D1217818

A Companion to
Television

BLACKWELL **COMPANIONS IN CULTURAL STUDIES**

Advisory editor: David Theo Goldberg, University of California, Irvine

This series aims to provide theoretically ambitious but accessible volumes devoted to the major fields and subfields within cultural studies, whether as single disciplines (film studies) inspired and reconfigured by interventionist cultural studies approaches, or from broad interdisciplinary and multidisciplinary perspectives (gender studies, race and ethnic studies, postcolonial studies). Each volume sets out to ground and orientate the student through a broad range of specially commissioned articles and also to provide the more experienced scholar and teacher with a convenient and comprehensive overview of the latest trends and critical directions. An overarching *Companion to Cultural Studies* will map the territory as a whole.

A Companion to
Television

Edited by

Janet Wasko

Blackwell
Publishing

CENTRAL MISSOURI
STATE UNIVERSITY
Warrensburg
Missouri

© 2005 by Blackwell Publishing Ltd

BLACKWELL PUBLISHING
350 Main Street, Malden, MA 02148-5020, USA
9600 Garsington Road, Oxford OX4 2DQ , UK
550 Swanston Street, Carlton, Victoria 3053, Australia

First published 2005 by Blackwell Publishing Ltd

2   2006

Library of Congress Cataloging-in-Publication Data

A companion to television / edited by Janet Wasko.
    p. cm. — (Blackwell companions in cultural studies ; 10)
  Includes bibliographical references and index.
  ISBN-13: 978-1-4051-0094-6 (hardcover : alk. paper)
  ISBN-10: 1-4051-0094-X (hardcover : alk. paper)
  1. Television broadcasting. 2. Television. I. Wasko, Janet. II. Series.

PN1992.5C615 2005
791.45—dc22

                                                          2005000692

A catalogue record for this title is available from the British Library.

Set in 11/13pt Ehrhardt
by Graphicraft Ltd, Hong Kong
Printed and bound in the United Kingdom
by TJ International Ltd, Padstow, Cornwall

The publisher's policy is to use permanent paper from mills that operate a sustainable forestry policy, and which has been manufactured from pulp processed using acid-free and elementary chlorine-free practices. Furthermore, the publisher ensures that the text paper and cover board used have met acceptable environmental accreditation standards.

For further information on
Blackwell Publishing, visit our website:
www.blackwellpublishing.com

Contents

Contents

Figures

Notes on Contributors

Jack Banks is Associate Professor in the School of Communication at the University of Hartford. At Hartford, he has also held positions as the Director of the Humanities Center and Distinguished Teaching Humanist. He teaches and conducts research in the areas of media ownership, media activist groups and gays and lesbians in popular culture.

David Buckingham is Professor of Education at the Institute of Education, London University, where he directs the Centre for the Study of Children, Youth and Media (www.ccsonline.org.uk/mediacentre). He is the author, co-author or editor of 17 books, including *The Making of Citizens: Young People, News, and Politics* (2000), *After the Death of Childhood* (2000), *Moving Images: Understanding Children's Emotional Responses to Television* (1996), *Children Talking Television: The Making of Television Literacy* (1993), and *Watching Media Learning: Making Sense Of Media Education* (1990). His work has been translated into 15 languages. He has recently directed projects on the uses of educational media in the home; young people's interpretations of sexual representations in the media; and the uses of digital media by migrant/refugee children across Europe. His most recent book is *Young People, Sex and the Media: The Facts of Life?* (with Sara Bragg).

Andrew Calabrese is an Associate Professor at the University of Colorado in Boulder. He has published many research articles on communication politics and policy, and he edited *Information Society and Civil Society: Contemporary Perspectives on the Changing World Order* (1994, with Slavko Splichal and Colin Sparks), *Communication, Citizenship, and Social Policy* (1999, with Jean-Claude Burgelman), and *Toward a Political Economy of Culture* (2004, with Colin Sparks). He won the Donald McGannon Award for Social and Ethical Relevance in Communication Policy Research, and he was a Fulbright scholar in Slovenia. He edits the book series "Critical Media Studies," serves on several editorial boards and is a founding member of the European Institute for Communication and Culture (EURICOM).

Notes on Contributors

Stuart Cunningham is Professor and Director of the Creative Industries Research and Applications Center (CIRAC), Queensland University of Technology. Known for his policy critique of cultural studies, *Framing Culture* (1992), and for the co-edited *New Patterns in Global Television* (1996) and the co-authored *Australian Television and International Mediascapes* (1996), his recent projects include a study of popular culture amongst Asian overseas communities (*Floating Lives: The Media and Asian Diasporas*, with John Sinclair, 2001) and the standard textbooks *The Australian TV Book* (2001) and *The Media and Communications in Australia* (2002) (both with Graeme Turner).

Peter Dahlgren is Professor of Media and Communication Studies at Lund University, Sweden. His research interests lie in the areas of media and social life in late modernity, and he has published widely on such themes as democracy, journalism, political participation, and identity. Among his publications is *Television and the Public Sphere: Citizenship, Democracy and the Media* (1995). At present he is working on a project about young citizens and new media. His forthcoming book is entitled *Media and Civic Engagement*.

Nabil H. Dajani is Professor of Communication at the American University of Beirut, Lebanon. He has been on the faculty for 35 years and has served as an Assistant Dean of its Faculty of Arts and Sciences, Chairperson of the Department of Social and Behavioral Sciences, and Director of its Mass Communication Program. Dr Dajani specializes in the study of the role of the media in society with an emphasis on the Lebanese and Arab media. Prof. Dajani is the author of three manuscripts and some 50 articles in professional journals. He has contributed to some 80 international and regional communication professional meetings. Between 1971 and 1975, Prof. Dajani served as a member of UNESCO's International Panel of Experts on Communication Research that contributed to initiating the international debate for a new world communication order.

Caren J. Deming is Professor of Media Arts at the University of Arizona. She teaches courses in film and television history, criticism, and writing. Her current research focus is the study of "Golden Age" television through the pioneering family comedy *The Goldbergs* and a critical biography of its creator Gertrude Berg. The essay in this volume is drawn from research at the Bird Library at Syracuse University, the Museum of Television and Radio in New York, the Museum of Broadcasting in Chicago, and the UCLA Film and Television Archive.

Bonnie J. Dow is Associate Professor of Speech Communication at the University of Georgia, author of *Prime-Time Feminism: Television, Media Culture, and the Women's Movement Since 1970* (1996) and former co-editor of *Critical Studies in Media Communication* (2002–4).

Albert Moran is Senior Lecturer at the School of Film, Media and Cultural Studies at Griffith University and Researcher for the Australian Key Centre for Cultural and Media Policy. His research interests include screen adaptation and film/TV and place. Recent publications include *Copycat TV: National Identity, Program Formats & Grundy* (2003), *Television Across Asia: TV Industry, Formats, Flows* (2003, co-edited with Michael Keane), *Wheel of Fortune: Australian TV Game Shows* (2003), and *Television Australia: Precedent, Period, Place* (Sydney: University of New South Wales Press, forthcoming).

Graham Murdock is Reader in the Sociology of Culture at Loughborough University. He has held visiting professorships in the United States, Belgium, Norway, and Mexico, and, most recently, in Sweden, as the Bonnier Chair at the University of Stockholm. He has long-standing interests in the political economy of the communications industries and in the social organization and impact of new communications technologies. He is currently co-directing a panel study of digital access and participation. His recent books include; as co-author, *Researching Communications* (1999) and as co-editor, *Television Across Europe* (2000). He is currently working on a book on the transformation of public broadcasting.

Horace Newcomb holds the Lambdin Kay Chair for the Peabody Awards in the Telecommunication Department at the University of Georgia, where he directs George Foster Peabody Awards Programs. He is editor of the *Museum of Broadcast Communications Encyclopedia of Television*, editor of *Television: The Critical View*, and author of numerous articles and essays about television.

Julianne H. Newton is Associate Professor of Visual Communication at the University of Oregon and editor of *Visual Communication Quarterly*. Her scholarship explores the interplay of the visual with society, our ways of knowing, and the integrity of the self. Newton's book *The Burden of Visual Truth: The Role of Photojournalism in Mediating Reality* (2001) won the 2003 Excellence in Visual Communication Research Award from the National Communication Association Visual Communication Division. Her publications on visual ethics and visual ecology span scholarly and public forums, and her documentary photographs of people and communities have been shown in more than 50 exhibitions. At the University of Oregon, she teaches photojournalism, visual ethics, visual communication theory, and ethnography.

Michael R. Real is Professor of Applied Communication at Royal Roads University in Victoria, BC, Canada. His books include *Exploring Media Culture* (1996), *Super Media* (1989), and *Mass-Mediated Culture* (1977). He has written scores of scholarly and general publications, directed local and international research projects, and hosted television and radio programs. The focus of his work is media, culture, and social responsibility.

Notes on Contributors

Paddy Scannell is a member of the Department of Mass Communication and Journalism at the University of Westminster where he has taught for many years. He is a founding editor of the journal *Media Culture & Society* and the author, with David Cardiff, of *A Social History of British Broadcasting, 1922–1939* (Blackwell, 1991). He is currently writing a review and critique of theories of communication and media in the last century.

Jane M. Shattuc is Associate Professor of Visual and Media Arts at Emerson College in Boston. She has written *Television, Tabloids and Tears: Fassbinder and Popular Culture* and *The Talking Cure: Women and TV Talk shows*.

John Sinclair is a Professor in the School of Communication, Culture and Languages at Victoria University of Technology, Melbourne, Australia. He has been researching the globalization of media for over 20 years, with special reference to the internationalization of the advertising and commercial television industries, particularly in developing regions such as Latin America and India. His published work includes *Images Incorporated: Advertising as Industry and Ideology* (1987), *Latin American Television: A Global View* (1999), and the co-edited works *New Patterns in Global Television: Peripheral Vision* (1996) (with Liz Jacka and Stuart Cunningham) and *Floating Lives: The Media and Asian Diasporas* (2001) (with Stuart Cunningham). He has held visiting professorships at the University of California, San Diego, and the University of Texas at Austin, and has been UNESCO Visiting Professor of Communication at the Universidad Autónoma de Barcelona.

Lynn Spigel is a Professor at Northwestern University, and author of *Make Room for TV* (1992), *Welcome to the Dreamhouse* (2001), and *High and Low TV* (forthcoming). She has edited numerous anthologies, including *Television after TV* (2004), and is the editor of the Console-ing Passion book series.

Ruth Teer-Tomaselli is a Professor of Culture, Communication and Media Studies at the University of KwaZulu-Natal, Durban, South Africa. She holds a UNESCO–Orbicom Chair in Communication. Her research interests include the political economy of broadcasting and telecommunications in Southern Africa; program production on television; radio, particularly community radio; and the role of media in development. She has served as a Director on the boards of the national public broadcaster, the South African Broadcasting Corporation, a commercial radio broadcaster, East Coast Radio, and a community radio broadcaster, Durban Youth Radio.

Sasha Torres teaches television studies and critical theory at the University of Western Ontario. An editor of *Camera Obscura* since 1993, she is the author of *Black, White and In Color: Television and Black Civil Rights* (2003) and the editor of *Living Color: Race and Television in the United States* (1998).

Janet Wasko is the Knight Chair for Communication Research at the University of Oregon, USA. She is the author of *Understanding Disney: The Manufacture of Fantasy* (2001) and *How Hollywood Works* (2003), and the editor or co-editor of collections relating to political economy of communication and democracy and media.

Shunya Yoshimi is a Professor at the Institute of Socio-Information and Communication Studies, University of Tokyo. His books in Japanese include: *Dialogue with Cultural Studies* (with Tasturo Hanada and others, 1999), *Birth of the News* (with Naoyuki Kinoshita, 1999), and *Cultural Studies* (2000). He also has edited *Media Studies* (2000), *Perspectives to Globalization* (with Kang Sang Jun, 2001), and *Introduction to Cultural Studies* (2001), as well as publishing numerous articles in English.

Yuezhi Zhao is an Associate Professor in the School of Communication at Simon Fraser University, Canada. She is the author of *Media, Market, and Democracy in China: Between the Party Line and the Bottom Line* (1998), the co-author of *Sustaining Democracy? Journalism and the Politics of Objectivity* (1998), and the co-editor of *Democratizing Media? Globalization and Public Communication* (forthcoming). She is currently working on a book manuscript on communication, power, and contestation in China.

Introduction

Janet Wasko

tel·e·vi·sion (pronunciation: tl-vzhn) *n.* [French télévision: télé-, *far* (from Greek tle-, *tele-*) + vision, *vision*]

1 The transmission of visual images of moving and stationary objects, generally with accompanying sound, as electromagnetic waves and the reconversion of received waves into visual images.

2 a. An electronic apparatus that receives electromagnetic waves and displays the reconverted images on a screen.

   b. The integrated audible and visible content of the electromagnetic waves received and converted by such an apparatus.

3 The industry of producing and broadcasting television programs.

   (*The American Heritage Dictionary of the English Language*, fourth edition, 2000)

Television? The word is half Greek, half Latin. No good can come of it.
                 C. P. Scott, English journalist (1846–1932)

What is television, how can we understand it, and why should we bother? Ultimately, these questions lie at the heart of this volume, which features original essays by an international collection of media scholars who have studied various aspects of television. But even these experts do not offer easy or conclusive answers to these key questions, for television presents a complex phenomenon that has become a ubiquitous feature of our modern world.

What is Television?

Television is a multifaceted apparatus. Most simply, it is a technological process, an electronic device, a system of distributing images and sounds. Although television as a form of mass communication did not emerge until the late 1940s and early 1950s, much of the technology of television was developed during the 1920s. As with many forms of media technology, the promises and expectations of the medium were optimistic and propitious. For instance, one of the often-overlooked inventors in the United States, Philo Farnsworth, was clearly hopeful about the future of television. One of his biographers explains:

> Philo began laying out his vision for what television could become. Above all else
> . . . television would become the world's greatest teaching tool. Illiteracy would be
> wiped out. The immediacy of television was the key. As news happened viewers
> would watch it unfold live; no longer would we have to rely on people interpreting
> and distorting the news for us. We would be watching sporting events and sym-
> phony orchestras. Instead of going to the movies, the movies would come to us.
> Television would also bring about world peace. If we were able to see people in
> other countries and learn about our differences, why would there be any mis-
> understandings? War would be a thing of the past. (Schwartz, 2002, p. 113)

Obviously, Farnsworth's full vision has not yet been realized, even though
some parts of his dream have been more than fulfilled. Television has become a
common household appliance that serves as a source of news, information, politics, entertainment, education, religion, art, culture, sports, weather, and music.
Television is an industrial system that produces and distributes products, as well
as (often) promoting other commodities and commerce. Hence, television is not
only a technical device, but also a social, political, economic, and cultural force.

Of course, the way television is produced and received has changed over the
years with changing political and economic climates, as well as the introduction
of newer technologies – VCRs, cable systems, pay TV, satellite systems, digital
and high-definition. In addition, other communication systems (such as computers and the Internet) increasingly challenge television's dominance as the primary mass medium. Television may also have a variety of meanings in different
parts of the world, as is evident from the discussions in this volume.

These variations and changes make television an enigmatic "moving target,"
its future uncertain and contested. Nevertheless, we must still attempt to define
its character and its influence.

Why Should We Bother to Understand Television?

Television continues to be a centrally important factor and an inescapable part
of modern culture. Many would still call it the most important of all the mass
media. As one television program about television concludes:

> From its public marketing in the 1940s to the present day, television can be listed
> as one of the most profound, if not the most profound, influences on human
> history. Television has affected every aspect of our lives including history, science,
> politics, culture and social mores. It is impossible to imagine a world without
> television, and most of us take for granted the way television has shaped and
> defined our society, and our lives. (The History Channel, 1996)

The pervasiveness of television is hard to ignore. For instance, in the United
States and Canada, 99 percent of households own at least one television set,

while the average number of sets is 2.93. In most cases, television is a central presence in individual homes – 66 percent of Americans supposedly watch television while eating dinner. But television sets are also prominent in other locations. We find them in schools, hospitals, prisons, bars, restaurants, shopping malls, waiting rooms . . . television seems to be (virtually) everywhere and often difficult to avoid. Obviously, television ownership and viewing may vary around the world – but the prevalence of television is a global, albeit varied, phenomenon.

We know that television is a fundamental part of everyday life for many people, although assessing television viewing is tainted with inevitable methodological problems. While computers may be luring some viewers away from the tube, it is claimed that the average American watches more than 4 hours of TV each day (that's 28 hours/week, or 2 months of nonstop TV watching per year). Of course, the American television diet may be more extensive than other countries. The point is that television often plays an important role in people's daily lives.

It might also be argued that television is central to the way that people learn about news and public events. Although the Internet may be increasingly providing citizens with news and information, television is still the primary source of news for many people. Events are now transmitted by television at the moment they are happening. In many countries, television is a key component in elections and campaigns, thus becoming part of the democratic process.

In addition to news and public affairs, television provides endless varieties of entertainment and diversion. Though the form and content may differ across time and space, the capacity of television to transmit sounds and images is potentially inexhaustible and seemingly unlimited. Thus, many have called television a storyteller, if not THE storyteller for society. As Signorelli and Bacue (1999, p. 527) explain:

> Television's role in society is one of common storyteller – it is the mainstream of our popular culture. Its world shows and tells us about life – people, places, striving, power and fate. It lets us know who is good and who is bad, who wins and who loses, what works and what doesn't, and what it means to be a man or a woman. As such, television has joined the ranks of socialization agents in our society and in the world at large.

Obviously, television systems and content exist within social contexts and are shaped by a variety of forces. Through its distribution of information, entertainment, education, and culture, television inevitably is a fund of values, ideals, morals, and ethical standards. In other words, television is an ideological source that cannot be overlooked in modern societies.

Nevertheless, there are differing opinions about television's fundamental value. (Note the sampling of opinions in the quotes about television by public figures included at the end of this introduction.) Television has been praised as a wondrous looking glass on the world, a valuable source of information, education, and entertainment. TV allows people to share cultural experiences, as well as allowing family members of all ages an opportunity to spend time together. Despite the disparaging comments about television's impact on print culture, some would point out that TV may serve as a catalyst for reading, as viewers may follow up on TV programs by getting books on the same subjects or reading authors whose work was adapted for the programs.

As envisioned by Farnsworth, television does indeed provide news, current events, and historical programming that can help make people more aware of other cultures and people. It is argued that "good television" can present the arts, science, and culture. Furthermore, good television can teach important values and life lessons, explore controversial or sensitive issues, and provide socialization and learning skills. Good television can help develop critical thinking about society and the world. More simply, many point out that television provides people with pleasure, as well as a welcome companion for lonely or isolated individuals.

The economic impact of television might also be noted. Manufacturers often depend on television to spread the word and encourage consumption of their products and services through commercial television. In 2001, total broadcast TV revenues in the United States were $54.4 billion. Revenues are also generated from programming production and distribution, as well as hardware sales. It follows that television also provides employment – not huge numbers, but certainly a significant workforce that obviously plays an important role in economic systems.

On the other hand, many commentators have also disparaged television as being valueless, vulgar, and vacuous. Indeed, the discussions of television as a negative force in society are so widespread and varied that they are difficult to summarize. Television is blamed for everything from passivity and obesity to stimulating aggressive and violent behavior. It has been singled out as leading an attack on literate culture, as well as shriveling public discourse (see Postman, 1986).

One of the most often-cited assessments of television acknowledged its potential value, but was damning of its current state. In 1961, Newton Minow, chairman of the Federal Communications Commission, proclaimed: "When television is good, nothing is better. But when television is bad, nothing is worse. I invite you to sit down in front of your TV set and keep your eyes glued to that set until the station signs off. I can assure you that you will observe a vast wasteland."

If television has become "a teaching tool," as envisioned by Farnsworth, this is not a positive development for many observers. For instance, John Silver, president of Boston University, recently declared "Television is the most important educational institution in the United States today." Silver went on to decry the

... degenerative effects of television and its indiscriminate advocacy of pleasure ... As television has ravenously consumed our attention, it has weakened the formative institutions of church, family, and schools, thoroughly eroding the sense of individual obedience to the unenforceable on which manners and morals and ultimately the law depend. (Silver, 1995, p. 2)

The role of television in promoting consumption has been widely attacked, because commercial systems are fundamentally ruled by advertising.

But even without advertising, some have argued that television cannot be transformed or altered, but is inherently destructive and detrimental. Former advertising executive Jerry Mander (1977) presented this viewpoint years ago, when he argued that television is not a neutral technology and its very existence is destructive to human nature.

It might also be noted that there may be different values and importance associated with television in different cultures. Nevertheless, television's key role in many societies, as well as its global prevalence and importance, is undeniable and makes it a significant issue for research and reflection.

How Do We Understand Television?

Since its inception, television has attracted a good deal of reflection and analysis. Within academia, television has been part of the ongoing study of mass media in general, which has been influenced by many disciplines, including political science, sociology, economics, psychology, and literary studies.

But scholarly research has also concentrated specifically on television, insisting that the medium itself is a worthy focal point for academic research. While general approaches to television research might be characterized as social scientific or humanistic, areas of research specialization have also evolved. Several chapters in this volume offer general overviews of television research, detailing different perspectives and approaches, while other contributors summarize specific areas of television research.

Much early television research adhered to a media effects orientation, searching for quantitative measures of television's impact on audiences, especially the impact of violent content on behavior. For instance, according to one estimate, approximately 4,000 studies have examined TV's effects on children. Still, no conclusive results have been found.

Meanwhile, other scholars focused attention on television content from the purview of literary or dramatic criticism. The growth of television studies in the 1970s and 1980s drew on this orientation, and has been characterized by work that focuses mostly on television texts and audiences, often integrating cultural studies, feminist analysis and drawing on a range of qualitative methodologies.

More recently, historical studies of television have blossomed, as well as work that examines television's structure, organizations, and ownership, its connections to the state and other media, and its role in influencing public opinion and the public sphere.

Indeed, debates continue to rage about what should be studied and what methods should be used to study television, as many (if not, most) studies of television still represent "single perspectives" or "specific agendas." However, numerous authors in this volume argue that interdisciplinary, multi-perspective approaches are needed. Horace Newcomb calls for "blended, melded research strategies," while Doug Kellner describes "multidimensional" or "multiperspectival" approaches to understand television from a critical perspective. As Newcomb argues: "we can best understand television not as an entity – economic, technological, social, psychological, or cultural – but as a site, the point at which numerous questions and approaches intersect and inflect one another."

Chapter Overview

The contributors to this volume offer a wide range of expertise on the study of television. They present overviews of the extensive research on television, as well as original insights into its development and significance in various regions of the world. Only a brief introduction to the chapters is presented here.

In the first section, Horace Newcomb traces the general development of television research and the growth of television studies, while Doug Kellner discusses critical perspectives on television from the 1930s through to the present day.

Perhaps surprisingly, historical dimensions of television are often overlooked in much of television studies. In the next section, Paddy Scannell discusses the histories of television, while Lynn Spigel explores television archives and the politics of television preservation.

Another neglected topic in typical television studies might be identified as the aesthetics of television. Julianne Newton considers television and "a moving aesthetic," while Caren Deming explores the "televisual," as exemplified in the Golden Age of American television. Meanwhile, Jane Shattuc examines the American TV production process and the question of authorship.

Analysis of structure and control is fundamental in examining television systems and a number of the contributors to this volume address these issues. Sylvia Harvey asks "Who Rules TV?" in her examination of the state, markets and the public interest. Graham Murdock looks at issues relating to public broadcasting and citizenship, while Stuart Cunningham analyzes changing television policies or policy regimes.

The prominence of American television also demands attention to the implications of commercial, privately owned television systems. Matthew McAllister discusses television advertising as a textual and economic system, while Eileen Meehan presents a political economic approach to the analysis of television viewing. Jack Banks looks at MTV as an exemplar of the development of media conglomerates, while Andrew Calabrese considers the trade in television news in the United States.

6

A good deal of research on television has focused on content, albeit using a variety of approaches and methodologies. In this section, Albert Moran introduces the new television landscape and explores the circulation of television formats. Reflections on specific types of programming or genres are presented by Christine Geraghty (soap operas), Jane Shattuc (talk shows), Michael Real (sports), and Gary Edgerton (historical programming). Meanwhile, issues relating to representation of specific social groups are considered by Bonnie Dow (women and feminism) and Sasha Torres (race).

Although a good deal of television research is devoted to audiences, a variety of approaches and methods have been used. Peter Dahlgren explores the reception of television in its broadest sense as he looks at its relationship to public spheres and civic cultures, while Justin Lewis examines television and public opinion. Specific audiences are considered in Annette Hill's discussion of audiences for reality television and David Buckingham's overview of the study of children and television.

In the final sections, discussions of the variety of television forms are presented. DeeDee Halleck outlines various alternative challenges to mainstream television, while television in different parts of the world is explored by John Sinclair (Latin America), Yuezhi Zhao and Zhenzhi Guo (China), Shunya Yoshimi (Japan), Ruth Teer-Tomaselli (South Africa), and Nabil Dajani (the Arab East).

Thus, contributors to this volume attempt to define television, consider why it is significant and present overviews of how it has been studied. Despite changes in television and in the world, no matter how difficult, we must endeavor to answer these questions. Welcome to the world of television at the dawn of the twenty-first century!

Acknowledgments

The editor would like to thank the contributing authors for their enthusiasm and professionalism in the preparation of this volume. Special thanks to the editors at Blackwell for their support and efficiency during this process, especially Jayne Fargnoli, Ken Provencher, Annie Lenth, and Nick Brock. More thanks to Christine Quail, Micky Lee, and Randy Nichols, for their research assistance during this project.

References

The American Heritage Dictionary of the English Language (2000), 4th edn., New York: Houghton Mifflin.

The History Channel (1996) *Modern Marvels*, "Television: Window to the World," accessed 1 June at http://www.historychannel.com/classroom/admin/study_guide/archives/thc_guide.0229.html.

Mander, J. (1977) *Four Arguments for the Elimination of Television*, New York: Harper Collins.

Postman, N. (1986) *Amusing Ourselves to Death: Public Discourse in the Age of Show Business*, New York: Penguin Books.

Janet Wasko

Schwartz, E. I. (2002) *The Last Lone Inventor: A Tale of Genius, Deceit, and the Birth of Television*, New York: HarperCollins.

Signorelli, N. and Bacue, A. (1999) "Recognition and Respect: A Content Analysis of Prime-time Television Characters across Three Decades," *Sex Roles: A Journal of Research* April, 527–44.

Silver, J. (1995) "The Media and Our Children: Who is Responsible?," accessed 10 June 2004 at http://www.johnsonfdn.org/winter96/media.pdf.

Quotes about Television

It is interesting how many public figures have commented on the nature and significance of television over the years. Included here is a sampling of these quotes (many by people deeply involved in television) that may provide amusement or reflection, but are also relevant to the discussions that follow in this volume.

Richard P. Adler: "All television is children's television."

Fred Allen: "Imitation is the sincerest form of television."

Lucille Ball: "Television is the quickest form of recognition in the world."

Clive Barnes: "Television is the first truly democratic culture – the first culture available to everybody and entirely governed by what the people want. The most terrifying thing is what the people do want."

Daniel J. Boorstin: "Nothing is really real unless it happens on television."

Ray Bradbury: "The television, that insidious beast, that Medusa which freezes a billion people to stone every night, staring fixedly, that Siren which called and sang and promised so much and gave, after all, so little."

David Brinkley: "The one function TV news performs very well is that when there is no news we give it to you with the same emphasis as if there were."

Rita Mae Brown: "Art is a moral passion married to entertainment. Moral passion without entertainment is propaganda, and entertainment without moral passion is television."

Art Buchwald: "Every time you think television has hit its lowest ebb, a new program comes along to make you wonder where you thought the ebb was."

Carol Burnett: "The audience is never wrong."

Prince Charles: "There are now more TVs in British households than there are people – which is a bit of a worry."

Paddy Chayefsky: "It's the menace that everyone loves to hate but can't seem to live without."

Paddy Chayefsky: "Television is democracy at its ugliest."

Imogene Coca: "Television is the only way I know to entertain 20 million people at one time."

Alistair Cooke: "When television came roaring in after the war (World War II) they did a little school survey asking children which they preferred and why – television or radio. And there was this 7-year-old boy who said he preferred radio 'because the pictures were better'."

Alan Coren: "Television is more interesting than people. If it were not, we would have people standing in the corners of our rooms."

Salvador Dali: "What is a television apparatus to man, who has only to shut his eyes to see the most inaccessible regions of the seen and the never seen, who has only to imagine in order to pierce through walls and cause all the planetary Baghdads of his dreams to rise from the dust."

Ani Difranco: "Art may imitate life, but life imitates TV."

Hugh Downs: "Television is the medium of the 20th century."

Dwight D. Eisenhower: "I can think of nothing more boring for the American people than to have to sit in their living rooms for a whole half hour looking at my face on their television screens."

T. S. Eliot: "It is a medium of entertainment which permits millions of people to listen to the same joke at the same time and yet remain lonesome."

Tony Follari: "Karl Marx is wrong. Television is the opiate of the masses."

David Frost: "Television is an invention that permits you to be entertained in your living room by people you wouldn't have in your home."

Larry Gelbart: "Television is a weapon of mass distraction."

Samuel Goldwyn: "Television has raised writing to a new low."

S. I. Hayakawa: "In the age of television, image becomes more important than substance."

Jim Henson: "Television is basically teaching whether you want it to or not."

Alfred Hitchcock: "Television is like the American toaster, you push the button and the same thing pops up everytime."

Alfred Hitchcock: "Seeing a murder on television . . . can help work off one's antagonisms. And if you haven't any antagonisms, the commercials will give you some."

Steve Jobs: "You go to your TV to turn your brain off. You go to the computer when you want to turn your brain on."

Nicholas Johnson: "All television is educational television. The question is: what is it teaching?"

Ernie Kovacs: "Television – a medium. So called because it is neither rare nor well-done."

Ann Landers: "Television has proved that people will look at anything rather than at each other."

Lee Lovinger: "Television is simply automated day-dreaming."

Mignon McLaughlin: "Each day, the American housewife turns toward television as toward a lover. She feels guilty about it, and well she might, for he's covered with warts and is only after her money."

Miriam Makeba: "People in the United States still have a 'Tarzan' movie view of Africa. That's because in the movies all you see are jungles and animals . . . We [too] watch television and listen to the radio and go to dances and fall in love."

Marya Mannes: "It is television's primary damage that it provides ten million children with the same fantasy, ready-made and on a platter."

Daniel Marsh: "If the television craze continues with the present level of programs, we are destined to have a nation of morons."

Groucho Marx: "I find television very educating. Every time somebody turns on the set, I go into the other room and read a book."

Marvin Minksy: "Imagine what it would be like if TV actually were good. It would be the end of everything we know."

Malcolm Muggeridge: "Television was not intended to make human beings vacuous, but it is an emanation of their vacuity."

Edwin Newman: "We live in a big and marvelously varied world. Television ought to reflect that."

Camille Paglia: "Television is actually closer to reality than anything in books. The madness of TV is the madness of human life."

Shimon Peres: "Television has made dictatorship impossible, but democracy unbearable."

Gene Roddenberry: "They say that ninety percent of TV is junk. But, ninety percent of *everything* is junk."

Rod Serling: "It is difficult to produce a television documentary that is both incisive and probing when every twelve minutes one is interrupted by twelve dancing rabbits singing about toilet paper."

Homer Simpson: "Television! Teacher, mother, secret lover."

Red Skelton: "I consider the television set as the American fireplace, around which the whole family will gather."

Harriet van Horne: "There are days when any electrical appliance in the house, including the vacuum cleaner, seems to offer more entertainment possibilities than the TV set."

Orson Welles: "I hate television. I hate it as much as peanuts. But I can't stop eating peanuts."

E. B. White: "I believe television is going to be the test of the modern world, and that in this new opportunity to see beyond the range of our vision, we shall discover a new and unbearable disturbance of the modern peace, or a saving radiance in the sky. We shall stand or fall by television – of that I am quite sure."

E. B. White: "Television hangs on the questionable theory that whatever happens anywhere should be sensed everywhere. If everyone is going to be able to see everything, in the long run all sights may lose whatever rarity value they once possessed, and it may well turn out that people, being able to see and hear practically everything, will be specially interested in almost nothing."

Frank Lloyd Wright: "Television is chewing gum for the eyes."

Unknown, from *New York Times* 1939: "TV will never be a serious competitor for radio because people must sit and keep their eyes glued on a screen; the average American family hasn't time for it."

Unknown: "A television is a device you can sit in front of and watch people do things that you could be doing, if you weren't sitting there watching them do it."

Unknown: "Sex on television can't hurt you unless you fall off."

Unknown: "TV. If kids are entertained by two letters, imagine the fun they'll have with twenty-six. Open your child's imagination. Open a book."

Famous last words: "I've seen this done on TV."

Sources: www.quotegarden.com, www.basicquotations.com, http://en.thinkexist.com/quotations, and Alison Bullivant, ed. (2003) *The Little Book of Humorous Quotations*, New York: Barnes & Noble Books.

PART ONE

Theoretical Overviews

CHAPTER ONE

The Development of Television Studies

Horace Newcomb

Since the 1990s "Television Studies" has become a frequently applied term in academic settings. In departments devoted to examination of both media, it parallels "Film Studies." In more broadly dispersed departments of "Communication Studies," it supplements approaches to television variously described as "social science" or "quantitative" or "mass communication." The term has become useful in identifying the work of scholars who participate in meetings of professional associations such as the recently renamed Society for Cinema and Media Studies as well as groups such as the National Communication Association (formerly the Speech Communication Association), the International Communication Association, the Broadcast Education Association, and the International Association of Media and Communication Research. These broad-based organizations have long regularly provided sites for the discussion of television and in some cases provided pages in sponsored scholarly journals for the publication of research related to the medium. In 2000, the *Journal of Television and New Media Studies*, the first scholarly journal to approximate the "television studies" designation, was launched.

Seen from these perspectives, "Television Studies" is useful primarily in an institutional sense. It can mark a division of labor inside academic departments (though not yet among them – so far as I know, no university has yet established a "Department of Television Studies"), a random occasion for gathering like-minded individuals, a journal title or keyword, or merely the main chance for attracting more funds, more students, more equipment – almost always at least an ancillary goal of terminological innovation in academic settings.

That the term could also potentially denote what some might call an "academic field," or, more aggressively, "a discipline," however, causes as many problems as it solves. Indeed, as Toby Miller cautions:

> We need to view the screen through twin theoretical prisms. On the one hand, it can be understood as the newest component of sovereignty, a twentieth-century cultural addition to ideas of patrimony and rights that sits alongside such traditional

topics as territory, language, history, and schooling. On the other hand, the screen is a cluster of culture industries. As such, it is subject to exactly the rent-seeking practices and exclusionary representational protocols that characterize liaisons between state and capital. We must avoid reproducing a thing called, for example, "cinema or TV studies or new media (urggh) studies," and instead do work that studies the screen texts and contexts, regardless of its intellectual provenance. (*Politics and Culture*, Issue 1, 2002, http://aspen.conncoll.edu/politicsandculture/arts.cfm?id=40)

It is, of course, significant that Miller is also editor of *Television and New Media* (2002), and elsewhere, in the preface to a collection of commentary (boldly entitled *Television Studies*), on various aspects of the medium, has written:

can anyone seriously argue against seeking to understand how and why television and its audiences make meaning? Of course, people can and do object, and one aim of this book is to convince doubting siblings, peers, and hegemons of the need for television studies. But the principal goal is to open up the field of thinking about television to students and show them how it can be analysed and changed. (BFI Publishing, 2002, p. vii)

I juxtapose these apparently varying statements not to "catch" Miller in "contradiction," much less to make light of comments from a scholar I consider a central contributor to whatever we choose to designate under the heading in question. Rather, I cite Miller's well-considered perspectives to indicate the troubling complexities encountered in any attempt to place this particular medium inside clearly defined boundaries. Miller's latter phrase in the introduction to his handbook, "show them how it can be analysed and changed," is indicative of a forceful motivation shared by many of us who have spent considerable time and effort in examining the complex phenomenon we call television. Indeed, that television needs changing is probably one of the most widely shared assumptions of the second half of the twentieth century, and certainly one that shows no signs of diminishing presence.

By contrast, the notion that television requires, or even that calls for change would somehow demand, "analysis," is widely considered silly. As Miller's comments indicate, the mere suggestion that television needs analysis itself requires supportive argument. "Everyone" knows how to think about, presumably how to "change" television. The sense that any change would either imply, or explicitly rely upon, *specific types* of analysis, *specific questions, particular bodies of knowledge*, flies in the face of our common and "commonsensical" experience of the ubiquitous appliance and its attendant "content." And if some of these bodies of knowledge, these questions, these strategies for analysis might be contradictory, or subversive of one another, or perhaps internally incoherent, the waters are muddied more thickly.

Moreover, there is yet another angle on this topic that is preliminary to any thorough description of the "development" of "Television Studies." It is important to recognize that "Television Studies" is not the same thing as "studying television." Even the most skeptical or hostile critic of the former may have no hesitation in supporting the latter. Indeed, the skepticism and hostility emerge precisely with attempts to extract television from other "studiable" topics and problems inside which television, while perhaps hugely significant, remains subordinate. It is with these varied approaches to "studying television," however, that any account of the development of the potentially institutionalized and focused designation must begin.

As I have indicated elsewhere, a number of those who paid early attention to the medium speculated in broad philosophical terms about its place in society and culture (see, for example, Newcomb, 1974). One example, Lee De Forest, will suffice. Best noted for contributions to the development of television technologies, De Forest was also deeply concerned – and broadly optimistic – about the sociocultural power of the medium. Television would, he believed, contribute to the rise of a particular social formation.

> A population which once more centers its interest in the home will inherit the earth, and find it good. It will be a maturer population, with hours for leisure in small homes, away from today's crowded apartments. Into such a picture ideally adapted to the benefits and physical limitations of television, this new magic will enter and become a vital element of daily life.

> This new leisure, more wisely used, welcoming the gifts, entertaining, cultural, educational, which radio and television will bestow, shall eventually produce new outlooks on life, and new and more understanding attitudes toward living. (De Forest, 1942, p. 356)

Embedded, rather remarkably, in this brief commentary, are multiple versions of possibilities and problems that continue to motivate a variety of topics related to television studies. The domestic nature of the medium, its range of offerings, its relation to time and space, its ability to affect attitudes and behaviors – all these observations lead to questions still open to exploration. And, of course, this last cluster of implied topics in De Forest's list, television's "effects" on behavior and attitude, quickly came to the fore in the early years of the medium's development as the "essential" questions to be addressed. But rather than exploring them within De Forest's optimistic frame, as "gifts," the effects were most often framed and examined as social problems. In this context, of television "as" social problem, a first wave of major studies of television came to prominence. And it is also the case that these questions were perceived as "essential" in two ways – as crucial questions for society, and as the "essence" of the medium itself. To try to think of "television" as other than the conduit for and/or cause of these problems required effort, if not audacity. One need only search under the keywords, "Television: Social Aspects," in library catalogs to discover large numbers of books, many of them bibliographies containing far larger numbers of essays, to survey the results of approaches to television from this perspective.

Still, it would be a mistake to suggest that these materials suggest an overly simple dichotomy between "the social sciences" and "the humanities," with the latter providing all the sources for newer uses of "television studies." Many examinations of television by social psychologists, sociologists, economists, political scientists, and others began early and continue to address questions and provide information, even "data," powerfully useful for any full understanding of the medium. It is also the case, as I shall suggest later, that "television studies" best understood implies (perhaps requires) the power of blended, melded research strategies that, while reshaping some of the issues and questions underpinning earlier work, profit by returning to them from new angles. Moreover, it is helpful to remember that much work from earlier periods was conducted by scholars for whom rigid divides among "fields," "disciplines," "approaches," and "methods" were less important than they may have become in harsher circumstances driven by the meager reward systems afforded by academic institutions – departmental resources, personal prestige, or narrow requirements for individual advancement and personal job security. Television, like film and radio before it, was a subject, a topic, and a source of great intellectual interest, attracting attention from many scholars from many fields as a result of a sensed obligation to acknowledge potential change of great import. The famous exchanges and collaborations between Paul Lazarsfeld and Theodor Adorno can be taken as exemplary struggles over appropriate questions and approaches without demand for final divisions, even though this is rarely the case when terms such as "administrative" and "critical" are attached to "research" as categories in conflict. And it is certainly worth recalling that Wilbur Schramm, often cited as one of the founders of social scientific media research, began his career with the study of literature. The foreword to his book, *Two Creative Traditions in English Poetry* (1939), was written by the great literary scholar Norman Foerster. And with Foerster and others, Schramm served as co-editor of *Literary Scholarship: Its Aims and Methods* (1941). It was hardly likely to be the case that all concern for expressive culture disappeared when he and his colleagues developed their work on children and television, or on the media as related to national development strategies.

In spite of these multiple connections and relations, however, there is no need to ignore the fact that television has most often been approached from single perspectives. Such precisely focused questions, and attendant methods of analysis or argument, generally reflect deep interests directed toward specific agendas. Thus, for the social psychologist concerned with the welfare of children, any study of television must gather data of a certain sort, capable of securing a voice in the arena of public policy, or at least in the appropriate bodies of academic literature that might be cited in public debate. For the economist focused on international flows of media, however, children's programming might be examined as a relatively inexpensive commodity best understood within the context of "public good" economic theory. Programming thus cited may be used as an example of why certain producing entities or nations have come to have particular influence in world markets. For the scholar of technology, the programs themselves might hold little or no interest, while processes of production and distribution could be fascinating. For the critic, whose approaches are grounded in a range of humanistic fields and who expresses interest in the history of fictional forms, the same body of programs might be "read" as versions of expressive culture, works that rely on familiar forms of narration, stories that can be placed within a very long tradition of "representation." Many of these focused agendas have resulted from a perceived need to "fill gaps," or to offer "new" perspectives on familiar phenomena. Thus, when humanities-based critics and scholars turned their attention to television's fictional programming it was often with the goal of "supplementing" (or, perhaps more arrogantly, "correcting"), analyses conducted by social psychologists, economists, or technologists, and social psychologists turning to issues of large social effects may have intended to "extend" or "expand" work focused solely on television and children.

More interesting questions begin to emerge, however, when the critic suggests to the social psychologist that it is impossible to study children's responses without some sophisticated notion of narrative theory, or when the economist is challenged by a political economist arguing that the relatively limited number of circulated forms and genres is the result of powerful interests in control of "storytelling" in all cultural and social contexts, or when a specialist in media technologies examines the roles of new media devices alter the processes and outcomes of producing works for children.

It is here, in my view, in the interstices of methodological facility and discipline or field grounded problematics that "Television Studies" begins to find its ground. But getting "here" can be mapped in a variety of configurations. In the introductory essay to *Television: The Critical View* (2000), I chart one pathway – typically, the one most influential in my own efforts – leading to current developments. In this account the first influential turn can be described as the rise of questions related to "popular culture studies," a movement primarily grounded in varieties of "literary" analysis and determined to take seriously works considered underappreciated because of structured hierarchies involving the sociology of taste and the aims of humanistic education as molder of citizenship. In higher education settings in the United States in the late 1960s those who decided to study popular expressive culture – popular literature, comics, sport, popular music – made particular choices that would involve struggles for place within university curricula and charges of triviality in the general press. Film Studies had secured a foothold by focusing on international cinema as art, but also faced uphill battles when the field turned to American popular movies. Television was among the last topics for which legitimacy was sought.

That these events, decisions, and movements began at that particular time is telling. My argument suggests the following motivations, with specific attention to other developments in the United States.

The choice to examine these "inferior" or "unappreciated" forms was motivated by a number of concerns. Philosophically, scholars in this movement often felt the works they wished to examine were more indicative of larger cultural preferences, expressive of a more "democratic" relationship between works and audiences than the "elite" works selected, archived, and taught as the traditional canon of humanistically valued forms of expression.

Politically, these same impulses suggested that it was important to study these works precisely because their exclusion from canonical systems also excluded their audiences, devalued large numbers of citizens, or saddled them with inferior intellectual or aesthetic judgment. (Newcomb, 2000, p. 2)

Despite the "political" motivation behind the study of popular culture, there was little overt analysis of "ideology." The sense of "rescuing" the materials from complete dismissal was considered a form of activism, and certainly led to substantial political conflict in academic settings. But it was the development of "Cultural Studies" in Britain that began far more thorough analyses of the medium, among other "cultural" topics, with a fundamental commitment to ideology critique. This work drew heavily on a range of Marxist social and cultural theory, as well as on other "continental" philosophies. In this setting culturalists also engaged in debate with those championing stricter applications of Marxist political economy, who viewed cultural studies as, at times, myopic regarding issues of ownership and control of media industries. The cultural studies perspectives, and sometimes the attendant debates involving political economy, were quickly taken up in the United States and were a second, if not parallel influence on the development of television studies there. It should be noted here that while there was comparatively little influence flowing from the United States to Britain regarding these matters, it remained the case that British and other European scholars – and later, Asian and Latin American scholars as well – often focused on television produced in the United States as sites for analysis or theory development. Indeed, the powerful presence of US television throughout the world became a central topic of discussion in the cultural studies literatures and that content has undoubtedly had its own influence on various approaches to the medium at large.

Cultural studies also blended easily with a third strain of influence in television studies – critical sociology. Here, scholars drew on the work of the Frankfurt School of sociocultural analysis, and often viewed television as the latest in a line of "culture industries" spreading false consciousness, turning masses of popular culture users into mere fodder for pernicious political control (see Horkheimer and Adorno, 1972).

Academic critics working both from this tradition and from sharper versions of cultural studies frequently critiqued what they considered to be a central weakness in the earlier "popular culture" approach, its apparent reliance on a naïve notion of "liberal pluralism" when examining many expressive forms. The arrival of "British cultural studies" required and enabled some scholars working within the tradition of critical sociology to sharpen their own critiques, to recognize weaknesses and gaps in their work, and to move toward a more complex perspective on television and other topics by recognizing greater textual complexity in industrially produced expressive culture.

As suggested earlier, a fourth influence in this account must be the array of film studies expanding in academic settings. "Art" films, "foreign" films, often constituted the subject matter in some earlier classes devoted to film studies, and, as with television, many analytical approaches were modifications of literary studies. "Film appreciation" classes were also popular among students (and, because they enrolled large numbers, equally popular with administrators and teachers in liberal arts literature departments), as were the offerings, relatively few in number, devoted to the technical production of films. The push to study popular American film – to study "Hollywood" – drew many of the same negative responses as those leveled at the study of television. Still, with a degree of "support" from European scholars and critics/filmmakers who praised the unrecognized "artistry" of Hollywood film and filmmakers, American film topics found their place in the academy. The entire body of film studies quickly developed subdivisions and an array of analytical approaches, methods, and theories. In some quarters and some journals, the field also developed its own specialized languages, often cited by beginning students, journalists, or "visitors" from other fields of study as unduly arcane. By the 1980s a number of film scholars were also attending to television. In some cases the turn to the newer medium enriched approaches that were already being applied. In others, film theory and analysis foundered in encounters with features fundamentally distinct from those for which they were developed.

One area in which film scholars encountered difficult problems involved actual settings and behaviors surrounding the practices of viewing the media. While "spectatorship" had become a major topic of film analysis, the domestic aspects of television viewing, combined with its role as advertising medium, repetitive or serialized narrative structures, and genres merged within the television schedule, led to serious reconsideration or revision of notions regarding actual viewer experiences. In somewhat fortuitous fashion, British cultural studies had posited the study of audiences as a major topic within the study of mass media. Drawing on the model developed by Stuart Hall, analytical strategies had developed around notions of "encoding and decoding" television "texts." By examining the professional/institutional/production process at one pole of this model and the activities of audiences at the other, emphasis on the "actual" audience became a central component of study of television. The notion of the "active audience" became a central tenet in much of this work, often used to counter earlier studies of "media effects" and a range of "ethnographic" approaches, drawn from anthropology replaced or amplified the "survey" and "experimental" methods of social psychologists.

This focus on audience activity became a major focus of the emerging television studies arena and was also central to yet another influential stream in the development of television studies – the development of a range of feminist approaches to media and culture. Focus on gendered distinctions has ranged from studies of production and performance involving women to theories of narrative. And the focus on active audiences has been a basic strategy for redeeming such denigrated forms as the soap opera. Television has even been defined, problematically, as a more "feminine" medium, in part because of its domestic setting and, in the US industries, its constant flow of advertising, often directed at women as primary consumers in households. Feminist theory has cut through and across almost all previous approaches to television, altering or challenging basic assumptions at every juncture.

A number of these factors came together in several works in the mid-1980s, most notably in the work of John Fiske. That analysis began in collaboration with John Hartley, *Reading Television* (1978), a significant study grounded in literary theory and semiotics, but pushing those approaches to the study of television in exciting new ways. By 1987 Fiske had articulated an overarching approach in *Television Culture*, a work that began to develop ideas considered radical, even in cultural studies circles. The most prominent concept, one developed further in later studies, suggested that the ability – indeed, the power and authority – of viewers could perhaps match or even override that of television "texts," and by implication the ideological authority in which those texts were grounded. In some instances Fiske suggested that viewers could perhaps subvert messages and, by creating meanings of their own, create a type of ideological response to dominant ideology. Fiske was soundly taken to task by those who found such a view far too "populist," too naïve. (See, for example, McGuigan, 1992 and 1996.) In my own view, however, Fiske never lost sight of the applied power afforded by access to production, control of discursive systems, and political policies. Rather, his work reminds us that the results of such power is always uneven in its effectivity, couched in multiple and varying contexts, and significant to individuals and groups in very different ways. The debates sparked by this body of work continue.

The account presented thus far suggests only one version of the development of television studies. In it, various emphases, on television programs, industries, audiences, remain, in varying degree, discreet. Or, better put, they remain fundamental starting points for applied work. Similar starting points are also found in another survey of the development of television studies constructed by Charlotte Brunsdon:

> Television studies emerges in the 1970s and 1980s from three major bodies of commentary on television: journalism, literary/dramatic criticism and the social sciences. The first, and most familiar, was daily and weekly journalism . . . The second body of commentary is also organized through ideas of authorship, but here it is the writer or dramatist who forms the legitimation for the attention to television. Critical method here is extrapolated from traditional literary and dramatic criticism, and television attracts serious critical attention as a "home theatre" . . .

Both of these bodies of commentary are mainly concerned to address what was shown on the screen, and thus conceive of television mainly as a text within the arts humanities academic traditions. Other early attention to television draws, in different ways, on the social sciences to address the production, circulation and function of television in contemporary society. Here, research has tended not to address the television text as such, but instead to conceptualise television either through notions of its social *function* and *effects*, or within a governing question of *cui bono?* (whose good is served?). Thus television, along with other of the mass media, is conceptualised within frameworks principally concerned with the main-tenance of social order; the reproduction of the status quo, the relationship be-tween the state, media ownership and citizenship, the constitution of the public sphere.

. . . Methodologies here have been greatly contested, particularly in the extent to which Marxist frameworks, or those associated with the critical sociology of the Frankfurt School have been employed. These debates have been given further impetus in recent years by research undertaken under the loose definition of cultural studies. The privileged texts, if attention has been directed at texts, have been news and current affairs, and particularly special events such as elections, industrial disputes and wars. It is this body of work which is least represented in "television studies", which, as an emergent discipline, tends towards the textual-isation of its object of study. (Brunsdon, 1997, pp. 1647–9)

Brunsdon goes on to discuss, as I have above, the move toward audience studies and the overarching influence of feminist approaches to the medium. She then concludes:

Television studies in the 1990s, then, is characterised by work in four main areas. The most formative for the emergent discipline have been the work on the defini-tion and interpretation of the television text and the new media ethnographies of viewing which emphasise both the contexts and the social relations of viewing. However, there is a considerable history of "production studies" which trace the complex interplay of factors involved in getting programmes on screen . . . Increasingly significant also is the fourth area, that of television history . . . This history of television is a rapidly expanding field, creating a retrospective history for the discipline, but also documenting the period of nationally regulated terrestrial broadcasting – the "television" of "television studies" – which is now coming to an end.

These same lines of influence are again reconfigured in John Corner's overview text, *Critical Ideas in Television Studies* (1999); Corner begins with a distinction between "Television as Research Object," (p. 6) and "Television and Criticism" (p. 7). As in other accounts he identifies the former with "anxiety about [tele-vision's] influence," focused on matters such as "a distortion of politics," or "the displacement of culture." With either concern the focus of "research" has been "the individual viewer." This approach, he suggests, misses two important aspects of the medium. First, he points out that television is itself "culturally constitutive, directly involved in the circulation of the meanings and values out of which a popular sense of politics and culture is made and which also then provides the interpretative resources for viewing" and, secondly, "that all of the television which we watch will bring about some modification in our knowledge and experience, however minor and temporary" (p. 6). Criticism, on the other hand, has a different set of concerns: "I take a defining feature of critical activity to be an engagement with the significatory organization of television programmes themselves, with the use of images and language, generic conventions, narrative patterns, and modes of address to be found there" (p. 7). The questions emerging from such matters foreground "the critic's own interpretive resources as a specialist in the medium and does not work with a notion either of 'data' or of 'method' in the manner conventional in the social sciences . . ." (p. 7). But "this does not stop the critic making inferences about the social relationships and configurations of value within which television's texts are placed . . . Television criticism has most often wanted to go beyond the textually descriptive and evaluative and to use its observations here as a route to a broader or deeper cultural diagnosis, either of the past or the present" (pp. 7–8).

Corner, like others, cites the influence of "European social thought," the Frankfurt School, and various strands of Marxism. But he also adds a key notion, the development of "postmodernist thinking" and its influence on the study of television.

> Not surprisingly, television, with those features of space-time manipulation, social displacement, and scopic appeal . . . has often been regarded as an agency of post-modern culture, despite its origins as a modernist cultural technology. It has been seen as the representational hub of a new pattern of knowledge and feeling and of new kinds of political organization, self-consciousness, and identity. (p. 8)

John Hartley (1999) quite succinctly sums up many of the sequence of issues addressed in these other accounts by clustering studies of television under four headings: television as mass society, television as text, television as audience, and television as pedagogy.

The problem faced by any scholar or student planning to study television is that all these questions, attendant "methods" or "approaches," all the lines of thought, bodies of information generated, remain in play. No single focus has replaced another. Despite scholarly arguments over epistemology or legitimacy of purpose, each can explain certain aspects of the medium, lead to identification and definition of new problems, overlap with other results. This is the stew of issues stirred by television. And while it would be a mistake to argue that there is no clear "progression," "refinement," or "development" of stronger and clearer approaches, it does remain the case that most studies of television (rather than "television studies") continue to deal with the medium and construct their questions from relatively discreet points of view. It is also the case that any developments in the field we might call television studies have been greatly complicated by changes and developments surrounding "television" itself. New technologies, alteration in policy arenas, varying business models, innovations in narrative strategies, revival of older strategies – these and other changes have made television something of a moving target. In turn, the changes have sharpened awareness of the fact that many "approaches," even "theories" of television were put forward in other contexts, very specific historical conditions and social formations. The degree to which questions framed and approaches developed in those contexts remain useful is a matter of some concern.

What these interactions suggest is that we can best understand television not as an entity – economic, technological, social, psychological, or cultural – but as a site, the point at which numerous questions and approaches intersect and inflect one another. For this reason television should also be thought of as "television," somehow "marked" to remind us that no single definition or set of terms can gather or control the power and significance of this entity. Indeed, in this tendency to confound singly focused approaches, television has also become the site at which various theories and methods, not to say larger systemic constructions such as "the social sciences" or "the humanities" or "critical theory," have been forced to recognize shortcomings and attempt conversation, if not always conjunction, with others.

At this point, we can say that television studies is a conflicted field of study in need of one or more controlling or guiding metaphors. Such terms should somehow acknowledge the "site-like" qualities of television, recognizing it as one of the most powerful such points of conjunction in human history. Yet any such recognition must not ignore knowledge generated by more specifically focused queries.

In this context, Corner's use of the term "hub" is useful. If "television" is at the center of structuring spokes, holding things together in order to roll on, we could perhaps account for intersecting influences by speculating about what might happen if a particular spoke were removed. Or we might explore the role of one spoke, acknowledging that its force and significance might be limited.

My own preference for metaphor would be that "television" is a "switchboard" through which streams of information, power, and control flow unevenly. Struggles for control of the switchboard occur at many sub-points. In the "creative communities" the struggles might be over the control of textual content, style, or even budgets. At the corporate level they are most likely focused on budgets, but even the dullest accountant employed in a media industry recognizes that it is impossible to predict the next "hit," and must therefore adapt a calculus allowing for failures. And these failures cannot be fully explained by research departments or demographers any more than they can by critics, political economists, or cultural historians.

The impossibility of fully analyzing, much less synthesizing such fluid activities should be clear. The task becomes one of recognizing the interplay and, when possible, mapping the lines of force and influence most pertinent to any case at hand. Some studies stand out as exemplary in this difficult process. In the early 1980s the collection of essays by Jane Feuer and colleagues, *MTM: Quality*

Television (1984), admirably linked certain shifts in the US television industry and various aspects of US sociology and culture to examine what seemed to be fundamental stylistic alterations in programming. They never lost sight of the connections of those newer programs to examples from previous periods in the brief history of the medium, but still made a convincing case for a set of intersecting influences shaping the changes they outlined. A cluster of important historical studies by William Boddy (1990), Lynn Spigel (1992), Christopher Anderson (1994), and Michael Curtin (1996) brought new sophistication to topics ranging from television as the site of policy struggles, to television's role in a new domestic context, to television's intersectional struggles with the film industry, to the role of network policies, government actions, and documentary production.

Studies of specific television programs have also been richly contextualized by scholars exploring a range of influences and affectivities of the medium. Julie D'Acci's *Defining Women: Television and the Case of Cagney and Lacey* (1994) is an outstanding work linking analyses of television industrial practices, production practices, texts, and audience responses. Jostein Gripsrud's *The "Dynasty" Years: Hollywood Television and Critical Media Studies* (1995) examines the ways in which a single American television program, thrown into the lake of another society and culture, sends ripples reaching to parliaments and political activist groups. John Thornton Caldwell's *Televisuality: Style, Crisis, and Authority in American Television* (1995) adds the layer of "redefining" television in light of specific developments in technologies and industrial history; Ron Lembo's *Thinking Through Television* (2000) explores audience relationships with television from a sociological perspective, incorporating a version of ethnographic study with a sophisticated sense of textual nuances and programming strategies; and Anna McCarthy's *Ambient Television* (2001) explodes the general conception that television is solely or primarily a domestic device by studying a range of sites in which the medium can be embedded.

Finally, in Hartley's *Uses of Television* (1999), I find what is, for me, the most challenging and from its own perspective explanatory treatment of television to date. Among other taxonomical gambits Hartley lumps the history of television studies into two large, crude clumps – The Desire School and The Fear School (p. 135), placing most of the work concerned with televisions presumed "effects" in the latter, most of the work treating television as an expressive form in the former. But the clustering is secondary to his own perspective that television primarily serves a "pedagogical" function in contemporary culture, spreading forms of broad knowledge and information into corners that might otherwise have missed such perceptions, or challenging received notions with purposeful provocations. In short, without focusing precisely on particular program "texts," or on specific analyses of overarching "ideology," on specific industrial formations or practices, or on details of audience response and activity, he returns to fundamental philosophical questions: What is television? How has it functioned? Why is it even important, or at least, why and how is it more important than the refrigerator?

I do not suggest here that Hartley, or the other works cited above, "explain" television in any total sense better than many earlier studies. Indeed, I am arguing that "television" is inexplicable. But it is no longer necessary for those who study television to remain bound by their own particular languages and strategies. Rather, it is necessary that they acknowledge one another more explicitly, incorporating those other strategies, topics, areas, and problems they find most pertinent, most forceful in modifying their own conclusions. In one sense, "television studies," as an intellectual accomplishment in itself, should best exercise a form of modesty. But the modifications should also lead toward a keen precision that might allow television studies to achieve a stronger voice in matters of policy, industrial practice, and viewer education. In both the modesty and the precision we can acknowledge that with regard to television from the mid-twentieth century to present day, this set of intersecting forces, practices, and influences has demanded attention and concern – and that at every turn of events it has refracted, prism-like, every light we bring toward its illumination. In the play of these bent, blended, and colored shadows we find the best repository for better questions.

References

Anderson, C. (1994) *Hollywood TV: The Studio System in the Fifties*, Austin: University of Texas Press.

Boddy, W. (1990) *Fifties Television: The Industry and its Critics*, Urbana, IL: University of Illinois Press.

Brunsdon, C. (1997) "Television Studies," in H. Newcomb (ed.), *Museum of Broadcast Communications Encyclopedia of Television*, Chicago and London: Fitzroy Dearborn Publishers, pp. 1647–9.

Caldwell, J. T. (1995) *Televisuality: Style, Crisis and Authority in American Television*, New Brunswick, NJ: Rutgers University Press.

Corner, J. (1999) *Critical Ideas in Television Studies*, Oxford: Oxford University Press.

Curtin, M. (1996) *Redeeming the Wasteland: Television Documentary and Cold War Politics*, New Brunswick, NJ: Rutgers University Press.

D'Acci, J. (1994) *Defining Women: Television and the Case of Cagney and Lacey*, Chapel Hill, NC: University of North Carolina Press.

De Forest, L. (1942) *Television: Today and Tomorrow*, New York: Dial Press.

Feuer, J., Kerr, P., and Vahimagi, T. (1984) *MTM: Quality Television*, London: British Film Institute.

Fiske, J. and Hartley, J. (1978) *Reading Television*, London: Methuen.

Gripsrud, J. (1995) *The "Dynasty" Years: Hollywood Television and Critical Media Studies*, London: Routledge.

Hartley, J. (1999) *The Uses of Television*, London: Routledge.

Horkheimer, M. and Adorno, T. (1972) *Dialectic of Enlightenment*, Translated by John Cuming, New York: Herder and Herder.

Lembo, R. (2000) *Thinking Through Television*, Cambridge: Cambridge University Press.

McCarthy, A. (2001) *Ambient Television*, Durham, NC: Duke University Press.

McGuigan, J. (1992) *Cultural Populism*, London: Routledge.

McGuigan, J. (1996) *Culture and the Public Sphere*, London: Routledge.

Horace Newcomb

Newcomb, H. (1974) *TV: The Most Popular Art*, Garden City, NJ: Doubleday Anchor.
Newcomb, H. (2000) *Television: The Critical View*, 6th edn., New York: Oxford University Press.
Schramm, W. (1939) *Two Creative Traditions in English Poetry*, New York: Farrar and Rinehart.
Spigel, L. (1992) *Make Room for TV*, Chicago: University of Chicago Press.

CHAPTER TWO

Critical Perspectives on Television from the Frankfurt School to Postmodernism

Doug Kellner

Paul Lazarsfeld (1942), one of the originators of modern communications studies, distinguished between what he called "administrative research," that deployed empirical research for the goals of corporate and state institutions, and "critical research," an approach that he associated with the Frankfurt School. Critical research situates the media within the broader context of social life and interrogates its structure, goals, values, messages, and effects. It develops critical perspectives by which media are evaluated and appraised.

Since the 1940s, an impressive variety of critical approaches to the media and television have developed. In this study, I will first present the Frankfurt School as an inaugurator of critical approaches to television studies and will then consider how a wide range of theorists addressed what later became known as the politics of representation in critical television studies, engaging problematics of class, gender, race, sexuality, and other central components of media representation and social life. Then, I discuss how a postmodern turn in cultural studies contested earlier critical models and provided alternative approaches to television studies. I conclude with some comments that argue for a critical approach to television and media culture and in this text sketch out a comprehensive critical model that embraces production and the political economy of television; textual analysis; and investigation of the effects and uses of television by audiences. As this study will indicate, such a multidimensional approach to critical media and television studies is found initially in the Frankfurt School and was developed by many other television theorists in diverse locations and from often conflicting perspectives, ranging from British cultural studies to critical feminism.

Doug Kellner

The Frankfurt School and the Culture Industries

From the classical Frankfurt School perspective, commercial television is a form of what Horkheimer and Adorno and their colleagues called the "culture industry." Moving from Nazi Germany to the United States, the Frankfurt School experienced at first hand the rise of a media culture involving film, popular music, radio, television, and other forms of mass culture.[1] In the United States, where they found themselves in exile, media production was by and large a form of commercial entertainment controlled by big corporations. Thus, the Frankfurt School coined the term "culture industries" to call attention to the industrialization and commercialization of culture under capitalist relations of production. This situation was most marked in the United States that had little state support of film or television industries.

To a large extent, the Frankfurt School began a systematic and comprehensive critical approach to studies of mass communication and culture, and produced the first critical theory of the cultural industries.[2] During the 1930s, the Frankfurt School developed a critical and transdisciplinary approach to cultural and communications studies, combining a critique of the political economy of the media, analysis of texts, and audience reception studies of the social and ideological effects of mass culture and communications. They coined the term "culture industry" to signify the process of the industrialization of mass-produced culture and the commercial imperatives which drove the system. The critical theorists analyzed all mass-mediated cultural artifacts within the context of industrial production, in which the commodities of the culture industries exhibited the same features as other products of mass production: commodification, standardization, and massification. The culture industries had the specific function, however, of providing ideological legitimation of the existing capitalist societies and of integrating individuals into the framework of its social formation.

Key early studies of the culture industries include Adorno's analyses of popular music (1978 [1932], 1941, 1982, and 1989), television (1991), and popular phenomena such as horoscopes (1994); Lowenthal's studies of popular literature and magazines (1961); Herzog's studies of radio soap operas (1941); and the perspectives and critiques of mass culture developed in Horkheimer and Adorno's famous study of the culture industries (1972 and Adorno, 1991). In their critiques of mass culture and communication, members of the Frankfurt School were among the first to systematically analyze and criticize mass-mediated culture and television within critical social theory. They were the first social theorists to see the importance of what they called the "culture industries" in the reproduction of contemporary societies, in which so-called mass culture and communications stand at the center of leisure activity, are important agents of socialization, mediators of political reality, and should thus be seen as major institutions of contemporary societies with a variety of economic, political, cultural and social effects.

Furthermore, the critical theorists investigated the cultural industries in a political context, conceptualizing them as a form of the integration of the working class into capitalist societies. The Frankfurt School was one of the first neo-Marxian groups to examine the effects of mass culture and the rise of the consumer society on the working classes, which were to be the instrument of revolution in the classical Marxian scenario. They also analyzed the ways that the culture industries and consumer society were stabilizing contemporary capitalism and accordingly sought new strategies for political change, agencies of social transformation, and models for political emancipation that could serve as norms of social critique and goals for political struggle. This project required rethinking the Marxian project and produced many important contributions – as well as some problematical positions.

The Frankfurt School provides television and media studies with a model that articulates the dimensions of production and political economy, text analysis, and audience/reception research. The Frankfurt School addresses all of these dimensions and at its best depicts their interrelationship. Indeed, Frankfurt School critical theory provides the "Big Picture," analyzing relationships between the economy, state, society, and everyday life (Kellner, 1989). Thus, a critical theory of television would articulate the relationships between the economy, the state, and television, analyzing television's production process, texts, and sociopolitical effects and audience uses within the context of its institutional role within specific types of social organization (see Kellner, 1990). I will accordingly discuss the classical Frankfurt School model of television and some specific attempts to provide analyses of television within the Frankfurt School tradition before turning to other critical approaches.

In their *Dialectic of Enlightenment*, Horkheimer and Adorno anticipate the coming of television in terms of the emergence of a new form of mass culture that would combine sight and sound, image, and narrative, in an institution that would embody the types of production, texts, and reception of the culture industry. Anticipating that television would be a prototypical artifact of industrialized culture, Adorno and Horkheimer wrote:

Television aims at a synthesis of radio and film, and is held up only because the interested parties have not yet reached agreement, but its consequences will be quite enormous and promise to intensify the impoverishment of aesthetic matter so drastically, that by tomorrow the thinly veiled identity of all industrial culture products can come triumphantly out into the open, derisively fulfilling the Wagnerian dream of the *Gesamtkunstwerk*, the fusion of all the arts in one work. The alliance of word, image, and music is all the more perfect than in Tristan because the sensuous elements which all approvingly reflect the surface of social reality are in principle embodied in the same technical process, the unity of which becomes its distinctive content . . . Television points the way to a development which might easily enough force the Warner Brothers into what would certainly be the unwelcome position of serious musicians and cultural conservatives. (1972, pp. 124, 161)

Following the model of critique of mass culture in *Dialectic of Enlightenment*, a Frankfurt School approach to television would analyze television within the dominant system of cultural production and reception, situating the medium within its institutional and political framework. It would combine the study of text and audience with an ideology critique and a contextualizing analysis of how television texts and audiences are situated within specific social relations and institutions. The approach combines Marxian critique of political economy with ideology critique, textual analysis, and psychoanalytically inspired depth-approaches to audiences and effects.

T. W. Adorno's article "How to Look at Television" (1991) provides a striking example of a classic Frankfurt School analysis. Adorno opens by stressing the importance of undertaking an examination of the effects of television upon viewers, making use of "depth-psychological categories." Adorno had previously collaborated with Paul Lazarsfeld on some of the first examinations of the impact of radio and popular music on audiences (see Lazarsfeld, 1941). While working on *The Authoritarian Personality* (Adorno et al., 1969 [1950]), Adorno took up a position as director of the scientific branch of the Hacker Foundation in Beverly Hills, a psychoanalytically-oriented foundation, and undertook examinations of the sociopsychological roots and impact of mass cultural phenomena, focusing on subjects as diverse as television (Adorno, 1991) and the astrological column of the *Los Angeles Times* (Adorno, 1994).

In view of the general impression that the Frankfurt School make sharp and problematic distinctions between high and low culture, it is interesting that Adorno opens his study of television with a deconstruction of "the dichotomy between autonomous art and mass media." Stressing that their relation is "highly complex," Adorno claims that distinctions between popular and elite art are a product of historical conditions and should not be exaggerated. After a historical examination of older and recent popular culture, Adorno analyzes the "multilayered structure of contemporary television." In light of the notion that the Frankfurt School reduces the texts of media culture to ideology, it is interesting that Adorno calls for analysis of the "various layers of meaning" found in popular television, stressing "polymorphic meanings" and distinctions between latent and manifest content. Adorno writes:

> The effect of television cannot be adequately expressed in terms of success or failure, likes or dislikes, approval or disapproval. Rather, an attempt should be made, with the aid of depth-psychological categories and previous knowledge of mass media, to crystallize a number of theoretical concepts by which the potential effect of television – its impact upon various layers of the spectator's personality – could be studied. It seems timely to investigate systematically socio-psychological stimuli typical of televised material both on a descriptive and psychodynamic level, to analyze their presuppositions as well as their total pattern, and to evaluate the effect they are likely to produce . . . (1991, p. 136)

Adorno's examples come from the early 1950s TV shows and he tends to see these works as highly formulaic and reproducing conformity and adjustment. He criticizes stereotyping in television, "pseudo-realism," and its highly conventional forms and meaning, an approach that accurately captures certain aspects of 1950s television, but which is inadequate to capture the growing complexity of contemporary television. Adorno's approach to "hidden meanings" is highly interesting, however, and his psychoanalytic and ideological readings of television texts and speculation on their effects are pioneering.

Adorno's study is one of the few concrete studies of television with the Frankfurt School tradition that addresses the sort of text produced by network television and the audience for its product. While Horkheimer, Adorno, Marcuse, Habermas, and other major Frankfurt School theorists never systematically engage television production, texts, or audiences, they frequently acknowledge the importance of television in their development of a critical theory of society, or in their comments on contemporary social phenomena. Following the Frankfurt School analysis of changes in the nature of socialization, Herbert Marcuse, for instance, noted the decline of the family as the dominant agent of socialization in *Eros and Civilization* (1955) and the rise of the mass media, like radio and television:

> The repressive organization of the instincts seems to be *collective*, and the ego seems to be prematurely socialized by a whole system of extra-familial agents and agencies. As early as the pre-school level, gangs, radio, and television set the pattern for conformity and rebellion; deviations from the pattern are punished not so much in the family as outside and against the family. The experts of the mass media transmit the required values; they offer the perfect training in efficiency, toughness, personality, dream and romance. With this education, the family can no longer compete. (97)

Marcuse saw television as being part of an apparatus of administration and domination in a one-dimensional society. In his words,

> with the control of information, with the absorption of individuals into mass communication, knowledge is administered and confined. The individual does not really know what is going on; the overpowering machine of entertainment and entertainment unites him with the others in a state of anesthesia from which all detrimental ideas tend to be excluded. (104)

On this view, television is part of an apparatus of manipulation and societal domination. In *One-Dimensional Man* (1964), Marcuse claimed that the inanities of commercial radio and television confirmed his analyses of the individual and the demise of authentic culture and oppositional thought, portraying television as part of an apparatus producing the thought and behavior needed for the social and cultural reproduction of contemporary capitalist societies.

Doug Kellner

Critical Perspectives from/after the Frankfurt School

While the classical Frankfurt School members wrote little on television itself, the critical theory approach strongly influenced critical approaches to mass communication and television within academia and the views of the media of the New Left and others in the aftermath of the 1960s. The anthology *Mass Culture* (Rosenberg and White, 1957) contained Adorno's article on television and many other studies influenced by the Frankfurt School approach. Within critical communication research, there were many criticisms of network television as a capitalist institution and critics of television and the media such as Herbert Schiller, George Gerbner, Dallas Smythe, and others were influenced by the Frankfurt School approach to mass culture, as was C. Wright Mills in an earlier era (see Kellner, 1989, p. 134ff).

From the perspectives of the New Left, Todd Gitlin wrote "Thirteen Theses on Television" that contained a critique of television as manipulation with resonances to the Frankfurt School in 1972 and continued to do research and writing that developed in his own way a Frankfurt School approach to television, focusing on TV in the United States (1980, 1983, 2002). A 1987 collection *Watching Television* contained studies by Gitlin and others that exhibited a neo-Frankfurt School approach to television, and many contemporary theorists writing on television have been shaped by their engagement with the Frankfurt School.

Of course, media culture was never as massified and homogeneous as the Frankfurt School model implied and one could argue that their perspectives were flawed even during their time of origin and influence. One could also argue that other approaches were preferable (such as those of Walter Benjamin (1969), Siegfried Kracauer (1995), Ernst Bloch (1986) and others of the Weimar generation). The original Frankfurt School model of the culture industry did articulate the important social roles of media culture during a specific regime of capital. The group provided a model, still of use, of a highly commercial system of television that serves the needs of dominant corporate interests, plays a major role in ideological reproduction, and in enculturating individuals into the dominant system of needs, thought, and behavior.

Today, it is more fashionable to include moments of Frankfurt School critique of television in one's theory than to simply adopt a systematic Frankfurt School approach. It would be a mistake, however, to reject the Frankfurt School tout court as reductive, economistic, and representative solely of a one-dimensional "manipulation theory," although these aspects do appear in some of their writings. Indeed, the systematic thrust of the Frankfurt School approach that studies television and other institutions of media culture in terms of their political economy, text, and audience reception of cultural artifacts continues to be of some use. Overcoming the divide between a text-based approach to culture and an empiricist social science-based communication theory, the Frankfurt School sees media culture as a complex multidimensional phenomenon that must be taken seriously and that requires multiple disciplines to capture its importance and complexity. Within the culture industries, television continues to be of central importance and so critical theorists today should seek new approaches to television while building upon the Frankfurt School tradition.

In recent decades other critical studies have researched the impact of global media on national cultures, attacking the cultural imperialism of Western media conglomerates or the creeping Americanization of global media and consumer culture (Schiller, 1971; Tunstall, 1977). Schiller and others focused on the political economy of television and its role, both nationally and globally, in promoting corporate interests. In *Mass Communications and American Empire* (1971), Herbert Schiller traced the rise of the commercial broadcasting industry in the United States, its interconnection with corporate capitalism and the military, and the use of communications and electronics in counterrevolution, such as Vietnam, and in promoting a global capitalist economic empire.

Political economy approaches to television charted the consequences of dominance of TV production by corporate and commercial interests and the ways that programming was geared toward concerns of advertisers and securing the largest possible mass audience. Herman and Chomsky (1988) presented "filters" by which corporate, advertising, media gatekeeping, and conservative control excluded certain kinds of programming while excluding less mainstream and conservative material. Scholars studying media imperialism traced how the importation of US programming and broadcasting institutions and structures impacted broadcasting on a global scale.[3]

Some critical approaches focused on the social effects of television, often decrying excessive TV violence. On television and violence, some literature continued to assume that violent representations in the media were a direct cause of social problems. A more sophisticated social ecology approach to violence and the media, however, was developed by George Gerbner and his colleagues at the Annenberg School of Communication. Gerbner's group has studied the "cultural environment" of television violence, tracking increases in representations of violence and delineating "message systems" that depict who exercises violence, who is the victim, and what messages are associated with media violence. A "cultivation analysis" studies effects of violence and concludes that heavy consumers of media violence exhibit a "mean world syndrome" with effects that range from depression to fearful individuals voting for right-wing law and order politicians, to the exhibition of violent behavior (Gerbner, 2003).[4]

Another approach to violence and the media is found in the work of Hans J. Eysenck and David K. B. Nias (1978) who argue that recurrent representations of violence in the media desensitize audiences to violent behavior and actions. The expansion of youth violence throughout the world and media exploitation of sensational instances of teen killings in the United States, Britain, France, Germany and elsewhere has intensified the focus on the interplay of media and violence and the ways that rap music, video and computer games, television and film, and other types of youth culture have promoted violence.[5]

In addition to seeing television as a social problem because of growing societal violence, from the 1960s to the present, left-liberal and conservative media critics coalesced in arguing that mainstream media promote excessive consumerism and commodification. In the 1960s FCC commissioner Newton Minow described TV as a "Vast Wasteland" and the term was taken up by both conservative and left-liberal critics to assail what was perceived as the growing mediocrity and low cultural level of television. This view is argued in sociological terms in the work of Daniel Bell who asserts, in *The Cultural Contradictions of Capitalism* (1978), that a sensate-hedonistic culture exhibited in popular media and promoted by capitalist corporations was undermining core traditional values and producing an increasing amoral society. Bell called for a return to tradition and religion to counter this social trend that saw media culture as undermining morality, the work ethic, and traditional values.

In *Amusing Ourselves to Death* (1986), Neil Postman argued that popular media culture – and, in particular, television – has become a major force of socialization and was subverting traditional literacy skills, thus undermining education. Postman criticized the negative social effects of the media and called for educators and citizens to intensify their critique of the media. Extoling the virtues of book culture and literacy, Postman called for educational reform to counter the nefarious effects of media and consumer culture.

Indeed, there is by now a long tradition of studies that have discussed children and media such as television (see Luke, 1990). Critics like Postman (1986) argue that excessive TV viewing stunts cognitive growth, creates shortened attention spans, and habituates youth to fragmented, segmented, and imagistic cultural experiences and thus television and other electronic media are a social problem for children. Defenders stress the educational benefits of some television, suggest that it is merely harmless entertainment, or argue that audiences construct their own meanings from popular media (Fiske, 1987, 1989a).

Negative depictions of the media and consumerism, youth hedonism, excessive materialism, and growing violence were contested by British cultural studies that claimed that the media were being scapegoated for a wide range of social problems. In *Policing the Crisis* (Hall et al., 1978), Stuart Hall and colleagues at the Birmingham Centre for Contemporary Cultural Studies analyzed what they took to be a media-induced "moral panic" about mugging and youth violence. The Birmingham group argued for the existence of an active audience that was able to critically dissect and make use of media material, arguing against the media manipulation perspective. Rooted in a classic article by Stuart Hall titled "Encoding/Decoding" (1973/1980), British cultural studies began studying how different groups read television news, magazines, engaged in consumption, and made use of a broad range of media. In *Everyday Television: Nationwide* Charlotte Brunsdon and David Morley (1978) studied how different audiences consumed TV news; Ien Ang (1985) and Liebes and Katz (1990) investigated how varying audiences in Holland, Israel, and elsewhere consumed and made use of the US TV series *Dallas*; and John Fiske (1987, 1989a and 1989b) wrote a series of books celebrating the active audience and consumer in a wide range of domains by audiences throughout the world.

Yet critics working within British cultural studies, individuals in a wide range of social movements, and academics from a variety of fields and positions, began criticizing the media from the 1960s and to the present for promoting sexism, racism, homophobia, and other oppressive social phenomena. There was intense focus on the politics of representation, discriminating between negative and positive representations of major social groups and harmful and beneficial media effects, debates that coalesced under the rubric of the politics of representation.

Oppositional Social Movements and the Politics of Representation

During the 1960s much television criticism was somewhat unsophisticated and underdeveloped theoretically, often operating with reductive notions of political economy; simplistic models of media effects; and one-dimensional models of media messages. Yet from the 1960s to the present, a wide range of critical theories circulated globally and many working within television studies appropriated the advanced critical discourses.

The ground-breaking work of critical media theorists within the Frankfurt School, British cultural studies, and French structuralism and poststructuralism revealed that culture is a social construct, intrinsically linked to the vicissitudes of the social and historically specific milieu in which it is conceived and that gender, race, class, sexuality, and other dimensions of social life are socially constructed in media representations (see Durham and Kellner, 2001). Media and cultural studies engaged in critical interrogations of the politics of representation, which drew upon feminist and gay and lesbian approaches, as well as critical race and multicultural theories, to fully analyze the functions of gender, class, race, ethnicity, nationality, sexual preference and other key issues in television and the media.

The social dimensions of media constructions of axes of difference and subordination are perceived by cultural studies as being vitally constitutive of audiences who appropriate and use texts. These approaches were strongly influenced by the social movements of the era. The feminist movement opposed media representation of women and criticized ones claimed to be sexist and inadequate, while calling for more positive representations of women and the participation of more women in the culture industries. Black and brown power movements criticized representations of people of color and militated for more inclusion in television and other media, as well as more realist and positive depictions. Likewise, gay and lesbian movements criticized the media for their neglect or misrepresentations of alternative sexuality and more representation.

All of these oppositional movements developed critical perspectives on television and often produced new forms of TV criticism, positioning the politics of representation as a crucial part of television studies.[6] Developments within British cultural studies are representative of this move toward a more inclusive politics of representation and TV criticism. While earlier British cultural studies engaged the progressive and oppositional potential of working-class and then youth culture, under the pressure of the social movements of the 1960s and 1970s, many adopted a feminist dimension, paid greater attention to race, ethnicity, and nationality, and concentrated on sexuality. During this period, assorted discourses of race, gender, sex, nationality, and so on developed within a now global cultural studies. An increasingly complex, culturally hybrid and diasporic global culture and networked society calls for sophisticated understandings of the interplay of representations, politics, and the forms of media.

Although a vigorous feminist film and cultural criticism had begun to emerge by the 1970s, little feminist TV criticism emerged until the 1980s.[7] As with feminist film criticism, early efforts focused on the image and representations of women, but soon there was more sophisticated narrative analysis that analyzed how television and the narrative apparatus positioned women and the ways that television constructed femininity and masculinity, as well as more sociological and institutional analysis of how TV functioned in women's everyday life and how the institutions of television were highly male-dominated and patriarchal and capitalist in structure.

Tania Modleski (1982), for instance, followed a ground-breaking essay by Carol Lopate (1977) on how the organization of the TV day followed the patterns of women's lives. Soap operas present a fragmented ongoing narrative that provides distraction and fantasies for women at home while ideologically positioning women in traditional stereotyped roles. The moral ambiguities and openness of the form provide spaces for multiple viewers, make possible varied readings, and provide predictable pleasures for its audiences. Addressing the alteration between the soap narratives and those of commercials, Modleski suggests that these modes address women's dual roles as "moral and spiritual guides" and "household drudges," thus reproducing the values and subject positions of patriarchal capitalism.

Many gay and lesbian theorists decried the ways that media representations promoted homophobia by presenting negative representations of gay sexuality. Larry Gross's "Out of the Mainstream: Sexual Minorities and the Mass Media" (1989) argues that corporate media culture defines and frames sexuality in ways that marginalize gay and lesbians, and "symbolically annihilate" their lives. Stereotypic depiction of lesbians and gay men as "abnormal, and the suppression of positive or even 'unexceptional' portrayals, serve to maintain and police the boundaries of the moral order" (1989, p. 136) in Gross's view. He argues for alternative representations – a call that has to a certain degree been heard and answered by gay and lesbian media producers coming to prominence in the contemporary era, with even US network television eventually presenting gay and lesbian characters.

A variety of critics of color have engaged racist representations in film, television, and other domains of media culture.[8] Herman Gray (1995), for example, scrutinizes the related trajectory of black representation on network television in an analysis that takes into account the structures and conventions of the medium as well as the sociopolitical conditions of textual production. Gray's examination of race and representation highlights the articulations between recent representations of blacks and much earlier depictions. He argues that "our contemporary moment continues to be shaped discursively by representations of race and ethnicity that began in the formative years of television" (1995, p. 73). Contemporary cultural production is still in dialogue with these earliest moments, he writes, and he is aware of the regressive as well as the progressive aspects of this engagement. Importantly, Gray identifies certain turning points in television's representation of blackness, situating these "signal moments" within the cultural and political contexts in which they were generated. His analysis brings us to a confrontation with the possibilities of mass cultural texts engaging the politics of difference in a complex and meaningful way.

Many critics emphasized the importance of connecting representations of gender, race, class, sexuality, and other subject positions to disclose how the media present socially derogatory representations of subordinate groups. bell hooks (1992) has been among the first and most prolific African-American feminist scholars to call attention to the interlocking of race, class, gender and additional markers of identity in the constitution of subjectivity. Early in her career she challenged feminists to recognize and confront the ways in which race and class inscribe women's (and men's) experiences. In "Eating the Other" (1992), hooks explores cultural constructions of the "Other" as an object of desire, tying such positioning to consumerism and commodification as well as to issues of racial domination and subordination. Cautioning against the seductiveness of celebrating "Otherness," hooks uses various media cultural artifacts – clothing catalogs, films, television, and rap music – to debate issues of cultural appropriation versus cultural appreciation, and to uncover the personal and political cross-currents at work in mass-media representation.

Elaine Rapping has written a series of books engaging dynamics of gender, race, and class while relating television to current social and political issues. *The Looking Glass World of Nonfiction Television* (1986) provides a study of local and national news, game shows, national rituals, beauty pageants, and presidential politics, as well as studies of TV documentaries, special reports, and soft news. Her studies of made-for-TV movies was expanded into *The Movie of the Week* (1992), a ground-breaking analysis of TV movies which had hitherto been somewhat ignored by both film and television scholars. Her recent *Law and Justice As Seen on TV* (2003) traces the history of crime drama and courtroom drama and the ways that actual crimes and problems of justice are represented in TV frames and dramas from the Menendez brothers trial, to the O. J. Simpson murder trials, and Timothy McVeigh's Oklahoma City bombing case.

TV representations often construct women, people of color, and members of various minorities and their social problems as victims and objects, and mainstream television rarely presents positive representations of women's movements or collective forms of struggle, rather focusing on women as individual examples of specific social problems like rape or domestic violence. Likewise, television series featuring people of color often appropriate groups such as African Americans or Latinos into typical white middle-class American behavior, values, and institutions, rather than articulating cultural specificity or showing oppressed groups voicing criticisms or organizing into political movements.

Just as critical television critics came to insist on the interaction of the politics of representation in race, gender, class, sexuality, and other key dimensions, so too did critical television scholars begin to integrate studies of the TV industry, texts, audiences, and social context into their work. For instance, in a groundbreaking work on *Cagney and Lacey*, Julie D'Acci calls for an "integrated approach" that analyzes how the politics of representation play out in the television production process, on the level of the construction and unfolding of TV texts and narratives, on the level of audience reception, and within the context of specific sociohistorical environments (1994, 2002). Such "modern approaches," however, were criticized by a postmodern turn in television and cultural studies.

The Postmodern Turn within Critical Television Studies

During the 1980s and 1990s, many noticed a postmodern turn toward cultural populism that valorized audiences over texts and the production apparatus, the pleasures of television and popular culture over their ideological functions and effects, and that refocused television criticism on the surface of its images and spectacle, rather than deeper embedded meanings and complex effects (see Best and Kellner, 1997; McGuigan, 1992; and Kellner, 1995). If for most of the history of television, narrative storytelling has been the name of the game, on a postmodern account of television, image and spectacle often decenters the importance of narrative. It is often claimed that in those programs usually designated "postmodern" – MTV music videos and other programming, *Miami Vice*, *Max Headroom, Twin Peaks*, high-tech ads, and so on – there is a new look and feel: the signifier has been liberated and image takes precedence over narrative, as compelling and highly artificial aesthetic spectacles detach themselves from the television diegesis and become the center of fascination, of a seductive pleasure, of an intense but fragmentary and transitory aesthetic experience.

While there is some truth in this conventional postmodern position, such descriptions are in some ways misleading. In particular, the familiar account that postmodern image culture is fundamentally flat and one-dimensional is problematic. For Fredric Jameson, postmodernism manifests "the emergence of a new kind of flatness or depthlessness, a new kind of superficiality in the most literal sense – perhaps the supreme formal feature of all the postmodernisms" (1984, p. 60). According to Jameson, the "waning of affect" in postmodern image culture is replicated in postmodern selves who are allegedly devoid of the expressive energies and individualities characteristic of modernism and the modern self. Both postmodern texts and selves are said to be without depth and to be flat, superficial, and lost in the intensities and vacuities of the moment, without substance and meaning, or connection to the past.

Privileging Jameson's category of the waning of affect, Gitlin (1987), for example, claims that *Miami Vice* is the ultimate in postmodern blankness, emptiness, and world-weariness. Yet, against this reading, one could argue that it pulsates as well with intense emotion, a clash of values, and highly specific political messages and positions (see Best and Kellner, 1997 and Kellner, 1995). Grossberg (1987) also argues that *Miami Vice* and other postmodern culture obliterate meaning and depth, claiming: "*Miami Vice* is, as its critics have said, all on the surface. And the surface is nothing but a collection of quotations from our own collective historical debris, a mobile game of Trivia. It is, in some ways, the perfect televisual image, minimalist (the sparse scenes, the constant long shots, etc.) yet concrete" (1987, p. 28). Grossberg goes on to argue that "indifference" (to meanings, ideology, politics, and so on) is the key distinguishing feature of *Miami Vice* and other postmodern texts which he suggests are more akin to billboards to be scanned for what they tell us about our cultural terrain rather than texts to be read and interrogated.

Against these postmodern readings, one could argue that *Miami Vice* is highly polysemic and is saturated with ideologies, messages, and quite specific meanings and values. Behind the high-tech glitz are multiple sites of meaning, multiple subject positions, and highly contradictory ideological problematics. The show had a passionately loyal audience that was obviously not indifferent to the series that had its own intense affective investments and passions. I have argued that *reading* the text of *Miami Vice* hermeneutically and critically provides access to its polysemic wealth and that therefore it is a mistake to rapidly speed by such artifacts, however some audiences may relate to them (Kellner, 1995, p. 238ff).

One-dimensional postmodern texts and selves put in question the continued relevance of hermeneutic depth models such as the Marxian model of essence and appearance, true and false consciousness, and ideology and truth; the Freudian model of latent and manifest meanings; the existentialist model of authentic and inauthentic existence; and the semiotic model of signifier and signified. Cumulatively, postmodernism thus signifies the death of hermeneutics; in place of what Ricoeur has termed a "hermeneutics of suspicion" and the polysemic modernist reading of cultural symbols and texts, there emerges the postmodern view that there is nothing behind the surface of texts, no depth or multiplicity of meanings for critical inquiry to discover and explicate.

From this view of texts and selves, it follows that a postmodern television studies should rest content to describe the surface or forms of cultural texts, rather than seeking meanings or significance. Best and Kellner (1997) have polemicized against the formalist, anti-hermeneutical postmodern type of analy-sis connected with the postulation of a flat, postmodern image culture and have delineated an alternative model of a "political hermeneutic" which draws on both postmodern and other critical theories in order to analyze both image *and* mean-ing, surface *and* depth, as well as the politics *and* erotics of cultural artifacts. Such an interpretive and dialectical analysis of image, narrative, ideologies, and meanings is arguably still of importance in analyzing even those texts taken to be paradigmatic of postmodern culture – though analysis of form, surface, and look is also important. Images, fragments, and narratives of media culture are satu-rated with ideology and polysemic meanings, and that therefore – against certain postmodern positions (Foucault, 1977; Baudrillard, 1981; and Deleuze/Guattari, 1977) – ideology critique continues to be an important and indispensable weapon in our critical arsenal.[9]

Another problematic postmodern position, associated with Baudrillard (1983a, b), asserts that television is pure noise and a black hole where all meaning and messages are absorbed in the whirlpool and kaleidoscope of the incessant dis-semination of images and information to the point of total saturation, where meaning is dissolved and only the fascination of discrete images glow and flicker in a mediascape within which no image any longer has any discernible effects. On the Baudrillardian view, the proliferating velocity and quantity of images produces a postmodern mindscreen where images fly by with such rapidity that they lose any signifying function, referring only to other images ad infinitum, and where eventually the multiplication of images produces such saturation, apathy, and indifference that the tele-spectator is lost forever in a fragmentary fun house of mirrors in the infinite play of superfluous, meaningless images.

Now, no doubt, television can be experienced as a flat, one-dimensional waste-land of superficial images, and can function as well as pure noise without referent and meaning. One can also become overwhelmed by – or indifferent to – the flow, velocity, and intensity of images, so that television's signifying function can be decentered and can collapse altogether. Yet people regularly watch certain shows and events; there are fans for various series and stars who possess an often incredible expertise and knowledge of the subjects of their fascination; people do model their behavior, style, and attitudes on television images and narratives; television ads do play a role in managing consumer demand; and many analysts have concluded that television is playing the central role in political elections, that elections have become a battle of images played out on the television screen, and that television is playing an essential role in the new art of governing (Kellner, 1990, 1992, 1995, 2001, 2003a and 2003b).

As British cultural studies have long argued, different audiences watch tele-vision in different ways. For some, television is nothing more than a fragmented collage of images that people only fitfully watch or connect with what goes before or comes after. Many individuals today use devices to "zap" from one program to another, channel hopping or "grazing" to merely "see what's happening," to go with the disconnected flow of fragments of images. Some viewers who watch entire programs merely focus on the surface of images, with programs, ads, station breaks, and so on flowing into each other, collapsing meaning in a play of disconnected signifiers. Many people cannot remember what they watched the night before, or cannot provide coherent accounts of the previous night's programming.

And yet it is an exaggeration to claim that the apparatus of television itself relentlessly undermines meaning and collapses signifiers without signifieds into a flat, one-dimensional hyperspace without depth, effects, or meanings. Thus, against the postmodern notion of culture disintegrating into pure image without referent or content or effects – becoming at its limit pure noise – many critics argue that television and other forms of mass-mediated culture continue to play key roles in the structuring of contemporary identity and shaping thought and behavior. Television today arguably assumes some of the functions traditionally ascribed to myth and ritual (i.e. integrating individuals into the social order, celebrating dominant values, offering models of thought, behavior, and gender for imitation, and so on; see Kellner, 1979 and 1995). In addition, TV myth resolves social contradictions in the way that Lévi-Strauss described the function of traditional myth and provided mythologies of the sort described by Barthes that idealize contemporary values and institutions, and thus exalt the established way of life (Kellner, 1979 and 1982).

Consequently, much postmodern cultural analysis is too one-sided and limited, in either restricting its focus on form, on image and spectacle alone, or in abandoning critical analysis altogether in favor of grandiose totalizing metaphors (black holes, implosion, excremental culture, and so on). Instead, it is preferable to analyze both form and content, image and narrative, and postmodern surface and the deeper ideological problematics within the context of specific exercises which explicate the polysemic nature of images and texts, and which endorse the possibility of multiple encodings and decodings.

Thus, I would conclude that critical perspectives developed by the Frankfurt School, British cultural studies, and other scholars who focus on dissection of television production and political economy, texts, audience reception, and sociopolitical context in a multiperspectivist framework provide the most comprehensive and flexible model for doing critical television studies. For some projects, one may choose to intensely pursue one perspective (say, feminism or political economy), but for many projects articulating together salient critical perspectives provides a more robust approach that helps to grasp and critique television's multifaceted production, texts, effects, and uses.

To avoid the one-sidedness of textual analysis approaches, or audience and reception studies, I propose that critical television studies itself be *multiperspectival*, getting at culture from the perspectives of political economy, text analysis, and audience reception, as outlined above. Textual analysis should utilize a multiplicity of perspectives and critical methods, and audience reception studies should delineate the wide range of subject positions, or perspectives, through which audiences appropriate culture. This requires a multicultural approach that sees

43

the importance of analyzing the dimensions of class, race, and ethnicity, and gender and sexual preference within the texts of television culture, while studying as well their impact on how audiences read and interpret TV.

In addition, a critical television studies attacks sexism, racism, or bias against specific social groups (i.e. gays, intellectuals, and so on), and criticizes texts that promote any kind of domination or oppression. In short, a television studies that is critical and multicultural provides comprehensive approaches to culture that can be applied to a wide variety of artifacts from TV series to phenomena like Madonna, from MTV to TV news, or to specific events like the 2000 US presidential election (Kellner, 2001), or media representations of the 2001 terrorist attacks on the United States and the US response (Kellner, 2003a). Its comprehensive perspectives encompass political economy, textual analysis, and audience research and provide critical and political perspectives that enable individuals to dissect the meanings, messages, and effects of dominant cultural forms. A critical television and cultural studies is thus part of a media pedagogy that enables individuals to resist media manipulation and to increase their freedom and individuality. It can empower people to gain sovereignty over their culture and to struggle for alternative cultures and political change. Cultural studies is thus not just another academic fad, but can be part of a struggle for a better society and a better life.

Notes

1 On the history of the Frankfurt School, see Jay (1973) and Wiggershaus (1994); for Frankfurt School readers, see Arato and Gebhardt (1982) and Bronner and Kellner (1989); for appraisal of Frankfurt School social and media critique, see Kellner (1989) and Steinert (2003).
2 For critical analysis and appreciation of the Frankfurt School approach to media and television studies, see Kellner (1989, 1995, and 1997), and Steinert (2003).
3 For useful overviews of political economy research in television studies, see Sussman in Miller (2002); for an excellent overview of discourses of media imperialism, including analysis of how the concept has become problematic in a more pluralized and hybridized global media world, see Sreberny in Miller (2002).
4 For a survey of studies of television and violence, see Morgan in Miller (2002).
5 See the studies depicting both sides of the debate on contemporary television and its alleged harmful or beneficial effects in Barbour (1994) and Dines and Humez (2003).
6 For examples of studies of the politics of representation, see Gilroy (1991), McRobbie (1994), Ang (1991), and texts collected in Durham and Kellner (2001) and Dines and Humez (2003).
7 For an excellent account of the genesis of feminist TV criticism by one of its major participants, see Kaplan (1992). For an anthology of feminist TV criticism, see Brunsdon, D'Accci and Spigel (1997), and for overviews of feminist TV criticism in the contemporary moment, see the studies collected under Gender in Miller (2002).
8 On race and representation in television, see Jhally and Lewis (1992), Hamamoto (1994), Gray (1995), the 1998 anthology edited by Torres, and Noriega (2000).
9 See Kellner (1995) for discussion of the issues at stake here and a program for combining ideology critique with formalist analysis, sociological interpretation and political critique. On ideology critique in television studies, see White (1986).

References

Adorno, T. W. (1941) "On Popular Music" (with G. Simpson), *Studies in Philosophy and Social Science*, *9*(1), 17–48.

Adorno, T. W. (1978 [1932]) "On the Social Situation of Music," *Telos*, *35* (Spring), 129–65.

Adorno, T. W. (1982) "On the Fetish Character of Music and the Regression of Hearing," in Arato and Gebhardt (eds.), *The Essential Frankfurt School Reader*, New York: Continuum, pp. 270–99.

Adorno, T. W. (1989) "On Jazz," in Bronner and Kellner (eds.), *Critical Theory and Society. A Reader*, New York: Routledge, pp. 199–209.

Adorno, T. W. (1991) *The Culture Industry*, London: Routledge.

Adorno, T. W. (1994) *The Stars Down to Earth and Other Essays on the Irrational in Culture*, London: Routledge.

Adorno, T. W. et al. (1969 [1950]) *The Authoritarian Personality*, New York: Norton.

Ang, I. (1985) *Watching Dallas*, New York: Methuen.

Ang, I. (1991) *Living Room Wars: Rethinking Media Audiences for a Postmodern World*, New York and London: Routledge.

Arato, A. and Gebhardt, E. (eds.) (1982) *The Essential Frankfurt School Reader*, New York: Continuum.

Barbour, W. (ed.) (1994) *Mass Media: Opposing Viewpoints*, San Diego, CA: Greenhaven Press.

Baudrillard, J. (1981 [1973]) *For a Critique of the Political Economy of the Sign*, St. Louis: Telos Press.

Baudrillard, J. (1983a) *Simulations*, New York: Semiotext(e).

Baudrillard, J. (1983b) *In the Shadow of the Silent Majorities*, New York: Semiotext(e).

Bell, D. (1978) *The Cultural Contradictions of Capitalism*, New York: Basic Books.

Benjamin, W. (1969) *Illuminations*, New York: Schocken.

Best, S. and Kellner, D. (1997) *The Postmodern Turn*, New York: The Guilford Press.

Best, S. and Kellner, D. (2001) *The Postmodern Adventure: Science Technology, and Cultural Studies at the Third Millennium*, New York and London: Guilford and Routledge.

Bloch, E. (1986) *The Principle of Hope*, Cambridge: MIT Press.

Bronner, S. and Kellner, D. (eds.) (1989) *Critical Theory and Society: A Reader*, New York: Routledge.

Brunsdon, C. and Morley, D. (1978) *Everyday Television: "Nationwide"*, London: British Film Institute.

Brunsdon, C., D'Acci, J., and Spigel, L. (eds.) (1997) *Feminist Television Criticism: A Reader*, Oxford: Oxford University Press.

D'Acci, J. (1994) *Defining Women: Television and the Case of Cagney and Lacey*, Chapel Hill, NC: University of North Carolina Press.

D'Acci, J. (2002) "Cultural Studies, Television Studies, and the Crisis in the Humanities," in J. Olsen and L. Spigel (eds.), *The Persistence of Television*, Durham, NC: Duke University.

Deleuze, G. and Guattari, F. (1977) *Anti-Oedipus*, New York: The Viking Press.

Dines, G. and Humez, J. M. (eds.) (2003) *Gender, Race, and Class in Media*, London and Thousand Oaks, CA: Sage.

Durham, M. G. and Kellner, D. (eds.) (2001) *Media and Cultural Studies: KeyWorks*, Malden, MA and Oxford, UK: Blackwell.

Eysenck, H. J. and Nias, D. K. B. (1978) *Sex, Violence and the Media*, New York: St. Martin's Press.

Fiske, J. (1986) "British Cultural Studies and Television," in R. C. Allen (ed.), *Channels of Discourse*, Chapel Hill, NC: University of North Carolina Press, pp. 254–89.

Fiske, J. (1987) *Television Culture*, New York and London: Routledge.

Fiske, J. (1989a) *Reading the Popular*, Boston: Unwin Hyman.

Fiske, J. (1989b) *Understanding Popular Culture*, Boston: Unwin Hyman.

Foucault, M. (1977) *Language, Counter-Memory, Practice*, New York: Cornell University.

Gerbner, G. (2003) "Television Violence: At a Time of Turmoil and Terror," in Dines and Humez (eds.), *Gender, Race, and Class in Media*, London and Thousand Oaks, CA: Sage, pp. 339–48.

Gilroy, P. (1991) *"There Ain't No Black in the Union Jack": The Cultural Politics of Race and Nation*, Chicago: University of Chicago Press.

Gitlin, T. (1972) "Sixteen Notes on Television and the Movement," in George White and Charles Newman (eds.), *Literature and Revolution*, New York: Holt, Rinehart and Winston, pp. 335–56.

Gitlin, T. (1980) *The Whole World is Watching*, Berkeley: University of California Press.

Gitlin, T. (1983) *Inside Prime Time*, New York: Pantheon.

Gitlin, T. (2002) *Media Unlimited: How the Torrent of Images and Sounds Overwhelms Our Lives*, New York: Metropolitan Books.

Gitlin, T. (ed.) (1987) *Watching Television*, New York: Pantheon.

Gray, H. (1995) "The Politics of Representation in Network Television," in *Watching Race: Television and the Struggle for "Blackness"*, Minneapolis: University of Minnesota Press, pp. 70–92.

Gross, L. (1989) "Out of the Mainstream: Sexual Minorities and the Mass Media," in E. Seiter (ed.), *Remote Control: Television, Audiences and Cultural Power*, New York: Routledge, pp. 130–49.

Gross, L. and Woods, J. D. (1999) "Introduction: Being Gay in American Media and Society," in *The Columbia Reader on Lesbians and Gay Men in Media, Society, and Politics*, New York: Columbia University Press, pp. 3–22.

Grossberg, L. (1987) "The In-Difference of Television," *Screen*, 28(2), 28–46.

Hall, S., et al. (1978) *Policing the Crisis: Mugging, the State, and Law and Order*, London: Macmillan.

Hall, S. ([1973] 1980) "Encoding/Decoding," in Centre for Contemporary Cultural Studies (ed.), *Culture, Media, Language: Working Papers in Cultural Studies, 1972–79*, London: Hutchinson, pp. 128–38.

Hamamoto, D. Y. (1994) *Monitored Peril: Asian Americans and the Politics of TV Representation*, Minneapolis: University of Minnesota Press.

Herman, E. and Chomsky, N. (1988) *Manufacturing Consent: The Political Economy of the Mass Media*, New York: Pantheon.

Herzog, H. (1941) "On Borrowed Experience: An Analysis of Listening to Daytime Sketches," *Studies in Philosophy and Social Science*, IX(1), 65–95.

hooks, b. (1992) "Eating the Other: Desire and Resistance," in *Black Looks: Race and Representation*, Boston: South End Press, pp. 21–39.

Horkheimer, M. and Adorno, T. W. (1972) *Dialectic of Enlightenment*, New York: Herder and Herder.

Jameson, F. (1984) "Postmodernism – The Cultural Logic of Late Capitalism," *New Left Review* 146.

Jay, M. (1973) *The Dialectical Imagination*, Boston: Little, Brown and Company.

Jhally, S. and Lewis, J. (1992) *Enlightened Racism: The Cosby Show, Audiences, and the Myth of the American Dream*, San Francisco: Westview Press.

Kaplan, E. A. (1992) "Feminist Criticism and Television," in Allen, Robert C. (ed.), *Channels of Discourse: Television and Contemporary Criticism*, Chapel Hill, NC: University of North Carolina Press, pp. 247–83.

Kellner, D. (1989) *Critical Theory, Marxism, and Modernity*, Cambridge, UK and Baltimore: Polity and Johns Hopkins University Press.

Kellner, D. (1990) *Television and the Crisis of Democracy*, Boulder, CO: Westview Press.

Kellner, D. (1992) *The Persian Gulf TV War*, Boulder, CO: Westview Press.

Kellner, D. (1995) *Media Culture: Cultural Studies, Identity, and Politics Between the Modern and the Postmodern*, London and New York: Routledge.

Kellner, D. (1997) "Critical Theory and British Cultural Studies: The Missed Articulation," in J. McGuigan (ed.), *Cultural Methodologies*, London: Sage, pp. 12–41.

Kellner, D. (2001) *Grand Theft 2000*, Lanham, MD: Rowman & Littlefield.

Kellner, D. (2003a) *Media Spectacle*, New York and London: Routledge.

Kellner, D. (2003b) *September 11 and Terror War: The Dangers of the Bush Legacy*, Lanham, MD: Rowman and Littlefield.

Kracauer, S. (1995) *The Mass Ornament*, Cambridge, MA: Harvard University Press.

Lazarsfeld, P. (1941) "Administrative and Critical Communications Research," *Studies in Philosophy and Social Science*, IX(1), 2–16.

Lewis, J. (2002) "Mass Communication Studies," in T. Miller (ed.), *Television Studies*, London: BFI Publishing, pp. 4–6.

Lopate, C. (1977) "Daytime Television: You'll Never Want to Leave Home," *Radical America*, 11, No. 1 (January–February).

Lowenthal, L. (1957) *Literature and the Image of Man*, Boston: Beacon Press.

Lowenthal, L. (1961) *Literature, Popular Culture and Society*, Englewood Cliffs, NJ: Prentice-Hall.

Luke, C. (1990) *TV and Your Child*, London: Angus and Robertson.

Marcuse, H. (1955) *Eros and Civilization*, Boston: Beacon Press.

Marcuse, H. (1964) *One-Dimensional Man*, Boston: Beacon Press.

McGuigan, J. (1992) *Cultural Populism*, London and New York: Routledge.

McRobbie, Angela (1994) *Postmodernism And Popular Culture*, London / New York: Routledge.

Miller, T. (ed.) (2002) *Television Studies*, London: BFI Publishing.

Modleski, T. (1982) *Loving with a Vengeance: Mass-Produced Fantasies for Women*, Hamden: Anchor.

Morley, D. (1986) *Family Television*, London: Comedia.

Noriega, C. (2000) *Shot in America: Television, the State, and the Rise of Chicano Cinema*, Durham, NC: Duke University Press.

Postman, N. (1986) *Amusing Ourselves to Death: Public Discourse in the Age of Show Business*, New York: Viking.

Rapping, E. (1986) *The Looking Glass World of Nonfiction Television*, Boston: South End Press.

Rapping, E. (1992) *The Movie of the Week*, Minneapolis: University of Minnesota Press.

Rapping, E. (2003) *Law and Justice As Seen on TV*, New York: New York University Press.

Rosenberg, B. and White, D. M. (eds.) (1957) *Mass Culture*, Glencoe, IL: The Free Press.

Russo, A. and Torres, L. (eds.) *Third World Women and the Politics of Feminism*, Bloomington: Indiana University Press, pp. 51–80.

Schiller, H. (1971) *Mass Communications and the American Empire*, Boston: Beacon Press.

Sreberny, A. (2002) "Media Imperialism," in T. Miller (ed.), *Television Studies*, London: BFI Publishing, pp. 21–3.

Steinert, H. (2003) *Culture Industry*, Cambridge, UK: Polity Press.

Sussman, G. (2002) "The Political Economy of Television," in T. Miller (ed.), *Television Studies*, London: BFI Publishing, pp. 7–10.

Torres, S. (ed.) (1998) *Living Color: Race and Television in the United States*, Durham, NC: Duke University Press.

Tunstall, J. (1977) *The Media are American*, New York: Columbia University Press.

White, M. (1986) "Ideological Analysis of Television," in Allen (1986), pp. 134–71.

Wiggershaus, R. (1994) *The Frankfurt School*, Cambridge, UK: Polity Press.

PART TWO

Television/History

CHAPTER THREE

Television and History

Paddy Scannell

I start with the worldliness of contemporary television. On the one hand, it is routinely experienced everywhere as part of the ordinary life-world of members of modern societies (watching TV is one of those things that most of us do in the course of an ordinary day).[1] On the other hand, and just as routinely, in daily news services the world over audiences experience, as a commonplace thing, their situated connectedness with what's going on elsewhere in the world. In exceptional moments people the whole world over are glued to their television sets as witnesses of celebratory or catastrophic events. In all this broadcasting has accomplished something quite unprecedented: the routinization of history on a worldwide basis. Television today makes the historical process visible. Through it we see the manifest truth of the claim that human beings do indeed make history; their own histories, the history of the country in which they live, the history of the world. But what is much harder to see is how to account for and understand these interlocking historical processes that are all embedded in each other. I have argued that the history of the world (world history) is an impossible narrative (Scannell, 2004b).[2] There is no point of view, no point of rest, from which it could be written by human beings. And the same is true, I think, for television. As a world-historical phenomenon it paradoxically appears as an impossible historical narrative. So in order to broach the world-historical character of broadcast television,[3] I begin with the perplexities of historiographies of broadcasting, communication and media technologies.

Broadcasting Histories

What is broadcasting history's natural subject matter? In the mid-1950s the British historian, Asa Briggs, embarked on a history of broadcasting in the United Kingdom which turned out to be the history of the British Broadcasting Corporation who commissioned him (Briggs, 1961–94). Fifty years and five volumes later, this is a still continuing history with Jean Seaton taking over from

Paddy Scannell

Lord Briggs to produce volume 6 (1974–86). This, the earliest scholarly history of broadcasting, was immensely influential and set the benchmark standard for subsequent histories of broadcasting in other countries. Briggs produced a meticulously researched history, based primarily on the BBC's huge written archive, which offered a rolling narrative of the development of the BBC as its activities grew and expanded over time. It was largely concerned with the internal history of the institution; its administrative structure, its hierarchy of policy and decision making, program production and delivery. At the same time it looked outwards to the external pressures that constantly impinged on the operational activities of the broadcasters from its two masters – the state on one hand, the audiences on the other. These pressures bore down on different aspects of the work of broadcasting, but together they helped to shape and define its universe of discourse, the limits of permissibility, of what could and could not be said or shown on radio or television, at any time. Radio broadcasting began everywhere on a local basis and sooner or later a process of consolidation and centralization took place that set in dominance a national system of broadcasting that remains intact today. This convergence took place very quickly in the United Kingdom, partly because of its small size, partly because of the rapid domestic uptake of radio by the population and partly because so much of British economic, political, and cultural power was already concentrated in the metropolitan capital, London. In other parts of the world, with much larger territories, with different sociopolitical geographies and a slower rate of uptake, the centralization of broadcasting took place more gradually and the central broadcasting authorities had less power over regional and local broadcasters.[4]

Briggs established a "first generation" history that put in place a narrative of the institutions of broadcasting. It served to generate further "social" and "cultural" histories, which focused on the output and impact of broadcasting or, in other words, the reception of broadcasting. Susan Douglas's engagingly readable history of "listening-in" to the radio in America is exemplary (Douglas, 1999).[5] Such histories, however, do not run in parallel with histories of the broadcasters. They are separate narratives whose concerns are with daily existence, the place of the radio or TV set in the spaces of domestic, family life, and their role (along with the movies and other elements of popular culture) in the lives of, say, girls growing up in America in the 1960s (Douglas, 1994). These histories have no necessary connection with the histories of the broadcasters because, as mass communication sociologists gradually learnt and as Stuart Hall (1980) argued, there is no direct correspondence between the outputs of broadcasting and their impact and effect on audiences.[6]

All these histories are embedded in national histories, for the nation-state remains the containing frame within which historiography operates, the world over, today. The possibility of comparative, international, or global histories has exercised historians for centuries.[7] It is an increasingly pressing issue today since all of us know that we are living in a single, common world. Broadcasting history, in response to this pressure, has tried to transcend its national boundaries.

A comparative study of Nordic television (Bono and Bondebjerg, 1994) brought together condensed histories of developments in Denmark, Finland, Norway, and Sweden, each drawing on its own, more comprehensive national history of broadcasting. Kate Lacey has made comparative studies of broadcasting in Germany, Britain, and the United States (Lacey, 2002). Michele Hilmes has argued the need for larger comparative broadcasting histories (Hilmes and Loviglio, 2002, pp. 1–19) and has brought together British and American broadcasting in *The Television History Book* (Hilmes, 2004). All these works proceed by setting national accounts alongside each other and considering their points of convergence and divergence. But what do we learn from them beyond the structural similarities of broadcasting's organization, mode of production and program service which are subject, inevitably, to national variations and differences determined by the size of available native audiences, and indigenous economic, political and cultural factors? The comparative study of national broadcasting certainly illuminates their idiosyncratic character – the Japaneseness of Japanese broadcasting, the Americaness of American broadcasting etc. – in a supranational historical context. But it does not bring us closer to the global character or impact of the spread of broadcasting in the twentieth century.

What of the history of *world* broadcasting? In this, the case of the BBC is exemplary. In the 1930s the BBC began overseas broadcasting first to white settler audiences in Britain's imperial outposts and then, in the late 1930s, with a European war imminent, to countries that the British government wished to influence. In the course of World War II the BBC developed a truly global broadcasting service that transmitted British versions of events, suitably inflected for reception in different parts of the world depending on their part in the global convulsion. Coming out of the war the BBC's now established World Service, funded by a grant-in-aid from the Foreign Office, played an important part in the Cold War, backed up by the government-funded Monitoring Service which eavesdropped on broadcasting transmissions from within the Soviet bloc and from many other parts of the world. It might be thought that this service, born out of raison d'état, should have begun to disintegrate as Britain gave up its empire in the decade after the war and to have disappeared completely following the fall of the Berlin Wall in 1989. It is remarkable then that, at present, the World Service's audiences continues to grow each year and not only for its English-language services. For example, the audiences for the Brazilian service, in Portuguese, have grown since September 11, 2001, and its staffing levels have doubled since then.[8]

The continuing existence and growth of the World Service indicates, I think, not only the overlooked global importance of radio as the parent broadcasting medium, but also the existence of a growing felt need around the world for reliable, authoritative news of the world that comes from one of its centers, from where the action is.[9] But what would the history of this service consist of? It is, inevitably, a history *of* the center; of the growth of the scale of its operations and of key historical moments such as Suez and Hungary in 1956 (Mansell, 1982).

What it cannot be is a history of its reception the world over, for that is historically irretrievable beyond the most fragmentary indications to be gathered from newspapers, magazines, and other sources in particular countries throughout the world. Thus, broadcasting historiography's natural limits are set by the situational geographies in which, and for which, broadcasting institutions exist – the territorial boundaries of nation-states. Moreover, it seems to be a one-sided history. Either you write about the institutional side, or you write about the reception side but between them there is a wall over which it is hard to see the other side. The narratives of institutions and their activities and the narratives of the social uptake of those activities are invisible to each other for good reasons, as we shall see.

Technological Histories

Broadcasting histories belong within the more encompassing history of the extraordinary growth in mediated forms of communication that underpin the modern, electronically wired-up and wireless world. Radio broadcasting is after all a by-product of an earlier technology (wireless telegraphy) conceived for different purposes and use. The same is true of the Internet and the worldwide web. Both were later applications of technologies that had, at first, a restricted military use as outcomes of earlier histories of scientific exploration and discovery. Communication technologies reach beyond national borders, and their histories are not constrained within them. Brian Winston (1998) has produced a sophisticated model of the complex transition from "pure" scientific experimentation, through the recognition of possible practical applications and the development of prototypes, to the invention of a new technology with a strong potential for use and profit. His magisterial narrative of developments from the early nineteenth-century telegraph to the late twentieth-century Internet is, throughout, a *technical* history of scientific discovery and commercial application. The same is true of Pawley's important history (1976) of the BBC's engineering division. In both books the concern is only with the scientific, technical process and its richly complex historical unfolding. The boundaries of technological histories are set by the moment of transition when the technology in question moves out of the laboratory, so to speak, and achieves social recognition and uptake. At that point different histories take over – the histories of their social application and use as discussed above, in the case of broadcasting.

It is important to note how this transition comes about. A technical thing comes out of the R&D laboratory and enters into the world. It ceases to be a technical thing and becomes a worldly thing. For this to happen it *must* present itself – if it is to be an ordinary, worldly thing – not as a complicated technological object but as a simple piece of equipment such that anyone can use. This point is clearly illustrated by the development of the radio set. In the aftermath of World War I radio had become a popular "scientific" hobby even before the British Broadcasting Company began to transmit a program service in November 1922. In garden sheds up and down the land, men and boys (it was very much a male pursuit) were building two-way radio transmitter-receivers or one-way receiving sets to scour the ether for sound signals. In either case the results were a naked display of valves, knobs, wires, and amplifiers. The scientific innards had yet to be encased and its operation required endless fiddling and twiddling. It was not yet a domestic object fit for family living rooms.[10] Adrian Forty describes three stages in the evolution of the first truly modern, mass-produced radio set in Britain, the Ekco AD65 receiver designed and manufactured by the E. K. Cole company and in the shops by 1934 (Forty, 1986, pp. 200–6; Scannell and Cardiff, 1991, pp. 356–62). The mediating stage in the transition from technology to domestic equipment is *design*. It is a basic mistake to think of design as style and aesthetics applied to mass-produced goods, as if it were some kind of value-added. In reality, design is essential to the transformation of user-unfriendly technologies that are only of use to trained experts into simple user-friendly devices. The famous Ekco set was designed by a leading architect of the time. Its scientific innards were concealed in a circular molded plastic case made of bakelite, with a chrome-plated grille and just three knobs for volume, wavelength, and tuning. It was not a piece of furniture, but a thoroughly new and modern piece of equipment suitable for any household with an electricity supply, and any child could use it.

The point is perhaps obvious enough; you do not need to know how a thing is made in order to understand how to use it. Nor do you need to know how programs are made in order to like or dislike or be bored by them. The labyrinthine complexities of the scientific-technical development of radio and television broadcasting and the production processes that lie behind their transmitted output are equally invisible in the design of the receiving equipment and in the design of programs. We are not aware of the manufactured character of either except when they malfunction. And yet it must be the case that the design of television sets and of television programs disclose, in different ways, how they are to be understood and used. How else would we know what to do with them? To study the hidden labor processes of technological innovation and application and of broadcasting institutions and their program making, is to begin to uncover the care-structures that are concealed and yet immanent in humanly made things.[11] More particularly, to attend to the design of receiving equipment and to the communicative design (or intentionality) of the programs they disclose is to begin to find answers to the question as to how something such as "television" appears in the world *as* a worldly thing; as an ordinary, available thing for use by each and all, anyplace, anytime.

Media Histories

A third approach to the historical study of communication was pioneered by the Canadian economic historian, Harold Innis, whose ideas were taken up and popularized by Marshall McLuhan.[12] McLuhan's fame has overshadowed and distorted the significance of Innis's late work which today needs some contextualizing in order to rescue it from the condescension of contemporary media historians (e.g. Curran, 2002, pp. 51–4). Outside Canada Innis is known primarily for two books written at the end of his life: *Empire and Communication* and *The Bias of Communication*. In these two works Innis developed what was then a startlingly original thesis about the *media* of communication, the material forms (and their technologies) through or upon which human communication is registered and moved. Today, as a result of their diffusion in McLuhan's writings, these ideas have become commonplace. They include the periodization of historical epochs according to their dominant form of communication (oral, manuscript and print cultures); the distinction between speech and writing (emphasizing the role of the latter in the management and maintenance of religious and political power); the communicative bias of different media of communication toward either time or space. Throughout, the emphasis is on the material forms of communication and not their particular content.

Innis's late work is hard to read today. It is written in an assertive, oracular style, employing a vast historical sweep and a high degree of abstraction: "Minerva's owl," the first chapter of *The Bias of Communication*, sweeps from ancient Babylon and Mesopotamia to the industrial revolution and the Communist Manifesto in little more than 20 pages. This kind of writing was more acceptable 50 years ago and in fact represented probably the last – and certainly the most original – attempt to write "world history," a genre which, even as Innis wrote, was in decline and has fallen out of favor ever since for reasons hinted at above. World history took its inspiration from Hegel's *Phanomenologie des Geistes* (The Phenomenology of the Spirit) in which the Enlightenment narrative of progress found its ultimate expression as the story of the Spirit of Humanity's long journey to self-understanding and reconciliation. The challenge to translate this from a philosophy of history into an actual historical narrative was taken up by historians in the nineteenth and early twentieth century. The most influential of these, in Innis's day, was Arnold Toynbee's multi-volume *Study of History* which started by tracing the history of the world first in terms of the rise and fall of civilizations and, later, of world-religions.

Innis's *Empire and Communication* used the same broad canvas as earlier world histories but painted a very different picture. The transcendental narrative of the movement of *Geist* in history via the rise and fall of civilizations was replaced by the movement and circulation of people, goods and information. To see how Innis arrived at this point we must return to his early historical work on the Canadian economy. In his detailed, empirical studies of Canada's export staples (fur, timber, and fish), Innis came to see them as key components of a front tier (frontier) economy heavily dependent on the "back tier" economies of Europe and its dominant American neighbor. More exactly, he found that his work was, in a fundamental way, a study of the movement and circulation of people and goods underpinned by available forms of transport and communication and all of which came up against the material exigencies of time and space. If his later work seems to operate at a high altitude, it is nevertheless grounded in the earthy, practical realities of his early empirical work. As part of his definitive study of the fur trade, Innis bought himself a canoe and paddled down the remote McKenzie River to the Hudson Bay (the route taken by nineteenth-century trappers) in order to understand how the pelts started on their long journey to the shops of London and Paris where they were sold as fashionable beaver hats.

It is customary to view "medium theory"[13] as being flawed by technological determinism; the view that technological innovation causes social change. The difficulties lie, to a considerable extent, in the way that the question is posed in terms of technology and its social effect. That formulation presupposes a dichotomy between the hidden processes of technical discovery, invention, application, manufacture, and distribution all on one side with "society" on the other side of the wall. It is as if human inventions are discovered outside society and then are suddenly parachuted into it. Furthermore, the question is posed in terms of a cause-effect relationship as if one could isolate and specify the particular change(s) that could be attributed to the technology itself and nothing else. Moreover what is almost completely overlooked in this analysis is that what begins, at the point of social uptake of modern technologies of communication, is the process of working out what can be done with them, the discovery of what in fact they are (good) for. Technologies do not arrive in the world with what Ian Hutchby calls their "communicative affordances" known and understood. Hutchby places this concept at the heart of his penetrating review of current approaches, in the sociology of science, to the question of technologies and their impact (Hutchby, 2001, pp. 13–33). The traditional deterministic interpretations of technology were largely negative, seeing technologies as the product of instrumental reason that exploited the natural environment and as instruments of social exploitation and domination. Recent sociology has challenged that view but, Hutchby argues, ends up by rejecting determinism completely. His own more nuanced position allows that technologies do indeed have constraining effects, but that these should be thought of as enabling rather than disabling. The question now becomes: What affordances do new communicative technologies open up? What are they good for? What difference, for instance, does television make to our lives? What does it do with us and what can we do with it?

The Historicality of Television

The historiographies of communication and media with which I have thus far been concerned all point to the difficulty of grasping the historicality of media and particularly the world-historical character of television. Histories of broadcasting, in which television's history is situated, turn out to have a one-sided institutional and national character that is difficult to transcend. Social and cultural histories are written on the other side of the wall. Narratives of the development of technologies of communication are similarly one-sided and stop at the point of social uptake. Finally, efforts to write the history of the world in terms of communication media appear today as discredited by our skepticism toward grand narratives. The wider question of the historical impact of communication technologies presents major hermeneutic difficulties. At the heart of these problems is an issue that medium theory highlights. Historiography is about history, but points in a different temporal direction. Historiography operates on the temporal axis of present and past, while history operates on the axis of present and future. History's subject matter is the history-making process. Both are situated in the present, the phenomenal "now." Historiography looks back to the past as a clue to the present situation. Meanwhile, however, the history-making process, in the very same phenomenal now, is moving forward into the future, is giving the world its future through its actions in the present. The writing of history and the making of history inevitably diverge. Broadcast television is part of the history-making process. That is what its *historicality* (its being historical) indicates. That is why historiography can never catch up with, can never quite grasp, its object of enquiry. As historiography looks back, history itself is moving forwards and away from it.

Historiography is about the *writing* of history. A much-debated crux in a number of disciplines is the status, in historiography, of the *event*. The influential *Annales* School (Burke, 1990) was deeply dismissive of *histoire événementielle* whose time was that of daily life and whose concern was with the kinds of event that show up in newspapers (Braudel, 1980, pp. 27–9). These historians argued that a preoccupation with historical actors (monarchs, statesmen, and military leaders) and with great events (politics and war) produced surface narratives which overlooked the underlying structural factors that produced both the events and their agents. The rejection of surface history, however, produced peculiarly motionless and abstract histories and the late twentieth century saw a return to narrative history, accompanied by vigorous debates about its reliability in relation to the "truth" of the event-as-narrated.[14] The event, for all the difficult issues it poses, is the bedrock of history. If nothing happens, there is nothing to tell. One elegant definition of daily life is precisely that there is nothing to say about it. It is uneventful *because* it has no storyable, tellable characteristics (Sacks, 1995, vol. 2, pp. 215–21).

History, however, is not *simply* the event. Events remain unhistorical unless or until they are narrated. History is the act of narrating the event. To narrate is not to chronicle. It is to find and tell the story of the event. The investigative process of finding and telling the story is the task of the historian and the journalist:

Yes (.) This just in (.)

You are looking at obviously a very disturbing *live* shot there

That is the World Trade Center and we have unconfirmed reports this morning that a plane has crashed into one of the towers of the World Trade Center CNN center right now is just beginning to *work* on this *story* obviously our sources are trying to figure out exactly what happened

But clearly something relatively devastating *happening* this morning there at the south end of the island of Manhattan. [emphases added]

This is the moment that the event breaks, live to air, into CNN news at 8:50 am on September 11, 2001. It is the moment of first sight, for viewers *and* the newsdesk, of a pall of smoke billowing from one of the towers of the World Trade Center, and these are the first words from the newsroom about what, coming out of the ad break, is now on screen with the strapline, BREAKING NEWS. It is immediately and naturally assumed, by the newscaster, that this – whatever it is – is a story. There is "something [. . .] happening" as viewers can see. What exactly, is unclear beyond "unconfirmed reports" of a plane crashing into the building. Though the situation presents itself as incomprehensible and inexplicable, it is spontaneously treated as self-evidently potentially meaningful and significant. The *work* of finding the story is the task of the CNN news center and it is now, off-screen and invisibly, working flat out on it. In the interface between its back-stage finding and its front-stage telling, the meaning and significance of the event-as-story will be uncovered. It was to be a long and terrible journey of discovery on that day (Scannell, 2004b).

Journalists are the historians of the present. To find and tell the story is to give structure, coherence and meaning to events-in-the-world and thereby historicize them. The world-historical character of life today shows up, like a bolt from the blue, in the world-historical event. Both are, in significant ways, an effect of television. To reiterate: it is not the event-in-itself that is historical. It becomes so only through the storytelling narratives of its historian(s). History is the sum of the relationship between event, story, and narrative. The attack on the World Trade Center in New York instantly became a world-historical event through its immediate uptake on television news programs round the world. Most news comes after the event. But on September 11, event and narrative were both in the same forward-moving, history-making, real-time now. The significance of television – its essential meaning, power and impact – is encrypted in its most fundamental communicative affordance as *live broadcasting*.

Live Television Broadcasting

"You are looking at obviously a very disturbing *live* shot there." To find and tell the story in the live, phenomenal now of television is to articulate a prospective, forward-looking narrative. This in contrast with written histories (including film and newspaper histories) that are backward-looking retrospective narratives. Innis and McLuhan drew attention to the fundamental communicative affordances of writing (inscribed in all its mediating technologies) and speech. But the force of this distinction was considerably vitiated by the terms in which it was made: the distinction between "oral" and "print" cultures has a curiously flattening and distancing effect (it is an academic distinction). We will have a more vivid grasp of its force if we think it in terms of the *living* and the *dead*. Historiography's subject matter (history) is in, as we say, the dead past. But history itself (the history-making process; the *a priori* of historiography) is in, as we also say, the living present. The past is dead because it is over and done with – "It's history" (it's finished). The perishability of news ("yesterday's news is dead news") reminds us of this each day. The present is alive because it is the now-becoming-future of the lives of the living. The liveness of television is not its technological effect but its existential basis, the condition of its existence in a double sense: its possibility and its manifest, expressed effect. It is because, and only because, television is live that it is inextricably implicated in the history-making process which today has long since been routinized by modern media (starting with the daily press) as news. Today's news is tomorrow's history.

The meaning of *live* has been much misunderstood in the academic literature on television. In most discussions it is pointed out that television was broadcast live to begin with but was, from the 1960s onward, replaced for the most part by recorded programs. But "recorded" is not the negation of "live." Jane Feuer's (1983) influential and much-cited essay on "The Concept of Live Television" conflates liveness with immediacy. Of course, in live broadcasting the moments of production, transmission, and reception are all in the same real-time now, but what Feuer neglects to consider is the temporal ontology of the immediate now and, crucially, what gives its possibility. As human beings we exist, at one and the same time, in many different and incommensurate orders of time. The immediate now, for instance, is radically different in digital and analog time. In digital time reckoning, we say: "*Now* it is 8:50. Now it is 8:51," etc. Time is manifest as an ever-present punctual moment that cannot ever be anything other than "now." In analog time reckoning we say: "Now it is ten *to* eight. Now it is ten *past* eight." Analog time's immediate now is expressed (both on the clock-face and in the way we say it) as being in a relationship with its before and after, neither of which exists in digital time. The now of analog time is the phenomenal now of our concern. It is the matter to hand in the "now" that matters. It is an immediate present that exists only by virtue of the *historic* and *future* present, which are the conditions of its possibility, of its coming-into-being. The possibility of live-to-air program transmissions, in which we experience liveness-as-immediacy, is given by the structure of the daily program schedule, which, in broadcasting, is attuned to the existential arc of days.

The two ontologies of time expressed in analog and digital time pieces are implicated in two temporal orders of the day. The day, in 24/7 news-time, exists in a continuous, never-ending succession of punctual moments that are always in the ever-present now. This strictly abstract, numbered, and sequential time overrides the natural temporality of the day with its immanent structure, rhythm, and tempo around which human life, even today, remains adjusted.[15] Light and darkness; waking and sleeping – the days of our lives have a natural arc of morning, noon, and night which is the storyable arc of our own existence, too. Life and days are inextricably folded into each other and show up in the schedules of the broadcast day in which the historic, immediate, and future present show up in relation to each other. *Good Morning America*, which Feuer briefly discusses, is a start-of-day program whose live-to-air unfolding format performs the task of orienting its audience to the day ahead and all its upcoming business. It is not just *at* that time of day, but *for* that time. For Feuer, liveness and immediacy are essentially ideological. She never sees either as matters of time or as time-that-matters.

Live *broadcasting*. The two terms must be considered together. We owe it to John Durham Peters for a corrective reminder of the communicative affordances of broadcasting, in his seminal discussion of Christ's parable of The Sower (Peters, 1999, pp. 51–62). To broadcast, before radio and television, meant to sow, to scatter seed abroad. In the parable the broadcaster is careless of where the seed falls. Some lands on stony ground and is pecked up by the birds of the air. Some falls among thorns and is choked as soon as it springs up. Some falls on shallow soil, springs up quickly and soon withers. And some falls on fertile soil and yields a good harvest; a hundredfold, sixtyfold, thirtyfold. This is inefficient communication that is indifferent to its success. It is inefficient because it is indiscriminate. It makes no effort to disseminate only to chosen, selected, and responsive audiences. It allows for rejection and indifference. It has no measure of its own success. It is a strictly one-way, or non-reciprocal form of communication. But whereas this has usually been regarded as its deficiency, Peters sees it is a blessing. To give (to broadcast) without any expectation of return is an unconditional communicative act that comes with no strings attached. Any recipient can make of it what they will, and that is allowed for. It is unforced, non-coercive communication that offers involvement without commitment. In all these ways broadcasting is deeply democratic. It is intrinsically non-exclusive and non-binding. Anyone can watch or listen and anyone can, if they so choose, disagree with what they see and hear. The generosity of broadcasting is strictly impersonal, but allows for persons and their personal opinions.

Television, History, and the World

The broadcast character of television indicates its spatiality. Its liveness is its particular temporality. Together, they yield an unprecedented historical here-and-now. History is no longer "then." It is "now." The event is no longer "there," but "here." The now-and-then, the here-and-there come together in the live immediacy of broadcast news and events which are structured in expectancy of what is to come. These real-time, real-world moments produce a spanned and gathered now in which, daily and routinely, countless individual lives and the historical life of societies intersect with each other the world over. In such moments each of us experiences the news-event as if it spoke to me-and-others now.[16] The world-event, through television, impinges directly and immediately, in each individual case, upon me and my life. In live transmissions individuals the world over are not so much spectators as witnesses of events.[17] *As* witnesses we become implicated in the events themselves. Witnesses have communicative entitlements and obligations by virtue of having been present at the event. As such we are not just entitled to our views and opinions, but we may be called upon to bear witness, to testify to what we saw and how we saw it (Peters, 2001).

BBC News, 11.09.01: 10.04 pm

Eyewitness, New York:

I wuz just standing here watching the World Trade Center after the first after the first plane hit (.) I just saw a second plane come in from the south and hit the whuh south (.) tower half way between the bottom and the top of the tower it's gotta be a terrorist attack I can't tellya anything more th'n that (.) I saw the plane hit the building . . .

To re-live a moment such as this testifies to the *pain* of witnessing. The anguish in the face and voice, in the whole body of this anonymous "man in the street" as he tells what he just saw is all caught in the recording. But what is our position, as viewers, in relation to what we witness on television?

Luc Boltanski has eloquently argued that, as "moral spectators of distant suffering" via television, we are unavoidably implicated in what he calls the *politics of pity*. In France, if you are an immediate witness of suffering, you have a legal obligation to come to the aid of the sufferer (Boltanski, 1999, pp. 7–17). What, then, is our obligation (if any) as television viewers in relation to what we witness? As *moral* spectators we cannot assume the indifference of an objective stance ("that's how it is") and turn away. We *feel* for what we see. The politics of pity requires that we take a stand and confront the choice between detachment or commitment, a choice made reflexively visible by broadcasting. We may be roused (politicized) to act; to protest, to demonstrate or at least to make a donation to an aid agency. At the very least we may be roused to speak; to express our indignation, pity, or even our malicious pleasure, to discuss with others, to form an opinion on the matter of the suffering of others. Through the communicative affordances of today's television, their suffering achieves a visibility and publicness which "presupposes an international public space" of discussion (Boltanski, 1999, p. 184), a global public sphere. This is how we, as viewers anywhere, encounter the world-historical character of life today. This is how we are implicated in what Boltanski calls "the politics of the present" which responds immediately to immediate events.

Critics of the politics of the present accuse it of a naïve humanitarianism, which merely responds to the victims of suffering without addressing its causes.

Boltanski replies that "to be concerned with the present is no small matter. For over the past, ever gone by, and over the future, still non-existent, the present has an overwhelming privilege: that of being real" (Boltanski, 1999, p. 192). It is the reality of suffering brought to presence by television everywhere that stirs us to present thought and action. Present actions have no guarantees of success. We cannot be wise before the event, though all of us can be wise in its aftermath. The CNN newsdesk and other broadcasters on the day had no such available wisdom as they wrestled with the unbelievable events unfolding live and in real time on their screens; yet, by the end of that day, newsrooms the world over, had digested, framed and interpreted their momentous significance. They had named Osama bin Laden as the likeliest perpetrator of the attacks on the United States and correctly anticipated an American-led attack on Afghanistan as its likeliest political consequence. Journalists, as historians of the present, face and anticipate the future that present events will bring about. They do this on behalf of their publics everywhere today.

Boltanski's meditation on the television news-viewer as moral spectator has a premise that this chapter shares – it is through television that we are implicated, day by day on a worldwide basis, in the history and politics of the present. The beginnings of that historical development was the theme of Jürgen Habermas's hugely influential account of the emergence of public opinion as the foundation of modern mass, democratic politics (Habermas, 1989). Habermas pinpointed the moment that the opinions of ordinary citizens became *historically* relevant as the moment that they became *politically* relevant. When the opinions of ordinary people began to impinge on the decisions and actions of those who exercised political power, the people themselves became, for the first time, involved in the process of making history. The role of media in making public the political-historical process was and remains crucial to the formation of critical public opinion as part of that process. In the last century the live and broadcast affordances of radio and television have drawn all of us into the history-making politics of the present which all of us experience normally, and normatively, as members of the societies in which we live. Our *own* situation and its attendant circumstances are understood by each of us as embedded in the world-historical framework of life today as disclosed, daily and routinely, in television news and events wherever and whoever we may be.

Notes

1 On the ordinariness of television, see Bonner (2003).
2 This history was, in the West, originally the Judeo-Christian narrative of humanity's fall and ultimate redemption. It was revised in the Enlightenment as the historical struggle for the kingdom of heaven on earth in the form of the perfectly free and just society. Postmodernism has proclaimed its incredulity towards such "grand narratives" (Lyotard, 1986).
3 "Given the overall mapping of the globe that today is taken for granted, the unitary past is one which is worldwide; time and space are recombined to form a genuinely *world-historical*

framework of action and experience" (Giddens, 1990, p. 21, my emphasis). I follow Giddens in thinking of "globalization" as the-world-as-a-whole experienced by each and all of us "embedded" in our own time and place.

4   Australia, France and the United States may serve as exemplary case studies. See, respectively, Johnson (1988), Meadel (1994), and Smulyan (1994).

5   Douglas has that rare ability to write as an academic (observing academic norms of scholarship, research etc.) for a non-academic readership and her books are widely reviewed and read outside academia. It is partly a matter of style but it is, more exactly, the narrative point of view that she assumes. She writes of radio in the way that it matters for listeners as part of their own lives and experience.

6   Except on very rare occasions. The Orson Welles' *War of the Worlds* scare in 1938 is an early and classic case of a single program with an immediate, dramatic effect on audience behavior.

7   Breisach (1983). See especially his discussion of "The Enigma of World History," pp. 319–22, 395–411.

8   In the early 1990s the Brazilian service was on the point of closure. It now has 40 staff, and is the one of the largest sectors in the BBC's foreign-language transmissions (see bbc.co.uk/brazil). I am grateful to Lorena Barbier of CBN (Central Brasilieras de Noticias) Recife, for this information.

9   The hegemony of English as the world's language is crucially important to the position of the World Service as the dominant global broadcaster today. In many countries people listen to improve their understanding of the English language.

10 For an account of this history in the United States, see Douglas (1999, pp. 55–82). See Douglas (1999, plate 1, opposite p. 192) for a photograph that vividly captures this moment.

11 See Scannell (2003) for a discussion of the broadcasting production process as a care-structure.

12 Notably in *The Gutenberg Galaxy*, that McLuhan describes in the preface as "a footnote to the observations of Innis on the subject of the psychic and social consequences, first of writing and then of printing" (McLuhan, 1962, p. ix).

13 The label attached to the approach of Innis and McLuhan by Joshua Meyrowitz (1994).

14 A useful review of history and narrative as discussed by historians, philosophers, and literary theorists is provided by Roberts (2001).

15 The time-of-day, like the lunar month and solar year, is a natural (non-human) order of time and is both linear and cyclical in its movement. Digital time is motionless and is a perfect example of Zeno's paradox of the arrow in flight. In any indivisible instant of its flight is a flying arrow moving or at rest? If the former, how can it move in an instant; if the latter, it is never moving, and therefore is at rest (Honderich, 1995, p. 922). The punctual moment of digital time, with no "before" or "after," appears trapped in the eternity of the ever-same now. *Groundhog Day* is a wonderful exploration of the paradoxes of digital and daily time.

16 For a fuller discussion of the complexities of how "we" are addressed by radio and television, see Scannell (2000).

17 There is a very basic issue at stake here. The witness has experienced something by virtue of having been there. Can the viewer lay claim to an experience having watched something on television? The various communicative entitlements of a witness derive from the assumed *authenticity* of their witnessing. That is presumed to be validated by the fact of their presence and their immediate, first-hand experience. If television offers mediated, second-hand experience, it is inauthentic. I have argued it is possible to have an authentic experience watching television and thus to be a witness (Scannell, 1996, pp. 93–116), a claim which underpins the whole of this chapter. See Ellis (2002, pp. 31–6) on television as "live witness realized."

References

Boltanski, L. (1999) *Distant Suffering*, Cambridge: Cambridge University Press.

Bonner, F. (2003) *Ordinary Television*, London: Sage.

Bono, F. and Bondebjerg, I. (1994) *Nordic Television: History, Politics and Aesthetics*, Copenhagen: University of Copenhagen, Dept. of Film and Media Studies.

Braudel, F. (1980) *On History*, London: Weidenfeld and Nicolson.

Breisach, E. (1983) *Historiography*, Chicago: University of Chicago Press.

Briggs, A. (1961–95) *The History of Broadcasting in the United Kingdom* (Volumes 1–5), Oxford: Oxford University Press.

Brunsdon, C. (1998) "What is the 'Television' of Television Studies?," in C. Geraghty and D. Lusted (eds.), *The Television Studies Book*, London: Arnold, pp. 95–113.

Bull, P. and Black, R. (2003) *The Auditory Culture Reader*, Oxford: Berg.

Burke, P. (1990) *The French Historical Revolution: The* Annales *School, 1929–1989*, Cambridge: Polity.

Cavell, S. (1976) *Must We Mean What We Say?*, Cambridge: Cambridge University Press.

Curran, J. (2002) *Media and Power*, London: Routledge.

Douglas, S. (1994) *Where the Girls Are: Growing up Female with the Mass Media*, New York: Times Books.

Douglas, S. (1999) *Listening In: Radio and the American Imagination*, New York: Times Books.

Dreyfus, H. and Rabinow, P. (1982) *Michel Foucault: Beyond Structuralism and Hermeneutics*, Brighton: Harvester Press.

Ellis, J. (2002) *Seeing Things: Television in the Age of Uncertainty*, London: I. B. Tauris.

Ellis, J. (2004) "Television Production," in R. Allen and A. Hill (eds.), *The Television Studies Reader*, London: Routledge, pp. 275–92.

Feuer, J. (1983) "The Concept of Live Television: Ontology as Ideology," in E. Kaplan (ed.), *Regarding Television*, Los Angeles: The American Film Institute, pp. 12–21.

Forty, A. (1986) *Objects of Desire: Design and Society 1750–1980*, London: Thames and Hudson.

Foucault, M. (1974) *The Archaeology of Knowledge*, London: Tavistock Publications.

Giddens, A. (1990) *The Consequences of Modernity*, Cambridge: Polity.

Habermas, J. (1989) *The Structural Transformation of the Public Sphere*, Cambridge: Polity.

Hall, S. (1980) "Encoding/Decoding," in S. Hall et al. (eds.), *Culture, Media, Language: Working Papers in Cultural Studies, 1972–1979*, London: Hutchison, pp. 128–38.

Heidegger, M. (1999) *Ontology – The Hermeneutics of Facticity*, Bloomington: Indiana University Press.

Heidegger, M. (1962) *Being and Time*, Oxford: Blackwell.

Hilmes, M. and Loviglio, J. (eds.) (2002) *Radio Reader: Essays in the Cultural History of Radio*, New York and London: Routledge.

Hilmes, M. (2004) *The Television History Book*, London: British Film Institute.

Honderich, T. (1995) *The Oxford Companion to Philosophy*, Oxford: Oxford University Press.

Hutchby, I. (2001) *Conversation and Technology*, Cambridge: Polity.

Innis, H. (1950) *Empire and Communication*, Oxford: Oxford University Press.

Innis, H. (1964) *The Bias of Communication*, Toronto: University of Toronto Press.

Jenks, C. (1995) *Visual Culture*, London: Routledge.

Johnson, L. (1988) *The Unseen Voice: A Cultural Study of Early Australian Radio*, London: Routledge.

Lacey, K. (2002) "Radio in the Great Depression: Promotional Culture, Public Service and Propaganda," in M. Hilmes and J. Loviglio (eds.), *Radio Reader*, New York and London: Routledge, pp. 21–40.

Lyotard, J. F. (1986) *The Postmodern Condition: A Report on Knowledge*, Manchester: Manchester University Press.

Paddy Scannell

Mansell, G. (1982) *Let the Truth be Told: Fifty Years of BBC External Broadcasting*, London: BBC.

McLuhan, M. (1962) *The Gutenberg Galaxy: The Making of Typographic Man*, Toronto: University of Toronto Press.

Meadel, C. (1994) *Histoire de la Radio des Années Trente*, Paris: Institut National de l'Audiovisuel.

Meyrowitz, J. (1994) "Medium Theory," in D. Crowley and D. Mitchell (eds.), *Communication Theory Today*, Cambridge: Polity Press, pp. 27–49.

Pawley, E. (1976) *BBC Engineering, 1922–1972*, London: BBC.

Peters, J. D. (1999) *Speaking Into the Air: A History of the Idea of Communication*, Chicago: University of Chicago Press.

Peters, J. D. (2001) "Witnessing," *Media, Culture & Society*, 23(6), 707–24.

Roberts, G. (2001) *The History and Narrative Reader*, London: Routledge.

Sacks, H. (1995) *Lectures on Conversation*, 2 vols, Oxford: Blackwell.

Scannell, P. (1996) *Radio, Television and Modern Life*, Oxford: Blackwell.

Scannell, P. (2000) "For-Anyone-as-Someone Structures," *Media, Culture & Society*, 22(1), 5–24.

Scannell, P. (2003) "The *Brains Trust*: A Historical Study of the Management of Liveness," in S. Cottle (ed.), *Media Organisations and Production*, London: Sage, pp. 99–113.

Scannell, P. (2004a) "Broadcasting Historiography and Historicality," *Screen*, 45(2), 130–41.

Scannell, P. (2004b) "What Reality has Misfortune?," *Media, Culture & Society*, 26(4), 573–84.

Scannell, P. and Cardiff, D. (1991) *A Social History of British Broadcasting, 1922–1939*, Oxford: Blackwell.

Smulyan, S. (1994) *Selling Radio: The Commercialization of American Broadcasting, 1920–1934*, Washington: Smithsonian Institution Press.

Winston, B. (1998) *Media, Technology and Society: A History from the Telegraph to the Internet*, London: Routledge.

CHAPTER FOUR

Our TV Heritage: Television, the Archive, and the Reasons for Preservation

Lynn Spigel

In the last decade of his life, Andy Warhol taped huge amounts of television programs. The remains of his television past – from *Father Knows Best* to *Celebrity Sweepstakes* – are preserved at the Andy Warhol Museum in Pittsburgh, and his collection has also been donated to other film and TV archives. In his usual fashion, Andy managed to create a counter practice out of popular culture by rearranging banal commercial objects under the banner of his trademark name. Indeed, there is nothing out of the ordinary about Andy's collection. It is the kind of stuff one might find – if saved under any other name – in the local thrift store bargain bins. What makes Andy's collection important is therefore not the programs, but the fact that they were saved by a unique collector; the programs are deemed worthy insofar as they shed light on Andy's viewing practices and, by extension, his psychic and artistic investments in the everyday commercial culture of twentieth-century America. To be sure, Andy's TV archive is a highly personal diary of the programs he taped – a kind of homemade archive – and for this reason it is a useful foil with which to begin a discussion of the institutional logics through which museums, academies, and the industry itself have historically deemed TV programs worthy of collection.

What is the logic of the TV archive? Why has TV been saved by public and private institutions? How have nostalgia networks like Nick at Nite and personal recording systems like the VCR and TiVO affected the canon of programs saved? And how does TV's preservation relate to public perceptions of television and to the kinds of questions that historians ask about the medium? In *The Archeology of Knowledge* (1972), Michel Foucault observes that history (as a narrative form and discursive mode) makes the archive. Rather than assume there is a pre-existing "collection" of facts waiting to be accessed, Foucault argues that the archive is preceded by a discursive formation that selects, acquires, and arranges words and things.[1] To be sure, film and television historians engage with a complex system of image classification that has its roots in modern archival systems. The evidence we find – whether paper or moving image – is saved and arranged according to technologies of filing, and, as John Tagg demonstrates in his history of photography, the archive files images according to the power dynamics and beliefs of the larger social system.[2] Historians enter the archive with fantasies and hunches; they search for something they imagine – or hope – was once real. That reality, however, turns out to be at best elusive, accessible mainly through deductions and interpretations of weak, incomplete evidence. Instead of finding the "truth" of the past, what we find in the end is the rationale (or lack thereof) for the filing system itself. For this reason it seems useful to think about the apparatus of the television archive – its strategies for collection, the reasons why people saved certain television programs, and the reasons why so many others are lost.

To this end, I want to trace the discursive formation, and corresponding institutions and bodies of power, through which a television archive has been formed over the past half century in the United States. When considering television preservation, archivists and museum librarians typically focus on pragmatic issues of space, financing, copyright laws, donors, and advances in recording technologies, and they also consider general methods of preservation, cataloguing, and selection. Yet, we know very little about the reasons why programs were saved in the first place.

The fact that I will be writing only about the United States is itself revealing. Nationalist rhetoric and the logic of governmentality have been a primary force in historical research on television, particularly because broadcasting was historically bound to nationalist agendas. Yet, despite the national character of broadcasting as a business and cultural form, the US government showed little interest in archiving television in its early and formative years. While the Library of Congress now holds the largest public television archive, it was slow to realize the medium's value. In a public statement, now on its website, the Library of Congress isolates three factors that led to its random collection procedures in the first two decades of television: (1) Most early live programs were lost to the ether because only a small percentage were recorded on kinescopes; (2) Television programs did not initially require copyright registration (see "Television . . . ," pp. 1–4)[3] and therefore the Library of Congress did not usually receive copies; and (3) Scholars and librarians typically did not consider TV programs worthy of preservation.

Expanding on this last point, the Library of Congress website explains,

> There was an attitude held by Library of Congress acquisitions officers toward television programming which paralleled that of the scholarly community in general. The Library simply underestimated the social and historical significance of the full range of television programming. There was no appreciation of television's future research value. So before the mid-1960s few TV programs were acquired for the Library collections.[4]

As this statement suggests, the eclectic nature of the TV archive is in part due to the fact that there was initially very little respect for television as a historical source. By 1966, the Library responded to a "broadening range of research needs," and TV acquisition expanded. It was not until the passing of the Copyright Act of 1976, "which gave the Library the awesome responsibility for establishing the American Television and Radio Archives" that a national center was created to "house a permanent record of television and radio programs" (Murphy, 1997, p. 13).[5] But if the national library was slow to show an interest in television in the formative decades of commercial television (the 1950s–60s), there were other people who did want to save TV.

Storing Waste: The First TV Archives

In US universities, it was in Journalism, Speech, and Mass Communication Departments – and secondarily in Theater Departments – that television was first studied and arranged as an historical object. These were purviews in which the first generation of historians (most notably, Eric Barnouw) worked. Given their institutional homes, it is perhaps no surprise that for this generation of TV historians, television history was very much based on print media and rhetorical models. The documents collected and arranged were largely paper (regulatory and censorship documents, network memos, scripts, performer biographies, etc.). The programs that comprised TV history were primarily documentary and news programs surrounding what the historians marked out as major political events (the Blacklist, the Korean War, Vietnam). The collection of such documents was governed by top-down and "great man"/exceptionalist views of history. For example, documents deemed worthy of collection included corporate memos of network presidents, statements by crusading newsmen like Edward R. Murrow or writers like Paddy Chayefsky, and Federal Communication Commission (FCC) materials. While entertainment programs were sometimes discussed as evidence for larger political mood swings, they were mostly archived in memory (the historian speaks from recollection) or else in scripts. Although some programs (especially news) were preserved on videotape[6] in university contexts, television's first generation of historians did not use textual analysis as a method; the programs weren't considered as narratives to be interpreted; rather, they were seen as documents to be cited and summarized. In general, then, this first archive is based on the written word, not on the moving image.

Television's first historians were writing in the aftermath of the Quiz Show Scandals and FCC Chair Newton Minow's 1961 "Vast Wasteland" speech, both of which had enormous effects on the national discourse on television at the time (Minow, 1964, pp. 45–64). Blaming the television industry for producing a "steady diet" of commercial pap, Minow set out to cultivate the wasteland, but even more importantly the speech itself established a way of speaking about television's failed promise and envisioning its higher national purpose for education and culture. In this regard, part of Minow's intent was to separate good TV from waste, but his judgments were highly personal and impressionistic. For example, even while he admonished sitcoms, westerns, and game shows, he championed the *Twilight Zone* and *The Bing Crosby Special* as examples of "wonderfully entertaining" TV (Minow, 1964, p. 52). In other words, Minow knew what he liked, but he offered few criteria.

The canon for TV art was developed elsewhere, and the project of saving television programs was very much part of the moment's preoccupation with memorializing TV's "Golden Age." The early 1960s witnessed the growth of an archive movement that sought to weed through trash in the wasteland and find the golden nuggets. This coincided with broader national initiatives in the Kennedy and Johnson Administrations to make art and culture a central federal concern.[7] In this context, Minow's speech formed a national imperative for debates around television as a historic and aesthetic object of national worth, and paradoxically in that sense, the history of television's preservation is very much bound up with its status as "waste."

Television's preservation was spearheaded by institutions that collected television driven by concepts of public service, art, commerce, and public relations. Among these groups the television industry itself was a major force, and for this reason the history of television's preservation is also a history of the industrial logic through which it was saved. In particular, that logic was rooted in public relations efforts to promote the industry by proving that television programs had an aesthetic and cultural value beyond crass commercial gain. Secondarily, although they developed synergies with the industry, art museums and universities took an interest in saving television in order to extend their own cultural authority. And, finally, the project of saving television was (and continues to be) intimately tied to urban planning and local tourism.

In this regard, what remains of TV today belies a set of strategies and statements made by groups that had particular investments in the medium. These strategies and statements are bound up in a complex web of belief systems and prevailing discourses about television's value as an object. Television has been a transitory figure in a shifting system of meanings that constitutes modern collecting practices.[8] It has variously been collected as (1) a commercial asset preserved and stored by studios and networks;[9] (2) an anthropological/historical artifact that sheds light on American history and culture; (3) an art object representing – in the Arnoldian tradition – the "best of man"; (4) an object of science and industry representing technological achievements; and (5) a souvenir representing both a marker of place (TV museums function as sites for the tourist industry) as well as a marker of time (TV is collected as memorabilia).

While there were a number of groups that sought to save TV in the early period, I want to isolate three of the most prominent institutions, each of which epitomized a particular founding vision for the TV archive. Notably in this regard, I am interested not simply in the success stories, but also in the spectacular failures.

1 TV archive as public relations: The Academy of Television Arts and Sciences

In the late 1950s and early 1960s, the Academy of Television Arts and Sciences (hereinafter referred to as the Television Academy) grew to become television's premier archivist. Its collection is now on permanent loan at the UCLA Film and Television Archive (the single largest non-governmental archive in the nation). The collection is primarily composed of programs nominated for the Television Academy's annual Emmy Awards, and in this respect it represents a particular ideal (or shifting ideals) for "quality TV" as selected by the industry itself. These taste standards and corresponding archival efforts are bound up in the Television Academy's own interrelated (if often conflicting) goals of public service and public relations. (For more on this, see Spigel, 1998, pp. 63–94.)

Founded in 1949, the Television Academy was, on the one hand, dedicated to models of culture based on enlightenment ideals of democracy, edification, and public service (ideals that had been integral to the rhetoric – if not the practice – of broadcast communications since the 1920s as well as to the rhetoric of modern museums). On the other hand, in order to sustain itself, it had to make itself a commercially profitable wing of the entertainment industry (most aggressively through its establishment of the Emmy Awards). The Academy's conglomeration of "founding fathers" already expressed the tensions inherent in these dual goals. The Academy's main visionary, TV reporter Syd Cassyd, shared the spotlight with a UCLA professor, an engineer at Paramount Studios, and the Television Academy's first president, ventriloquist and popular radio personality, Edgar Bergen. The Television Academy's link with educators served a strategic role in its ability to promote itself as the leading industry organization for television.

The effort to build what was alternatively referred to as a library or museum was part of the Television's Academy's non-profit educational wing known as the Academy Foundation. Formed in 1959, the Foundation was masterminded as a solution to the problems that the Television Academy had getting tax-exempt status for donations, especially donations of copyrighted materials (such as kinescopes, tapes, and scripts) that could be considered corporate assets. As I have detailed elsewhere, the Foundation's public service/educational programs (including the library and its journal *Television Quarterly*) were devised in the context of the Television Academy's wider public relations efforts to polish television's tarnished image after the Quiz Show Scandals and in the context of 1960s "wasteland" criticism (Spigel, 1998). Along these lines, at an April 1966 meeting, board members noted that the press was "generally hostile to television," and recommended that "through its public relations, the Academy should establish itself as the industry organization speaking for its highest ideals through such activities as the publication of TELEVISION QUARTERLY, its many forums and seminars, its fellowship and scholarship program, the National Library of Television and such services as the ETV Committee . . ." [emphasis in original].[10] Not surprisingly in this regard, the Television Academy hired public relations executive Peter Cott to direct the Foundation, and virtually all Foundation board members had connections with the television industry. At best, educators sometimes served as liaisons, consultants, or editors/writers for the journal. Most typically, however, the Foundation viewed educators as the grateful recipients of the public services (grants, scholarships, lectures) that it supplied.

Industry insiders likewise established the criteria for program collection. In 1959, Cott outlined a plan that served as the basic architecture for the library. Cott's plan was also the basis of a set of discursive rules – that is to say, a "canon" – for generating notions of what was "collectable" from an educator's standpoint. He told Academy members of the need to "establish criteria" for selecting programs to preserve.[11] He also spoke of the need to court "networks, agencies, producers, etc." for program sources.[12] In this sense, far from being established in some ivory tower of "art for art's sake" critical distance, the canon of Golden Age programs (at least as represented by the Academy collection) is a product of the marriage between public service and public relations; it was established in relation to wider industry practices of copyright, ownership, and tax exemptions, as well as the need to bolster the public image of the television industry and the Television Academy itself.

Most important, the Television Academy's archival efforts were influenced by coastal warfare between its Los Angeles and New York chapters. Almost from the start, there were various battles over the way this Los Angeles-based organization was representing itself as the official site for the production of national standards for the television arts and sciences. Its Hollywood locale angered the New York newspaper critics, especially Ed Sullivan, host of the popular Sunday night show, *Toast of the Town*, and an influential Broadway columnist in his own right.[13] Sullivan fought to get control of the organization away from Hollywood, and he rallied support from local chapters across the country that were equally bitter about what they perceived to be the Hollywood bias of the Emmy Awards. By 1957, the original Academy of Television Arts and Sciences had been transformed into the National Academy of Television Arts and Sciences (NATAS), and Sullivan was elected as its first president. Despite the national reorganization, the Television Academy continued to be embroiled in bitter battles between the New York and Los Angeles chapters, which domineered all other local chapters.

More than just a geographical split, the coastal wars were perceived as a culture war between Hollywood's crass commercial tastes and New York's cultural refinement. For example, when commenting on Sullivan's initial takeover, Dick Adler of the *Los Angeles Times* wrote,

> There have always been rumblings of discontent inside the television academy since it began as a Hollywood-based organization . . . New York appears to have always looked upon Hollywood as the sausage factory, the place where canned comedy and cop shows come from. Hollywood's attitude toward its eastern colleagues was equally derisive: They were late snobs who thought that their

involvement in news and live drama gave them special status. (cited in O'Neil, 1992, p. 9)

To be sure, the Hollywood–New York culture war was never so simple in practice. Even while Cassyd developed the pomp and pageantry of the Emmy Awards, he was committed to educational pursuits and in fact argued against hiring public relations companies to promote the Television Academy and the TV industry. Meanwhile, Sullivan was the quintessential showman, hawking Lincoln sedans and featuring popular performers (Elvis, the Beatles), even as he adorned his stage with the Russian ballet and Italian opera. The Hollywood–New York split, then, was more myth than reality. It came to represent a pleasing explanation for the very messy and chaotic indeterminacy surrounding critical judgments of television and its national purpose. Nevertheless, the struggles between New York and Los Angeles were an immediate site of contention in the plans to build a library.

Insofar as the Academy collection is primarily culled from Emmy nominations, the award selection process had an enormous impact on what was saved. To resolve, or at least temper, potential conflicts, in 1963, Cott recommended that the Academy Foundation's library committee split off into two "entertainment Program Criteria Sub-Committees" – one operating in New York and one in Hollywood.[14] In addition to this sub-committee plan, the more general Emmy Award process already suggested a compromise between New York and Hollywood notions of quality TV. Over the course of the 1950s and 1960s, programs produced in New York such as the critically acclaimed *Playhouse 90*, *Omnibus*, and *See It Now*, shared the honors with Hollywood telefilm fare such as the highly popular and critically esteemed *Dick Van Dyke Show*, *Gunsmoke*, and *Disneyland*. Yet, as all of these programs suggest, the prime-time Emmys favored national network series, and in this respect the selection process undermined the achievements of local television producers across the country. Despite the fact that many local broadcasters construed "quality" to mean "in the public interest" of local markets (a criteria firmly established in the 1946 *Blue Book* distributed by the FCC[15]), Hollywood board members often expressed disdain for local productions. In the 1970s, when local academy chapters grew in number and influence, the Hollywood chapter became increasingly hostile. Larry Stewart, president of the Hollywood Chapter, complained that "a news cameraman in Dayton, Ohio, had a vote equal to the cinematographer on *Roots*," and that "Bobo of Seattle's morning children's show was voting for best actor" (O'Neil, 1992, p. 11). By 1977, these disputes resulted in the formal division of the Television Academy into the Hollywood-run ATAS (which presides over prime-time Emmys) and the New York-run NATAS (which is responsible for sports, news, documentaries, international, local, and daytime Emmys). For contemporary researchers, therefore, the Television Academy's "Golden Age" collection is based on a national network bias and rooted in the internal struggles between Los Angeles, New York, and the "rest of the country."[16]

The Academy Foundation's efforts to build a library were also connected to its search for an actual archival site, which was also based on big city bias. In the early 1960s, the Foundation considered partnerships with museums and universities in three urban centers: New York, Washington, DC, and Los Angeles.[17] Because of UCLA's historical links to Television Academy founders, Los Angeles was the firmest connection. In 1960, UCLA began holding materials for the Television Academy, and the Chancellor appointed Cassyd as Acting Curator. The archive was formally established in 1965, and by 1968 UCLA had become the central university television archive in the nation.[18] Nevertheless, at the time, the Television Academy considered UCLA to be a temporary holding site, and it planned to build its own library for public use. Although this did not happen, the important point is that the Television Academy envisioned the collection as a long-term asset, and not just a public service for educators. The collection was indeed always imagined foremost in relation to tax benefits and PR, and, as I will demonstrate further on, the Television Academy held on to that vision of the archive through the mid-1990s.

2 TV archive as art museum: The Museum of Modern Art

A second model for the TV archive is best represented by New York's Museum of Modern Art (MoMA), which displayed, arranged, and aimed to preserve television in the context of the fine arts. While its efforts did not pan out, during the 1950s and 1960s, MoMA aggressively sought to incorporate the new medium of television into the museum. Whereas the Television Academy's archival efforts were bound up with its attempts to radiate an aura of public service, MoMA's interest in the new medium was rooted in the museum's desire to extend its cultural authority over modern art at a time when New York City was becoming the center of the art world (see Spigel, 1996, 2000, 2004).

In 1952, when television was becoming an everyday reality, MoMA received a three-year grant from the Rockefeller Brothers Fund to explore the uses that the museum might make of the new medium. Unlike the Television Academy, which was led by a consortium of industry insiders, MoMA's "Television Project," as it was called, was spearheaded by people in the fine arts, and in this respect it was an attempt to merge the art world with the world of commercial television. To that end, the museum hired avant-garde filmmaker Sidney Peterson to direct the Television Project. Peterson's primary goal was to build a commercially viable, self-supporting TV production company at MoMA and to produce regular series of what the museum called "experimental telefilms" that would appeal to television's mass audiences but still warrant a museum label. Peterson imagined television as an egalitarian art form, and he wrote a lengthy dissertation on its prospects – even likening the television image to the Mannerist movement in painting.[19] The Television Project also extended MoMA's previous attempts to use television to publicize museum collections and shows by, for example, having museum officials appear on women's programs; and finally, the Television Project included plans for a TV library comparable to MoMA's film library.[20]

Despite the enthusiasm for the new medium, anxieties regarding "vulgarization" underwrote all of MoMA's efforts to engage with television. Indeed, whenever MoMA toyed with the terrain of the popular (and the museum was famous for its exhibits on everyday objects and industrial design), critics launched charges of vulgarization (Lynes, 1973, p. 233; Marver, 2001, Introduction). The "mass" nature of television exacerbated the issue. Despite MoMA's success with Peterson's television film *Japanese House* and its children's TV series *Through the Enchanted Gate* (produced by the Education department), MoMA director Rene D'Harnoncourt grew weary of Peterson's experimental telefilm pilots that tried to combine avant-garde practices of poetic montage with popular broadcast aesthetics and stars.[21] The Museum directors rejected a number of Peterson's pilots on basis of their potential offense to patrons and the directors' sense that "by affixing its signature to . . . the films the Museum would be thought to be . . . lowering its intellectual level."[22] By 1955, MoMA shut down its plans for in-house commercial TV productions.

MoMA's efforts to build a TV library were similarly plagued by conflicts between aesthete and popular dispositions. As James Clifford argues, commercial objects cannot travel directly from the sphere of what is deemed "inauthentic" mass culture to the status of an "authentic" piece of art worthy of display in a fine art museum. To make this journey, the commercial object first has to be authenticated by moving through some other system of display. (For example, it has to be elevated to an artifact of anthropological or scientific importance before it can be re-evaluated as an artwork [Clifford, 1988, p. 224].) In this sense, it is no surprise that in the early 1950s, when museum officials were considering collecting television as a work of art, feathers were ruffled.[23] Perhaps because their medium was closest to television, the people in the film library were particularly concerned to distance themselves from the new medium. Richard Griffith, director of the Film Library, openly expressed his desire to keep television distinct from film. One internal memo, written in 1952, noted that Griffith said the film people "hate TV" and would tolerate its inclusion only "as long as it is not [in] the same department as the film library."[24]

Despite this initial antipathy, the Film Library, and MoMA officials more generally, recognized the economic value that television might have for the library, particularly with regard to potential rental requests for footage. Moreover, the Film Library had to respond to the wishes of museum directors, board members, and, most importantly, the museum's founding family, the Rockefellers. As early as 1952, Nelson Rockefeller wrote to D'Harnoncourt suggesting that the museum put on an exhibit featuring "the best in TV (films or kinescopes)."[25] Upon hearing of Rockefeller's suggestion, Griffith acknowledged that the museum might consider using some television films or kinescopes in a film retrospective to be held at MoMA. But, in keeping with his fear of vulgarization, Griffith specifically suggested that films and kines selected should be "only the

75

best, very short, and constituting the museum's explicit endorsement of the kinds of art film-making they represent, with an implicit denouncement of other kinds."[26] In other words, Griffith wanted to make sure that television would not pollute the Film Library's image as a "tastemaker."

By 1955, Griffith had tempered his views, and in a museum report he even spoke of the "obvious need for a central television archive analogous to the Film Library."[27] Yet he also pointed to a number of obstacles, including copyright permissions[28] and funding.[29] Speculating that the library's mission would lie in "preserving the history of the art" and "disseminating representative films of the past to qualified educational institutions throughout the country," he noted that the selection of programs for the collection would require (as it had for film) the "active help and guidance of the television industry."[30] In the 1956 museum *Bulletin*, Griffith continued with these themes, although he noted, "some progress has been made."[31] He reported that networks used the library to rent source material and rare footage and that "as an experiment, the Film Library has acquired for its collection the kinescope of a single 'live' television production, Horton Foote's *The Trip To Bountiful*, with Lillian Gish, which later was translated to Broadway with the same star."[32] This first acquisition is characterized by a set of qualities that are consistent (at least in some combination) with more general critical hierarchies already established by the leading East Coast critics of the 1950s. These criteria included "live" production; the presence of well-known stage talent (especially playwrights and Broadway stars); and/or an indigenous relation to New York.

In 1962, these "Golden Age" criteria – as well as Griffith's general views on the selection process – resurfaced in MoMA's *Television USA: 13 Seasons*, the first museum retrospective of television programming. Mounted by the film library, *Television USA* established criteria by which MoMA would endorse programs worthy of collection and display. The retrospective appeared just one year after Minow's "Vast Wasteland" speech and resonated with the prevailing evaluative discourses about when and how the degraded medium of television might ever approximate art. But unlike Minow, who just knew what he liked, the film library at MoMA formed a committee of what Griffith referred to as "television artists" to select programs for the show.[33] Headed by Jac Venza, a television director and set designer, the committee consisted entirely of television producers and executives.[34]

The program book for *Television USA* began with a mission statement from Griffith who outlined criteria for aesthetic judgment. Admitting that they "immediately faced a problem of policy" when establishing standards for selection, Griffith asked, should the museum "attempt to represent every aspect of the intricate pattern of television programming today?" Or should the programs be selected "for quality alone, even if this meant scantily representing or even omitting altogether certain categories of television material?" Answering his own rhetorical question, Griffith stated,

We are unanimous in deciding for the second course . . . It seemed to us what the Museum could most usefully provide would be a look at first-class work which many of our public had missed because they are in the habit of looking at television only at certain hours (or not at all). With the exception of historical milestones, included to make the record as complete as possible, every program in the exhibition has been selected because we thought it used the medium to the top of its capacity.[35]

In this respect, Griffith reinforced the "Golden Age" discourses of the times, but did so in the context of MoMA's larger struggles to valorize its own tastes. First, the criteria are stated in relation to the taste proclivities of his presumed high-brow patrons – those who watch a little or no TV at all. Second, Griffith defines artistic worth in relation to fine art values of media specificity (i.e. programs that "use the medium to its capacity").

Despite the seeming clarity of his aesthetic criteria, however, the actual programs chosen for the retrospective belie a much less consistent evaluative scheme. *Television USA* exhibited not just TV, but also a number of competing ideas and discourses about television's status as an art form. As in the case of the Television Academy, MoMA's selections overwhelmingly demonstrate a geographical bias. Given MoMA's New York location, and the presence of New York critics and talent on the selection board, it is not surprising that almost all of the programs chosen for the exhibit were produced in New York. These included live anthology dramas such as *Goodyear Playhouse*; documentaries and public affairs programs like *See It Now*; programs on the arts such as *NBC Opera Theater*; variety shows like *Your Show of Shows*; and a spattering of programs chosen for their formal experimentalism (an attribute that both resonated with Golden Age critical discourses concerning media specificity and with MoMA's own bias towards various modernisms). The only Hollywood-produced telefilm series that MoMA included in the exhibit was an episode of *Gunsmoke*, and it "was chosen because it is an almost perfect adaptation of the genre to the medium" rendered in "almost a classical manner."[36] (In other words, it too fit with the fine art criteria of media specificity.)

Although East Coast-centric, and steeped in a preference for live theater over telefilm fare, MoMA's aesthetic criteria were not entirely coherent. Instead, judgments shifted between extreme investments in theatrical realism ("If there were a Golden Age, it was when television drama concerned itself with real problems, real issues, and real people"[37]) and a desire for modernist experimentation (programs such as *Danger*, the *Ernie Kovacs Show*, and *Adventure* are praised for their formal innovations[38]). So too, MoMA's attitude toward commercialism was inconsistent. On the one hand, the program catalog for *Television USA* defined the industry as art's enemy, stating that television was divided in "two camps": the industry that is concerned with money and "artists and journalists whose standard of 'success' is the degree to which television realized its potentialities as an art form."[39] On the other hand, *Television USA* embraced commercials as the apex of TV art. The program catalog stated, "Almost everything has been tried to create original commercials. As a result, radical avant-garde experiments which would be frowned upon in other areas of television are encouraged in this field."[40] Consequently, *Television USA* exhibited everything from Brewer's beer ads to Rival dog food ads as proof of television's potential avant-garde status.

Why did MoMA reject commercialism, but honor commercials? It seems likely that MoMA's embrace of commercials was based not only on its historical willingness to display industrial design, but also on the emergence in this period of Popism and Assemblage Art (a broad term for three-dimensional collage or collage sculpture, often featuring "junk" castoffs like neon signs, ads, abandoned car parts, and hollowed out TV sets). In 1961 MoMA mounted William C. Seitz's "Art of Assemblage" and by 1962 MoMA held a symposium on Pop. (See Lippard, 1966, for a discussion of New York Pop, especially pp. 69–90.) In this respect, although MoMA and the Film Library still operated on enlightenment ideals of cultural edification, the museum also responded to the shifting nature of art discourses and practices, particularly the leveling of "high" and "commercial" genres that was so important to Pop aesthetics. *Television USA* reflected the museum's competing claims to Popism's aesthetic embrace of the commercial and the Wasteland Era's anti-commercial ideals.

After *Television USA*, the museum's Junior Council[41] continued to explore the possibilities of a television archive. However, their vision culminated in a much more narrowly defined collection aimed at art historians proper. In 1964, the Junior Council began to collect television documentaries on artists, and with the help of the three major networks, New York's Channel 13 (WNDT), and NET, in 1967, they established a Television Archive of the Arts.[42] In 1968, CBS Chairman William S. Paley (who had previously sat on the board at MoMA) became president of the museum's Board of Trustees. However, the presence of the CBS chairman seems to have had little impact on the growth of the television archive. Instead, in the early 1970s Paley turned his attention to establishing his own privately funded TV museum. Meanwhile, MoMA's early conception of TV as art, and even its more limited vision for an archive devoted to art documentaries, quickly vanished. Instead, by the early 1970s, MoMA had embraced the emerging world of video art, engaging a more narrowly defined "art" public. In 1972, MoMA staged "Open Circuits: An International Conference on the Future of Television," which brought together artists, museum officials, and critics who spoke almost exclusively about emerging video art forms. Indeed, despite the conference's subtitle, MoMA might as well have left television out of the future altogether, because after that time video art had usurped TV as the preferred object for preservation and display in the art world.[43]

3 TV archive as tourist site: The Hollywood Museum

A third model for the TV archive is rooted in tourist trade. In fact, even at MoMA, where the TV exhibit and archive were premised on appeals to art, tourism was a key consideration. *Television USA* was staged in the context of urban planning for the 1963–64 New York World's Fair, and MoMA officials thought they would attract international interest in this context. However, by far the most sustained efforts to launch a TV tourist attraction took place back in Hollywood where industry people got the city of Los Angeles to front seed money for the never-to-be-realized Hollywood Museum.

The Hollywood Museum was intended as a pleasing tourist experience that would promote television, movies, radio, and the recording arts. In this respect, like the Television Academy, the Hollywood Museum was conceived as a public relations arm for the entertainment industry. In 1959, the Los Angeles Board of Supervisors formed the Museum Commission; the Advisory Council included industry insiders Desi Arnaz, Jack Benny, Frank Capra, Walt Disney, William Dozier, Jack Warner, Arthur Miller, Ronald Reagan, and Harold Lloyd.[44] According to the founding document, "The goal is to portray these four communicative arts as having a justification not only as entertainment media but also as important contributions to humanity . . . the Museum will be of aid in a positive way in overcoming the damaging effect of the constant and growing criticisms of the industries by numerous private and public groups."[45] In other words, the museum was intended to counteract the bad press that was generated in the wake of radio's Payola Scandals[46] and Quiz Show Scandals of the decade; by the early 1960s, this mission resonated against the backdrop of Minow's famous speech. In 1963, when the county staged ground-breaking ceremonies for the building, the planning committee promoted the museum not simply as a local attraction, but instead as a national event of international importance. In line with President Kennedy's use of art as a strategic force in "free world" rhetoric, the museum promoters sent a telegram to Secretary of Defense Robert S. MacNamara, which stated that the museum and the mass media more generally would help create better understanding among nations.[47]

Rather than just collecting programs or films, the Hollywood Museum collected objects (kinescopes, early TV sets, costumes, etc.). In this regard, television was imagined not simply as a text, but as a technological artifact like those exhibited in a museum of science and industry. Moreover, the museum's designers imagined the exhibits as an immersive experience akin to the kind offered at a theme park. The then-wondrous technologies of rear screen projection and hi-fidelity sound were the prime technologies on display at the museum, and the promoters represented the museum experience as a kind of thrill ride for patrons.

This experiential aspect of the museum is particularly evident in a short promotional film titled "Concept" that was made to attract donors by demonstrating the architectural plans as well as the overall vision. Narrated by Edward

G. Robinson, who first appears in front of a wall of books, the film depicts the museum as what Robinson variously calls "one of the most exciting showplaces of the world"; the "first international center of audiovisual arts and sciences"; and a "research center," that will serve as a "living memorial to [the] media." In other words, the museum is divided across a number of touristic, artistic, industrial, and scholarly functions, which also are integrated into its architectural plan. In the opening sequence Robinson shows a small-scale model of the museum. A study in California "moderne," the building comprises an "education tower" (which houses "extensive materials for study"), a four-level pavilion (the main building that houses displays, visitor participation shows, exhibits, a restaurant, a library, and a theater), and two fully equipped sound stages. The promotional film then segues to more detailed sketches of the museum interiors, and a bevy of Hollywood stars explain the various displays.

The sketches mostly show a tourist class of patrons interacting with elaborate World's Fair- and Disneyland-inspired exhibits (in fact, Mickey Mouse himself narrates a sequence, and at the end of the film Mary Pickford specifically says the museum is designed in "World's Fair fashion"). Gregory Peck shows tourists experiencing the "synthesis show" which, for example, uses the "magic of rear screen projection" to transport visitors back to a virtual Imperial Rome. Next, Bing Crosby takes a walk on the "Discovery Ramp," where visitors learn about entertainment history. Bette Davis leads a tour through the elaborately designed museum restaurant where a replica of the Hollywood canteen transports visitors back to the USO shows of World War II. Demonstrating the interactive exhibits in the Wardrobe Hall, Doris Day models a gown she wore in one her recent films.

Although the museum is primarily imagined as a tourist pavilion, because it is aimed at donors, the promotional film emphasizes the economic use value that the museum will have for the entertainment industries that support it. For example, Jack Benny explains that the museum's two sound stages will not only serve as exhibition halls where visitors can see how movies or TV shows are made; they will also serve as laboratories for advertising agencies that will use the visitors gathered there as test markets and perform studies in audience psychology, motivation research, and ratings. The film creates a subtle balance between these direct industry uses and the museum's aura of art and education. From the wall of books in the opening sequence, through the numerous professions of art, history, and international understanding, the promotional film suggests that the museum's educational aura will have great PR value for the industry. All of this comes to a climax in the final sequence when silent film star and one-time industry mogul Mary Pickford displays a space age "information center" that will link the Hollywood Museum via computer to universities and museums worldwide. The research library is rendered in the style of a "world of tomorrow" World's Fair pavilion; the sketches show families (not researchers) walking through the planets while futuristic music plays on the soundtrack. In the tradition of the Kennedy Era competition for worldwide technological and cultural supremacy, Pickford promises that the museum's "computer oriented library" will be a "truly international center."

Due to a series of local disputes, funding problems, and the failing health of its main visionary, the grandiose plans of the Hollywood Museum never came to fruition.[48] But even while it failed, the vision for an entertainment theme park lived on – most obviously in the opening of Universal Studios – just a mile down the road – in 1964. This vision also underwrote the future of the television archive in the next decade.

The Nostalgia Mode and the Postmodern TV Museum

The Academy, MoMA, and the Hollywood Museum demonstrate a number of historical contexts through which television came to be collected, organized, and displayed. While each represents a dominant vision of the collection (public relations agency, art museum, and tourist attraction), as I have suggested, they each also contained elements of all three, and all three institutions justified their efforts with claims to public service. In this respect, the goals of public relations, artistic achievement, and tourism all served to form a context for collecting TV that influenced not only the canon for preservation, but also the archive experience itself.

Today, there are two kinds of archival imaginations. One is aimed at a research model and mostly situated in universities and public museums (e.g. the UCLA Film and Television Archive, the University of Georgia's Peabody Award Archive, the Wisconsin Center for Film and Television, Vanderbilt University's News Archive, and the Library of Congress). Unlike the early archival efforts at MoMA or the Hollywood Museum, these institutions are aimed almost exclusively at an intellectual class of researchers (a public that formed itself through the institutionalization of television and media studies at universities) and at industry-oriented research. Changes in copyright laws, recording and preservation technologies, and especially (as the Library of Congress website suggests) attitudes toward TV have contributed to the growth of these institutions. These public archives arrange their spaces for "serious" study in quiet library settings, and they typically require prior arrangements with archivists. Paradoxically, in this respect, while the 1960s-era museum and library was steeped in enlightenment ideals of public service, these institutions have become the purview of intellectual and industry elites. That said, these archives are not purposefully exclusive; rather, as the 1996 Congressional report on television and video preservation determined, these public archives are often under-funded and they have to abide by copyright interests and restrictive usage policies (Murphy, 1997, p. 9). Meanwhile, private museums such as the Museum of Television and Radio (MTR) in New York and Los Angeles and the Museum of Broadcast Communications (MBC) in Chicago have positioned themselves as tourist sites for a general public. As the MTR's former president the late Robert M. Batscha stated, "We're effectively the first public library of the work that's been created for television and radio. You don't have to be an academic. You don't have to be in the television business. Anybody can have access to the collection" (cited in Weintraub, 1996, p. C11).

These private museums appeal to the general public through contemporary strategies of museum exhibition, including blockbuster festivals, celebrity signings and star-studded panels, interactive "touristic" exhibits, and, most of all, nostalgia. The nostalgia mode still contains the earlier era's public relations, aesthete, and touristic functions but arranges them differently. Rather than preserving TV (and thus producing its value) within coastal battles around geographical place (and taste), the nostalgia mode encourages a ritualistic relation to time. Audiences are convened around generational memories, mythic pasts, and "retro" aesthetics, and although these museums do claim to endorse "quality" TV, they are not constituted around the high seriousness of Wasteland Era disputes. Like the Academy or Hollywood Museum, their trustee boards are composed mostly of industry insiders, but the nostalgia mode is rooted in corporate synergies that far surpass the PR functions of the previous decade.

The first and most successful institution to build on this new vision was the Museum of Broadcasting that was founded in 1975 and opened its New York City location in 1977. The brainchild of William S. Paley, the museum combined all three previous ideals for the TV archive: it was (and continues to be) a public relations arm for the industry; a museum that promised to establish a canon of TV art; and a tourist site for visitors. Serving as the board chairman, Paley guaranteed funding for the museum, which was augmented by industry donations, membership fees, and an admission fee for the general public. Robert Saudek, the executive producer of the critically acclaimed 1950s program *Omnibus*, served as the museum's first president and gave the museum the proper Golden Age aura (Shepard, 1979). The collection was – and still is – secured through contractual agreements with networks, studios, producers, and, in some cases, private donors. In 1991, in response to the rise of cable TV, the museum changed its name to the Museum of Television and Radio.

Like the earlier archival efforts, Paley's museum was a finely crafted balancing act between public service and public relations. Of all the three major networks, CBS had always been the most invested in the apparatus of television criticism and preservation. Not only did Paley himself sit on the board of MoMA, but the network more generally sought to raise television's reputation. As with the Television Academy, CBS's efforts in this regard were aimed at undermining negative criticism that circulated in the popular press. As the only network indicted for its questionable practices during the Quiz Show Scandals, CBS was particularly prone to critical attacks of crass commercialization. For this reason network executives often referred to the newspaper critics as "hacks," and they also tried to discredit them by establishing their own in-house army of "quality" critics who often had distinguished university credentials (Spigel, 1998, pp.

67–70). Just like the Academy, in the early 1960s CBS devised a plan to publish a scholarly television journal. Although the journal never came to fruition (CBS dropped its plans when it heard that the Academy was publishing *Television Quarterly*), it did materialize in 1961 in book form as *The Eighth Art* (Shayon, 1962).[49] Throughout the 1960s, CBS remained wed to the idea of creating its own "in-house" critical apparatus, and from 1967 to 1971 the William Paley Foundation commissioned Dr William B. Bluem to study the possibility of creating a master collection of broadcast programs. The Bluem Report found that "there is an urgent and vital need to create a master plan and a centralized collecting institution to prevent destruction and loss."[50] That master plan and centralized institution became Paley's museum.

The museum's mission was to provide an interpretation of the broadcast past, and "Paley himself saw museum interpretation as one of the greatest benefits for the general public."[51] As the Library of Congress report on television and video preservation suggests, at the MTR "the act of interpretation manifests itself in the collections," which are organized to establish a balance of significant programming that represents all of the important genres.[52] For this reason the collection does not typically feature whole series but rather samplings with diverse appeal. From the start, the museum was a popular success. In 1979, the *New York Times* reported that the museum was so crowded that in the two years since it opened it had to turn away "three-quarters of its visitors" (Shepard, 1979).[53] The *Times* also spoke of the eclectic cultural sensibilities of the museum's patrons, commenting on the mix of people watching such "high" Golden Age performances as Toscanini's NBC orchestra and those watching the commercial likes of *The Making of Star Wars*. Despite the "something for everyone" ethos, the MTR has historically appealed mainly to middlebrow family publics, a constituency courted through the museum's posh Manhattan and Beverly Hills shopping districts. So too, while the MTR boasts of its popular appeal, it nevertheless arranges and displays television according to legitimizing discourses of art collecting. Museum events cater to connoisseur values of "rarity," "authorship," and "private screenings." For example, the museum has variously mounted exhibits on the lost live *Honeymooners* sketches and TV "auteurs" like Dennis Potter, and it began to feature private screenings when the museum expanded in 1979 to include a 63-seat auditorium.

In 1991, the museum took this combination of ballyhoo showmanship and aesthete connoisseurship one step further when it moved to its current West 52nd Street location where it occupies a somber and richly appointed limestone-clad tower designed by the famous modern architect, Philip Johnson. The relocation to the Johnson building is a symbol of the museum's own cultural movement from an urban curiosity to an established, and at least quasi-respectable, part of the New York City museum circuit. Paley's office (still preserved in the building) is a testimony to his eclectic vision where high culture and commercial culture live in harmony. A television set and Paley's many awards are on prominent display, but the office is also lined with examples from Paley's collection of modern paintings. However, the MTR is not just a product of Paley's tastes. The MTR's relocation to the Johnson building is consistent with a more general postmodern logic where the museum building is every bit as (or perhaps more) spectacular than the objects on display.

When it comes to the TV museum, this postmodern logic is tied to the peculiarities of the larger TV collector's market. The growth of cable, home video, VCRs, and now DVDs makes it much easier for the public to see and tape old TV shows at home. In this context, the TV museum no longer has a unique value in relation to programs. Instead, the museum's value is extra-textual. Its blockbuster annual festivals, star-studded panels, University Satellite Seminars (with industry leaders beamed into classrooms), and especially the museum's own architecture have been the sites of investment and prestige. This logic became even more apparent over the course of early 1990s when the MTR began to plan its sister museum in a new Beverly Hills location that opened on March 17, 1996.

Designed by architect Richard Meier (who also designed the Getty Center in Brentwood, California and the Museum of Contemporary Art in Barcelona), the Beverly Hills MTR is finished in enameled white metal panels with expansive walls of glass and a circular-shaped glass-plated rotunda entranceway. Harking back to the modernist sentiments of Le Corbusier, Meier claimed, "The main purpose of the building, like of the media it celebrates, is communication. We made it as open and transparent as possible, and devoid of mystery, so that people passing by can plainly see what happens inside" (cited in Whiteson, 1996, p. K-5). Although Meier's claims to transparency are rather suspect (as with all of the archives I have discussed, there are many business deals going on in the museum that are not seen by the pedestrian eye), his use of the media as metaphor for the building suggests the extent to which the museum itself is the main attraction on display. Upon its opening, Hollywood insiders praised the museum not only – or even primarily – for its collection, but rather for the building. Admitting that a museum of television would probably have to contain some "garbage," producer Larry Gelbart (of M*A*S*H) claimed that the building was, nevertheless, "the best piece of architecture to come up in Beverly Hills in years" (cited in Weintraub, 1996, p. C-12). In fact, the museum boasts no new acquisitions; its holdings are identical to those in New York, available instantaneously through electronic links so that users can access most programs within minutes. The exhibits are typically the same as those in New York, and many of the Golden Age programs housed in the MTR are highlights of the more voluminous collection held down the road at UCLA. Clearly, then, the museum's most distinguishing feature is the space and location itself.

Indeed, the MTR's cultural worth is not really demonstrated by its documents, but rather by the monument – the building – that contains them. The museum interior is subdivided to memorialize its various star/industry donors (e.g. the Danny Thomas Lobby, the Aaron Spelling Trustee Reception Area, the Steven and Barbara Bochco Scholar Room, the Gary Marshall swimming pool (yes, swimming pool), and so on). The building itself is named for ABC founder Leonard Goldenson who donated $3,000,000 (Murphy, 1997, p. 60). The star names confer value onto the experience of being in the museum space while the elegantly designed spaces bestow value onto the people for whom they are named. As an architecture critic for the *Los Angeles Times* wrote at the time of its opening, "The building's severe Modernist style, and the choice of Meier – rather than a more funky designer like, say, Frank Gehry – comes from the cultural aspirations of the people who create television and radio. It reflects their desire to be taken seriously as artists rather than as mere entertainers" (Whiteson, 1996, pp. K1, 5). In this sense, it is not surprising that even while MTR publicity boasts about the museum's popular appeal, the museum directors and industry people still promote the museum through discourses of art, education, and public service. At the Beverly Hills MTR opening, Diane English, the creator/producer of *Murphy Brown*, said, "Being able to look at television is as important for our culture and our history as looking at a great painting" (cited in Weintraub, 1996, p. C12).[54]

While the MTR is certainly the most successful private TV museum of its time, its vision is in no way idiosyncratic. The more modest Museum of Broadcasting Communications in Chicago copied its success in 1987. Like the MTR, the MBC was founded through industry investment (it is the brainchild of Bruce Dumont, the son of TV tycoon Alan B. Dumont). Currently relocating to a newly renovated building with a glass façade that mimics the transparent look of the Beverly Hills MTR, it now claims to be the "12th most visited cultural tourist attraction in Chicago."[55] Meanwhile in the 1980s, the Academy of Television Arts and Sciences renewed its efforts to build a public library, but this time the library was slated to be part of a $350 million redevelopment plan for North Hollywood. The Academy Project, as it was called, was the highlight of the investment and resulted in the construction of the Academy Plaza, which opened in the early 1990s – at about the same time that North Hollywood dubbed the area the NOHO Arts District (named after New York's SOHO).[56] The Academy Plaza contained not only the Television Academy business office and space for a library (which was originally intended to serve as a public archive for its program collection), but also an apartment complex (adjacent to, but not owned by, the Television Academy) and a "Hall of Fame" courtyard adorned with bronze statues of beloved TV stars and a 15-foot-tall golden replica of an Emmy Award looming in the center of a huge fountain. Once again, the real estate was more the attraction than the TV programs themselves.

Unfortunately for the Academy, however, the MTR's Beverly Hills opening and Universal's opening of City Walk (which is just down the road) trumped its more modest NOHO location. As one LA tourism website warns, "It isn't easy to find this new Hall of Fame . . . it's in a slightly seedy part of town, and it isn't visible from the street."[57] Having proved itself unsuccessful, the Academy left its TV collection at UCLA (where it remains on permanent loan), and donated its documents to USC's Doheny Library (which it had partnered with since 1988).

Today, the Academy Foundation still runs an archive, but its "Archive of American Television" is arranged as a holding repository for oral histories of TV pioneers. And while LA tourism websites like seeingstars.com still promote the Academy headquarters with glossy pictures of the giant Emmy and bronze stars, in reality Academy Plaza is a TV ghost town.

TV is its Own Archive: Nostalgia Networks

While I have so far discussed the TV archive as a physical and public site, it is also true that since the 1980s, television has increasingly become its own archive. With the rise of multi-channel cable systems, and sweeping changes in broadcast television, a number of networks have found a new "vintage" use value for re-runs and begun to amass particular kind of archives that appeal to narrowcast demographics.[58]

Two of the most successful of these cable ventures – Nick at Nite and TV Land – brand themselves as nostalgia networks, and in this regard they have taken on the role and function of the TV museum. Their business practices are based on corporate synergies with sister companies in their larger "umbrella" parent corporation Viacom. Originally a syndication company that specialized in the market-by-market sale of off-network re-runs, Viacom is now one of the top five multinational conglomerates in the business.[59] Among its many holdings, Viacom owns the children's cable network, Nickelodeon, which since the late 1980s has filled its prime-time and late night hours with its Nick at Nite line-up of vintage TV programs targeted to the 18–49 demographic. In 1996, Nick at Nite spawned TV Land, which is devoted to the same re-run fare, but on a full-time basis and geared toward a slightly older skewing demographic (24–54).[60] By 1997, Nickelodeon established links with the MTR, and today Mel Karmazin (president and CEO of Viacom) is a vice president of the MTR's Board of Trustees. Indeed, the nostalgia industry is now a revolving door of electronic and physical sites.

More generally, nostalgia networks amass audiences not only by taking them back in time, but also by promising them a fantastic sense of shared place. As befitting the moniker "TV Land," these networks blur distinctions between the physical world of the viewer and the diegetic (storyworlds) of old TV characters, storyworlds that are not only familiar, but also affectively meaningful for the generations of viewers the networks try to attract. Nostalgia networks transport viewers back to the Brady Bunch's ("oh so 70s") brown and orange kitchen, Mary's Minneapolis newsroom, and Cosby's posh but cozy Brooklyn brownstone, places that have become an imaginary geography for viewers. This nostalgic relation to TV space – the sense of locality and community it provides – is a particularly interesting twist on what David Harvey has explained to be a hallmark characteristic of postmodern geography. Harvey argues that in a global culture where electronic communications have made space more abstract, a local sense of place and tradition has become more important (Harvey, 1989).[61] To achieve this, numerous cities promote tourist and business trade by creating nostalgic places that refer back to a mythic past. (For more on this topic, see Hannigan, 1998.) For example, 1970s festival markets like the Faneuil Hall Marketplace in Boston or new urban festival malls like The Grove in Los Angeles are nostalgic throwbacks to old town squares. Viacom has taken this strategy one step further in merchandizing tie-ins and promotional ventures that evoke a sense of nostalgia not for actual places, but rather for television's diegetic places.

For example, Viacom has strategically blurred the lines between TV places and real places by creating tourist and shopping venues that allow people to interact with TV storyworlds. In May of 1997, Viacom opened its Viacom Entertainment Store on Michigan Avenue in Chicago, a two-level 30,000-square-foot store that offered some 2,500 branded products and promoted itself (in five languages) as a "fun house, an architectural treat, a museum, a theater, a concert hall, a boutique."[62] Like the Hollywood Museum, the store used the wonders of contemporary technologies of illusion to build 60 different types of interactive entertainment experiences that focused on six of the corporation's divisions: MTV, VH1, Nickelodeon, Nick at Nite, Paramount and *Star Trek*.[63] Just like the Hollywood Museum, the store was a spectacular failure; it closed within 18 months.[64] Nevertheless, the experience of "touring" through "televisionland" survived in the TV Land Landmarks campaign that places bronze statues of TV stars in place-appropriate sites. To date, Minneapolis has a Mary Tyler Moore statue; Raleigh, North Carolina has TV father and son statues of Andy Griffith and Opie; and the New York City Port Authority Bus Terminal has a Ralph Kramden statue which, the TV Land website reports, has become the confidante of "several crazy people" who talk with the bronze Ralph.[65]

Meanwhile, back on their cable networks, both Nick at Nite and TV Land create a particular kind of archival experience for viewers by scheduling old programs in new ways. Rather than simply stripping re-runs across a weekly schedule in daytime or fringe dayparts (the strategy common to off-network (re-run) syndication, the nostalgia networks create what can be called "themed flows" targeted at specific demographics and tastes. Nick at Nite and TV Land create themed flows by repackaging old TV according to a camp (or mass camp) sensibility that is registered in station identifications, promotional ads, marathons, and specials. As I have argued elsewhere, these networks appeal to a TV-literate generation by suggesting that their viewers are somehow more sophisticated and "hip" than the naïve audiences of the past.

Of course, because Viacom's archive is arranged solely for entertainment value and includes mostly prime-time genre hits and "retrommercials," it omits the vast amount of local television, public affairs, and news and documentary shows of the past. Nevertheless, with their trademark campy wink, Nick at Nite and TV Land present themselves as television's premier historians. For example, in the early 1990s, Nick at Nite ran promos featuring vintage sitcom star Dick Van

Dyke who informed viewers of the network's mission to preserve "our television heritage." More recently, in 2003 and 2004, Nick at Nite and TV Land simulcast an awards show for the all-time best TV programs – a strategy that recalls the Television Academy and MoMA's attempts to institutionalize themselves as television's best critics. But in this case, the ceremony was in many ways a camp rendition of the whole idea of "quality" TV, capped off with Eric McCormack (of *Will and Grace*) singing "Mary," a cheesy, over-the-top tribute to Mary Tyler Moore.

If Viacom has successfully cornered the market for camp, other nostalgia networks have positioned themselves as the "family values" networks. In the mid-1980s, the Christian Broadcast Network (CBN) initiated this trend in prime-time hours by creating a themed flow of TV re-runs that harked back to wholesome 1950s family life. The schedule included *Father Knows Best*, *The George Burns and Gracie Allen Show*, and more obscure programs like *I Married Joan*. In 1998, the PAX TV network took up a similar strategy. Launched by media magnate and born-again Christian Lowell White "Bud" Paxson, PAX TV bills itself as the network for "family entertainment" and brands itself with a God theme and Midwestern "heartland" values.[66] While its schedule includes first-run syndicated programs, paid programs, and original programs, over the years it has mixed these with family-oriented re-runs (*Flipper*, *Bonanza*, *Eight is Enough*, *Big Valley*) and re-runs of CBS's old-skewing *Diagnosis Murder* and religious-themed hit *Touched By an Angel*. The CBN and PAX strategy for a TV archive is rooted less in Nick at Nite's mission of preserving "our television heritage" than in a more extensive mission to preserve "core" American family and religious values.

This same preservationist mission is now aggressively marketed on GoodLife TV, a Washington, DC-based cable network that has ownership ties to the controversial Unification Church. Launched in 2001, GoodLife bills itself as "The Boomer Network – the nation's only full-time cable channel dedicated to providing lifestyle, entertainment, and information programming for the baby boomer generation."[67] With library deals involving Warner Brothers and other syndicators, GoodLife airs vintage series from *77 Sunset Strip* to *Welcome Back Kotter*, and schedules these alongside original series, including informational shows and lifestyle genres.

GoodLife distinguishes itself from Nick at Nite and TV Land not only through its somewhat older-skewing target audience (38–55) (*Multi-channel News*, 2002, p. 1), but also through its unique mission. Calling GoodLife "the go-to net for this boomer generation," Network President Lawrence R. Meli says:

> We're giving them the classic TV they grew up with, but we also want to give them information to help them cope with key issues. Because many families of this generation have elderly parents living longer and kids going to college, there's a big financial as well as emotional strain on them, and they want guidance over what to do about that.[68]

In this sense, GoodLife's program schedule not only works to preserve traditional core values (as in the case of CBN or PAX) but, in the fashion of boomer-era self-help marketing, the network also strives to have "therapeutic" value.

To this end, GoodLife schedules family-oriented re-runs alongside originally produced informational programs. For example, the network has scheduled 1950s sitcoms such as *Make Room for Daddy* next to its original program *American Family*, which takes viewers into the homes of real life families who discuss their problems. And it places vintage war programs like *Combat!* or westerns like *Maverick* in a line-up with *Homefront America*, an original series inspired by 9/11 that tries to "inform families . . . what dedicated Americans all across the country are doing to enhance the security of themselves and their fellow citizens."[69]

As the different strategies of Viacom and Goodlife indicate, the nostalgia network has imagined through two alternative "themed flows" – one camp, the other family values – that each create a brand identity for their respective markets of "hip" and "square" viewers. As TV becomes its own archive, then, the archive is increasingly "niche."

The Archive, the VCR, and DIY TV History

Although television networks like Nick at Nite, TV Land, and GoodLife have arranged a TV archive according to the demographic profiles of their corporate "brands," at the same time the TV archive has gone in the opposite direction – toward home-modes of collecting and increased personalization. As early as the 1950s people took photos off their television sets as a form of hobby art (see Spigel, 1992, pp. 97–8),[70] while others were recording TV on audiotape as a way to preserve and exchange programs (Anderson, 2003). While this popular attempt to save TV was at that time a specialized practice, the advent of the VCR in the 1970s, the mainstreaming of that technology by the mid-1980s, and the growth of the home video and DVD market have meant that more and more people are now able arrange their own TV archives in what might be called a new practice of "Do-It-Yourself" TV history.

These recording technologies have also had an impact on the canon itself. For example, when I first bought a VCR in 1985, I began to tape numerous programs off-air and to purchase others from cottage industry companies specializing in old TV. By the late 1990s, I had amassed a rather sizable archive, and after roughly 12 years of teaching with my personal archive of TV tapes, I realized that a number of my graduate students (now also university professors) were duping the programs I taught and using them to teach their own classes. Suddenly, the sitcoms, soaps, and game shows that I had collected were beginning to form a canon that actually had very little in common with (and in fact were often anathema to) the "Golden Age" canon formed by the first generation of TV museums, critics, and scholars. I suspect this has happened to other TV professors in the VCR generation and will continue to happen so long as teachers amass personal archives. (As if to verify this assumption, in my teaching evaluations last year, a student wrote, "Lynn is a TV archive!")

Despite the historian's stake in canon formation, however, teaching TV is very much in dialog with the wider "archive" amassed by TV museums, nostalgia networks, home video and DVDs, and Internet sites. Because students often come into contact with old TV programs on television, their relation to TV history is formed through the media industry itself. In turn, professors often try to offer students a classroom experience that draws upon familiar and popular ways of thinking about TV. But, at the same time – and this is the crucial point – professors also have to ensure that education is distinct from the kind of TV history students can find in popular culture. To put it another way, in the context of the popular nostalgia market for television history, university professors have to teach through what sociologist Pierre Bourdieu called strategies of "counter-distinction" (Bourdieu, 1984). This has become harder to do as camp, irony, and parody (that is, all the means intellectuals traditionally had at their disposal to distinguish themselves from mainstream culture) are increasingly mainstreamed on networks like Nick at Nite. Part of the problem for TV historians, then, is to find a way of teaching history that isn't just a rerun of the "fun" sensibilities and syndication holdings of Viacom, while at the same time not being so dull as to turn students off entirely.

For these reasons, it seems to me, television studies in the university is an ambivalent mix of the popular and the serious. In our attempts to find a space for counter-distinction from the contemporary nostalgia industry, our work is often partly governed by the high seriousness and anti-commercial rhetoric that governed the formation of TV's first archives in the 1960s Wasteland Era. The typical TV history syllabus still includes the great moments of the "Golden Age." Celebrated dramas like *Marty*, news shows like *See It Now*, and live variety shows like *The Texaco Star Theater* are still in the archive. Yet, as people who live in the context of the nostalgia industries, TV scholars often mix camp/ironic humor with the high seriousness of negative critique. I realize the "we" for whom I speak is rather royal, but in many ways I believe this generation of TV scholarship is caught in the contradictory sway of these two TV archival sensibilities. Just look at the spate of university press book covers with their campy images of old TV sets and fifties housewives, and you see immediately the way scholarly work is packaged according to marketing imperatives of camp, cult, and nostalgia.

Perhaps, however, this particular problematic of the scholarly vs the popular is itself becoming history. We might well be entering a different phase of the archival imagination altogether, a phase not caused by, but certainly facilitated through, new modes of digital storage. Although this deserves attention in its own right, I'll offer some preliminary thoughts by way of conclusion.

Ephemerality, Storage, and the Fantasy of Total Accumulation

In his work on early television and recording technologies, Christopher Anderson notes the tension between the ephemeral and the artifact in television studies and television culture more generally.[71] Certainly, the TV archive is itself exemplary of this tension as it attempts to store and give pattern to an extremely ephemeral medium and cultural form. The TV archive tries to arrest the "present-tenseness" of television, turning TV's aesthetics of liveness, flow, and channel surfing into a document of the past. Despite the fact that archivists cannot possibly amass the sheer amount of volume on TV, the archive – as a modern system of identification and classification – always suggests the possibility of metonymic representation (that the documents it holds are a representation of a larger abstract whole). Museum curators have long grappled with these issues of representation, and, not surprisingly, the problem was on center stage at the 1996 Congressional Hearings which sought to explore the future of film and television preservation. But if the archive is just a representation of the past, then we always need to ask what lies outside the archive. What isn't arranged and collected there? What can't the archive contain? This line of inquiry – the one that Foucault posed most brilliantly in all his studies of madmen and hermaphrodites – gets harder to sustain as the archive (or at least the archival imagination) enters the age of digital storage.

Today, with the Internet, digital systems like TiVO, and the general proliferation of technologies of storage, we are confronted by a fantasy of total accumulation – an encyclopedic fantasy that promises that we have accounted for and arranged every object. This fantasy is rooted in a desire to see all pasts, all presents, and presumably all futures – and, in that sense, this is the fantasy of "tele-vision" at last realized! Indeed, one of the central questions that the Internet raises is the status of the visible itself. In other words, although the Internet promises to make all knowledge visible on demand, the ephemeral nature of the Net makes the actual institution of the archive – the site of enunciation – less visible. The Internet makes it appear that facts have been collected and arranged by some data god, and not by the human sciences. With their graphical displays and endless links to other sites, websites create an aura of disembodied "truth" that can be quite seductive. This poses serious questions as to the project not just of TV history, but history more generally. As should be obvious, I am not saying that the library/physical archive is necessarily a less ideological space than the electronic archives of TV or the Internet. Nor am I dismissing the Internet's positive value as a research tool. Instead, I am suggesting that we think about the contemporary discursive formation in which electronic technologies play a central role in how we arrange, interpret, and ultimately use the past.

Television history is a favorite subject on numerous fan and museum websites. Yet, because of the slow transmission, memory capacity, and copyright laws,

Internet archives are for the most part unable to transmit images of TV shows. As Andrew Lange points out, this means that the "Internet as a tool for the history of television is then a paradox: as you cannot use illustrated material – video or still picture – you have as the only solution to come back to the written text" (or else, he adds, JPEG pictures of things in the public domain) (Lange, 2001, p. 44). In other words, in a strange circle of events, the Internet in some ways returns to the archival imagination of television's first historians, an imagination based mostly on writing as the technology for preservation. But, not entirely.

TV history and archive sites are also products of the contemporary nostalgic archival imagination. Some, like tvtome.com, are more like TV encyclopedias with extensive summary guides and production credits; others, like jumptheshark.com (which features information about old and new TV series that have failed), are more like roadside museums that contain offbeat TV facts; others, such as mztv.com, are more like a museum of science and industry, devoted to visual and aural displays of TV artifacts; others, like tvthemesongs.com, allow users to listen to soundbites of favorite hits; and still others, like internetarchive.com, are archives of archives which, for example, display historical websites of TV networks that users can access on a "wayback machine" (which is really a sophisticated search engine). As the "wayback machine" suggests, these sites are not aimed at scholars but rather at a broad-based public of TV fans; they include gimmicky graphics, audio effects, and trivia games, and many are linked to shopping sites where fans can buy TV memorabilia.

In all of these cases, the Internet – both for pragmatic but also for aesthetic reasons – rearranges the archive to suit its communicative and commercial form. Indeed, even if the Internet is not ordered on the classificatory systems and spatial arrangements of the traditional library/archive, it is the still the case that the Internet is a human arrangement. As Jacques Derrida argues with respect to the storage of images, "Today, we can at least pretend (in a dream) to archive everything, or almost everything . . . But because it is not possible to preserve everything, choices, and therefore interpretations, structurations, become necessary." And for this reason, "whoever is in a position to access this past or to use the archive should know concretely that there was a politics of memory, a particular politics, that this politics is in transformation, that it is a *politics*" (Derrida and Stiegler, 2002, pp. 62–3).

Finally, although I have highlighted the various rationales and discursive systems that govern TV's archive, there is an important factor that this kind of inquiry leaves out. Namely: much of what remains of our TV past remains largely through accidents. Given its ephemeral nature, television is still largely viewed as disposable culture, and what is saved is in large part based on what happens to be recorded, what happens to be in someone's basement, a thrift store, flea market, someone else's flight of fancy. So, once again, we are back to Andy's Archive. With the advent of the VCR and the newer digital TV systems, much of what remains of the TV past is really just what someone else recorded.

92

And in this sense, television history may well just be an attempt to give reason to – to arrange and systematize – these recording accidents of the past.

For example, the promotional film for the Hollywood Museum that I discussed in this chapter was not saved because anyone thought it was particularly important for TV or media history. Instead, I found it in an underground LA video store called Mondo Video A-Go-Go which, for the most part, sells porn videos catalogued in categories like Gummy Grandmas, and arranges these alongside cult movies and a spattering of TV programs. I found the Hollywood Museum film preserved on a film reel tape in the "shockumentary" section. According to the owner, it was stored there not because of its content, but because of its context; the film had been found at the murder site of an octogenarian porn producer, amid his rotting corpse and rumpled in a mound of old *Hustler* magazines. I will end there, hoping to convince you that despite the archive's search for reason, the reason things are saved are never as reasonable as they appear.

Notes

1   In rather oblique reference, Friedrich A. Kittler (1997) argues that Foucault's notion of the archive and archeological method does not work for media technologies other than writing. He claims that Foucault's "analyses end immediately before that point in time when other media penetrated the library's stacks. For sound archives or towers of film rolls, discourse analysis becomes inappropriate" (p. 36). Although Kittler doesn't amplify this argument, it seems to be entirely wrong to say that Foucauldian-inspired discourse analysis is somehow inappropriate for sound or image technologies. John Tagg (1988), for example, has brilliantly demonstrated the power of Foucauldian method in his analysis of crime photographs and their relation to record keeping among police and other social institutions of the nineteenth century.

2   Tagg (1988) discusses the relationship of photographic record to the identification of criminals by the police, law, and penal system. See especially chapters 2, 3, 4. See also Ginzburg (1980, p. 25).

3   Copyright registration was voluntary, and because many early television producers did not see long-term commercial value in their product, they did not seek copyright protection. In addition, "owners who did wish to obtain copyright projection for early television transmissions encountered the legal concepts of 'fixation' and 'publication.' As had been previously established with film, performances by broadcasting did not per se constitute publication; publication came at a later point, when the material had been fixed and offered for sale, lease, or rental" (p. 1). As the website further explains, this created a "legal morass" around the interpretation of who actually owned the production. Some legal advisors favored the syndication date as the first "publication date," but since many shows (games shows, variety shows, sports and talk shows) did not go into off-network syndication, the interpretation of publication as syndication favored the registration of prime-time entertainment series. Even when these were copyrighted, however, the Library of Congress was not in the practice of saving every series or even whole series. Of the series registered for copyright before the mid-1960s, "the library chose only an occasional sample of entertainment series . . . and the so-called "quality programs" (p. 2). Importantly, in his regard, the Library depended on the quality "canon" largely developed by the institutions I discuss in this paper. In the early 1970s, new

FCC syndication and ownership rules as well as the belated copyright registration of many older TV series filled out the collection. In July of 1986 NBC donated 18,000 programs preserved mainly by the network, and the library has a large collection of educational television from National Educational Television (NET) and the Public Broadcast Service (PBS). See the website for details.

4 In 1966 the Motion Picture Section of the library's reference staff became responsible for deciding which TV copyright deposits would be retained. Responding to a broadening range of research interests, the Motion Picture reference staff expanded TV acquisitions. Then, in the early 1970s, new FCC syndication and ownership rules as well as the belated copyright registration of many older TV series filled out the collection. In July of 1986 NBC donated 18,000 programs preserved mainly by the network, and the library has a large collection of educational television from National Educational Television (NET) and the Public Broadcast Service (PBS). For details, see the website lcweb.loc.gov. For more on the early history of the Library of Congress preservation efforts, see Murphy (1997), hereinafter referred to as "Television and Video Preservation."

5 The report nevertheless admits "Educational access remains largely unattainable for a variety of reasons," including the under-funding of public archives (p. 9).

6 In 1951, Bing Crosby Enterprises' Electronic Division introduced a method by which to record television electronically rather than photographically on kinescopes. Electronic recording on video eliminated optical distortions of kinescopes and could be replayed instantly, recorded over, and reused. In 1956 Ampex Corporation set an industry-wide standard for video recording. See Jacobs (2000, p. 24) and Winston (1986, p. 90). The use of video as a means to store programs became more widespread in universities in the 1970s.

7 In 1962, President Kennedy appointed August Heckscher as Special Consultant on the Arts to the President, and after Lyndon Johnson assumed the presidency he appointed a Special Assistant to the President on the Arts – the first full-time arts consultant in US history. This set the stage for significant arts legislation. Three landmark events occurred in 1965: (1) the publication of the Rockefeller Panel Report, *The Performing Arts: Problems and Prospects* (1965), which articulated the arts cause to the public; (2) the passage of the Elementary and Secondary School Act that authorized schools to develop innovative projects that utilized the services of arts groups and cultural resources in their communities; and (3) the passage of the National Foundation on the Arts and Humanities Act of 1965 that allowed the federal government to become a small but official patron of the arts. See Taylor and Barresi (1984, pp. 25–30) and Reiss (1972).

8 I am drawing on Clifford (1988, pp. 215–52).

9 The major Hollywood studios have assets protection programs aimed at preserving extensive inventories of television programs, which are preserved for economic reasons of domestic and foreign syndication, video and DVD sales, and other possible profit venues. Although these archives sometimes cooperate with public archives and TV museums, they are not open to the public. Networks often retain film or broadcast tapes of shows (even when they do not own them), and news divisions at networks keep extensive documents of historic public events as well as news footage. See Murphy (1997, pp. 35–41).

10 National Minutes, April 15, 16, 17, 1966, p. 14, Box 1: Folder 1963–67, Academy Archives, on permanent loan at Doheny Library, University of Southern California, Los Angeles, CA (hereafter referred to as Academy Archives).

11 Minutes: National Board of Trustees Meeting, September 13, 14, 15, 1963, p. 36, Box 1: Folder 1963–67, Academy Archives.

12 Ibid.

13 In the early 1950s, Sullivan tried to establish a rival award for the Emmys that he called the "Michaels."

14 Minutes: National Board of Trustees Meeting, September 13, 14, 15, 1963, p. 36, Box 1: Folder 1963–67, Academy Archives.

15 The "Blue Book" is the colloquial name for the FCC report titled, "Public Service Respons-ibility of Broadcast Licensees," March 7, 1946, reprinted in Kahn (1968, pp. 125–206).

16 The Peabody Award Archive, which opened in 1976 (but whose collection dates back to 1948), contains a more diverse selection of local programs, especially in the public affairs and documentary genre. Today, UCLA holds the Academy of Television Arts and Sciences/UCLA Collection of Historic Television, but it also includes donations from other Golden Age network series included in the ABC collection (1950s–70s), the Hallmark Hall of Fame and Jack Benny collections, holdings from the DuMont collection, commercials, and a variety of news-related collections. Although the Academy collection represents mostly national network fare, UCLA also has an extensive collection of local Los Angeles programs. At the present moment, each year the UCLA Archive receives tapes of the Primetime and Los Angeles Area Emmy Award nominees and winners from the Academy of Television Arts and Sciences, and tapes of the Daytime Emmy nominees and winners from the National Academy of Television Arts and Sciences.

17 In 1963 the Television Academy tried (unsuccessfully) to nationalize the library efforts through what the Foundation called a "National Literary Subcommittee" that operated out of New York. In the early 1960s, the Television Academy also considered partnerships with the Museum of Modern Art in New York, Hollywood Museum, and the Cultural Center in Washington (which was typically more interested in the fine arts than in mass media). But these partnerships never crystallized. Instead, the Academy Foundation more aggressively sought liaisons with universities, including New York University, George Washington University and American University (both in DC), and the University of California in Los Angeles. See Spigel (1998, p. 73).

18 Minutes: National Board of Trustees Meeting, September 13, 14, 15, 1963, p. 36, Box 1: Folder 1963–67, Academy Archives. Cassyd's appointment by the chancellor is discussed in ATAS (Los Angeles Chapter) Minutes, April 6, 1960, p. 2, Box 1: Folder, Minutes Los Angeles, January 1960–March 1968, Academy Archives.

19 Peterson discussed this in his report, *The Medium* (1955), p. 46, Series III. Box 14: Folder 14, Museum of Modern Art Library, New York, New York (hereafter referred to as MoMA Library); Peterson's assistant Douglas Macagy also wrote a long report on the Television Project titled *The Museum Looks in on TV* (1955), Series III. Box 14, MoMA Library. For more about Peterson and Macagy, see Spigel (2000, 2004).

20 Even before this, in 1939, MoMA became the first museum to experiment with using television to promote the arts. For more on the TV project at MoMA see Spigel (2000, 2004).

21 For example, Peterson's pilot "Manhole Covers" (in his never-to-be aired "Point of View" series) was a montage of footage that showed New York City from the point of view of a sewer worker. Peterson mixed this footage with ragtime music, silent film shorts, and narration from the popular radio host, Henry Morgan. A print of the film is located in the Museum of Modern Art Archive, New York. It was written by Peterson and directed by Ruth Cade.

22 "Point of View," report, n.d., Series III. Box 20: Folder 16b.b, MoMA Library; and Macagy, *The Museum Looks in on TV*, p. 205. Museum directors similarly rejected a series of animated films about artists that Peterson scripted and made in conjunction with NBC and the anima-tion company UPA (the creators of *Mr. Magoo* as well as television commercials). See report on "They Became Artists," Series III, Box 18: Folder 3 ca. December 1954, MoMA Library.

23 It is interesting to note in this regard that Peterson himself felt that the idea of collecting TV as an art object would be a sad substitute for producing it. Considering the problems he had with his various TV productions at MoMA, he said, "These problems are so basic . . . that the museum may well decide to forgo the whole project in favor of a non-productive anti-quarian interest in collecting – when enough time has passed for a few by then antiquated kinescopes to have acquired both charm and significance, to become, as it were, museogenic." See Peterson, *The Medium*, p. 131.

24 See Betty Chamberlain, memo to Rene d'Harnoncourt, April 11, 1952, Series III. Box 18: Folder 3, MoMA Library. However, by 1954 the Film Library did cooperate with a producer at NBC when the network put film librarian Iris Barry on the "NBC payroll as official agent and coordinator in Europe." See Richard Griffith, memo to the Coordinating Committee, October 22, 1954), Series III, Box 18: Folder 7, MoMA Library.

25 Richard Griffith, memo to Douglas Macagy, July 24, 1952. Series III, Box 18: File 2a, MoMA Library.

26 Ibid. Note that the idea for a film and TV retrospective was stated again in a letter from Macagy to Griffith dated March 6, 1953, Series III, Box 18: Folder 2a.

27 Richard Griffith, "Appendix 3: Prospect for a Television Archive," in Macagy, *The Museum Looks in on TV*, p. 291.

28 Ibid., p. 293.

29 Ibid., pp. 299–300.

30 Ibid., pp. 296 and 293.

31 Richard Griffith, "A Report on the Film Library, 1941–1956," *Bulletin*, *XXIV*(1) (Fall 1956), p. 14.

32 Ibid.

33 Richard Griffith, "Television and the Museum of Modern Art," Introduction in *Television USA: 13 Seasons*, Museum of Modern Art Exhibition Catalogue, designed by Mary Ahern (New York: The Museum of Modern Art Film Library and Doubleday, 1962), p. 3.

34 Ibid., p. 3.

35 Ibid., pp. 3–4.

36 *Television USA: 13 Seasons*, p. 23.

37 Lewis Freedman, *Television USA: 13 Seasons*, p. 17. Lewis Freedman was the producer of *Twentieth Century* and a member of the exhibit's selection committee. In this same statement he compares the Golden Age live dramas to the Hollywood filmed series, which he says have "escaped into a world of adventure, suspense, and melodrama" (p. 17).

38 *Television USA: 13 Seasons*, pp. 19, 30, 32.

39 Jac Venza, *Television USA: 13 Seasons*, p. 15.

40 Abe Liss, *Television USA: 13 Seasons*, p. 38. Liss was president of a production company for television commercials, Elektra Film Productions, Inc. and he was a member of the exhibit's selection committee representing the genre of commercials.

41 Founded in 1949, the Junior Council was a group of young volunteers who were concerned with extending MoMA's services to the community. In 1964, James Thrall Soby, chairman of the Committee on the Museum Collections, spearheaded the idea for the Television Archive of the Arts. See Press Release, January 19, 1967, Box 14; Folder 14, MoMA Library.

42 Ibid. See also Lynes (1973, pp. 381–2). MoMA officials did hope to expand this collection into the full archive of American television they imagined in the 1950s, and in line with the national agenda for the arts, museum officials even envisaged "a time when such archives are computerized and connected in a nation-wide – or perhaps world-wide network of museums" ("Preserving Our Artistic Heritage," *TV Guide*, July 1, 1967, n.p.). This is a press clipping in Box 14: Folder 14, MoMA Library.

43 The conference proceedings appeared in book form as Davis and Simmons (eds.), 1977. It should be noted that in the early years, there was some popular interest in video art, especially as the work was publicized on NET, PBS, and also some commercial stations.

44 See Sol Lesser, untitled document, 1962, n.p., Papers of August Heckscher, White House Staff Files, John F. Kennedy Memorial Library, Boston, MA. (Hereafter referred to as JFK Library).

45 Ibid.

46 The Payola Scandals of the late 1950s involved radio disc jockeys that were bribed by record manufacturers to plug records. A House Congressional investigating committee heard testimony from many famous disc jockeys.

47 Schumach telegram to Robert S. MacNamara, 1963, n.p., Papers of August Heckscher, White House Staff Files, JFK Library.

48 For a complete history of the Hollywood Museum and the reasons for its demise, see Trope (1999).

49 As the first network-sponsored foray into humanities-oriented television criticism, the book was edited by a widely respected Golden Age critic and contained essays by both leading critics and professors (and the biographical notes made a point to reference their university backgrounds at Cambridge, Harvard, and the like). For more about the book and CBS's interests in television scholarship in this period, see Spigel (1998, pp. 67–70).

50 The Bluem Report cited in Murphy (1997, p. 16). In the early 1970s the American Film Institute also decided to include television in its preservation interests. See Murphy (1997, pp. 16–17).

51 Ibid., p. 60.

52 The museum's criteria include historical importance, social relevance, and artistic excellence as evidenced in awards. See ibid., pp. 60–1.

53 To meet the popular demand, in 1979 the West 52nd location expanded by taking over two more floors of its building and adding new viewing and listening facilities.

54 In the same article, Dick Wolf, the creator of *Law and Order*, commented, "You spend three hours in this museum, and it's the best short course in the social history of the country" (p. C-12). Beamed in via satellite from Washington, President Clinton offered congratulatory remarks about the MTR's historical importance. Meanwhile, corporations that sponsor the museum similarly trade on the "art" value of the product to promote their own corporate image. For example, General Motors sponsors the MTR's University Satellite Seminars and boasts of its "Mark of Excellence" television presentations, claiming "At GM, we believe every good education should include a few hours of prime-time." See museum catalogue for "Television: The Creative Process," Fall 1996, p. 2 and back inside cover.

55 Donor solicitation letter and museum brochure from Bruce DuMont, November 20, 2003. While MBC houses a special collection of Chicago TV, it also has highlights (in smaller volume) of much of the same network "Golden Age" fare that can be found at the MTR and UCLA.

56 See www.noko.org.

57 See www.seeingstars.com.

58 In addition to the nostalgia networks I discuss herein, a number of networks form themselves around particular genres aimed at niche markets and air both original and re-run fare in these genres (for example, the Sci-Fi channel, The Game Show channel, and Lifetime). Trio, a fairly recent network, positions itself as the TV connoisseur channel by showing never-aired pilots and rare series.

59 For discussions of conglomerates and their holdings see Aufterheide (1997), McChesney (1997), Bagdikian (2000), Alger (1998).

60 These demographics are listed on the TV Land website press releases, "TV Land and Nick at Nite See Record-Breaking Viewership Levels in First Quarter '03," April 1, 2003. See www.tvland.com.

61 These nostalgic places (what Harvey calls the "atmosphere of place and tradition"), however, wind up being extremely monotonous reproductions – what Harvey (following M. Christine Boyer) calls "molds" of each other that are almost identical in ambiance (p. 295).

62 Martin Peers, "Nick Dents Retail," variety.com, posted May 22, 1997, p. 1. The promotional rhetoric is cited from the store's promotional flyer.

63 The interactive environments are pictured on the store's promotional flyer and also described from my personal observation at the store.

64 In its public explanation, a spokesperson for Viacom said that licensing "offers significantly greater flexibility and potential to expand faster with very limited use of capital" than retailing. See Martin Peers, "Viacom Gets Out of Studio Store Biz," variety.com, posted December 14,

1998, p. 1. Peers also gives a number of other explanations for the store's demise. See also Chandler (1998) (posted on Proquest, pp. 1–4); Chandler (1999) (posted on Proquest, pp. 1–3).

65   See www.tvland.com.

66   As Victoria E. Johnson argues, while the programming strategy might seem counter-intuitive in an industry bent on getting younger, urban demographics, PAX (like CBS) has fared remarkably well (see Johnson, 2004). PAX is the largest owned and operated group of stations in the country, and as of 2004 reaches 88 percent of US households.

67   See the GoodLife official website, www.goodlifetvnetwork.com.

68   "New Job on Cable TV Will be Fun," posted on www.georgeclooney.org., p. 2.

69   See the program descriptions on www.goodlifetvnetwork.com.

70   Since writing *Make Room for TV*, I received a number of early photographs that viewers took off their screens in the 1950s. Sometimes, viewers wrote ironic commentary on the photos or had balloons that attributed inappropriate dialog to characters in the pictures.

71   Personal email from Christopher Anderson to Lynn Spigel, March 22, 2004.

References

Alger, D. (1998) *Megamedia: How Giant Corporations Dominate Mass Media, Distort Competition, and Endanger Democracy*, New York: Rowman & Littlefield.

Anderson, C. (2003) "A Season in a Box," paper for Society for Cinema and Media Studies conference, Minneapolis, Minnesota, March.

Aufterheide, P. (ed.) (1997) *Conglomerates and the Media*, New York: New Press.

Bagdikian, B. H. (2000) *The Media Monopoly*, 6th edn., Boston: Beacon Press.

Bourdieu, P. (1984) *Distinctions: A Social Critique of the Judgment of Taste*, trans. R. Nice, Cambridge, MA: Harvard University Press.

Chandler, S. (1998) "Less-Than-Magnificent Year for Viacom Entertainment Store," *Chicago Tribune*, March 10, posted on Proquest, pp. 1–4.

Chandler, S. (1999) "Mag Mile's Changing Face While Official Vacancy Remains Low," *Chicago Tribune*, January 23, posted on Proquest, pp. 1–3.

Clifford, J. (1988) "On Collecting Art and Culture," in *The Predicament of Culture: Twentieth Century Ethnography, Literature, and Art*, Cambridge, MA: Harvard University Press, pp. 215–51.

Davis, D. and Simmons, A. (eds.) (1977) *The New Television: A Public/Private Art*, Cambridge, MA: MIT Press.

Derrida, J. and Stiegler, B. (2002) *Echographies of Television: Filmed Interviews*, London: Polity Press.

Foucault, M. (1972) *The Archeology of Knowledge*, trans. A. M. Sheridan Smith, New York: Pantheon.

Ginzburg, C. (1980) "Morelli, Freud and Sherlock Holmes: Clues and Scientific Method," *History Workshop*, 9 (Spring), 1–36.

Hannigan, J. (1998) *Fantasy City: Pleasure and Profit in the Postmodern Metropolis*, London: Routledge.

Harvey, D. (1989) *The Condition of Postmodernity*, Oxford: Basil Blackwell.

Jacobs, J. (2000) *The Intimate Screen: Early British Television Drama*, London: Oxford.

Johnson, V. (2004) "Welcome Home? CBS, PAX-TV, and 'Heartland' Values in a Neo-Network Era," in R. C. Allen and A. Hill (eds.), *The Television Studies Reader*, New York: Routledge, pp. 404–17.

Kahn, F. J. (ed.) (1968) *Documents of American Broadcasting*, New York: Meredith Corporation, 1968.

Kittler, F. A. (1997) *Literature, Media, Information Systems* (introduced and edited by J. Johnson), Amsterdam: G & B Arts International.

Lange, A. (2001) "The History of Television Through the Internet: A Few Notes on the Project www.histv.net," in G. Roberts and P. M. Taylor (eds.), *The Historian, Television and Television History*, Luton: University of Luton Press, pp. 39–45.

Lippard, L. R. (1966) *Pop Art*, New York: Praeger.

Lynes, R. (1973) *Good Old Modern: An Intimate Portrait of the Museum of Modern Art*, New York: Scribner.

McChesney, R. (1997) *Corporate Media and the Threat to Democracy*, New York: Seven Stories Press.

Marver, A. (2001) "New York Eyeline," unpublished PhD dissertation, University of California, Irvine.

Minow, N. N. (1964) "The Vast Wasteland," in *Equal Time: The Private Broadcaster and the Public Interest*, New York: Atheneum, pp. 45–69.

Multi-channel News (2002) "GoodLife Wants Slice of Boomer Pie," *Multi-channel News*, June 21, posted on www.georgeclooney.org, p. 1.

Murphy, W. T. (1997) "Television and Video Preservation: A Report on the Current State of American Television and Video Preservation, Volume 1," prepared for the Library of Congress, October, Washington, DC: Government Printing Office.

O'Neil, T. (1992) *The Emmys: Star Wars, Showdowns, and the Supreme Test of TV's Best*, New York: Penguin.

Reiss, A. H. (1972) *Culture and Company: A Critical Study of an Improbable Alliance*, New York: Twayne Publishers.

Rockefeller Panel Report (1965) *The Performing Arts: Problems and Prospects*, New York: McGraw-Hill.

Shayon, R. L. (ed.) (1962) *The Eighth Art*, New York: Holt, Rinehart and Winston.

Shepard, R. F. (1979) "Tune in Soon for Best of Yesteryear," *New York Times*, July 22.

Spigel, L. (1992) *Make Room for TV: Television and the Family Ideal in Postwar America*, Chicago: University of Chicago Press.

Spigel, L. (1996) "High Culture in Low Places: Television and Modern Art, 1950–1970," in C. Nelson and D. P. Gaonkar (eds.), *Disciplinarity and Dissent in Cultural Studies*, New York: Routledge, pp. 314–46.

Spigel, L. (1998) "The Making of a TV Literate Elite," in C. Geraghty and D. Lusted (eds.), *The Television Studies Book*, London: Arnold, pp. 63–94.

Spigel, L. (2000) "Live From New York: Television at the Museum of Modern Art, 1948–1955," *Aura*, 6(1), 4–25.

Spigel, L. (2004) "Television, The Housewife, and the Museum of Modern Art," in L. Spigel and J. Olsson (eds.), *Television after TV: Essays on a Medium in Transition*, Durham, NC: Duke University Press, pp. 349–85.

Tagg, J. (1988) *The Burden of Representation: Essays on Photographies and Histories*, Minneapolis: University of Minnesota Press.

Taylor, F. and Barresi, A. L. (1984) *The Arts and the New Frontier: The National Endowment for the Arts*, New York: Plenum Press.

"Television in the Library of Congress," lcweb.loc.gov.

Trope, A. (1999) "Mysteries of the Celluloid Museum: Showcasing the Art and Artifacts of Cinema," unpublished PhD dissertation, University of Southern California.

Weintraub, B. (1996) "Museum of TV Goes Bicoastal," *New York Times*, March 18, p. C11.

Whiteson, L. (1996) "TV Museum Both Formal and Inviting," *Los Angeles Times*, June 2, p. K5.

Winston, B. (1986) *Misunderstanding Media*, Cambridge, MA: Harvard University Press.

PART THREE

Television/ Aesthetics and Production

Television as a Moving Aesthetic: In Search of the Ultimate Aesthetic – The Self

Julianne H. Newton

It is the mark of our period that everything can be regarded as a work of art and seen in textual terms. . . . Contemporary art replaces beauty, everywhere threatened, with meaning.

Arthur Danto, The Madonna of the Future (2000, p. xxx)

The etymology of all human technologies is to be found in the human body itself: they are, as it were, prosthetic devices, mutations, metaphors of the body or its parts.

Marshall and Eric McLuhan, Laws of Media (1988, p. 128)

I recently saw the most amazing sight – on television.

In the debut of a Fox television program, what may be the ultimate aesthetic unfolded before my eyes. The classic book-form fairy tale *The Ugly Duckling* transformed into the video-form *The Swan*, in the same breath manifesting the wisdom of the ages, the fantasies of the many, the belief that we can become whomever/whatever we decide to become, a feast (or orgy) of spectacle to be devoured by the voyeur or condemned by the critical gaze, and the twenty-first-century realization of the virtual made incarnate.

I could not believe my eyes. Two women self-described as Ugly Ducklings (but termed average by one of their significant others) were gifted by a visit from the Swan team (twenty-first-century fairy godmothers?) and whisked away for a three-month bout of plastic surgery, dental work, self-esteem therapy, dieting, exercise, and expert makeup, hair styling and costuming. All mirrors – those symbols of narcissism and self-confrontation – were removed from the transformees' environments, as were family members and significant others – our less-obvious but all-too-real embodied mirrors (or co-determinants?) of self. At season's end, a group of semi-Swans competed, not only against each other, but also against the virtual presence of their past selves cast beside them through the magic of television reality, to be adorned with the title and crown of The Swan.

The debut of *The Swan* manifested a shift in our cultures of the aesthetic, the ultimate blending of reality and fantasy into a purposeful physical transformation toward a beauty deemed worthy of the gazes of millions and cast within the mythology of fairy tale through the miraculous intersection of wishful thinking, culture, technology, media, and commoditization. No longer content to express our feelings and perceptions solely with brush, camera, or sculptor's hand, we have combined them with a surgeon's scalpel to mold the medium of the body – an ultimate aesthetic at once horrifying yet strangely appealing to our human desire to improve and control our destinies. Following *Extreme Makeover*, *Queer Eye for the Straight Guy*, and *I Want a Famous Face*, *The Swan* magnified issues of aesthetic judgment of the body by introducing the competitive runway – that apparatus for parading the body as an aesthetic for the mass gaze. But *The Swan* was the Ultimate Aesthetic for this season only – the current bulls-eye of a perpetually moving target and an apt exemplar for my exploration of an aesthetic of television.

While watching the swollen, bruised faces of the women visually documented in the physical and psychological agony of their performances, I could not help but think about Orlan. For years, the French performance artist has been the embodied manifestation of the "anti-aesthetic," a term which Hal Foster (1983) wisely explicates as a continual "practice of resistance" (p. xvi). Orlan's unique, videotaped performances of purposeful, conscious transformation into a physical form of her choosing challenge all who gaze upon her or hear her story to consider the idea of an ultimate aesthetic. "I can observe my own body cut open, without suffering! . . . ," Orlan (2004) says in her "Carnal Art Manifesto." "I see myself all the way down to my entrails; a new mirror stage."

The Swan moved Orlan's unique performance of the anti-aesthetic into mass culture, at once conveying and betraying her sacrificial expression. Her message that we can choose our own bodily aesthetic was conveyed through *The Swan*. Yet the aesthetic the women in the program pursued was an assumed aesthetic, a stereotype of Western, light-skinned female beauty characterized by svelte bodies with large breasts, carefully made-up faces with smooth skin and full lips, perfect bright-white teeth, flowing and coiffed long hair, revealing costumes and newly proclaimed self-confidence evidenced through exhibition in swimsuits, evening gowns, and – gulp! – lingerie (an overt bow to female sexuality perhaps influenced by the soft-porn aesthetic of lingerie advertising).

Yet, as Orlan (2004) declares, "Carnal Art is anti-formalist and anti-conformist." In reference to Orlan, Swift (2004) writes, "We need a reminder that beauty isn't always pretty. Beauty can also be painful, shocking, controversial, and even fatal. . . . Some people give their bodies to science when they die; Orlan has given her body to art while still alive."

The challenge of understanding the paradoxical appeal of both Orlan and *The Swan* is to comprehend the competing forces at play within our aesthetic explorations of the inner and outer worlds of human existence. And therein lies the key to understanding the *moving aesthetic of television.*

The Moving Aesthetic

In this chapter I argue for the concept of a moving aesthetic, one that dances between convention and the transgressive, between established codes and the challenging of codes through an anti-aesthetic. The moving aesthetic is a kind of expanded frame, not a relative aesthetic, but one that nevertheless shifts and sways in the breezes of time and perception as part of the dialogic process of sensing, interpreting, and knowing that is human communication, regardless of medium, message, or intent. Until the Internet began to command significant attention as the technology of the moment, television epitomized that moving aesthetic through its ability to reinvent itself at will – and quickly. As the Internet gathers steam as the carrier of our increasingly global conversation – and debate – television enters a new era of aesthetic exploration . . . as art did when photography came along, as photography did when movies came along, as radio, movies, and print publications did when television came along. It is the way of growing, dynamic entities, be they driven by economic, political, artistic or basic human needs, emotions, expressions and exertions of self.

Asking the Right Question

Over the course of the last 70 years, many scholars and critics have argued about the rights of television to term itself an art form. "Television is a relative of the motorcar and airplane," wrote psychologist Rudolph Arnheim in 1935 (in Adler, 1981, p. 7), continuing, "To be sure, it is a mere instrument of transmission, which does not offer new means for the artistic interpretation of reality – as radio and film did." Yet, predating Marshall McLuhan's extensions-of-man concept, Arnheim recognized the importance of television to helping humankind extend one's "range of interest . . . beyond the reach of his senses" (p. 7). "Like the transportation machines," Arnheim wrote, "television changes our attitude to reality: it makes us know the world better and in particular give us a feeling for the multiplicity of what happens simultaneously in different places" (p. 7).

Rather than ask, "Is there an aesthetic of television?," I want to begin by joining those scholars who assert television aesthetics as a given. In developing his now-classic theory of applied television aesthetics, Herbert Zettl (1981) argued that the television medium "has precise and decisive aesthetic require-ments that can make or break a message, regardless of the significance and integrity of the initial intent of the 'communicators'" (p. 116). In her synthesis of "The Aesthetic Aspects of Television," Ruth Lorand (2002) points out that television's "influencing power . . . for better or worse – has something to do with its aesthetic qualities" (p. 5). We may not understand how those qualities work; we may even fear "the uncontrollable aesthetic power to convey implied messages" (Lorand, 2002, p. 5). Yet the fact that a "theory of TV aesthetics is undoubtedly at its very inception" does not preclude the benefits of reflecting on its qualities (Lorand, 2002, pp. 29–30).

Arthur Danto (2000) puts it another way – somewhat tautological but nevertheless useful: "What does it mean to live in a world in which anything could be a work of art? . . . It is to imagine what could be meant by the object if it were the vehicle of an artistic statement" (p. xxix). So, one of our challenges in a discussion of television aesthetics is to shift our conception of television in terms of its functions and forms toward a consideration of interface – our interplay with that which we perpetually create and which perpetually creates us.[1] That is the core of understanding the aesthetic.

Among the questions we might ask are:

- What are the key qualities of television aesthetics? What are the core aesthetic characteristics of television as a medium?
- How have political and economic forces shaped those aesthetics?
- How have the aesthetics of television affected the way we perceive and act in the world, in terms of personal psychology, visual perception, ideology, and personal and public power?

A number of scholars before me have addressed those questions in more detail than I can approach here. Zettl (1990) explores the production and perception of "a number of aesthetic phenomena, including light, space, time-motion, and sound" (p. 14). Metallinos (1996) offers a useful synthesis of key ideas inherent in classical Western debates of "philosophical aesthetics" ranging from Plato and Aristotle to Kant to Nietzsche to Dewey, movements shaping contemporary artists (Marxism, Freudianism, Existentialism, and Semantics), and four influential "media aesthetics theories that emerged from the literature of contemporary media": traditional (philosophical), formalist, contextualist, and empiricist (pp. 2–9). Metallinos asserts that, while the latter four theories bridged "aesthetic concepts of the arts with those of the media products," they do not address "the processes of perceiving visual and auditory images in motion, recognizing or interpreting such images, and synthesizing, or composing moving images with sound" (p. 9). Metallinos advocates that study of television aesthetics focus "on the analysis of three factors: perception, cognition, and composition of television images" (p. 9).

In their edited book on television aesthetics, Agger and Jensen (2001) emphasize three theoretical areas: the medium, the genres, and the aesthetics (p. 11). Of particular interest in our investigation is the book's first chapter, in which Jorgen Stigel uses his assessment of television's unique strengths and limitations, which are founded "on proximity, participation and immediacy" (p. 28), to explicate his concept of the "aesthetics of the moment." "The central dimension of the aesthetics of television," Stigel (2001) writes, "has become the aesthetics of expediency which make a virtue of making and communicating contingence and circular, or cyclical, recurrence" (p. 47). This "instant aesthetic," which "makes things literally accessible at a glance, so that the viewer is given an immediate experience" ("'it is as though I were being directly spoken to' and, 'things are shown to me as though I were actually present'") plays against formal, recurring frameworks with lengthy periods (pp. 47–8). "This mix of lengthiness and momentary intensity exists in a form of symbiosis with the everyday lives of the viewers," Stigel notes (p. 48).

Within the current handbook, Carol Deming (2004) offers a particularly insightful synthesis of five categories of the "televisual": temporality, spatiality, aurality, femininity, and hybridism. "Viewing televisuality as a synthesis of stylistic, technological, and ideological characteristics," Deming's chapter "reveals the concept's resistance to being fixed in time or in relation to other media" (p. 126).

Rather than offer additional summary of available literature, I want to focus on two aspects of these issues:

- What more do we need to consider in our effort to define an aesthetic of television?
- What happens when we experience television aesthetics?

I believe we will find the answer to these questions by reconceptualizing television aesthetics, often described as flat-fielded and two-dimensional, as multi-dimensional, with infinite ramifications for human life and our perception of that life. I hope to enhance that reconceptualization by suggesting that we need to study these issues within the context of human visual behavior grounded in an ecology of the visual.

Definitions

Aesthetics in popular usage refers to theories of art and people's responses to art. Often associated with beauty and the judgment of what is beautiful, an aesthetic might be defined as a property or set of properties characteristic of a particular kind of artistic expression. But defining aesthetics is not that simple, of course. The task has challenged the best thinkers and artists over the course of millennia. Twentieth-century scholars were particularly occupied with trying to determine the relation (or non-relation) between aesthetics and ideology, with valid concern about issues of power and the use of aesthetics to inculcate ideology and reinforce social control. Some scholars now call for the elimination of aesthetics as a separate field, arguing that "conceptual categories themselves manifest and reinforce certain kinds of cultural attitudes and power relationships" (Feagin, 1995, p. 11). "They favor instead a critique of the roles that images (not only painting, but film, photography, and advertising), sounds, narrative, and three-dimensional constructions, have in expressing and shaping human attitudes and experiences," Feagin writes (p. 11).

My position on aesthetics resonates with concerns about ideology, power, cultural attitudes, and critiquing the roles of images of all kinds in human experience. However, I do believe that examining these issues under a centralizing rubric termed *aesthetics* is crucial to understanding what happens when we experience a particular aesthetic or combination of aesthetics in a particular medium or group of media.

Television aesthetics encompass concerns ranging from artistic processes through mass persuasion techniques. Aesthetics include such creative concerns as frame, composition, proxemics, movement, color, and sound; such technical issues as film versus video versus digital media, single or multiple cameras, light placement and quality, high-definition TV, stereo versus surround sound, studio versus location shooting, recorded versus live performance; such cultural issues as monitor location, frequency of use, programming and content; such content issues as genre, violence, sex, pornography, representation, manipulation; such economic issues as the relationship between commercials and programming, cable versus satellite transmission, and product placement; and such cognitive and psychophysiological issues as television effects on brain waves, long-term and subconscious memory, information retention, social learning and stereotyping. As you undoubtedly thought while reading this list just now, each of the processes and techniques involves the other. Color and sound, for example, resonate between creative and technical concerns. Economic concerns of advertising and programming resonate with content issues of representation. And so on.

As noted earlier, I bow to the many excellent works addressing these varied and interrelated aspects of television aesthetics in more complete and specific ways. My discussion here focuses on the broader problems of defining an aesthetic of television and illuminating the significance of that aesthetic to our lives.

Approaches

One way to begin our analysis is to consider the ways different fields of inquiry might pose the question before us. This immediately raises flags of complexity. Communication scholars, for example, might begin with the classic messenger-message-receiver model. To that we must add perspectives of the Birmingham, American, and Australian schools of cultural studies. Critical scholars might focus on the political and economic dynamics of television as a commodity and means of social control. Cognitive and perceptual scholars might examine the neurological and physiological bases of practices of watching television and the ensuing effects on mind and body. Psychologists might study the personal and interpersonal factors affecting ways people engage television imagery. Sociologists, anthropologists, and political scientists might focus on the social and psychological patterns of viewing television as art – or not – and of using television, whether owner, creator, participant, or viewer.

One particularly astute approach that invites inclusion of the above perspectives is David Thorburn's (1987) call for an anthropology of television aesthetics. In his discussion of American television, Thorburn suggests:

> The best understanding of television . . . will be reached by those among us who can achieve something of the outsider's objectivity or partial neutrality but who can remain also something of a native informant: alive to the lies and deceptions inscribed in and by the medium, aware of its obedience to advertising and the ideology of consumption, yet responsive also to its status as America's central institution for storytelling. (p. 172)

Thorburn's call echoes Sol Worth's important 1980 article defining a shift in visual anthropology as a field of study using the camera for illustration and information gathering toward an anthropology of visual communication. To fully comprehend the complexities of television in our lives, we need to make a similar shift, and to adopt a participant observation mode, combining methods and reflecting upon "televisual aesthetics" as part of the larger environment in which humans participate.

Piccirillo (1986) asserts that the "study of television will be enriched greatly as technological and transcendentally aesthetic biases give way to practical consideration of everyday televisual experience" (p. 353). Piccirillo concludes not only that television is capable of originality, but also that television experience is authentic (p. 352). He suggests television can be understood in terms of "rhetorical aesthetics," which facilitates the study of television "in terms of the aura in which program and viewer are united" (p. 347):

> "Good" and "bad" television can be identified, if it is essential to risk such transcendental judgments; but good and bad cannot be abstract aesthetic criteria associated with such primitive art forms as painting and theater. Television needs an aesthetic developed from analysis of that aura which arises in consequence of everyday aesthetic experience. (p. 347)

In an entry to the 2004 edition of the *Encyclopedia of Television*, Thorburn (2004) writes of American television:

> Though we are still too close to the Broadcast Era for a definitive verdict, it is probable that American television of the second half of the twentieth century will be recognized as a significant aesthetic achievement, the result of a never-to-be-repeated confluence of social, technological and historical forces, a unique precursor to the digital entertainment future now impending. It would not be the first time that popular diversions scarcely valued by the society that produced them were judged by the future to be works of art. (p. 13)

Based on literature about the aesthetics of television we might conclude that television aesthetics have emerged from a resounding triumph of the popular over the elite, a redefining of the social, and an unavoidable merging of the technological and the ideological (or abstract) through the fires of capitalism, mass audiences, global corporatization, and individual perception. This chapter explores those claims – not with a presumption of thorough synthesis, but rather with a challenge to explore the messy, murky geography of a "moving aesthetic" manifested in television form.

Consider a typical evening of watching television. In a period of a mere four hours – should one be inclined to watch the screen with conscious attention for four full hours – one would witness and choose from among an astounding array of images, in terms of both quantity and content. During that time, a purposeful viewer might see scripted programs, reality shows, commercials and PSAs, news, sports events, historical and scientific documentaries, civic meetings, and videotaped classroom events. The key word in the above scenarios is "conscious." For usually we watch with an eye for distraction, seeking to leave the world that requires us to act and think in purposeful ways for a world that requires virtual and – we assume – minimal participation instead.

But what holds us is *movement* – not our own through physical shifting of body through space, for we often are "vegging" in front of the television, but of the flickering of lines on a screen and our ability to combine those lines into patterns we can identify. The television aesthetic constantly shifts objects of viewing, forms of presentation, ease of interpretation, as well as sounds, cadences, and levels of reflexivity. Just as the moving lines constantly refresh the image in order to maintain and convey the images, so do the characteristics of television visual and aural stimuli constantly refresh in order to engage our perceptions.

I do not claim to be the first to note this movement and its metaphorical implications for understanding the nature of television perception. Television studies pioneer Herb Zettl (1981) writes, "The television image is composed of electrical energy, a rapidly scanning electron beam or series of beams which we perceive as variations in light and color. . . . The *material* [author's italics] of television are not illuminated objects and people, but constantly changing patterns of light and color whose very existence depends upon the fluctuating energy of the electron beam" (p. 117). Noting that "the scanning beam is constantly trying to complete an always incomplete image," Zettl writes, "the basic television unit is ephemeral, forever fleeting . . . It is in a continual process of becoming, regardless of whether the screen image has at its electronic base the television camera, the videotape, or any other electronic storage device" (p. 130).

One excellent recent analysis of this phenomenon is Thorburn and Jenkins's (2003) "aesthetics of transition":

> We must resist notions of media purity, recognizing that each medium is touched
> by and in turn touches its neighbors and rivals. And we must also reject static

definitions of media, resisting the idea that a communications system may adhere to a definitive form once the initial process of experimentation and innovation yields to institutionalization and standardization. In fact, as the history of cinema shows, decisive changes follow upon improvements in technology (such as the advent of sound, the development of lighter, more mobile cameras and more sensitive film stock, the introduction of digital special effects and editing systems); and seismic shifts in the very nature of film, in its relation to its audience and its society, occur with the birth of television. (pp. 11–12)

These processes of imitation, self-discovery, remediation and transformation are recurring and inevitable, part of the way in which cultures define and renew themselves. Old media rarely die; their original functions are adapted and absorbed by new media, and they themselves may mutate into new cultural niches and new purposes. The process of media transition is always a mix of tradition and innovation, always declaring for evolution, not revolution. (p. 12)

My emphasis on a moving aesthetic necessarily encompasses the transition and evolution of aesthetic forms. But I want to focus on an aesthetic of television as a process of "imitation, self-discovery, remediation and transformation" that is rooted in the human organism more than in culture, technology, or any external form of expression. We know from studies of the human visual system that our eyes are drawn to movement. This is part of our vision instinct, a key to how we have survived as a species. When something moves in our field of vision, it draws our attention, demanding that we determine whether it is a threat to our bodies or merely something we notice.[2] Similarly, our ears are quick to note differences in aural stimuli, an instinctual response television advertising has exploited in order to arouse our attention to their interruptions in programming.

A key function of mass media is to meet our need for information. The first thing many of us do when disaster strikes, as when the planes flew into the World Trade Center buildings, is to turn on the television, simultaneously tending our fears and participating in a collective consciousness made possible by signals transmitted through mass technology. In this way television supplies the information we are instinctively driven to seek in order to survive as a species. We have learned over time that the world is larger than the field of view we can directly scan with our own eyes. This is one reason we find the world of images so compelling – in all their forms, whether a painting at the Metropolitan Museum of Art or a row of plasma screens at a sports bar – they offer points of view and content we would not otherwise see.

But certainly, you say as you read these words, watching *The Swan* is not essential to our survival. Yet in a fundamental way, watching *The Swan* or *American Idol* or even *Friends* is indeed essential to our survival, as essential as touch is to the survival of an infant. The issue of survival is about more than physical safety. It encompasses the state of an organism, for better or worse, which exists in relation to other organisms. The state of the human organism depends upon the millions of living cells, each possessing its own consciousness and function, that compose the body. In the same way, society and culture depend upon millions of bodies, each possessing its own consciousness and function, that compose the various groupings into which humans find or put themselves. In the process of living, we humans constantly attempt to refresh ourselves, much as the television signal refreshes the images on the screen as it creates them, by seeking information from sources we deem important to the state of our organisms.

So, watching *The Swan* informs some of those who watch about the potential for altering themselves literally and metaphorically. *The Swan* is a twenty-first-century fairy tale made incarnate through the bodies of the real-life actors on the stage of television, yet safely virtual for mass consumption, contemplation, critique, and even shock. As DiTommaso (2003) writes,

> To witness the incomprehensible possibility of the play of light and movement in a "life-less" object is to witness the sublime event of life being created.... We simply need to switch on our set to encounter and appreciate this continual event of becoming and creation. Indeed, it is precisely at the moment of instantiation that we become confronted with our aesthetic experience of television as sublime. It is at the very threshold where we turn on the TV, in the moment of tension – where we are consumed with the *anticipation* of television's capacity and delighted by television's ability to *satisfy* this anticipation – that we are engaged in an aesthetic experience. We are in awe, if only for a moment, enraptured by the sublime and unthinkable movement of life in an inorganic object. It is this encounter with the TV at the very threshold of instantiation that permits us to think of television as capable of promoting and inducing an aesthetic experience. (npn)

DiTommaso stresses the importance of conceptualizing television "as a medium of light and movement," rather than critiquing the mundanity of broadcasting: "The aesthetic experience of television is available to all who sit in front of the TV, and in the moment of turning on the box, we experience our postmodern identity; an identity that is perpetually in flux" (npn).

We need not agree about whether television evokes a sublime aesthetic experience in order to appreciate DiTommaso's point that the very process of encountering televised stimuli captures our attention enough to habituate us to turning on the set. That process is an aesthetic interface – that which causes, enables, provokes, stimulates, annoys, and draws a person to experience something outside the self, to experience something that evokes a response within the self . . . that commands our attention . . . resonating, articulating, enunciating, mesmerizing, prodding, challenging, threatening, obfuscating, cloaking.

Aesthetics and Survival

What is the role of art – and, therefore, of aesthetics – in our survival? Beyond the issue of the physical safety function of surveying our environments rests the distinction humans have assumed – rightly or wrongly – that makes them unique among creatures of the earth – the ability to consciously reflect upon the nature of our existence. We may debate our so-called distinction from other species; consider, for example, that Koko the gorilla paints self-portraits, and that an elephant at the Portland, Oregon, Zoo creates art that sells for thousands of dollars. My discussion here is not intended to assert human superiority over other organisms of the earth, but rather to spark discussion of a form of activity in which many of us regularly engage. Most important is the need to express, to find and share common experiences – or communications – within the projected space between living organisms. Forest algae and microbial entities communicate via cellular conveyance of chemical substances, from the base of the forest floor to the tips of the leaves on the tallest tree. Humans communicate within their bodies in much the same way, creating and sending forth chemical messages from cells in the neural pathways of the brain to cells in far reaches of the toes.

Those are all a kind of aesthetic consideration. We can imagine, and, through the technologies of magnetic resonance imaging, PET scans, CAT scans, angioplasty, X-rays and the tiniest of optical probes, we can see inside the body to observe the spectacle of its internal media system.

A human being sitting before a television screen can be likened to a cell at the tip of a toe – at once a fully conscious entity capable of independent and unique action, yet also dependent upon the stimuli received via the device that collects and transmits signals from the larger self of the world. In that way, the aesthetics of television are linked to the well-being of that human as information, in both form and content, to be dealt with – whether absorbed or rejected – for the improvement or to the detriment of the human organism.

Can we identify aesthetic codes of television? As noted earlier, I do not have space to explore the full range of codes.[3] However, two holistic points are important here. One, television's aesthetic codes are as complex as any expressive form we can imagine, except perhaps holography. Two, because they are mediated through electronic signals collected and transmitted to us via a boxed frame, they are always a multiple-dimension translation of stimuli created by humans in hundreds of other spaces and times, who performed, manipulated, reflected, and dripped their light and sound waves for collection and transmission through space in a kind of quantum transmigration of energy not unlike a *Star Trek* "beaming up" of bodies from a planet to the *Starship Enterprise*. The manner in which the human sitting before the screen collects the quanta is unique to that human's cellular programming and memory. And yet that human, too, is part of a collective consciousness that is larger than the one body.

So the aesthetic codes of television encompass the stimuli of the real and the fantastic, the translation of those stimuli into electronic form for transmission through space and time, the reformulation of the transmitted stimuli onto a cathode ray tube screen, the human brain's perception via ear and eye of those stimuli projected by the screen, the interpretation of perceived stimuli by the brain, and the encoding into memory and/or action by the body.

113

An example

To help explain the process of aesthetic experience of television, I offer the example of a fascinating email found on the website for *Queer Eye for the Straight Guy* (2004). According to the site, the email is from the mother of two boys who watch the program:

> I wanted to express how thankful I am for your show. I have two young boys, Tyler age 10, and Kevin age 8. They get picked on because Kevin is heavy and Tyler is in the enrichment program in school and has red hair and a mole on his face.
>
> It is so sad to see this, and hard as a mom to encourage them to be strong. Then your show came along and it opened up a whole new world for them. They watch it every week and they enjoy every part of it. They love Carson when he jokes, especially since he is from our area. Kevin tries to pick up hints on how to dress. Tyler loves to watch Ted cook, he tries to remember the desserts. He is a very thoughtful young man. He knows I work two jobs and he wants to cook for me. They also panic when I threaten them that I will write to all of you about the mess they call a room.
>
> You are such role models, all of you, for my wonderful sons. You have such a gift. God bless you all. Now my boys know it's ok to be creative and caring and still be men.
>
> Kevin asked for his birthday in October for a portable CD player and a new copy of your CD. He now plays and listens to my copy of your CD on this old player, but he wants his own copy. He is also looking for a poster of the Fab 5 to hang in his room, he is hoping someday to meet all of you and get you to sign his poster. I tried to explain how busy you all are, and he understands. He's a good eight year old. You would be right in the middle of the Eagle's and Giant's football teams! Thank you again for all you have done for my family.
>
> Sincerely and with lots of love to all of you,
>
> Susan

Let me explain how I interpret the mother's email as an aesthetic experience of television in terms of an ecology of the visual rooted in human visual behavior. *Human visual behavior* refers to "all the ways human beings use seeing and images in everyday life" (Newton, 2001, p. 19). Visual activity can be either external, "meaning that people can observe something outside of themselves, such as someone else or a photograph" or internal, in experiences such as imagining or dreaming.

> Visual behavior includes how people act in front of cameras, as well as behind them. It includes seeing of every kind: looking at photographs, watching a sunset, noting the way a cat slowly stalks a bird, absorbing the beauty of a sleeping child, scanning the galaxy for changes through a telescope. It includes witnessing the enactment of countless deaths in the movie *Die Hard II*, watching in mesmerized numbness the real-time bombing of Baghdad via a medium that is more often

about make believe, consumerism, and entertainment than about attempts to convey truth. It includes police mug shots, family albums, roadside billboards, and Internet zines. It includes all the ways people use these various visual artifacts, both consciously and subconsciously. It includes the ways people pose, mask their intimate personalities, project false personae, take on roles in order to manipulate opinion, model clothing, and unconsciously reveal that they are lying. It includes an editor's decision to use one photograph over another, a judge's decision to forbid cameras in the courtroom, a school board's decision to use video cameras on school buses, the military's decision to use a satellite to spy on another country. It includes an artist's decision to use bright red and yellow acrylic paints, a teenager's decision to sport purple hair, or an aging person's decision not to color graying hair. (Newton, 2001, pp. 20–1)

One way to understand the complexities of human interaction with television stimuli, then, is in the context of human visual behavior. I have been working on this idea for some time, having developed it from thinking about Stanley Milgram's (1977) work with photographic behavior, and about the work of non-verbal communication theorists Jürgen Streeck and Mark Knapp (1992). I mean the term to include the larger whole of human creation, interaction without and within, and responses to imaging systems. These include interior imaging systems of dreaming, imagining, self-imaging, and unconscious memory, as well as more obvious exterior imaging systems such as painting, photographing, filming, gesturing, and video recording. The term encompasses not only looking behaviors, but also performing, interacting, perceiving, and remembering behaviors.

I have found the concept particularly helpful in ameliorating the challenges of translating visual activities into inevitably inadequate verbal interpretations because it keeps before us the fact that, although we can observe some behaviors and things, we can never explain them fully through words. Visual behavior has non-translatable, organic roots, whether the behavior is external – caused in part by responses to other organisms or stimuli – or internal – perhaps caused by chemical interactions within the human nervous system. Our very process of observing and explaining changes our understanding of the behavior or thing. As Heisenberg observed, the very act of observing something changes what is observed.[4] Edmund Carpenter (1975), much criticized for his anthropological experiments of introducing such imaging technologies as mirrors and cameras to New Guineans in the 1970s, noted their quick adaptability to the act of "seeing themselves." Evidence of the effect of videotaping on the culture of the New Guineans was their immediate decision to discontinue the ritual of scarification through which young men were painfully admitted to manhood – once they saw the ritual on tape (p. 457).

Eric Michaels (2000) recounts a similar response of the Warlpiri in Australia after the filming of their Fire Ceremony in 1972: "Remarkably, the ceremony lapsed shortly after this film was made" (p. 708). When the Yuendumu community decided to perform the ceremony again in 1984, the 1972 film was considered "Law," a script for shooting the new videotape (though Michaels notes,

"When this did not happen, no one in fact remarked on the difference" [p. 709]). But that is not the whole story. The main point of Michaels' article is that, through the leadership of Warlpiri broadcaster Francis Jupurrurla Kelly, the Yuendumu community worked to insure its "cultural future" by using television as political resistance.

Human visual behavior, then, includes the decision of Los Angeles policemen to beat Rodney King under cloak of darkness, King's bodily movements during the beating, the opportune videotaping of a chance observer while witnessing the beating, public broadcast of the video as news, breaking the video into still frames for print media and for courtroom analysis, scholarly analysis of the video, and the images in your own mind called to your attention by reading these words just now.[5] Human visual behavior is also manifest in the horrific images from Abu Ghraib prison in Iraq: making the men strip naked or masturbate in front of others, sodomizing them in front of others and the camera, taking pictures of the activities, sharing the images with each other and via the Internet, the publication of the images, public viewing of the images, public response to the images, public memory of the images, the impact of the images on world perceptions of the United States, the use of the images as evidence against military personnel, and so on, are all encompassed in the concept of human visual behavior.[6] Among the most extreme examples of human visual behavior are Buddhist monk Quang Duc's 1963 visual statement in protest against religious oppression through burning himself to death in front of cameras, as well as the 2004 beheading of Nick Berg in front of a video camera and for mass distribution.[7]

A visual act deemed outrageous by some, but naïve compared with the above examples, was the baring of rock star Janet Jackson's breast (was the behavior hers, that of co-star Justin Timberlake, or an act of collusion between them?) during the 2004 Super Bowl half-time show and the subsequent outcry calling for the censoring of live broadcasts.[8] Though we can analyze these behaviors in terms of Blumer's (1969) formal explication of symbolic interaction theory (preceded by McLuhan's *Mechanical Bride*, 1951), conceptualizing them as behaviors rather than as "symbolic interactions" encourages us to remember the organisms who produce, enact, respond to and change as a result of encountering the stimuli.

Note that I focus here on "visual" behavior, as opposed to "aural" behavior. Because 75 percent of the information humans process is visual, because visual stimuli are so influential on memory, and because the visual is the dominant mode of television, specific attention to human interaction with visual stimuli is essential. Research indicates that when visual and aural stimuli send different messages via televised media, we direct our attention, comprehension and memory to the visual, not the aural. One notorious example of this phenomenon was CBS News correspondent Lesley Stahl's critical analysis of discrepancies between then President Ronald Reagan's actual policies and televised images about the president's activities. Lance Bennett (1986) writes:

Stahl was nervous about the piece, because of its critical tone and the practice of the White House Communications Office to call reporters and their employers about negative coverage. The phone rang after the report was aired, and it was a "senior White House official." Stahl prepared herself for the worst. In her own words, here is what happened:

And the voice said: Great piece.

I said: What?

And he said: Great piece!

I said: Did you listen to what I said?

He said: Lesley, when you're showing four and a half minutes of great pictures of Ronald Reagan, no one listens to what you say. Don't you know that the pictures are overriding your message because they conflict with your message? The public sees those pictures and they block your message. They didn't even hear what you said. So, in our minds, it was a four-and-a-half minute free ad for the Ronald Reagan campaign for re-election.

I sat there numb. I began to feel dumb because I'd covered him four years and I hadn't figured it out. Somebody had to explain it to me. Well none of us had figured it out. I called the executive producer of the Evening News . . . and he went dead on the phone. And he said, Oh My God. (Smith, 1988, in Bennett, 1996, p. 98)

Repeated, controlled experiments have reliably documented the validity of the phenomenon illustrated by Stahl's experience. Especially important to note is that not only are visuals – even subtle facial expressions – more likely to grab and hold our attention and frame our understanding of what is before us, but they also are what we are most likely to remember (Graber, 1990; Mullen, 1986; Schultz, 1993). In addition, neuroscientists have documented compelling evidence that visual information stored in the subconscious mind is a key determinant of how we respond to subsequent stimuli we encounter. One other element is essential to connecting mediated imagery with human behavior: we have strong evidence that our memory galleries do not differentiate between images we obtained via media and images we obtained in real life.

An *ecology of the visual* encompasses human visual behavior within an integrated cultural and physiological system, simultaneously core and primal to human organisms and evolving even as the organisms evolve.[9] Applied to our current discussion, it is as if the aesthetic of television is both creating and showing us our entrails. By addressing our deepest fears, anxieties, and desires through the experiences of real people (however constructed their video presentations may be) and fictional personae, we cut open the raw innards of the human psyche for mass view. Like it or not, think it to be art or trash, aesthetic experience compels us. We hunger for the aesthetic because it offers a touch of the experience of feeling, seeing, and hearing that may otherwise be absent in our lives. In this way the aesthetic of television draws and repels us, informs, fools, reforms us.

This is all part of "human visual behavior." As Hill (2004) writes in her own chapter in this volume:

> Viewers of reality programming are most likely to talk about the truth of what they are seeing in relation to the way real people act in front of television cameras. The more ordinary people are perceived to perform for the cameras, the less real the program appears to be to viewers. Thus, performance becomes a powerful framing device for judging reality TV's claims to the real. And, television audiences are highly skeptical of the truth claims of much reality programming precisely because they expect people to act up in order to make entertaining factual television. (p. 449)

Hill (2004) notes that the reality television phenomenon is global: "After the 'smash hit' of *Survivor*, the networks scrambled to glut the market with a winning formula of game show, observational documentary and high drama." In her earlier research on *Big Brother*, she "noted that the tension between performance and authenticity in the documentary game show format invites viewers to look for 'moments of truth' in a constructed television environment" (Hill, 2002, cited in Hill, 2004).

We seek these "moments of truth," which we need not define metaphysically but rather as a kind of resonant knowing evoked by the recognition of something positively or negatively meaningful to us, in everything we watch – that is the vision instinct, in part the surveillance function. Yet good theater, a good painting – and good television, whether fictional or so-called reality based – also offer us opportunities for connecting with these moments of truth, whether in a conscious moment of profound realization or in a casual moment of everyday watching.

The tetrad

Marshall and Eric McLuhan's (1988) concept of the tetrad can help us comprehend the complexities of television activity as one form of human behavior within an ecological system of the visual. The tetrad expresses the McLuhans' ideas of the Four Laws of Media: Enhancement, Obsolescence, Retrieval and Reversal. Eric McLuhan (personal communication, 2004) conceives the tetrad as a resounding chord, through which media play their music. The tetrad is a "heuristic device, a set of four questions, that can be asked (and the answers checked) by anyone, anywhere, at any time, about any human artefact [*sic*]":

> What does it enhance or intensify?
> What does it render obsolete or displace?
> What does it retrieve that was previously obsolesced?
> What does it produce or become when pressed to an extreme? (McLuhan and McLuhan, 1988, p. 7)

	(Enhance)	(Reverse)
	the multisensuous	
	using the eye as	inner trip:
	hand and ear	exchange of inner and outer
	(Retrieve)	(Obsolesce)
	the occult	**radio,**
		movie,
		point of view

Figure 5.1 A Tetrad for TV, adapted from McLuhan and McLuhan, 1988, pp. 158–9

In *The Global Village* (McLuhan and Powers, 1989), a second book on the tetrad and also published after Marshall McLuhan's death, television is presented as a resonating Mobius strip expressing figure-ground ebb and flow of conscious attention:

> After the Apollo astronauts had revolved around the moon's surface in December of 1968, they assembled a television camera and focused it on the earth. All of us who were watching had an enormous reflexive response. We "outered" and "innered" at the same time. We were on earth and the moon simultaneously. And it was our individual recognition of that event which gave it meaning.
>
> A resonating interval had been set up. The true action in the event was not on earth or on the moon, but rather in the airless void between, in the play of the axle and the wheel as it were. We had become newly aware of the separate physical foundations of these two different worlds and were willing, after some initial shock, to accept both as an environment for man. (McLuhan and Powers, 1989, p. 4)

McLuhan and McLuhan (1988) expressed a tetrad for television (pp. 158–9) as shown in figure 5.1.

If we return now to our email example, the tetrad becomes especially helpful in understanding the aesthetic significance of the two boys' experiences of *Queer Eye for the Straight Guy*. For the boys, the program enhanced creative aspects of masculinity, obsolescing their frustrations with trying to adapt to a conventionalized peer masculinity. The program facilitated the boys' retrieving their own personal sense of aesthetic value of the self and reversed into a space through which not only could they validate themselves, but their mother could also find support for validating her sons' nascent individuality.

In this way, television is a powerful medium in our arsenal of extensions of self, of efforts to be more than self, and of efforts to understand self. And that aesthetic, while always moving, is the ultimate we seek. What many consider to be a dominating ideological weapon of the corporate elite can, with conscious effort, be reversed into an instrument for self-actualization. Visual communication theorist Rick Williams has developed a number of techniques to help people achieve this. Williams (in press) asserts, "It is critical to our survival as self-aware, self-determining individuals and to the survival of our planet, that we learn to reverse the effects of these messages of consumerism on the psyche and to reverse the subsequent, unbridled development of the consumer culture that is, itself, consuming our self identities, our resources, and our environments". We can, then, learn to understand contemporary aesthetic life by pondering the nature of our interactions with television, by paying closer attention to its aesthetic power. Television is both of the world and in the world – and so are we.

The intersection of multiple gazes (Lutz and Collins, 1993) prevails – viewer, camera person, producer, actor, sponsor, network, corporation, earth – all within a framed box, a moving painting if you will, in which the strokes are lines of video and images that seem to move too fast to ponder. Yet we do ponder . . . both consciously and unconsciously. This moving aesthetic is like a fugue, a kind of "mosaic that results from the collaboration" (Arnheim, 1981, p. 4), or a "mosaic logic" (Barry, 1997), resonating through time and space, physical and virtual realities, through us and surrounding us and emanating from us. The television equivalent of the future may project a holographic image far more compelling than any we now encounter, projecting the visual and aural signals, in turn inviting ever-more-real-seeming projection of self, into the shared communication space between us and the reception apparatus.[10] McCarthy (2001) wisely calls our attention to the idea that television is far more than a living room or bedroom presence; rather, it is an "ambience" surrounding us in such public spaces as sports bars and airline terminals. Many of us have experienced the startling realization that we have transported our minds out of our bodies, which are sitting on stadium bleachers at a major sporting event, via watching the virtual (and closer-appearing) version on a giant arena screen. With the increasing popularity of wireless videophones, a wristwatch-sized television cannot be far behind.

Rather than decry our aesthetic involvement with technologies such as television, we would do well to embrace Haraway's (1991) cyborg manifesto. "It is not clear who makes and who is made in the relation between human and machine" (p. 177), Haraway writes, adding, "The machine is us, our processes, an aspect of our embodiment. We can be responsible for machines; *they* [author's italics] do not dominate or threaten us. We are responsible for boundaries; we are they" (p. 180). It is time, as Williams (2004, personal communication) says, to mandate "a paradigm change in the ways we ponder and understand the illusive images of television." For the Warlpiri, the path to control over their cultural future means television that reaches "forward and backwards through various temporal orders," a political resistance conceived "in terms of the convoluted temporalities" of the present (Michaels, 2000, p. 714). Media scholars should be so courageous. In a call for the serious study of media aesthetics in Europe, Wolfgang Schirmacher (1991) argued, "In media we are challenged to write our own lives. . . . Mouse and remote control are only the beginning of inter-active features in media which allow us to edit and cut, stop and go, break and flow whatever situation we encounter. . . . In media we write our autobiography – and if we don't, somebody else will do it for us."

Conclusion

I want to conclude in a place similar to where I began, but with a different "beauty" pageant. While Rick Williams and I were surfing with the remote recently, the opening moments of *The Miss Universe Pageant* caught our attention. One after another, the women deemed most beautiful in the world introduced themselves. "I'll bet we won't see a Miss Iran or Miss Iraq," Williams said. Nor did we see a Miss Pakistan or a Miss Afghanistan. We made an easy conjecture – that those countries of Islam would not want to be represented by scantily clad women put on public display since Islam reserves viewing of women's beauty for their husbands. Westerners are quick to criticize that religion-based norm. Yet many of us also criticize the public voyeurism and objectification of women's bodies in beauty pageants. The point here is not to determine what is morally or religiously acceptable, but to note that through the "geographic phenomenon" (as one of the announcers termed it) of the Miss Universe Pageant broadcast live from Quito, Ecuador, we participated in a global aesthetic experience in which we observed absences and found ourselves in a thoughtful discussion about public and private displays of female beauty. When we stop to consciously consider what is happening through the aesthetic of television, we learn.

In the final chapter of *The Transfiguration of the Commonplace*, Danto (2000) discusses the etymology of the *stilus* [*sic*], noting that its "specific inscriptional use" redeems the term from the "certain sexual hilarity" of overtones connoted by its "near of kin *stimulus* (point, goad) and *instigare* (to goad or prick)" (p. 197):

> It is as an instrument of representation that the stilus has an interest for us and, beyond that, its interesting property of depositing something of its own character on the surfaces it scores. I am referring to the palpable qualities of differing lines made with differing orders of stiluses: the toothed quality of pencil against paper, the granular quality of crayon against stone, the furred line thrown up as the drypoint needle leaves its wake of metal shavings, the variegated lines left by brushes, the churned lines made by sticks through viscous pigment, the cast lines made by paint dripped violently off the end of another stick. It is as if, in addition to representing whatever it does represent, the instrument of representation imparts and impresses something of its own character in the act of representing it, so that in addition to knowing what it is of, the practiced eye will know how it was done. (p. 197)

One of the challenges in understanding television aesthetics is that most of us who view "television art" do not think about "how it was done," a process that encompasses the infinite possibilities of dots, lines, frames, and forms the television stylus and its users employ to represent various views of the world – and of ourselves – to us. Therein lies the source of the aesthetic of television and answers to what happens when we experience television aesthetics. Just as television is technologically possible because of the way the eye, as part of the brain, puts together bits of visual stimuli to interpret patterns of meaning, so we can draw upon our brains to reflect upon the content the television stylus represents to us as a profoundly evocative, moving aesthetic that simultaneously entreats, repels, enchants, horrifies, soars, falls, and moves forward and backward along the winding path by which we seek the ultimate aesthetic of the self. As Schirmacher (1991) wrote, "It is in aesthetics, when we are open to the phenomenon itself, that we discover media's authenticity as mediation" (npn).

Television is a container, a framed box, that gathers visual and aural stimuli in concentrated form for our perusal. The space between our bodies and the "set" of reality with which we choose to engage at any viewing period is the aesthetic. We are in a constant state of change, in which any object perceived, including the self, is at once known and knower, author and work. We are simultaneously pushing the limits of conscious understanding of the known world and creating new spaces in which to connect. What some scholars describe as para-social relationships, what I have called mass-interpersonal communication (Newton, 2001), is the conceptual experience of the aesthetic self and the aesthetic other meeting in the spaces of the mind and heart. Whether the stylus of television continues to draw precisely articulated narratives, such as genres of carefully crafted situation comedies or dramas, or more loosely conceived stages on which reality plays such as *American Idol* and *The Swan* are celebrated, the visual/acoustic aesthetics of television will continue to engage collective and individual yearnings to experience . . . to experience.

A stained-glass window in the Cathedral of St. John the Divine in New York City features the evolution of media from quill-holding scribes to a then-futuristic medium: television. As hard as it is to believe, the turn-of-the-century (nineteenth to twentieth) rendition of a glowing rectangular screen, along with an entranced viewer, looks all too familiar. Is it a television? Or is it a computer? If we want to attempt to isolate the aesthetic qualities exhibited by television, the distinction is significant. But in the larger scheme of the ecology of human visual behavior, it

Figure 5.2 *A Prototype of the Television*, stained glass panel, communications bay, the Cathedral Church of Saint John the Divine, New York, NY. © by The Cathedral Church of Saint John the Divine.

matters little. Television is one means through which we experience and become aware of aspects of living along the path of our search for the Ultimate Aesthetic – the Self.

Notes

1 I am indebted here to Marshall McLuhan's idea that "Truth . . . is something we make in the encounter with the world that is making us" (McLuhan and Powers, 1989).

2 For an exploration of the Vision Instinct, see Newton's (2001), Chapter 3, "The Burden of Visual Truth." For an exploration of surveillance theory applied to news, see Shoemaker (1996) "Hardwired for News."

3 See Fiske and Hartley's (1989) chapters on "The Signs of Television" (pp. 37–58) and "The Codes of Television" (pp. 59–67) for their seminal explication of "bardic television" within a semiotics framework.

4 See Babbie (1986), for a discussion of applying Heisenberg's uncertainty principle, as well as the implications of the Hawthorne effect, to social science research.

5 See, for example, Gerland (1994) for an analysis of defense attorneys' frame-by-frame deconstruction of George Holliday's videotape "in order to dismantle the judgment to which it 'naturally' gives rise: that the police officers are guilty" (p. 306).

6 See Newton (2004) and Sontag (2004) for analyses of visual behaviors related to the images.

7 See Goldberg (1991, pp. 212–15) for the story of the protest event and the uses and misuses of Associated Press photographer Malcolm Browne's images. See *USA Today* (2004) for a visual/verbal report on the Berg slaying.

8 See Drudge (2004) for verbal and visual description of the media event.

9 See Newton (2001, chapter 9) for a full explanation of the ecology of the visual.

10 See Winston (1996) for an excellent exposition on technological possibilities and the forces contributing to and constraining their diffusion in mass culture.

References

Adler, R. P. (ed.) (1981) *Understanding Television: Essays on Television as a Social and Cultural Force*, New York: Praeger Publishers.

Agger, G. and Jensen, J. F. (eds.) (2001) *The Aesthetics of Television*, Aalborg, Denmark: Aalborg University Press.

Arnheim, R. (1981) "A Forecast of Television," in R. P. Adler (ed.), *Understanding Television*, pp. 3–9, New York: Praeger Publishers, pp. 3–9. (Original article published in 1935.)

Babbie, E. (1986) *Observing Ourselves: Essays in Social Research*, Belmont, CA: Wadsworth.

Bennett, W. L. (1996) *News: The Politics of Illusion*, 3rd edn., White Plains, NY: Longman.

Blumer, H. (1969) *Symbolic Interactionism: Theory and Method*, Englewood Cliffs, NJ: Prentice Hall.

Caldwell, J. T. (2000) "Excessive Style: The Crisis of Network Television," in H. Newcomb (ed.), *Television: The Critical View*, 6th edn., New York: Oxford University Press, pp. 649–86. (Originally published in 1995.)

Carpenter, E. (1975) "The Tribal Terror of Self Awareness," in P. Hockings (ed.), *Principles of Visual Anthropology*, The Hague: Mouton, pp. 452–61.

Danto, A. (1981) *The Transfiguration of the Commonplace: A Philosophy of Art*, Cambridge: Harvard University Press.

Danto, A. (2000) *The Madonna of the Future: Essays in a Pluralistic Art World*, New York: Farrar, Straus and Giroux.

DiTommaso, T. (2003) "The Aesthetics of Television," *Crossings: a Journal of Art and Technology*, *3*(1), npn, accessed on May 28, 2004 from http://crossings.tcd.ie/issues/3.1/DiTommaso/#25.

Drudge, M. (February 1, 2004) "Outrage at CBS After Janet Bares Breast During Dinner Hour; Super Bowl Show Pushes Limits," accessed on June 11, 2004 from http://www.drudgereport.com/mattjj.htm.

Feagin, S. L. (1995) "aesthetics [*sic*]," in R. Audi (ed.), *The Cambridge Dictionary of Philosophy*, Cambridge: Cambridge University Press, pp. 10–11.

Fiske, J. and Hartley, J. (1989) *Reading Television*, London: Routledge. (Originally published in 1978 by Methuen & Co.)

Foster, H. (1983) *The Anti-Aesthetic: Essays on Postmodern Culture*, Port Townsend, WA: Bay Press.

Gerland, O. (1994) "Brecht and the Courtroom: Alienating Evidence in the 'Rodney King' Trials," *Text and Performance Quarterly*, *14*, 305–18.

Goldberg, V. (1991) *The Power of Photography: How Photographs Changed Our Lives*, New York: Abbeville Press.

Graber, D. A. (1990) "Seeing Is Remembering: How Visuals Contribute to Learning in Television News," *Journal of Communication*, *40*(3), 134–55.

Haraway, D. J. (1991) *Simians, Cyborgs, and Women: The Reinvention of Nature*, New York: Routledge.

Hill, A. (2004) "Reality TV: Performance, Authenticity and Television Audiences," in J. Wasko (ed.), *A Companion to Television*, Malden, MA: Blackwell.

Lorand, R. (ed.) (2002) *Television: Aesthetic Reflections*, New York: Peter Lang Publishing, Inc.

Lutz, C. A. and Collins, J. L. (1993) *Reading National Geographic*, Chicago: University of Chicago Press.

Metallinos, N. (1996) *Television Aesthetics: Perceptual, Cognitive, and Compositional Bases*, Mahwah, NJ: Lawrence Erlbaum Associates.

McCarthy, A. (2001) *Ambient Television: Visual Culture and Public Space*, Durham, NC: Duke University Press.

McLuhan, M. (1951) *The Mechanical Bride: Folklore of Industrial Man*, New York: Vanguard.

McLuhan, M. and McLuhan, E. (1988) *Laws of Media: The New Science*, Toronto: University of Toronto Press.

McLuhan, M. and Powers, B. R. (1989) *The Global Village: Transformations in World Life and Culture in the 21st Century*, New York: Oxford University Press.

Michaels, E. (2000) "For a Cultural Future," in H. Newcomb (ed.), *Television: The Critical View*, 6th edn., New York: Oxford University Press, pp. 701–15. (Originally published in 1987.)

Milgram, S. (1977) *The Individual in a Social World: Essays and Experiments*, Reading, MA: Addison-Wesley.

Mullen, B. (1986) "Newscasters' Facial Expressions and Voting Behavior of Viewers: Can a Smile Elect a President?," *Journal of Personality and Social Psychology*, *51*, 291–5.

Newcomb, H. (ed.) *Television: The Critical View*, 6th edn., New York: Oxford University Press.

Newton, J. H. (May 16, 2004) "Indelible Images: Photos Reveal Truth, Shake Us From Complacency," *The Anniston Star*, accessed on June 11, 2004 from http://www.annistonstar.com/opinion/2004/as-insight-0516-0-4e14q2316.htm.

Newton, J. H. (2001) *The Burden of Visual Truth: The Role of Photojournalism in Mediating Reality*, Mahwah, NJ: Lawrence Erlbaum Associates.

Orlan (2004) Official ORLAN WebSite, accessed on May 27, 2004 from http://www.orlan.net/

Piccirillo, M. S. (1986) "On the Authenticity of Televisual Experience: A Critical Exploration of Para-Social Closure," *Critical Studies in Mass Communication*, *3*, 337–55.

Schirmacher, W. (1991) "Media Aesthetics in Europe," accessed on May 23, 2004 from http://www.egs.edu/faculty/schirmacher/aesth.html.

School of Fine Art, Duncan of Jordanstone College of Art, University of Dundee, Scotland, Orlan (1997) "Woman with Head . . . Woman with Head, 1993," *Transcript Magazine*, *2*(2), accessed on May 27, 2004 from http://www.dundee.ac.uk/transcript/volume2/issue2_2/orlan/orlan.htm.

Shoemaker, P. (1986) "Hardwired for News," *Journal of Communication*, 46, 32–47.

Schultz, M. (1993) "The Effect of Visual Presentation, Story Complexity and Story Familiarity on Recall and Comprehension of Television News," Unpublished doctoral dissertation, Indiana University, Bloomington.

Smith, H. (1988) *The Power Game: How Washington Works*, New York: Ballantine.

Sontag, S. (May 23, 2004) "The Photographs Are Us, Regarding the Torture of Others," *The New York Times Sunday Magazine*, accessed on May 23, 2004, from http://www.nytimes.com/pages/magazine/index.html?8dpc.

Stigel, J. (2001) "Aesthetics of the Moment in Television: Actualisations [*sic*] in Time and Space," in G. Agger and J. F. Jensen (eds.), *The Aesthetics of Television*, Aalborg, Denmark: Aalborg University Press, pp. 25–52.

Streeck, J. and Knapp, M. L. (1992) "The Interaction of Visual and Verbal Features in Human Communication," in F. Poyatos (ed.), *Non-Verbal Communication*, Amsterdam: John Benjamins Publishing Co., pp. 3–23.

Susan (May 25, 2004) Viewer Email. Accessed on May 28, 2004 from http://www.bravotv.com/Queer_Eye_for_the_Straight_Guy/Community/Viewer_Email/2004.05.25.shtml.

Swift, E. (March 2, 2000) "Skin Deep: Orlan Takes Beauty to A Whole New Level," Jolique, accessed on May 27, 2004 from http://www.jolique.com/orlan/skin_deep.htm.

Thorburn, D. (1987) "Television as an Aesthetic Medium," *Critical Studies in Mass Communication*, *4*, 161–73.

Thorburn, D. (2004) "Television Aesthetics," in H. Newcomb (ed.), *The Television Encyclopedia*, 2nd edn., London: Taylor and Francis.

Thorburn, D. and Jenkins, H. (2003) *Rethinking Media Change: The Aesthetics of Transition*, Cambridge, MA: MIT Press.

USA Today (May 11, 2004) "The Tragic Death of Nick Berg," accessed on June 11, 2004 from http://www.usatoday.com/news/gallery/2004/nick-berg/flash.htm.

Williams, R. (in press) "Theorizing Visual Intelligence: Practices, Development and Methodologies for Visual Communication," in Diane Hope (ed.), *Visual Communication and Social Change: Rhetorics and Technologies*, Cresskill, NJ: Hampton Press.

Winston, B. (1996) *Technologies of Seeing: Photography, Cinematography and Television*, London: British Film Institute.

Worth, S. (1980) "Margaret Mead and the Shift from 'Visual Anthropology,' to the 'Anthropology of Visual Communication,'" *Studies in Visual Communication*, *6*, 15–22.

Zettl, H. (1981) "Television Aesthetics," in R. P. Adler (ed.), *Understanding Television: Essays on Television as a Social and Cultural Force*, New York: Praeger Publishers, pp. 115–41.

Zettl, H. (1990) *Sight, Sound, Motion: Applied Media Aesthetics*, 2nd edn., Belmont, CA: Wadsworth.

CHAPTER SIX

Locating the Televisual in Golden Age Television

Caren Deming

What the televisual names then is the end of the medium, in a context, and the arrival of television as the context. What is clear is that television has to be recognised as an organic part of the social fabric; which means that its transmissions are no longer managed by the flick of a switch.

Tony Fry, *R/U/A/TV?: Heidegger and the Televisual* (1993, p. 13)

Introduction

American network television's apparent decline following the rise of cable, videogames, and the Internet fuels an intensifying debate over the definition of an aesthetics of the televisual. Serious engagement with "the problem" of defining a televisual aesthetics unsettles long-standing assumptions about the technology that enables "seeing at a distance" and what it means to do so by watching television. Assumptions about the nature of the medium, production practices, industry contexts, and the larger social forcefield in which television operates – all come to bear on the problem.

This chapter organizes prevalent claims about the televisual into five categories: temporality, spatiality, aurality, femininity, and hybridism. Such a list reflects the variability of the television literature in purpose, method, and critical orientation. The concepts subsumed in these categories comprise a formalist, economic, discursive, and ideological mix seasoned with a little each of phenomenology and physics. Viewing televisuality as a synthesis of stylistic, technological, and ideological characteristics, this study reveals the concept's resistance to being fixed in time or in relation to other media.

This exploration of the constituents of the televisual is motivated by curiosity about the "Golden Age" of television in the United States. Ultimately, my purpose is to search for the televisual there. If the traits identified and described begin to distinguish television as a medium, and if those traits can be seen, in retrospect, to have emerged in the early days of television, we may view the Golden Age as the beginning of the televisual. Recognizing the cultural freight borne by the term and tracing its origins to the earliest days of television, then, dispels nostalgia for some prelapsarian state preceding commercial, aesthetic, and sociopolitical degradation. There is everything to gain by looking back now that television is old enough to afford a longer view. In television years – where series longevity may be defined in a few episodes – 50-something is very long indeed.

The case I have selected is *The Goldbergs* (1949–56), a series whose production run is virtually homologous with the Golden Age. In the decade from 1948 to 1958, television drama went from "apprenticeship to sophisticated anthologies to series, from New York to Los Angeles, and from live dramas to recording on film or videotape" (Hawes, 2001, p. 2). *The Goldbergs* reflects the interest in experimentation and innovation characteristic of a period when new technologies attract the attention of major artists of all fields. Series creator Gertrude Berg worked with Lee J. Cobb, Cedric Hardwicke, and Sidney Lumet, among others. The programs she produced, wrote, and starred in contain the residue of theater and film experience these figures brought to the new medium of television. The productions of the time were seldom beautiful. They were, as Hawes (2001, p. 1) points out, the product of a period of experimentation, rather than of a mature period of achievement. Nonetheless, as the study of *The Goldbergs* reveals, those productions can contain moments of astonishing televisuality.

After a brief review of the categories of the televisual, I will deploy those categories in an illustrative reading of *The Goldbergs*. This reading demonstrates the complexity of the concept and challenges the popularly-held notion that the convergence of television with other technologies and their associated styles is a recent development. It argues instead for the longitudinal study of entertainment technologies in relation to one another rather than in the technologically defined divisions that have tended to characterize the academy's encounter with them.

Constituents of the Televisual

Before embarking on a brief review of the concepts constituting the televisual, it is important to be clear about the term. I mean "televisual" to refer to a complex of formal tendencies that shape television works and their reception. For the purposes of this project, I focus on narrative television. Other modes of presentation (news and advertising, for example) obviously participate in the televisual, though they cannot be dealt with here. I am not limiting my use of "televisual" to a narrowly defined aesthetic, such as Caldwell's (1995) "excess of style over substance" or Redmond's (2004) "videographic frames," inasmuch as my purpose is to identify traits synthetically and to look for roots or precursors in early television.

Temporality: Commoditized Flow

Television is inescapably about time. The sense of immediacy originating from simultaneous "seeing at a distance" arose from genuine excitement about television as a new technology. Television's capacity (if not its dominant practice) to deliver events in real time remains its most salient claim to importance. As the unfolding events of September 11, 2001 demonstrated, it does deliver the real thing often enough to keep the claim to immediacy viable. Sobchak (1996) observes, however, that television's capacity for liveness is managed so that what is (increasingly) simultaneous is not the event and the experience of it but rather the event and its representation (and, ultimately, immediacy and its mediation). The prescience of Sobchak's observation is driven home by the fictional series *24*, which claims to render its narrative in "real time," complete with scenes running simultaneously on a divided screen. The fact that a represented hour is not an hour long is glibly elided along with the "missing" time needed for advertising and promotion. In the context of contemporary television, "real time" is a construct that, like liveness, grows increasingly surreal.

Like radio before it, initially television was broadcast live. Telefilms quickly became common because of their promise of efficiency (repeatability), image quality, and quality control. Hawes (2001, p. 1) traces the preference for live productions and the mystique about them to nineteenth-century stage productions, a preference carried to television by the radio interests who developed and nurtured the new medium. The proportion of "transcribed" material increased over the years, but even in the video age a fair amount of television is created before a live audience, whether in a studio or at home, and nothing inherent in the medium allows viewers to detect the difference between live and videotaped images.

Ontologically, the video image is "always becoming," as it requires a pattern of encoded electro-magnetic signals to be recreated continually. Even at a time when digitization has stabilized a paused "frame" of video, it still requires an epistemological leap to imagine frames of video at all. Phenomenologically, televisual liveness

> is related to the strong sense of distant seeing which the medium generates, together with the fascination of seeming close. A medium unable to produce anything but recorded images does not produce the temporal alignment (happening-as-you-watch) upon which the special magic of distant seeing is premised. (Corner, 1999, p. 2)

Although the relationship between the continuity of the signal and the experience of liveness remains largely conjectural, the "special magic of distant seeing" is highly contingent. As explained by Fry (1993, p. 42) the televisual possesses, paradoxically, "a presence of perpetual absence," something that is always arriving and being received but which "can never come to be." For Williams, the continuity of the signal is "the first constitution of flow" (Heath and Skirrow, 1986, p. 15). Although the notion of televisual flow applies at several levels – ranging from the atomic to programs and the "continuity" connecting them – more important to Williams is the structuring of endlessly flowing program and interstitial matter by television programmers hoping to keep viewers tuned in.

Surfing with a remote control alters the flow envisioned for viewers by any given broadcast or cable source, though the capacity to surf emphasizes the experience of multiple flows auditors can enter and leave at will. Awareness of other flows continuing even when not intended is common in the experience of television in a way that it is not in the experience of a movie in a theater. This is not to say that spectators at a multiplex are never distracted enough to wonder how the show next door is going, but that moving between continual flows of images is a more difficult mental and physical proposition in the theatrical experience of film.

John Ellis (1982) observed that television's program flow is more segmented than continuous. The segmentation occurs at various levels, including the common division of programs into acts separated by commercial breaks (themselves highly segmented) and series divided into episodes. The divisions between segments are marked emphatically, perhaps none more so than the fade to black before and after commercials placed between the acts of dramatic programs. The segments manifest varying degrees of closure, but none as emphatic as the closure of the classical Hollywood narrative, even allowing for that medium's current proclivity for sequels and prequels.

In stark contrast to the classical Hollywood narrative (often characterized as seamless), the televisual text is all about seams, or segment markers, which don't interrupt the programs so much as help to constitute them (Williams, in Heath and Skirrow, 1986, p. 15). Television narratives use classical Hollywood rendering (such as shot–reverse shot patterns) within acts to inject television's highly elliptical narratives with natural illusion (see Olson, 1999). More often than disguising divisions, then, the televisual flow manifests a preoccupation with division at the expense of continuity. Going beyond Olson's account of ellipticality as a factor in the narrative transparency that makes Hollywood film and television globally exportable, Caldwell claims that since the 1980s television has "deontologized" its own focus on liveness in favor of "style and materiality." In Caldwell's view, "hypostatized time and massive regularity comfort the viewer by providing a rich but contained televisual spectacle, an endless play of image and sound" (1995, p. 30). In keeping with Caldwell's characterization, then, we might say that television's heavy marking of time divisions serves the embodiment of managed, commoditized time.

Indeed, the segmentation of most television into regular and repeatable temporal units bespeaks the dominance of commerciality in the United States and, increasingly, elsewhere. Time is television's commodity form: units of time are bought and sold even in most so-called noncommercial settings. The prominence of regular temporal units in American television underscores the American interest in the exportability of film and television. In a formulation reminiscent of Carey's account of the creation of time zones in the United States in service to the need for the trains carrying commodities to run on regular schedules, global export of commercial television requires temporal regularity and repetition as well as narrative redundancy. In other words, the extra-diegetic features of televisual time also contribute to negentropy, one of the constituents of the narrative transparency that Olson (1999, p. 98) claims makes American media exportable. If, as Caldwell argues, contemporary television has relegated liveness to secondary status, it does not do so at the expense of temporality. To the contrary, television relegates the rendering of space to secondary importance in the interest of time.

Spatiality: The Window Itself

Televisual space privileges the two-dimensional space of the screen's x and y axes. As observed by Herbert Zettl (1989), televisual space is increasingly a graphical space in which computer-generated graphics and overlays emphasize the electronic image's absence of depth. This depthlessness is in stark contrast to cinema, in which creating the illusion of depth beyond the plane of the screen has been a perennial ideal. The contemporary practice of incessantly flattening the appearance of the televisual space with overlaid graphics suggests that collapsed space is as much a matter of televisual style and convention as it is of the focal length of lenses. The practice of emphasizing the flat, overlapping planes parallel to the screen (known in computer parlance as "windowing") is a televisual phenomenon easily adapted to the computer and the flat screens now regarded as desirable in both. For even freed of the television studio apparatus' limitations on mobility, television continues to respect the proscenium's immutable limitation of the performance space and, concomitantly, on the sphere of viewing positions.

Television's predisposition for emphasizing two-dimensional space accounts for its preference for proscenium-style shooting through the period when virtually all prime-time television narratives (both drama and comedy) have been shot on film and into the age of digital post-production. Barker (2000) illustrates the "highly utilitarian" approach of telefilm style with *Leave It to Beaver*, wherein primary movement is used only as a means to move characters in or out of the shot and tertiary movement consists of repetitive, predictable sequences of alternating medium shots. Clearly, more than technological and budget limitations are implicated here.

The predominance of medium shots of people in television suggests that a qualification of the mythology of television as "a close-up medium" is in order. The epithet, apparently derived from the habit of showing products close up in commercials, applies to things more than to people. In the world of the televisual aesthetic, it's as difficult to see a person close-up in a window as it is to enter the space beyond it without getting hurt.

More important than television's supposed inability to render deep space is its *denial* of space by repeatedly and intentionally flattening it through the use of blocking and superimposed graphics. As Morse (1998, p. 94) points out, the ancestral metaphor for the framing of television news subjects is the cartoon balloon, not the window. By extension, the space behind the television screen is also flat. It is an immaterial space in which camera, graphics, editing, and sounds "transform 'the world' into picture and we watch this picture which appears to be, but never is, the world we are *in*" (Fry, 1993, p. 30). Indeed, as Morse's discussion of the virtual subject positions constructed for humans by machines such as television and "more completely interactive and immersive technologies" suggests, it may be that the televisual locates viewers more precisely than it locates the people and objects it presents for viewing.

Aurality

The presentational formats of much television suggest that television viewers also are expected to be listeners. Most messages are verbalized by the medium's infamous talking heads and voice-overs, but fictional characters seldom miss opportunities to articulate the moral import of narratives. Morse (1998, p. 6) names television as the first machine to mediate stories and also to "simulate the act of personally narrating them in a shared virtual space." The simulation thus extends to news and other "reality" formats, which depend on the "enunciative fallacy" identified by Greimas as a feature of any speech act. In electronic media, immediacy and aurality conspire to foster the idea that the person speaking is a subject "in the here and now" and not twice removed – once by the mediation itself and again by the fact that speaking is already a first-order simulation of the enunciating subject and the time and place of enunciation. The speaking voices of television are disembodied in various ways, not the least of which is the medium's flattened visual space described in the section above.

Music and sound provide punctuation and emphasis, as well as the all-important signature theme, to representational and presentational television forms. The addition of stereo and even surround sound to broadcasting replaces visual depth with aural depth. This development complements Altman's (1986) analysis of television's discursive character, which invites viewers into dialogue and conveys the suggestion that the images television delivers have been collected "just for us." One of his prime examples is the signaling of the replay in a televised football game. Sound cues call viewers to attention so they can see images assembled just for their viewing (Altman, 1986, p. 50). The television industry's preoccupation with improving picture quality is ironic in the light of the importance of sound, which only in recent years has been a priority for technological improvement.

Television "speaks" in a variety of modes, its flow constituted by a series of presentational and representational moments nearly filled with talk. Its narrative forms, dominated as they are by melodrama and family comedy genres, contain an abundance of social commentary. Deriving their content from the quotidian and the topical, these narratives are every bit as important and powerful at bearing television's social meaning as more clearly presentational forms such as news, so-called "reality" forms, and commercials. The (oft) spoken morality of television befits a medium that takes its social role seriously. Indeed, the wave upon wave of entertaining (or annoying) advocacy for consumption is perhaps the most eloquent testimony that television knows and embraces its role as social arbiter. Thus, it is fitting to note television's reflexivity in association with its aurality and commerciality. Intertextual promoter, huckster, and zealot, television seems ever aware of itself as such (as well as its auditors' awareness of these roles). It is no wonder that it seeks to fill the very air with repetitive sounds and endless talk.

Femininity: Talking (Back)

Television's domesticity (its location in the home and preoccupation with family matters), its acknowledged role as sales agent for commodities and consumption, and its openness to women as creators and performers are among the factors that some writers have used to characterize television as more feminine than other media. Corner (1999, p. 26) maintains that television's visual scale, domestic mode of reception, and forms of spoken address provide the medium with the grounds for "a relaxed sociality" that contributes to its sense of co-presence with the outside world. Corner traces the association with gender and domesticity to the medium's durable preoccupation with the housewife and the equally durable regulatory concern with television's impact on children and the family.

The characterization of mass culture as woman has been a persistent theme in critical theory since Marx wrote of the "elusive, illicit, femininity of the commodity" that "seems to found the very idea of possession and production itself" (Zucker, 2002, p. 178). As feminist critics have observed, the language for analysis may have changed since Marx and the Frankfurt School authors who adopted it, but the association often continues as a tacit assumption. The larger domain of literature and the arts stands in decided contrast to modernism's masculinist preoccupation with action, enterprise, and progress. Huyssen (1986, p. 190) points out that for art, repudiating the feminine, whether implicitly or explicitly, "has always been one of the constitutive features of a modernist aesthetic intent on distancing itself and its productions from the trivialities and banalities of everyday life." The modernist valorization of the abstract amounts to relegation of popular, realist forms to inferior, feminine status, despite the fact the production of mass culture has been under the control of men (see Huyssen, 1986, p. 205).

Television's appeals to women as consumers are complicated, too. Lynn Spigel (1992, p. 159) has argued convincingly for early television's self-conscious and paradoxical appeals to women in the family comedy, which "transforms everyday life into a play in which something 'happens.'" Spigel sees that play enacted in the "prefabricated social setting" of 1950s suburbs, which she and others such as Mellencamp (1986) and Feuer (2001) find implicated in the strategic containment of women characters such as Gracie, Lucy, Roseanne, and *Absolutely Fabulous'* Patsy and Edina. The televisual view of women thus manifests another paradox: the avatars of consumer culture must be taught how to behave and to keep their place. Hatch (2002) demonstrates this paradox in her analysis of the selling of soap to the American housewife interpellated by postwar soap operas.

Michele Hilmes (1997) has uncovered the blatant efforts of network executives to "masculinize" a medium that paid too much attention to women because it had so much soap to sell and so much air (time) to fill. Though powerful women on both sides of the cameras have been targets of misogynist degradation, women, their sensibilities, and their buying potential remain influential in television. Perhaps the most telling evidence of a truth that won't go away is the fact that television still works hard at masculinization. Advertising executives are still talking about addressing television ads to women, and the topic is exceptional enough to elicit comment in the trade press. In that context, a masculine voice addresses a presumably masculine reader over the perennial topic of what "to do about" women.

Hybridism: Messing with the Borders

Television's resistance to modernist analytical categories has made critical engagement with it nettlesome, too (see Deming, 1989). Indeed, television's postmodernist characteristics reinforce its characterization as feminine. As emphatic as television is about time divisions, it flaunts its fluidity where genre is concerned. Although most individual episodes of any series are formulaic by design, television is determinedly recombinate at the series level. In part, television's generic hybridism is attributable to the paradoxes built into its need to be familiar and centrist while claiming to cut away at the edges – simultaneously exercising its penchants for recycling and topicality, nostalgia and immediacy. Critics noticed television's generic hybridism in the 1980s, the decade in which terms such as "dramedy" found their way into the critical vocabulary. However, television's resistance to formal categories is inveterate.

Television also flirts with the borders of reality and fantasy. Spigel concludes her discussion of family comedies with the observation that they transport viewers to a "new electronic landscape where the borders between fiction and reality were easily crossed" (1992, p. 180). If fictions may be said to blur the borders of reality, then "reality" genres (from news to talk, games, contests, and makeovers) may be said to blur the boundaries between "primary experiences" (such as conversation or other interpersonal relations) and constructed social realities (see Morse, 1986, p. 74). Stars appear as themselves in narratives or in commercials, and characters morph into hawkers with ease. Such fluidities combine with other postmodernist characteristics: intertextuality, pastiche, "multiple and collaged presentational forms," textual messiness (more textural than transparent), and reflexivity (Caldwell, 1995, p. 23). Though not the first to observe

Figure 6.1 Olga Fabian plays Mrs. Bloom on *The Goldbergs*, NBC-TV.
Source: Library of American Broadcasting, University of Maryland.

television's unrelenting postmodernism, Caldwell observes in that very observation postmodern critics' inability to distinguish the modern from the postmodern
in television.

Televisuality in the Golden Age: Gertrude Berg and *The Goldbergs*

The critical literature on television locates the elements of the televisual explored
here – peculiar patterns of temporality, spatiality, aurality, femininity, and
hybridism – in various historical periods. The question I wish to pose now is:
how successfully can these traits be deployed to illuminate work produced when
television was just coming into its own? If the televisual was invented in the
celebrated Golden Age in the United States, it ought to be possible to find at
least traces of it in those texts that have survived.

Gertrude Berg (1899–1966) was a pioneer broadcaster and prolific creator of
theater, radio, television, and film. After more than 5,000 radio programs (including *The Goldbergs*, 1931–4, 1936–50), she moved to television to reinvent the

series, transforming it from soap opera to domestic comedy. Berg is one of the inventors of radio drama, of television drama, and of television comedy. Her oeuvre is significant because of her contributions to the development of the television domestic comedy (especially the ethnic comedy), her unusually powerful industry status, the broad popularity of her programs, and her timing.

The Goldbergs is regarded as exemplary television from the Golden Age, and its history incorporates momentous developments in television. The series was carried live on CBS (1949–51) until controversy over the blacklisting of actor Philip Loeb ended the relationship. The series reappeared on NBC under new sponsorship for the 1952–3 season and then moved to the Dumont network in 1954. A filmed version ran in first-run syndication in 1955–6. Some kinescoped broadcasts from the early live years, and the whole of the 1954–5 and 1955–6 seasons (more than 80 episodes all together) have survived in various archives.

The Goldbergs portrayed the trials and tribulations of the eponymous Jewish family living in a Bronxville tenement. Over the years, the series included over two hundred characters, though it was sustained by five central figures. Molly Goldberg (Gertrude Berg) was a powerful and benevolent mother absorbed with finding sensible solutions for family and neighborhood problems. Her humor, derived primarily from malapropisms and Yiddish dialect, was lovingly authentic, never patronizing or condescending. Molly's husband Jake Goldberg (Philip Loeb, Harold J. Stone, Robert H. Harris) worked in a dress shop, though audiences knew him as the reliable husband and father, and as the perfect foil to Molly. Jake could be impatient or critical. He also could seem a little silly or irrelevant, by comparison with his wife, especially when the schemes he criticized proved beneficial.

The Goldberg children, Rosalie (Arlene McQuade) and Sammy (Larry Robinson, Tom Taylor), typified first-generation Americans trying to make sense of their heritage. "Rosie" and "Sammily" (as Molly familiarly referred to them) were dedicated to modernizing their parents and correcting their pronunciation. Molly's Uncle David Romaine (Eli Mintz) rounded out the Goldberg family regulars. Uncle David was integral to the household, often cooking or washing the dishes. He wore a ruffled apron with as much nonchalance as his *yarmulke*. A ready enlistee in Jake's sideline commentaries on Molly's activities, Uncle David would echo Jake's protests and judgments. The beloved Uncle David's eccentricities were taken in their stride by the rest of the family.

In the surviving episodes of *The Goldbergs*, features such as the theatricality of the set, the creative use of a proscenium shooting style, and high-quality acting are all prominent. The episodes also contain visual and narrative treatments that eventually became conventions of the domestic comedy. The focus of the reading that follows is on spatiality. This is so not only because there is not enough room in these pages for a detailed exploration of all of the elements of the televisual. Analysis of the treatment of space in *The Goldbergs* reveals the extent to which the features of the televisual interpenetrate one another. In other words, analysis of one implicates the others.

Televisual Space: Sample Synthetic Reading

Molly is typical of the 1950s television mother in that she cooks wearing attractive dresses, takes care of everybody, and never seems to get dirty. In 1955, she moves her family to the suburbs and seems always ready to shop or entertain. The obvious difference from other television mothers of the time is that she is neither thin nor glamorous. Moreover, she is in charge. This *mother* knows best, and her narrative dominance is matched by her visual dominance.

Molly dominates both the television frame and the conversation. From the beginning of the series, she tends to be "downstage" of other characters (closer to the camera). When the Goldbergs form the tableau reminiscent of soap opera (and film melodrama), Molly typically is centered in the lower, "heavier" portion of the screen. Even in shot-reverse shot sequences cut from film in the final year of the series, Molly's close-up is tighter, and her body is literally larger on the screen. Berg uses her big bones, big hair, and big features to advantage. Her gestures often call attention to her matronly form – which grows bigger every season – as she smoothes her apron or plants her hands on her hips. She fills the famous window frame from which she hails her neighbors with a musical "Yoohoo" or presents the virtues of her sponsors' products in direct address to the camera.

In one scene from 1955, Molly appears in the center of a pyramid formed by the other characters and the dining room table. She stands behind the table, hands on hips, larger and taller than the others. Even more dramatic effect comes from a scene in the same episode in which Molly, Jake, and Uncle David have been tied up by robbers. Molly's body looks especially large as a result of her central placement and the perspective of the shot.

In scenes with the other actors, Berg's body often is nearly still, leaving the bustling movement associated with television domestic comedy largely to the rest of the cast. These blocking techniques underscore Berg's character as axial to the narrative. This is in contrast to Carroll O'Connor's performance of Archie Bunker on the day of Florence and Herbert's wedding, when Archie is constantly in motion, "orchestrating" the movements of other characters. Barker (2000, p. 176) interprets this activity as the visual manifestation of Archie's axial role in the narrative. Molly achieves her status as the visual anchor of the action largely by standing still, a blocking technique that is echoed in the performance of Kelsey Grammer as Dr. Frasier Crane in the contemporary domestic comedy *Frasier*.

Typical of domestic comedy, any of the Goldbergs can become the focus of an episode. Nonetheless, Molly is always the focalizer, the moral center of the form's discursive universe. By contrast, the male roles are largely superfluous. When, in the episode with the young robbers, Jake and Uncle David don't want Molly to be a hero attempting to take down the robbers, she replies, "It's no time for a faint heart." A woman of action as well as a minder of manners, Molly references her own centrality. The men's place is on the sidelines, where they make fun of Molly's projects, including matchmaking. She ultimately triumphs in spite of them. Molly teaches them a lesson and then preaches the moral of the story at the close of each narrative, a lesson that carries over to the closing commercial pitch of the live episodes.

Unlike radio, which had to overcome critical and regulatory resistance to commercialism (over which there was at least a public debate), television was born commercial. From the beginning, television programs were seen as devices to secure the attention of eyes and ears for the commercial pitch, even if programmers worried over how much blatant selling Americans would tolerate. In *The Goldbergs*, Molly is the only character privileged to interact with the audience by addressing the camera directly. She does so only during her commercial pitches, which she does in character, often using the "problems" of other characters in the narrative as an excuse to talk about the calming properties of Sanka decaffeinated coffee or Rybutol vitamins.

Berg relishes her intimate minutes with the camera, earnestly pitching "her" Sanka, Rybutol, or the television set itself (for RCA). The commercials lend authority to Molly's character, as she is part of the narrative and (at beginning and end of the broadcast) is also central to its framing device. In the commercials, viewers get their special time with Molly, in effect, more important time even than she spends with her fictive family. Viewers get the undivided attention of "the mother audience members had or wished they had" [in] "an era in which Jewish mothers were the models of perfect mothers, sacrificing all for the children's happiness" (Epstein, 2001, pp. 72–3). Leaning out of her window toward the camera, Molly shares her advice on how to live well by consuming. She speaks in an intimate, almost conspiratorial, tone as she talks about other characters with the audience at home. Assertive and authoritative, Molly is the axis upon which commercial *and* narrative imperatives turn.

The window frame functions to contain the narrative visually, but not Molly's body. She fills the window frame, and her gestures take place in the space between the "window" and the camera as she leans forward. Spigel points out that the convention of making the commercial pitch seem "closer" to the audience than the narrative is most pronounced in *The Goldbergs* (in contrast to *Burns & Allen* or *I Love Lucy*, which had actual or animated theatrical curtains to mark this strategy). The transition from presentation in the commercial to representation in the narrative creates "the illusion of moving from a level of pure discourse to the level of story, of moving from a kind of unmediated communication to a narrative space" (Spigel, 1992, p. 168).

If the pitch enacted television's enunciative fallacy, the dramatic transition from it to the narrative (when Molly literally rotates 180 degrees and begins to speak with her family, and the director cuts on the action) also calls attention to the seam where the two forms of address meet. The convention of marking the transition between narrative and commercial with a fade-out and fade-in occurs in live broadcasts of other series when the commercial is done by someone located elsewhere in the studio. As soon as the shows are no longer broadcast live, but recorded on film or videotape, the fade to black becomes the standard delineator of narrative time and space, a convention which continues to the present day. Musical "stingers" also mark segments when they can be added in post-production. Gone from the final, syndicated season are Molly's famous sales pitches, indicative of her eroding visual and discursive power.

In the early years of *The Goldbergs*, theatricality is suggested by the look of the set and graphics more often than by blocking. What today's students would call "cheesy," primitive-looking graphics open the show. The titles are hand lettered, almost as though by the hand of a child. The signature geranium on Molly's window sill is crudely drawn. The flower grows in an empty Sanka can, announcing the sponsor's product (and the Goldbergs' postwar frugality) from the opening moment of the show. The window Molly leans from is cut out of a "wall" of painted-on bricks. Such painted-on or outlined set elements are common, and they complement the graphic style of the titles.

In a clear nod to its radio predecessor, *The Goldbergs* on television has a character who is neither seen nor heard. Pinky the dog is "spoken to" and "stepped over," but he is the product of sheer imagination. The impracticality of having a live dog in the television studio (not necessary for radio) no doubt explains why Pinky is never seen. Having the actors pretend that he is there barking and wagging his tail is a startlingly reflexive gesture. The "presence" of Pinky in the television episodes also assumes an audience that followed the Goldberg family from radio to television. The ubiquity of theatricality is not surprising in that New York (Manhattan) is American television's place of birth and the "salad days" of its Golden Age. Together with the visual elements, the "endless self-referentiality" Spigel (1992, p. 169) describes in *The Goldbergs* signals the characteristic reflexivity of the domestic comedy genre and of television.

In the early episodes of *The Goldbergs*, the shooting style is designed to create the illusion of depth more than to create a proscenium effect, however. Movement often occurs along the z axis; and characters are composed in patterns of overlapping planes. (Molly does a lot of overlapping!) It has not yet become conventional to move characters parallel to the screen plane in the family comedies. Rather, the actors are choreographed in deep space before nearly stationary cameras. Sometimes they are stacked in depth for simultaneous reaction to events happening in the foreground, even though few variations in angle, distance, and lighting are available. The medium focal length of the lens allows as many as three rooms of the Goldberg apartment to be visible (and in focus) in a single shot, and characters are busy talking and moving in as many as four planes at once. Thus, the multiple planes parallel to the screen defined by framing devices incorporated into the set (window frames and archways outlined in paint, for example) define interior space in depth traversed by actors moving on the z axis.

Often the actors execute a complex dance in which they must hit their marks accurately in space and time. Set movement patterns are evident from episode to episode, and misses are admirably infrequent given the exigencies of live studio production. However, the episodes are not without moments of experimentation. In a 1949 episode, the camera placed outside the window to shoot Molly during the commercials is used to frame characters inside the apartment with the window, half-drawn shade and all. (This voyeuristic moment occurs five years before the release of Hitchcock's *Rear Window*.) In another early episode, we see the visiting Uncle Beirish (Menasha Skulnik) through a tank full of live, swimming fish! In addition to the reflexivity generated by people having their fun with the new technology of television, such shots call attention to the frames themselves and concomitantly emphasize the flatness of the screen plane and the space behind it. Thus, the live years of *The Goldbergs*, the era before telefilm values assert their dominance over the visual style of television comedy, reveal a more elastic approach to the rendering of space. By the final season, the telefilm style (not something more cinematic) characterizes the episodes.

The early *Goldbergs* shows manifest narrative television's proclivity for the long and medium shot. Even when shot on film in the final season, shot–reverse shot sequences utilize the medium shot much more than the close-up. The predominance of medium shots persists, with the exception that later sitcoms use more limited (though more elaborate) sets and more limited kinds of movement by actors. In sum, *The Goldbergs'* seven seasons manifest signs of the solidifying televisual conventions as they incorporate techniques from radio, theater and the cinema. Experimentation, the use of different directors for different episodes (still common practice in series television), and rapidly evolving conventions – all confound static definitions of production style even at the series level. Long-running series such as *The Goldbergs* are particularly difficult to describe without an even more dynamic approach stemming from close analysis of a substantial number of episodes and an appreciation for variation and change over the life span of the series.

Conclusion

The cross-currents of visual style evident in *The Goldbergs* demonstrate that easy dichotomies between media have been challenged by television from the beginning. They also suggest that the convergence of cinematic and televisual styles needs to be understood in evolutionary terms and contextualized accordingly. Berg's work demonstrates that televisuality, like genre, is an unstable construct to be applied with the utmost delicacy, especially when approaching bodies of work traditionally regarded as the most formulaic.

Ultimately, the constituent technologies of the televisual need to be seen more broadly and in more integrated ways than most past scholarship has done. The televisual is at once technology, style, and ideology. It is at once art, economics, and politics. Such complexity demands no less than to place television (along with its precursors and predecessors) in the history of entertainment technologies writ large. Fry's declaration, that the ability to identify the televisual signals the end of television as medium and the acknowledgement of television as environment, is chastening. For, in the light of the televisual, the window metaphor becomes painfully apt. It places the medium and all its related, overdetermining formations behind the distancing glass that reveals television as the computer waiting to happen. Does that recognition obviate the need to study television series as texts? Not at all, because defining the televisual is a project only just begun. The secrets still locked in television's past are crucial grounds upon which its future may yet be understood and contested.

References

Altman, R. (1986) "Television/sound," in T. Modleski (ed.), *Studies in Entertainment: Critical Approaches to Mass Culture*, Bloomington, IN: Indiana University Press, pp. 39–54.

Barker, D. (2000) "Television Production Techniques as Communication," in H. Newcomb (ed.), *Television: The Critical View*, 6th edn., New York: Oxford University Press, pp. 169–82.

Caldwell, J. T. (1995) *Televisuality: Style, Crisis, and Authority in American Television*, New Brunswick, NJ: Rutgers University Press.

Corner, J. (1999) *Critical Ideas in Television Studies*, Oxford: Oxford University Press.

Deming, C. J. (1989) "For Television-centered Television Criticism: Lessons from Feminism," in M. E. Brown (ed.), *Television and Women's Culture: The Politics of the Popular*, Paddington, NSW, Australia: Currency, pp. 37–60.

Ellis, J. (1982) *Visible Fictions: Cinema, Television, Video*, London: Routledge.

Epstein, L. J. (2001) *The Haunted Smile: The Story of Jewish Comedians in America*, New York: PublicAffairs.

Feuer, J. (2001) "The Unruly Woman Sitcom*: I Love Lucy, Roseanne, Absolutely Fabulous*," in G. Creeber (ed.), *The Television Genre Book*, London: BFI, pp. 68–9.

Fry, T. (ed.) (1993) *R/U/A/TV?: Heidegger and the Televisual*, Sydney: Power.

Hatch, K. (2002) "Selling Soap: Postwar Television Soap Opera and the American Housewife," in J. Thumin (ed.), *Small Screens, Big Ideas: Television in the 1950s*, London: I. B. Tauris, pp. 35–49.

Hawes, W. (2001) *Live Television Drama, 1946–1951*, Jefferson, NY: McFarland.

Heath, S. and Skirrow, G. (1986) "An Interview with Raymond Williams," in T. Modleski (ed.), *Studies in Entertainment: Critical Approaches to Mass Culture*, Bloomington: Indiana University Press, pp. 3–17.

Hilmes, M. (1997) *Radio Voices: American Broadcasting, 1922–1952*, Minneapolis: University of Minnesota Press.

Huyssen, A. (1986) "Mass Culture as Woman: Modernism's Other," in T. Modleski (ed.), *Studies in Entertainment: Critical Approaches to Mass Culture*, Bloomington: Indiana University Press, pp. 188–208.

Mellencamp, P. (1986) "Situation Comedy, Feminism, and Freud: Discourses of Gracie and Lucy," in T. Modleski (ed.), *Studies in Entertainment: Critical Approaches to Mass Culture*, Bloomington: Indiana University Press, pp. 80–95.

Morse, M. (1998) *Virtualities: Television, Media Art, and Cyberculture*, Bloomington: Indiana University Press.

Olson, S. R. (1999) *Hollywood Planet: Global Media and the Competitive Advantage of Narrative Transparency*, Mahwah, NJ: Lawrence Erlbaum.

Redmond, D. (2004) *The World Is Watching: Video as Multinational Aesthetics, 1968–1995*, Carbondale, IL: Southern Illinois University Press.

Sobchak, V. (ed.) (1996) *The Persistence of History: Cinema, Television, and the Modern Event*, New York: Routledge.

Spigel, L. (1992) *Make Room for TV: Television and the Family Ideal in Postwar America*, Chicago, IL: University of Chicago Press.

Zettl, H. (1989) "Graphication," in G. Burns and J. R. Thompson (eds.), *Television Studies: Textual Analysis*, New York: Praeger, pp. 137–63.

Zucker, L. (2002) "Imagism and the Ends of Vision: Pound and Salomon," in W. S. Wurzer (ed.), *Panorama: Philosophies of the Visible*, New York: Continuum, pp. 169–84.

Television episodes

Composition. *The Goldbergs*. CBS, October 10, 1949.
Desperate Men. *The Goldbergs*, November 3, 1955.
The Goldbergs (untitled episode). NBC, August 7, 1953.

CHAPTER SEVEN

Television Production: Who Makes American TV?

Jane M. Shattuc

Who makes American television? "Created by Michael Crichton" punctuates *ER*'s opening credits. "Executive Producer Dick Wolf" portentously materializes at the end of all episodes of *Law and Order*. Rod Serling slips out of the shadows of a *Twilight Zone* to explain his definitive meaning of each episode. "Gene's vision" still lingers in the rhetoric of the production staff to maintain Roddenberry's ideas in the newest *Star Trek*. Popular magazines profile executive producers – David Chase of *The Sopranos*, Larry David of *Seinfeld* or Josh Whedon of *Buffy the Vampire Slayer* – as "creators." Seemingly, we need an inspired source to make sense of American television. Commercial producers consider themselves "creators" as they continually speak of holding onto their "original idea." But one might wonder: why should this be the case? American television is mass-produced; a series of stages serve as the assembly line where workers put together a similar product weekly with the production of over 200 like-products in the case of a successful series. This process is organized along rationalized lines, with as many as 300 people having some influence over the production of one program. So why do we need to have the agency of an individual, a source, or a creator to understand television?

European state networks have also nominated "creators" for dramatic programming for decades in the name of their national culture. The German ARD network has traditionally depended on adaptations of known "German" authors as the source for its television dramas (*Fernsehspiele*). Such a policy paved the way for the German New Wave where filmmakers such as Rainer Werner Fassbinder and Volker Schlondorff began their careers adapting great literature for TV. Their success led to television support for their original works – films seen on television as *Fesehspiele* (TV plays) and exported as *Autorenfilme* (theatrical art films).

A parallel also exists with the BBC, which has a similar mandate to produce "British culture" in the face of the dominance of American television. According to Glen Creeber (1998, 2001), this form of "public service" led the system initially to become an extension of British theater, producing the great works of famed British playwrights such as Shakespeare and Shaw. This penchant for the writer as the "creator" has remained the logic at the BBC, even as it has moved into producing original works for TV. There is no more famous example of this history than the origination of *The Singing Detective* (1986). This work has become synonymous with the writer Dennis Potter, even though well-known producers Kenith Trodd and John Harris and the director Jon Amiel, as well as a production team of several hundred people, were responsible for creating the series (Creeber, 1998, 2001, p. 167).

So why must television have an originator? A simple answer is: people make television, therefore there has to be a human source. A counterargument would be that every product has a source. In 1955 someone or some team invented *Crest* toothpaste in 1955 for Proctor and Gamble, but we do not consider their inspiration when brushing our teeth. Yet TV is different: it is culture. It also tells stories and creates images that are related to aesthetic traditions associated with artists. TV programs are descended from popular literature (the dime novel and theatrical melodrama), painting and photography or cinematography ("writing with light"). No two products are the same. Each is individuated. Originality, however marginal, is one of TV's attractions and why it has become classified as an art form.

Applying the concept of the "author" or "artist" to television is always difficult, but it is particularly thorny when applied to American television, which is based on profit and mass production – not on an aesthetic or national mandate. Martha Woodmansee (1994) reminds us that in contemporary usage, an author "is an individual who is solely responsible – and thus exclusively deserving of credit – for the production of a unique, original work" (p. 35). American television programming involves long-term series of possibly several hundred like-products, in the case of a successful series, as opposed to the British and European model of a single work or a short-term series. Clearly, American commercial success is constructed around the pleasure of familiarity and repetition rather than originality, as we tune weekly into variations on a basic norm, a format or, ultimately, a formula. This chapter attempts to resolve this dilemma: how and why the concept of the author – that unique creative source – can be applied to television in general and to American commercial TV in particular.

Before Television: Authorship in History

Authorship is not a simple term; it is the subject of continual academic debate. The belief that an individual is the source of meaning or originality is relatively new. But the application of the title to a mass-produced product is a development of the late twentieth century. Michel Foucault (1994, 1995) argues that a culture needs to pull back and consider how easily it ascribes authorship to a cultural work: "It might be worth examining how the author became individualized in a culture like ours, what status he has been given, at what moment studies of authenticity and attribution began." So when did the notion of an individual creative source become assigned to an American television program and, more importantly, why? The definition of authorship or artist as a singular individual has important ramifications for how we understand creativity in the American TV industry.

Consider why we know who created one program and not another. Here, we might question why Norman Lear is known for the creation of *All in the Family* (1971–83) and Paul Henning is not regarded as the creator of *The Beverly Hillbillies* (1962–71) and *Green Acres* (1965–71). Is it simply that one had greater social aspiration than just entertainment? Perhaps Paul Henning intended *The Beverly Hillbillies* as a critique of capitalism and social class – a plausible interpretation (Marc, 1984/1997). Why is Henning not seen as a "creative" individual in the mold of Lear? To understand the complexity of human origin, Foucault also asks for a consideration of "what kind of system of valorization the author was involved, at what point we began to recount the lives of authors rather than of heroes, and how this fundamental category of 'the-man-and-his-work criticism' began" (Foucault, 1994, 1995). Seemingly, we need to identify with "heroes" in television production, but again, we must still pose the question: why?

In its early years American television functioned with little or no allusion to TV makers as a reference point. Much of early television was filmed theater and the playwright became a key figure in establishing the importance of the program. Dramatic anthology series – *Kraft Television Theater* (1947–58), *Philco Television Playhouse* (1948–55), and *Playhouse 90* (1956–60) – was one of the earliest American television fictional forms. Glenn Creeber (1998, 2001) points out that the primitive nature of the medium (live productions seen on a small screen) emphasized the spoken word over its visual representation. The limited virtuosity of the medium caused the director's work to be more ephemeral and technical than "creative" (p. 20).

As American TV moved to producing original series, the works stood simply as entertainment: a commercial pastime. One simply watched *The FBI*; there was no need to conjure up a creative individual in order to understand it. Yet by the late 1970s, producers' names – Grant Tinker, Steven Bochco, Michael Mann and even Quinn Martin (the producer of *The FBI*) – began to be ascribed to programs as makers. And around this time, "man-and-his-work criticism" about television started to appear in magazines, newspapers, and academic books in the United States. The nomination of a source allowed critics to parallel TV culture with the traditional arts. In what may be one of the most rationalized of visual forms, critics isolated TV "heroes" fighting for the originality of their vision over the networks' constant drive for profit. Meaning was no longer the result of only a program (a product), a network, or a star; there was now a maker. This change resulted from a number of different discursive changes in how American culture redefined art and commerce in the late twentieth century. Media critics and studies programs in universities began to isolate creators in 1970s and 1980s commercial television as they sought to propose that TV should be considered a serious cultural form.

Prior to this time, authors and artists were relatively anonymous figures; they earned their livelihood at the behest of the church, the court, or wealthy patrons. Woodmansee (1994) argues that there were two parallel and competing ways of understanding authorship in the Renaissance and earlier – craftsmanship and inspiration. The writer-as-craftsperson was "the master of a body of rules or techniques, preserved and handed down in rhetoric and poetics, for manipulating traditional materials in order to achieve the effects prescribed by the cultivated audience of the court to which he owed both his livelihood and social status" (p. 36). This person was a "skilled manipulator" of predetermined rules, at best – not guided by the individual inspiration that one associates with genius or artistry.

This legacy has left its stamp on the television industry as production teams are divided by "crafts" or "craft unions," such as the Writer's Guild of America, Director's Guild of America, and IATSE (the International Alliance of Theatrical Stage Employees, Moving Picture Technicians, Artists and Allied Crafts of the United States). This language is associated with the Medieval and Renaissance tradition of craftsmen and guilds. It underlines the fact that television making is based on artisan labor – a skilled worker who practices some trade or craft. Traditionally, craft unions are divided off from industrial unions because they are organized around a particular skill or occupation that adds to the concept of mastery. In fact, nearly all television labor works on a model of apprenticeship, both for producing and technical labor. Even though the organization of a TV production staff – above the line (performers and producers) and below the line (technical personnel) – echoes a creative hierarchy, the division is still based on knowledge of rules (abstract or concrete).

However, another definition of authorship – the inspired writer – evolved during the Renaissance and involved a more spiritual understanding of artistry. Reacting to the prosaic concept of the writer as a mere vehicle for established rules, this new assumption was based on the following premise:

> [T]here are those rare moments in literature to which this [craftsmanship] concept does not seem to do justice. When a writer managed to rise above the requirements of the occasion to achieve something higher, much more than craftsmanship seemed to be involved. To explain such a moment a new concept was introduced: the writer was said to be inspired – by some muse, or even by God. (Woodsmansee, 1994, p. 36)

This belief in divine intervention affirms the underlying authority found in the TV credit "created by."

Nevertheless, this inspired creator remains at the behest of a larger force or set of rules. Woodsmansee argues that these two definitions coexisted, but were ultimately an uncomfortable marriage – dutifully following rules as opposed to being infused with a higher meaning. In both cases, the writer was not the source of the ideas in his/her work – it came from outside the individual – either from a set of rules or from a muse or god-like figure. He/she was still an instrument or a medium, but in the latter instance was one infused with divine insight. He/she was still not held directly responsible for the work.

The concept of the "creator" or "auteur" in television also owes a debt to Romanticism and the rising literate middle class, as well as the consequent marketplace for books and art in the eighteenth century as the craftsmanship sensibility diminished. In this period, the source of creativity shifts from outside the writer/artist to within. Creative inspiration emanated from one's own genius, not from a preconceived or spiritual source: "That is, from a (mere) vehicle of preordained truths – truths as ordained either by universal human agreement, or by higher agency – the *writer* becomes an *author* (Lat. *Auctor*, originator, founder, creator)." The writer and the artist became portrayed as expressive individuals who produce original works. Woodmansee (1994) quotes William Wordsworth as saying in 1815: "a genius is someone who does something utterly new, unprecedented, or in the radical formulation that he prefers, produces something that never existed before" (p. 39). Slowly, Romantic writers established ownership of their work – copyright – based on their own originality. Consider Goethe's dictum in the early eighteenth century for "constructive criticism" of writing – "What did the author set out to do? Was his plan reasonable and sensible, and how far did he succeed in carrying it out?" Such a view of authorship corresponds to the rhetoric of present-day newspaper articles, which highlight the work of executive producers, such as John Chase of *The Sopranos*, as their *own* work.

In the 1950s, critics began to ask if film art could, and whether it did, take place within a commercial context. Prior to this period American movies had one of three possible organizing influences: the studio ("MGM's *Wizard of Oz*"), the actor ("a Judy Garland movie"), or the genre ("I am going to see a musical"). At no point did the director play a vital role in the understanding of the film. The rethinking about artistic vision took place initially in *Cahiers du Cinema* in France, but also later in *Movie* in Britain and *The Village Voice* in the States. Edward Buscombe (1980) writes: "*Cahiers* was concerned to raise not only the status of the cinema in general, but of American cinema in particular, by elevating its directors to the ranks of the artists" (p. 23). It is only in the postwar period that the director as the central creator has come to the fore as an auteur or someone who could leave a personal signature on a commercial process.

More a system for evaluating a film's worth than an intellectual framework, the auteur theory involved three assumptions. According to John Caughie (1981), they are:

1. a film, though produced collectively, is more likely to be valuable when it is essentially the product of the director ("meaningful coherence is more likely when the director dominates the proceedings": Sarris);

2. the presence of a director who is genuinely an artist (an *auteur*) a film is more likely to be an expression of his individual personality;

3. that personality can be traced in a thematic and/or stylistic consistency over all (or almost all) the director's films. (p. 9)

The critic's and the viewer's job, then, became ferreting out the marks of the auteur/artist and tracing them across a number of films and themes. In fact, Hollywood auteurs were often considered to have an even more powerful artistic presence than their European art cinema counterparts, since they were able to break through the constraints of the studio system of mass production by sheer strength of personality and leave their personalities on a film – thereby overcoming obstacles not known to the independent European filmmaker.

Todd Gitlin's *Inside Prime Time* (1985) – a study of the decision process at the networks by which programs "rise and fall" in American television – fits this sensibility. The book centers on the epic of executive producers Steven Bochco and Michael Kozoll, who – in their efforts to hold on to their conception of *Hill St. Blues* against the pressures of mass production and corporate meddling – destroyed the program's originality (pp. 273–324). In *Television's Second Golden Age* (1996), Robert Thompson further pursues Gitlin's focus on creative individuals in American TV. He argues that the 1980s produced a wave of great television because of the work of visionary executive producers (such as Grant Tinker), who "had the courage . . . to gather talented creative people together and leave them alone" (p. 47). Producers echo this aesthetic individualism discourse. In my 15 years of interviews in television, producers have consistently told me how they must hold onto their creative vision against the pressures of profit. Even the executive producer of *COPS* has described his aesthetics and the battle with network pressure to keep the program original.

This Romantic definition of an artist has led to a clash between auteur-based critics with cultural studies scholars and other academics who see commercial artistry as a naïve Romantic construction. Most traditional academics are suspicious of any attempt to apply the concept of artistry to commercial television, contesting that the workings of television are ultimately driven by profit rather than by the artist's inner necessity. The auteur theory when applied to TV is a highly romanticized worldview and a naïve account of the dictates of commercial production. Art historians have traditionally defined art in opposition to commerce: true art emanates from a higher inspiration. When art and commerce mix, it becomes a craft (architecture) or an applied art (e.g. graphic arts).

This opposition between art and industry has been increased further by cultural studies and its Marxist logic, which sees American television as an agent of American capitalism and ideology. Cultural studies is more interested in how the viewer appropriates the ideas or dominant ideology of TV than to understand the complexity and contradictions of the institution. At best, art can only be produced outside TV or by an outsider subverting its logic. Richard Caves (2000, p. 4) argues that there is an impossible underlying assumption here:

147

"Imagination and passion carry their own warrant and should not compromise with reason and established practice. Successful imitation of a master, once considered a worthy achievement, becomes an act of cowardice and sloth." Such idealism leads to an impasse in the production of art:

> Asked to cooperate with humdrum partners in some production process, the artist is disposed to forswear compromise and to resist making commitments about future acts of artistic creation or accepting the limitations on them. The rub is that resources are scarce, and compromise is hence often unavoidable. Rejecting it on principle distracts one's mind from making the best deal available. (p. 4)

As a result, art has always claimed a purity of vision and ultimately superiority over craftsmanship in graphic arts, advertisements, and the film and television world.

The understanding of TV production remains caught between the naiveté of the auteur theory, which belies the constraints of an industrial form, and the absolutism of a purist definition of art. Upon first thought, comparison between our traditional image of television makers as designer water-sipping dealmakers and this Byronic image of creativity seems farfetched. Just the disparity in aesthetic aspirations and income between a poet and an executive producer makes such comparisons near impossible.

Raymond Williams (1995) offers a less quixotic explanation of creativity based on the growth of art in a marketplace economy. But instead of tracing the discourse of the artist as individual, he outlines the stages of the loss of the artist's control of the artwork under capitalism. Williams describes the independent writer of the eighteenth century as an artisan who had a measure of creative control. He was "wholly dependent on the immediate market, but with its terms his work remains under his own direction, at all stages, and he can see himself, in this sense as independent" (pp. 44–5). However, as corporations such as publishing houses and newspapers grew in the 1800s, the content and form of the artwork began to be prescribed by the needs of the market. Slowly, the artist moved from directly selling the work him or herself, to taking on a "distributive intermediary" – a firm that distributes the work. In this post-artisanal phase, Williams maintains that this company became his/her factual or occasional employer, shaping the content and form of the work. By the mid-1800s, the writer (less so, visual artists) had evolved into a market professional. Copyright not only established an "individual" as the creator, but, along with royalties, it created a contracted and dependent relationship with a corporation. Williams argues: "the newly typical relationship was a negotiated contract for a specific form or period of publication, with variable clauses on its terms and duration." Royalty replaced outright purchase. Now the writer was directly involved in the salability of her book by receiving a percentage of the profit for each book sold. The creation of the work was more likely to be framed by the needs of the market. Not only did the press dictate the length and type of work, but the writer's income was also dependent on the popularity of the work.

This dependency on the market presaged the final and current phrase of artist in the marketplace – the corporate professional – the term Williams applies to television makers. Here, the writer or artist is wholly an employee of a corporation. This relationship is characteristic of the magazine writer, the graphic artist, the photographer, and the TV writer and producer. In prior markets, the work originated in the commission. "But in the corporate structure this has become very much more common – the direct commissioning of planned saleable products has become the normal mode" (p. 52). Williams maintains that this situation – the artist within the corporation – dominates cultural production in the twenty-first century with the growth of highly capitalized forms of art production, such as filmmaking, recorded music, new media, and, centrally, television. "The scale of capital involved, and the dependence on more complex and specialized means of production and distribution, have to an important extent blocked access to these media in older artisanal, post-artisanal and even market professional terms, and imposed predominant conditions of corporate employment" (p. 53). The older arts – painting, sculpture, poetry, and orchestral music – have maintained a more traditional production process but one dependent on the largess of government and foundation grants. Nevertheless, these backers have become increasingly scarce and narrow in their definition of acceptable art.

The "creative industries" (as they are often referred to these days; see for instance, Caves, 2000) have absorbed much of artistic practice and have become powerful social institutions. Williams points to the advertising agency – the creative corporation par excellence – that once stood on the margin of culture. But we also can parallel television to advertising as dominant social agencies where creative individuals exert unprecedented authority. Today, they are powerful capitalist agencies exerting a central influence on the arts, politics, and the economy. They not only set consumer trends but also exert considerable influence upon both politics and our social priorities. The narrow definition of art as "art for art's sake" does not apply to advertising or television. But one might consider how, in a commercial world where thousands of programs air and are never seen again, the concept of artistry or authorship produces an aura of originality and a greater social value for the single episode as a one-of-a-kind creation not to be missed or, to quote NBC's fame slogan, "Must See TV."

Prime-Time Drama and Authorship: *Law and Order*

So how might we understand the creative origins of commercial television while recognizing its manufactured basis? The genre that is most closely associated with authorship is prime-time drama. There is no clearer instance of a prime-time drama being both turned into a commercial franchise and also still associated with an individual maker than Dick Wolf's *Law and Order*. The series and spin-offs (*Law and Order: Criminal Intent*, *Law and Order: Special Victims Unit*, and *L.A. Dragnet*) exemplify the tension between commercial imperatives. The original *Law and Order* – one of the longest-running series, with over 300 episodes – was clearly conceived as a product for American corporate television in 1988. The series depends on no "name" creative individuals. Dick Wolf – the "creator" – functions as the business executive producer, as a former advertising executive who conceived of the "concept." His genius was that the program did not depend upon a traditional creative name. *Law and Order*'s makers are interchangeable – actors, writers, and directors have rotated in and out of the program for over a decade. They are the embodiment of classic corporate professionals.

Wolf's status as "creator" originated in his ability to "pitch" a remake of a 1960s series for a 1980s milieu. He presented CBS with the idea of a one-hour self-contained crime program that was split into two parts – law (police investigation) and order (the court system). This formal conception was based on the extended profitability of repeat viewing and in the second market of syndication of cable, where it could be shown in any order and in half-hour segments. This second market appeal led the NBC network to take over the program in 1990, with Universal Studios as the production studio. It built its audience through repeat showings and thus succeeds as one of the Nielson's top ten after nearly ten years.

Law and Order results from its rewriting of popular genres – the conventions on which commercial television draws its formats and series' structure. Foremost, the detective genre – a staple of prime-time television – serves as the pattern. The actual two-part structure was lifted from a 1963–66 detective series called *Arrest and Trial*, in which Ben Gazzara as a police officer caught the criminal during the first half, and Chuck Conner as a public defender went to court to defend the perpetrator during the second half. This sober detective style, with its focus on the detection and prosecution (rather than the private lives of the officers and lawyers), has a long history dating back to Sherlock Holmes to *Dragnet* to *The FBI*.

Historically, prime-time dramas get their status and structure from the early TV anthology series, when TV drama was seen as an adjunct of the theater and worthy of respect and the affixing of an author. Its status as the most respected or "serious" of fictional TV offerings is underscored by its placement before the news in the program schedule. In fact, the public and academic writers have taken TV dramas quite seriously, with entire critical studies being devoted to individual programs. Among the most notable of recent years have been Robin Roberts's *Sexual Generations: "Star Trek, the Next Generation" and Gender* (1999), Stephen Holden's *The New York Times on The Sopranos* (2000), Toby Miller's *The Avengers* (1998), and Julie D'Acci's *Defining Women: Television and the Case of Cagney & Lacey* (1994).

These prime dramas draw from a number of established literary and filmic genres for their larger logic – melodrama, hospital, detective, police, and law. Critics have also labeled them "professional dramas" because they combine highly detailed renderings of a professional setting and the drama inherent in that profession. Yet these TV series draw from the larger cultural tradition of paraliterature or popular fiction. These are the books that populate the shelves of today's airport and drugstore book counters – the work of authors such as John Grisham, Sue Grafton, Barbara Cartland, Tom Clancy, and Anne Rice. It is not surprising that *ER* advertises "Michael Crichton" – one of the most prolific paraliterature producers – as one its creators, even though he has done little more than write a novel and co-write the pilot on which the series is based. He serves as a mark of novelistic "quality" and generic conventions associated with paraliterature that pull in an audience.

Even though Dick Wolf maintains that the *Law and Order* series' success depends on its writing ("It's always the writing. There has been 17 actors in the cast, and they're all really good actors, but they don't make up the lines."), the writers function as interchangeable parts – the hallmark of mass production. Much like para-novels, TV dramas are variations on a standardized narrative pattern that the author initiated in his or her first works (or pilot). Most often they deal with the workings of a profession (legal, medical, military), a historical period, or a scientific world. The pleasure comes from the familiarity of their structure, but more so from the precision by which they describe and give us insights into the workings of a specific world. Charles Elkins and Darko Suvin (1979) see paraliterature as a consequence of the rise of the commercial market-place for writers in the nineteenth century. Much like the corporate television writer, who either lives a gypsy-like existence moving from series to series or does not work at all, the nineteenth-century writers became "driven by market demands": "Some integrated into affluent bourgeois life; others lived in garrets and eked out livings as hack writers for firms bent on capturing the mass market with its insatiable mechanisms of ephemerality and quick turnover." The popular writer had lost control of the creation of his/her work; its core came from the needs of profitability.

Like all of fictional TV, *Law and Order*'s writing comes from a steady diet of new writers, who turn over yearly as the series searches for new ideas and the deviation necessary to keep the series from falling into sameness or developing as a true mass-produced product. The executive producer (or the head writer, not Wolf) serves to maintain the series' logic, as each episode becomes a classic example of the definition of a generic work: a variation on a norm.

Much like early popular writers, such as film scenarists who turned to popular magazines, newspapers, and traveling theater dramas (such as *Uncle Tom's Cabin*) as models, *Law and Order* uses familiar newspaper narratives with its "ripped from the headlines" content. According to Kristin Thompson (2003), the method involves "simplified versions of classical notions of what constitutes a story." In particular, this technique follows Aristotle's strictures concerning beginnings, middles, and ends, as well as his views on unity. Slowly, these techniques became codified within the Hollywood studio system in the 1920s through 1950, and this was the model inherited when television began to evolve in the 1940s and 1950s. Television drama adapted its storytelling from the Hollywood system because the method has "been so suited to telling straightforward, entertaining stories" (p. 19). Bordwell et al. (1985) have called this normative system the classical Hollywood narrative. It depends on a unified narrative where events happen in a clear cause-and-effect manner. Action proceeds from a goal-oriented character that motivates the causes inherent in the action through the character's desires. In television, these conventions become even more formalized and normative. Consider how easy it is to render the template for a *Law and Order* script:

> Act 1: Before the first 15 minutes pass, two homicide detectives . . . are asking questions and taking down names. In the first segment, the search for suspects has begun in earnest. Cut to the commercial.
> Act 2: Solving the crime is never easy, and the detectives generally run into some hurdle, some complication, some aggravation (this is *drama*, after all). At least one visit with the boss . . . helps clarify things, so by the half-hour, an arrest has been made. Before going to the next break, the "Law" segment has concluded.
> Act 3: It's in the hands of the lawyers now. Executive Assistant D. A. McCoy and assistant district attorney Southerlyn are preparing for trial. But typically moral or ethical matters need to be resolved. D. A. Branch sometimes provides the impediment, sometimes the solution. Either way, by the end of the third segment the case seems in jeopardy.
> Act 4: Any lingering issues from the previous act are brought to resolution. Perhaps McCoy and the more impulsive Southerlyn are at odds, or there could be something Briscoe or Green did or learned in their investigation . . . but by now it's time for the trial. By the time the hour's over, a judgment has been rendered. "Order" has been restored. (Lowry, 2003, p. E25)

Innovation – the changes in each script – constitutes the "art" of television writing. But one must factor in the constraints, even those beyond the popular conventions, which allow a drama such as *Law and Order* to be so highly profitable. There is the continual threat of censorship from advertisers and viewers. Network executives send "notes" or suggested changes for each episode, thus enforcing popularity and profits. Union rules control who can do what and when; the assembly line is highly regulated and rationalized. Network scheduling affects the length of an episode, but also the environment in which it is received and therefore the content. Consequently, the degree for creative variation is narrow. The program's talent revolves, maintaining a returning audience through the familiar frame yet subtly crafted differences. Although highly skilled, this work is much more akin to a craft than modern definitions of art, which are informed by concepts of genius, originality and expressive individualism. An idiosyncratic authorial signature would undercut the logic of the pleasure of television, a game of predictability and slight deviation.

American television is the child of an industrial system, not art. It emanates out of the crafts tradition where skilled workers or corporate professionals produce works that are based on "a body of rules or techniques" much like their

Renaissance counterparts. Commercial TV is in a grand tradition of custom production where items are "individually crafted for the purchaser, made singly to discrete specifications" (Scranton, 1997, p. 10). Parallels can be found in the jewelry, cabinetry, tailoring, bakery, and catering industries. This manufacturing category involves a tension between standardization and attention to specific customer needs – similar to a genre work as a variation on a norm. Although there is creativity and thought involved in the variations we call episodes and series, we misread its logic if we understand it as a form of personal expression or originality. Nor is this the bulk production associated with consumer goods, such as canned goods or even autos. But what *Law and Order* – and now its spin-offs -make clear is the power of a brand name. These works are commodities, which are designed in a highly regulated manner, without any background story, to be seen repeatedly to maximize their lives and profits (residues). Once we jettison the belief in the Romanticized artistry of these works, we can get down to the business of mapping the constraints and understanding of the possibilities of innovation under the commercial imperative of television.

References

Alexander, D. and Bradbury, R. (1994) *Star Trek Creator: The Authorized Biography of Gene Roddenberry*, New York: Roc.

Allen, R. (ed.) (1992) *Channels of Discourse, Resembled*, Chapel Hill, NC: University of North Carolina Press.

Boddy, W. (1990) *Fifties Television*, Urbana, IL: University of Illinois Press.

Bordwell, D., Staiger, J., and Thompson, K. (1985) *The Classical Hollywood Cinema: Film Style and Mode of Production to 1960*, New York: Columbia University Press.

Burke, S. (ed.) (1995) *Authorship: From Plato to the Postmodern*, Edinburgh: Edinburgh University Press.

Buscombe, E. (1980) "Creativity in Television," *Screen Education*, 35 (Summer), 5–17.

Buscombe, E. (2001) "Ideas in Authorship," in J. Caughie (ed.), *Theories of Authorship: A Reader*, London: British Film Institute, pp. 22–34.

Caughie, J. (ed.) (1981) *Theories of Authorship: A Reader*, London: British Film Institute.

Caves, R. (2000) *Creative Industries: Contracts between Art and Commerce*, Cambridge, MA: Harvard University Press.

Coward, R. (1987) "Dennis Potter and the Question of the Television Author," *Critical Quarterly*, 29(4), 79–87.

Creeber, G. (1998) *Dennis Potter: Between Two Worlds*, London: Macmillan.

Creeber, G. (ed.) (2001) *The Television Genre Book*, London: British Film Institute.

D'Acci, J. (1994) *Defining Women: Television and the Case of Cagney and Lacey*, Chapel Hill, NC: University of North Carolina Press.

Elkins, C. and Suvin, D. (1979) "Preliminary Reflections on Teaching Science Fiction Critically," *Science Fiction Studies*, 6(3), November. http://www.depauw.edu/sfs/backissues/19/elkinssuvin19art.html.

Foucault, M. (1994) *The Order of Things: An Archeology of Human Science*, New York: Vintage.

Foucault, M. (1995) "What Is an Author," in S. Burke (ed.), *Authorship: From Plato to the Postmodern*, Edinburgh: University of Edinburgh Press, pp. 247–62.

Geertz, C. (1973) *The Interpretations of Cultures*, New York: Basic Books.

Jane M. Shattuc

Gitlin, T. (1985) *Inside Prime Time*, New York: Pantheon.

Holden, S. (2000) *The New York Times on The Sopranos*, New York: Ibooks.

Kuney, J. (1990) *Take One: Television Directors on Directing*, New York: Greenwood.

Lowry, B. (2003) "How 'Law and Order' Rewrote the Rules," *Los Angeles Times*, May 18, pp. E1 and E25.

Marc, D. (1984) *Demographic Vistas: Television in American Culture*, Philadelphia, PA: University of Pennsylvania Press.

Marc, D. (1997) *Comic Visions: Television Comedy and American Culture*, Malden, MA: Blackwell.

McGrath, C. (2003) "Law & Order & Law & Order & Law & Order & Law & Order . . . ," *New York Times Magazine*, September 21, 48–51.

Messenger-Davies, M. and Pearson, R. (forthcoming) *Screen, Big Universe: Star Trek as Television*, Berkeley: University of California Press.

Miller, T. (1998) *The Avengers*, London: British Film Institute.

Newcomb, H. (1974) *TV: The Most Popular Art*, New York: Anchor Press.

Roberts, R. (1999) *Sexual Generations: "Star Trek, the Next Generation" and Gender*, Champaign, IL: University of Illinois Press.

Rosaldo, R. (1989) *Culture and Truth: The Remaking of Social Analysis*, Boston: Beacon Press.

Sandeen, C. A. and Compesi, R. J. (1990) "Television Production as Collective Action," in R. J. Thompson and G. Burns (eds.), *Making Television: Authorship and the Production Process*, New York: Praeger, pp. 161–74.

Scranton, P. (1997) *Endless Novelty: Specialty Production and American Industrialization, 1865–1925*, Princeton, NJ: Princeton University Press.

Self, D. (1984) *Television Drama: An Introduction*, Houndmills, England: Macmillan.

Sewell, W. H. Jr. (1992) "A Theory of Structure: Duality, Agency, and Transformation," *American Journal of Sociology*, 98 (July), 1–29.

Thompson, K. (2003) *Storytelling in Film and Television*, Cambridge, MA: Harvard University Press.

Thompson, R. (1996) *Television's Second Golden Age*, Syracuse: Syracuse University Press.

Tomashevsky, B. (1995) "Literature and Biography," in S. Burke (ed.), *Authorship: From Plato to the Postmodern*, Edinburgh: Edinburgh University Press, pp. 81–9.

Williams, R. (1995) *The Sociology of Culture*, Chicago: University of Chicago Press.

Woodmansee, M. (1994) *The Author, Art, and the Market: Rereading the History of Aesthetics*, New York: Columbia University Press.

PART FOUR

Television/The State and Policy

CHAPTER EIGHT

Who Rules TV?
States, Markets, and
the Public Interest

Sylvia Harvey

The answer to the question "Who rules television?" depends partly upon the empirical observation of particular television systems and partly upon the conceptual approach adopted by the observer. The word "rule" suggests the exercise of power, the activity of controlling, governing, or dominating. It is a problematic term in this democratic era when "rulers" are expected to be appointed by and answerable to the "ruled," but where a widespread skepticism about "who runs the show" follows from the observation that accountability is more honored in the breach than in the observance.

It is an equally problematic term in the field of media studies where the word "rule" can be little more than a kind of metaphor for thinking about the complex interplay of freedom and constraint in the making, showing and watching of television programs. Depending upon who is asking the question, the answer to "Who rules?" may be the government, the investor, the owner, the manager, the scheduler, the commissioning editor, the program maker, the spectator, or the customer.

This chapter will consider the role of the state as "ruler" in the sense that the state normally provides a legislative framework for television and sometimes also a regulatory body that – formed under statute – may have responsibilities in respect of media ownership, structure and content. However, it is important first to situate the idea of "rule" by government in the context of other explanations about "how television works" or "how television should work." Two alternative explanations will be considered. The first examines the power of spectators considered as customers or consumers, and the second emphasizes the role of owners and investors.

While a rounded view of television as industry and as culture requires an appreciation of these different explanations, there is also a sense in which we are considering competing paradigms and different (often opposed) political views. From one viewpoint television is seen as an institution that contributes (or should contribute) to the good of society as a whole: television programs are thought to enhance cultural expression and to strengthen informed political participation. From this perspective some programs will be thought to exhibit "non-market" qualities. From a contrary viewpoint programs are seen primarily as entertainment or leisure commodities. And the objective of creating a mature market is designed to ensure that all production revolves around the choices made by individual consumers in selecting commodities and in deciding whether to consume larger or smaller amounts of them. This second viewpoint also has implications for theories of ownership, since owners of the meaning-making machine may be thought to make only what consumers want. A contrary view asserts that owners, in setting agendas and in exercising editorial control over content, provide only a limited range of choices for viewers. Most "political economies" of culture argue that there is a link between ownership and the control of content, while theories of consumer sovereignty argue the opposite: that owners only make what customers want.

The following sections of this chapter explore the propositions that "consumers rule" and that "owners rule" before returning to the main issue – namely, the extent to which and the ways in which the state can be said to rule television.

The Power of the Consumer?

For some commentators, the freedom of consumers requires the absence or minimal presence of government. The needs of consumers are better served, it is argued, by competition between suppliers than by any form of government intervention designed to maintain standards, quality, or diversity. The marketplace itself is believed to be virtuous in providing sufficient choice and in ensuring customer satisfaction, and competition between companies is thought to ensure that owners will be sensitive to the demands of consumers. For other commentators, the state has a duty to act to protect the interests of consumers and citizens, in all spheres – and particularly in the fields of information and culture. The defense of the interventionist state in a market economy relies upon arguments about the public interest and the general good. But it may also, paradoxically, draw upon the philosophy of individualism and the proposition that individual consumers have, in practice, relatively little power in the marketplace and that large corporations exercise overwhelming and unaccountable power. The basic argument about the role of the state in such a context was clearly outlined by an early twentieth-century American president. In his autobiography Theodore Roosevelt suggested that:

> [a] simple and poor society can exist as a democracy on the basis of sheer individuals. But a rich and complex industrial society cannot so exist; for some individuals, and especially those artificial individuals called corporations, become so very big that the ordinary individual is utterly dwarfed beside them, and cannot deal with

them on terms of equality. It therefore becomes necessary for these ordinary individuals to combine in their turn . . . through the biggest of all combinations called the government. (quoted in Tracey, 1998, p. 286)

This notion of government considered as the proxy for a combination of individuals – banding together to obtain better outcomes in their dealing with large corporations – remains an interesting one, although the high success rate of corporate lobbyists in influencing new legislation might suggest that the opposite is the case. Nonetheless, the general pertinence of the argument remains whether the corporations concerned supply fruit, electricity, oil, audiences (to advertisers) or television programs (to viewers). But the application of the model in the field of communications soon finds itself entangled in the issue of free speech rights, since for some the free speech rights of corporations must be defended as vigorously as those of individuals. Moreover, within an economic frame of reference, the emergence of a spatial metaphor for free speech – the "marketplace of ideas" – has tended to reinforce the role and rights of corporations rather than those of individuals. For individuals frequently exchange ideas outside of the market (families and friends inform and entertain each other, at no cost, on a daily basis) as well as making purchases within it.

Nonetheless, the concept of the virtuous marketplace – the "marketplace of ideas" in the fields of information, culture and even political debate – remains influential and opposition to government intervention remains strong. For some the mechanism of the market remains sufficiently robust to ensure choice and diversity without external intervention or assistance. In Europe the institution of "public service broadcasting" might be thought to contravene the norms of competitive market provision since it is often based on government underwriting of costs, or on government support for a universal "licence fee" payment to support the service. An addendum to the European Union's Treaty of Amsterdam (1997) establishes the right of public service broadcasting to exist, but acknowledges the principle that it should not be permitted to have negative effects on the conduct of free trade in an open market. The complex wording of this short "Protocol" tries to combine free market principles with the right of state intervention:

> The provisions of the Treaty establishing the European Community shall be without prejudice to the competence of Member States to provide for the funding of public service broadcasting insofar as such funding is granted to broadcasting organisations for the fulfilment of the public service remit as conferred, defined and organised by each Member State, and insofar as such funding does not affect trading conditions and competition in the Community to an extent which would be contrary to the common interest, while the realisation of the remit of that public service shall be taken into account. (Goldberg, Prosser, and Verhulst, 1998, p. 19)

The Protocol is the outcome of a long and intense lobbying process in which business and citizen advocates often found themselves on different sides. Thus we may see it as both an important and authoritative statement and as the expression of a political compromise. For advocates of a free market in broadcasting the statement is taken to recognize and uphold their principles. For advocates of government intervention, and of public service broadcasting, it is seen as a major policy achievement designed to protect indigenous forms of communication and culture in an increasingly global television market. The differences of opinion embodied in the Protocol continue to be reflected in the outcomes of public policy and in court judgments affecting the audiovisual sector, just as the tensions around the propriety and extent of state intervention continue to be addressed by politicians, voters and scholars (Tongue, 2002).

It may be useful at this point to draw out in more detail the links between free market philosophy and its neo-liberal principles and the concept of choice in the "marketplace of ideas." It has been a key tenet of liberal individualism that the press should remain free of government control, and it is possible to trace in various countries the point at which the licensing of the press was ended (in 1694 in Britain, for example, and from 1791 in the United States, where the new republic enshrined the principle of press freedom in the Bill of Rights). However, the right of owners to establish publications free from government interference is sometimes assumed to guarantee the freedom of consumers, and to imply that readers or viewers will be provided with exactly that range of ideas and stories that they wish to find.

The principles of free speech and of a free press were enshrined in the famous first amendment to the American Constitution: "Congress shall make no law . . . abridging the freedom of speech or of the press" (Wilson, 1993, p. 48). This wording prohibits any negative intervention by the state, but remains silent on the issue of how to promote such freedoms, thus giving rise to a couple of centuries of debate about "positive" (how to enable) as opposed to "negative" (how not to interfere) freedoms.

Of course, a written document defending freedom does not mean that all individuals are equally able to disseminate their views or to persuade others of their value, and there are both economic and cultural explanations for such imbalances in the field of cultural "self-representation." Over the last two centuries significant sections of the population have at times been excluded from education and from literacy – from the "right to read," the time to read and the education required to understand what is read. Despite the struggles of the autodidacts, there have been whole classes of people – slaves, workers, and most women – who have found themselves routinely excluded from the world of learning. And many were to discover that if the freedom of the consumer to choose in the "marketplace of ideas" depended upon the ability to read, it also depended upon the ability to purchase publications and to be aware of their existence.

In the richer countries, the availability of universal, free secondary education from at least the middle of the twentieth century (and, arguably, the availability of public service broadcasting) has improved levels of education and of meaningful access to ideas. But, nonetheless, financial constraints have continued to play a part in the acquisition of a good education and of "cultural capital." Moreover, we can extend this argument about the freedom to read and thus to enter the "marketplace of ideas" to the issue of the freedom to write or, more exactly, to the freedom to represent oneself in the wider public sphere.

How do individuals enter the marketplace of ideas as writers as well as readers? Entry to public platforms (newspapers, television stations) is often jealously guarded for market or political reasons. By contrast, the development of the Internet is bypassing some of these "gate-keeping" constraints and reviving some of the older forms of non-commodity communication. However, the costs of marketing new information or other cultural commodities (in order to build a market that can cover the costs of production) can be prohibitive, even with the assistance of the net. The purpose of these comments about the role of "writing" as well as of "reading" is to underline one of the blind spots of the debate about cultural consumption – namely, the relative absence of consideration of those factors that assist or impede the activity of cultural production. The marketplace of ideas must fail as a guarantor of freedom of expression if it excludes certain sorts of individuals and certain sorts of ideas.

To return to the debate about cultural consumers and consumption, it is also the case that the choice between cultural commodities needs to be meaningful. In this respect, it has been suggested by some commentators that the emergence of multi-channel television, for example, provides an instance of "more of the same" rather than a meaningful choice. The idea has been most famously expressed in the title of Bruce Springsteen's song *57 Channels and Nothin' On* and by other pessimistic observers of the American scene: "Something touted as 'new' is usually a variation on what has been done before . . . rather than a carefully crafted artistic advance" (Sterling and Kittross, 2002, p. 720). This may be unfair to those programs and series that do "break the mold," though these in turn may be seen as the exceptions that prove the rule. Researchers who have looked closely at the operation of choice in practice have emphasized the point that choice occurs where there are distinctive alternatives on offer. As the British legal scholar Mike Feintuck (1999, p. 72) notes: "meaningful choice must be between a range of attractive, desirable, differentiated options; the choice between fifty or a hundred remarkably similar options is scarcely worthy of the name."

Thus, the specifics of choice in a given marketplace require closer examination. The economists Andrew Graham and Gavyn Davies have analysed the television market and noted that its high fixed costs and low marginal costs (one program of high quality is very expensive to produce, but very cheap to distribute) are "the natural creators of monopolies." In their view this tendency to monopoly results in lack of adequate competition, market failure and the absence of meaningful choice. Government support for the provision of public broadcasting services of high quality is seen as one way of ensuring adequate choice and of recognising the sophisticated information needs of citizens in a democratic society (Graham and Davies, 1997, pp. 16–17). The final section of this chapter will explore in more detail the positive role that might be played by the state in this regard.

Graham and Davies developed their analysis in the context of television in Britain, with Davies going on to serve as the chair of the governing body of the BBC, until his resignation in 2004. But the American scholar, Edwin Baker, develops a comparable theorization of distortions in the television market. In his analysis of the actual and potential tensions between media markets and the political and cultural requirements of a democracy, Baker refines the definition of choice to include the key variable of program *quality*. He defines quality in terms of the wider social benefits that may be delivered by television (from an economist's point of view these are "significant positive externalities") and argues that "market-based firms will produce and deliver drastically inadequate amounts of 'quality' media content." In addition he finds that the market: "devotes insufficient resources to creating diverse, quality media desired primarily by the poor and by smaller groups, especially marginalized or disempowered groups" (Baker, 2001, p. 115).

In this Baker reflects and extends the critique developed by other American analysts. Writing in 1978, Erik Barnouw had noted the negative effects of advertiser funding and influence, arguing that "The pre-emption of the schedule for commercial ends has put lethal pressure on other values and interests" (Barnouw, 1978, p. 95). More recently, Sterling and Kittross recorded their concern that "Most programs are . . . produced as inexpensively as practicable" in order to maintain levels of profitability (Sterling and Kittross, 2002, p. 720).

Baker's contention that there is insufficient high-quality programming on American television leads him to challenge the belief that the buying or watching habits of consumers necessarily reflect satisfaction with the product. Instead, he suggests: "it is plausible to conclude that audiences now believe they get too much junk even though they continue to buy it" (Baker, 2001, p. 116). In the light of identified market failures, he feels able to support "interventions to preserve or increase diversity over that which the market would provide" (Baker, 2001, p. 244).

The purpose of intervention, however, needs to be clarified and this is especially true for those who argue that television has a social role and function that takes it well beyond the realms of personal entertainment. James Napoli's recent study of the work of the American Federal Communications Commission (FCC) demonstrates some regulatory blind spots in this regard. Napoli's work suggests that even where there has been regulatory oversight by agencies appointed by the state, these agencies themselves may seem partial or undiscriminating in their approach to the issue of consumer choice, and in their analysis of the actual workings of the market.

Napoli pays close attention to the Commission's use of the term "marketplace of ideas" over time, and notes some changes of emphasis in the use of the term considered as a kind of standard-setting objective or benchmark for regulatory decisions. The phrase has become, he suggests, "contestable terrain in communications policymaking" and he finds that current uses tend to favor deregulatory approaches. Considering the phrase to be a metaphorical description of a complex process he finds a "narrowing" in the use of the term and demonstrates that the Commission increasingly emphasizes "the economic theory dimension of the metaphor over the democratic theory dimension" (Napoli, 2001, pp. 122–3). It follows from this that where "democratic theory" might call for regulatory intervention, "economic theory" appears to support non-intervention and a deregulatory agenda that promotes the perceived interests of broadcasting corporations above those of consumers.

One final point needs to be made about the powers of consumers. In the case of advertiser-funded television, broadcasters clearly have an interest in attracting the audiences sought by advertisers. And in an increasingly fragmented television environment there is, in some respects, a move away from serving a "mass market" or a "family market" and toward the construction of various niche markets. This has been one of the hard lessons learnt by large, mainstream channels in both the United States and the United Kingdom. Although the overall situation is complicated, and perhaps improved, by the presence of other revenue streams: by the development of the subscriber base for cable in the United States and by the large audience share enjoyed by the non-advertising-based BBC in the United Kingdom.

But the issue of "advertiser power" serves to remind us of one peculiar feature of the economics of one kind of television, namely that the true economic consumers are the advertisers (who pay for the airtime) and not the viewers who watch the programs. For some 40 years (from 1955 to the 1990s), the monopolistic, highly regulated, extremely wealthy system of commercial television in the United Kingdom resulted in fairly "light touch" control by advertisers, and in the production of richly resourced and generically varied programs. But this has begun to change as the audience fragments, channels proliferate and competition intensifies. In the United States the regulatory system was never designed to place a kind of "cordon sanitaire" between advertisers and program-makers and here most programming decisions have been responsive to advertiser requirements and pressures, since the emergence of television.

From this short review of the efficacy of the concept of consumer sovereignty in the media field, we may conclude that the choices available to consumers, and the power they exercise over the production process, is limited by a variety of factors. Among these is the fact that it may be the advertisers and not the viewers who are the true economic consumers. It is the advertisers who "pay the piper and call the tune." And it follows from this that the programs transmitted may reflect the objectives and priorities of advertisers at least as much as the wishes of viewers. Although viewer satisfaction may appear to be guaranteed since advertisers wish to fund programs that are successful at attracting particular kinds of viewers, economically unattractive viewers may have little choice but to switch off or to "listen in" on a language that is not really meant for them.

The Power of the Owner?

There is an extensive literature on the role of ownership and investment in the media industries and only a brief summary of some key issues will be offered here. Advocates of private ownership and of minimal state intervention tend to argue not so much in favor of profitability and the payment of high dividends to investors, as in favor of consumer choice in a free market. But the two phenomena – profitability and consumer satisfaction – are connected by an underlying theory. This is the theory or concept of the "invisible hand" of the market, developed by the eighteenth-century philosopher and economist Adam Smith, and it proposes the virtuous reconciliation of these two sets of interests (owners and consumers).

This theory proposes that the functioning of a competitive free market ensures four connected and beneficial outcomes: the absence of any barriers to new entrants; the most efficient use of resources leading to the most cost-effective forms of production; the best services to consumers; and the highest levels of profit for successful companies. The theory also assumes that consumers will act in informed and rational ways in maximizing benefits to themselves. Competition between suppliers, it is argued, ensures both the lowest possible prices and a constant supply of innovation as newcomers enter the market with new and better products. In many areas of economic activity it can be seen to be the case that the most cost-effective forms of production bring the most benefits to consumers. Although the theory is tested and even undermined whenever individual consumers suffer some bruising encounter with an unsatisfactory corporate supplier, or when producers who are also consumers lose their jobs as the corporation searches for a supply of cheaper labor.

Notwithstanding the general theory that efficient outcomes are ensured by the "invisible hand of the market," when we turn to consider the "market in meanings" – the market for symbolic goods – we find that the production of these goods is not so easily brought within the overarching rubric of the rationality and efficiency of the market. There are at least four reasons for this. First, since each television program is unique (these are not substitutable commodities even where there are shared generic features), it is difficult for consumers to know in advance exactly what they are looking for; this is an industry of pleasant and unpleasant surprises. Secondly, since individual programs are still, for most spectators, embedded in a portfolio of channel offerings, the spectator-as-consumer remains relatively unaware of the cost of individual programs. Thirdly, programs may have a social and political impact that goes far beyond their "consumption" by an individual; this is not a factor unique to the consumption of television programs, as the effects of cigarette smoking and the global warming debate demonstrate. And fourthly, the owners themselves have, in some cases, as much interest in the "meanings" as in the "money"; that is to say that the motivations of owners may in part include the wish to exercise social and political influence as well as the desire to make a profit.

It is important to note these reservations about the extent to which the "invisible hand" of the market reconciles the interests of owners and consumers. However, it is also the case that television ratings themselves demonstrate an active process of selection and choice as viewers watch one program and reject another. This is so even where channel controllers operate on the principle of transmitting "what they think the audience wants to hear" and not "what a variety of creative people want to say," thus excluding certain choices at an early stage in the process.

The previous section of this chapter has already suggested that television cannot be thought of as the sum total of individual viewer interactions, but rather, like a pebble dropped into a pool, that it is a medium whose operation has consequences for the whole of society. On the basis of this proposition, it was also suggested that the deficiencies of television may include: a failure to serve the interests of particular audiences, a failure to serve the cultural needs of society considered as a whole, and a failure to serve the information requirements of a functioning democracy. Moreover, it follows from the observation that television has a broad social and cultural role that programs cannot be thought of as exclusively entertainment commodities. They have a wider cultural impact, and in the political arena they may inform, misinform or fail to inform viewers considered in their other roles as citizens and voters.

If these issues are taken into account then this becomes a demanding industry for both private owners and public service providers. Much of the literature on media ownership traces the impact of ownership on content, the consequences of the tendency to concentration of ownership, the problems of cross-media ownership (where the same company might be the dominant provider of newspapers, television and radio stations in one geographical area) and the implications of unequal audiovisual trade flows between countries. As long ago as 1977, the American scholar Erik Barnouw cited the critical observations of a Guyanese writer: "A nation whose mass media are dominated from the outside is not a nation" (Barnouw, 1977, p. 470). Some countries, including the United States, seek to counter this by operating nationality controls so that the owners of television stations must be also be citizens and not foreign nationals.

As regards media concentration and cross-media ownership, most market-based democracies have enacted basic ownership controls in the interests of content pluralism, although it is clear that the deregulatory policies pursued over the last two decades have facilitated mergers and the control of the market by a few very large companies. The US Telecommunications Act of 1996 and the British Communications Act of 2003 both reflect this tendency. Even where there is economic competition between a number of players, Gillian Doyle has pointed out that this may not result in adequate choice for audiences:

> Sometimes, markets that raise no concerns in terms of competition may nonetheless lack the range and diversity of independent voices needed to safeguard pluralism. So, although promoting competition sometimes overlaps with promoting pluralism, these are fundamentally different objectives. (Doyle, 2002, p. 179)

And if technical competition between companies does not always produce pluralism of content, the internal governance of media companies may also result in the suppression of particular voices and values. It is difficult to collect evidence of this, but there are a few published sources that indicate the working of a process, as the following observations indicate:

> Programme-makers feel strongly that real control, artistic as well as financial, has moved further and further away from themselves . . . The people at the networks say they want something fresh, they want something new, they want something different. You come in with something new, fresh and different. You work on it a little more and they say, wait a minute – that's a little too different. (Gallagher, 1982, pp. 168–9)

From these various examples and arguments, it becomes clear that the media industries do not always operate in ways that adequately reconcile the interests of owners, of journalists or creative people, and of audiences. Moreover, while owners and investors may find themselves at quite some distance from the detail of what is made and how it is made, they must also take responsibility both for lack of choice and for negative social impacts.

The Power of the State

If consumers (and program makers) are not able to exercise sufficient leverage on the content commissioning process, and if private owners have interests that cannot always be reconciled with the broader communicative requirements of society, then there may be a role for the state, acting in the public interest to correct the deficiencies of the market. But there are sharp differences of analysis and of political opinion on this issue.

From the perspective of economic liberalism, the state should be no more than a "night watchman," ensuring effective competition and policing the laws of property, but otherwise intervening as little as possible in the affairs of the market. From the perspective of communism or state socialism all resources should be held by and used in the interests of working people, private capital should be abolished and the link between personal wealth and political control broken. A middle-of-the road position, associated with social democracy, tries to facilitate a good standard of living for the majority including public education and healthcare, seeks to diminish the hold of big business on the political class, and acts to enlarge the scope of participatory democracy. The presence of all three positions can be discerned in a wide variety of nation-states in the twenty-first century, along with the complicating factor of religious fundamentalism in the case of Christianity, Hinduism, Islam, and Judaism.

Much, of course, has changed since the emergence of classical liberalism in the eighteenth century. These changes include widespread acceptance of the principle of universal adult suffrage in politics, opposition to slavery, a philosophy of "equal opportunity" (with some continuing disagreements about the role of women) and the payment of income and other taxes to provide public services. But the philosophy of the market itself – as the primary means for allocating and distributing resources – has become deeply and extensively entrenched most especially since the end of the Cold War between communism and capitalism. On the other hand, state intervention in the form of taxation (at between 30 percent and 50 percent of gross domestic product) has become much more economically significant, not only because of military spending but also because of the view that the state should play a more extensive and positive role in minimizing risk and enhancing quality of life for all members of society. This view, of course, remains in sharp contrast to the belief that individuals should "stand on their own two feet" and that the family and private property are the key institutions of modern life.

At stake in the debate about public communication is the link proposed by some between the quality of available public information, the quality of public life, and the ability of citizens to maintain the historical experiment of democratic politics. And this debate is not only about factual communication since fictional representations and what has been called "entertainment" also embody cultural and political values and therefore impact upon the political process and the prospects for democracy.

Since the beginnings of broadcasting most nation-states have reserved the right to control the use of the airwaves by licensing those who use spectrum space. These fixed-term licenses, sometimes justified on the basis of spectrum scarcity and the need to prevent more than one organization using the same wavelength, have constituted a radical limitation on ownership. In addition, the granting of a licence has sometimes been conditional upon the fulfilment of certain public benefit or public interest principles. Only in conditions of civil war (for example in Lebanon or in Rwanda), or where a country is extremely poor, or where the state has claimed exclusive control of the airwaves, has there been an absence of this licensing process according to principles established in law.

With the development of digital compression and of cable, satellite, and Internet services, the amount of communicative "space" has greatly extended and some have argued that the relative end of spectrum scarcity provides, at last, the opportunity for a more complete form of private ownership of the means of communication. In part, this advocacy of an extension of ownership principles to the airwaves has taken the form of a debate about spectrum trading. Individuals and companies, it is argued, should be able to buy and sell frequency space in the way that land and other property has been able to be traded, relatively free of state involvement, for several centuries. There is clearly a strong logic at work here. However, the proposal does not, of itself, address the communicative needs of society, though it may address the issue of economic efficiency in the use of spectrum space. And the tendency to concentration of ownership in the media sector (and, indeed, the takeover of media companies by much larger non-media conglomerates) does not inspire confidence in the prospects for a pluralistic public sphere, accessible to all citizens and capable of meeting their information needs.

Given the extensive commodification of information and entertainment over the last century and a half, it is perhaps not surprising to find in some quarters the assumption that people will buy the knowledge that they want and need. By contrast, the principle of public service broadcasting or of public television was developed in order to ensure universal access to high-quality news and information for all, largely on a non-commodity basis. These services have been operated in different ways in different countries but always at a distance from the market, sometimes without reliance on advertising income and with any surpluses generated reinvested in production.

The advocacy of spectrum trading and enhanced private ownership of the means of communication, like the argument in favor of the deregulation of telecommunications, has tended to emphasize the issues of efficiency and cost in the transmission or carriage of messages. And it has tended to avoid the issue of *content*. This omission may also go hand in hand with an emphasis on quantity (many channels to choose from) not quality (high levels of expenditure on individual programs, wide generic range, pluralism of ideas and freedom for creative producers).

Where content, quality, and social impact are considered to be significant issues then it becomes important to ask of television output: "Who makes it and who controls it?" There is a role here for the democratic state not just as "night watchman," but as a noonday umpire, ensuring and enabling vigorous public debate well beyond the interests of the party in power. This, however, requires the unleashing of new energies and new thinking as well as a rejection of the "bad histories" of dictatorial or monopoly state control.

So, how has the state exercised control in the past? Apart from the almost universal application of a licensing regime, in some countries broadcasting has been directly controlled by the state whether dictatorial or democratic. In the case of India's Doordarshan, for example, radio and television stations were located within a government ministry and broadcasters were employed as civil servants (Rajagopal, 1993, pp. 98–9). Ministerial control was also a feature of French broadcasting up until the 1980s. In Europe, prior to World War II, there were instances of dictatorial, not just ministerial control (for example, in Hitler's Germany and in Franco's Spain), while in Britain the government established a single broadcasting organization, the BBC, in the 1920s providing it with a Royal Charter in 1927 and a measure of independence from the government of the day. (See Graham Murdock's chapter in this volume.)

In the United States, the government established a regulatory regime for largely privately owned broadcasting services giving supervisory responsibility first to the Federal Radio Commission and, from 1934, to the Federal Communications Commission (FCC). Both agencies had the duty to issue licences and the power to uphold the "public convenience, interest or necessity" (Barnouw, 1968, p. 321). In the early days public debate included the expression of some reservations about the use of advertising revenue. In 1924 the Secretary of Commerce, Herbert Hoover, declared: "If a speech by the President is to be used as the meat in a sandwich of two patent medicine advertisements, there will be no radio left" (Barnouw, 1966, p. 177). The tension between advertiser interests, audience needs, and the public interest remains very much a live feature of debates about broadcasting regulation.

State intervention or control has often been seen in negative terms as an unwelcome interference in freedom of speech, and there have been examples of this in a number of countries where journalists have been terrorized, imprisoned and even killed. In China and also in a number of Arab states, governments have maintained a strict control over the content of television. As Naomi Sakr remarks: "While state controls over broadcasting were being removed in other parts of the world during the 1990s, state broadcasting monopolies and strict government censorship remained the norm in most Arab states and in Iran." However, the development of cross-border satellite services has begun to erode these forms of control, introducing what Jordan's information minister referred to in a pithy phrase as "offshore democracy" (Sakr, 2001, pp. 3–4). (See Nabil Dajani's chapter in this volume.) In South Africa, under the apartheid regime, the government maintained a similarly strict control over the content of broadcasting. But the introduction of the first democratic elections in the country in 1994 was preceded by some complex, brave, and imaginative reconfiguring of the South African Broadcasting Corporation. And in 1993 new legislation established an Independent Broadcasting Authority designed to remove broadcasting from the day-to-day control by the government of the day (Maingard, 1997; Teer-Tomaselli, 1995; and Ruth Teer-Tomaselli's chapter in this volume). It is the creation of such an institutional space between the state, considered as the generally legitimating authority, and the government, led by the party in power, that provides a basis for the hope that the state can – where there is the political will – enable the process of democratization, through broadcasting.

We may find a kind of equivalent in the famous 1969 "Red Lion" judgement of the American Supreme Court. Here the Court found in favor of the right of the FCC to continue to implement its "Fairness Doctrine" (designed to support an even-handed coverage of controversial issues) and supported the view that: "It is the right of the viewers and listeners, not the right of the broadcasters, which is paramount . . . It is the right of the public to receive suitable access to social, political, esthetic, moral and other ideas" (Kahn, 1973, p. 426).

By the 1980s the regulatory winds were blowing in an opposite direction and, despite some expression of Congressional alarm, the FCC itself suspended the operation of the Doctrine in 1987. In the strongly neo-liberal climate of the time, the FCC argued that, in part, the Doctrine was defective since it interfered with the free speech rights of broadcasting owners. In this regard, the principle of private property and of corporate free speech rights appeared to weigh more heavily than the principle of public access to reliable and balanced information. On the other side of the Atlantic, however, the British government of Margaret Thatcher – a strong supporter of neo-liberalism in most respects – was legislating to continue the principle of "due impartiality" in British broadcasting. Thus, in Britain, both the Broadcasting Act of 1990 and the "New Labour" Communications Act of 2003 have enshrined the professional practice of impartiality reflecting, in a sense, the principles embodied in the 1969 American Supreme Court judgment.

The due impartiality rule requires British broadcasters to present a range of opinions on the controversial issues of the day and thereby obliges them to serve the interests of society as a whole rather than the political interests of their proprietors or senior managers (Harvey, 1998, 2004). Such a form of regulation may limit the free speech rights of owners but it serves, arguably, as one of the key guarantors of pluralism and diversity in television. Although the requirement tends to be applied more systematically to standards in factual reporting than to the field of fictional representations – where key social groups and experiences are often missing.

A broader approach to the principle of pluralism, and one that also addresses the issue of international trade in cultural services, may be found in some recent work by the United Nations Educational, Scientific and Cultural Organization (UNESCO). In its 2001 *Declaration on Cultural Diversity*, the member states have agreed to the proposition that cultural goods "must not be treated as mere commodities," that "cultural diversity is as necessary for humankind as biodiversity is for nature" and that public radio and television have a responsibility to promote "diversified contents in the media." The Declaration asserts that "market forces alone cannot guarantee the preservation and promotion of cultural diversity" and that, therefore, public policy, "in partnership with the private sector and civil society" has a key role to play in encouraging this diversity (UNESCO, 2001, pp. 2–5). Early in 2004 the organization also adopted a plan to develop a legally binding and enforceable international convention on cultural diversity. This convention or "international cultural instrument" is designed – if it succeeds in coming into existence – to limit the powers of the World Trade Organization (WTO) in enforcing an exclusively market-based approach to the issue of cultural production and cultural trade (Despringre, 2004, p. 6).

These actions taken by UNESCO, as well as the work of non-governmental organizations like the International Network for Cultural Diversity (INCD, 2004), indicate the growth of concerns about both the opportunities for and the threats to pluralism of expression and cultural diversity. The issue, therefore, of "rule" by the state must be seen in the light of these international or supranational developments.

To some extent, the outcome of such developments will be affected by the political complexion of national governments and by the shifting fortunes of the United Nations itself. But it is worth noting that the government of the United States has accorded some recognition to the principle of cultural diversity in its endorsement of the final communiqué issued by the summit of G8 nations in 2000. The "Cultural Diversity" section of this communiqué recognizes "the importance of diversity in linguistic and creative expression" and endorses the view that "cultural diversity is a source of social and economic dynamism which has the potential to enrich human life in the 21st century, as it inspires creativity and stimulates innovation" (G8, 2000, Clauses 39–42). This rhetoric endorsed by the leaders of eight rich nations may be viewed with some skepticism. But we also know that changing rhetorics sometimes emerge in response to shifting balances of power. The devil, as always, will be in the detail of international trade agreements and in the varied prerogatives exercised by media owners and national governments.

Conclusion

The role of the state in regulating television, as well as the content of television output, varies considerably from country to country. International media studies is barely beginning to catch up with this richness and diversity on the one hand and with the enormous difficulties facing developing countries in sustaining the most basic television services on the other. This chapter has tried to address the issue of the role of the state in promoting the public interest within television. In all countries, rich or poor, it is important to ask the question "where is the space for original creative voices?" and "where is the space for dissent?" If the end of spectrum scarcity means that these voices are not heard on the "big media" but are side-tracked onto the small backroads of the Internet (in those countries where citizens can afford access), then the potential for informed voting, well-being and democratic participation is undermined. The role of the state, in democratic societies, must be to ensure that this potential is not threatened, that the biggest and loudest voices are not always the richest voices, that there is shared communicative space for all and that there can be "new entrants" into the world of politics as well as into the world of the market.

References

Baker, E. (2001) *Media, Markets and Democracy*, Cambridge: Cambridge University Press.
Barnouw, E. (1966) *A Tower in Babel: A History of Broadcasting in the United States. Vol. I – to 1933*, New York: Oxford University Press.
Barnouw, E. (1968) *The Golden Web: A History of Broadcasting in the United States. Vol. II – 1933 to 1953*, New York: Oxford University Press.

Barnouw, E. (1977) *Tube of Plenty: The Evolution of American Television*, Oxford and New York: Oxford University Press.

Barnouw, E. (1978) *The Sponsor*, New York: Oxford University Press.

Despringre, C. (2004) "Next Steps in the UNESCO Process to Develop a Convention on Cultural Diversity," *Coalition Currents* 2:1 (February), 5–7. http://www.cdc-ccd.org/coalition_currents/Fev04/coalition_currents_en.html.

Doyle, G. (2002) *Media Ownership: The Economics and Politics of Convergence and Concentration in the UK and European Media*, London and Thousand Oaks, CA: Sage.

Feintuck, M. (1999) *Media Regulation, Public Interest and the Law*, Edinburgh: Edinburgh University Press.

G8 (2000) *Final Communiqué from the G8 Summit at Okinawa.* http://www.g8usa.gov/g8usa/24714.htm.

Gallagher, M. (1982) "Negotiation of Control in Media Organisations and Occupations," in M. Gurevitch, T. Bennett et al. (eds.), *Culture, Society and the Media*, London and New York: Methuen, pp. 151–73.

Goldberg, D., Prosser, T., and Verhulst, S. (1998) *EC Media Law and Policy*, London and New York: Longman.

Graham, A. and Davies, G. (1997) *Broadcasting, Society and Policy in the Multimedia Age*, Luton: University of Luton Press.

Harvey, S. (1998) "Doing It My Way – Broadcasting Regulation in Capitalist Cultures: The Case of 'Fairness' and 'Impartiality,'" *Media, Culture and Society*, *20*(4, October), 535–56.

Harvey, S. (2004) "Living with Monsters: Can Broadcasting Regulation Make a Difference?," in A. Calabrese and C. Sparks (eds.), *Toward a Political Economy of Culture*, Lanham, MD, and New York: Rowman & Littlefield, pp. 194–210.

International Network for Cultural Diversity (INCD) (2004) *Home Page* http://www.incd.net/incden.html.

Kahn, F. (ed.) (1973) *Documents of American Broadcasting*, 2nd edn., Englewood Cliffs, NJ: Prentice-Hall.

Maingard, J. (1997) "Transforming Television Broadcasting in a Democratic South Africa," *Screen*, *38*(3, Autumn), 260–74.

Napoli, P. (2001) *Foundations of Communications Policy: Principles and Process in the Regulation of Electronic Media*, Cresskill, NJ: Hampton Press.

Rajagopal, A. (1993) "The Rise of National Programming: The Case of Indian Television," *Media, Culture and Society*, *15*(1, January), 91–111.

Sakr, N. (2001) *Satellite Realms: Transnational Television, Globalization and the Middle East*, London and New York: I.B. Tauris.

Sterling, C. and Kittross, J. (2002) *Stay Tuned: A History of American Broadcasting*, 3rd edn., Mahwah, NJ: Lawrence Erlbaum Associates.

Teer-Tomaselli, R. (1997) "Moving Toward Democracy: The South African Broadcasting Corporation and the 1994 Election," *Media, Culture and Society*, *17*(5, October), 577–601.

Tongue, C. (2002) "Public Service Broadcasting: A Study of 8 OECD Countries," in P. Collins (ed.), *Culture or Anarchy? The Future of Public Service Broadcasting*, London: The Social Market Foundation, pp. 107–42.

Tracey, M. (1998) *The Decline and Fall of Public Service Broadcasting*, Oxford and New York: Oxford University Press.

UNESCO (2001) *Declaration on Cultural Diversity* http://www.unesco.org/confgen/press_rel/021101_clt_diversity.shtml.

Wilson, V. (1993) *The Book of Great American Documents*, Brookeville, MD: American History Research Associates.

Suggested further reading

Bagdikian, B. (1992) *The Media Monopoly*, Boston: Beacon Press.

Garnham, N. (2000) *Emancipation, the Media and Modernity: Arguments about the Media and Social Theory*, Oxford: Oxford University Press.

Herman, E. and McChesney, R. (1997) *The Global Media: The New Missionaries of Corporate Capitalism*, London and Washington: Cassell.

Hesmondhalgh, D. (2002) *The Cultural Industries*, London and Thousand Oaks, CA: Sage.

Murdock, G. (1982) "Large Corporations and the Control of the Communications Industries," in M. Gurevitch, T. Bennett et al. (eds.), *Culture, Society and the Media*, London and New York: Methuen, pp. 118–50.

Sassen, S. (1996) *Losing Control? Sovereignty in an Age of Globalization*, New York: Columbia University Press.

Schiller, H. (1989) *Culture Inc.: The Corporate Takeover of Public Expression*, New York: Oxford University Press.

Schudson, M. (1998) *The Good Citizen: A History of American Civic Life*, New York and London: The Free Press.

Stevenson, Veronis Suhler (2004) Home page: http://www.veronissuhler.com/

CHAPTER NINE

Public Broadcasting and Democratic Culture: Consumers, Citizens, and Communards

Graham Murdock

Re-imagined Communities

In the spring of 2004 the BBC, the world's best-known public service broadcaster, announced its program plans for the coming year. Faced with mounting criticism from critics who accused the Corporation of moving "down market" to compete effectively with commercial channels, it took the opportunity to reaffirm its commitment to its core principles of underpinning active and informed citizenship, enriching the cultural life of the nation, contributing to education for all, connecting communities, and helping to create a more inclusive society (BBC, 2004, p. 5).

The history of public service broadcasting (PSB) is in large part the story of how these ideals have come to be understood, how they have been institutionalized through a variety of organization forms and mixes of funding, and how they have been continually argued over, contested, and challenged. To fully understand the dilemmas currently facing public broadcasters around the world, however, we need to go back before the age of television and re-examine the origins of modern broadcasting in the years following World War I. The key decisions taken then, the institutions and systems they produced, and the arguments used to justify them continue to define the framework for contemporary policy and debate in fundamental ways.

The idea of broadcasting is underpinned by an image taken from agricultural labor of a sower walking a plowed field dipping her hand into a basket of seeds held in the crook of the arm and throwing them out in a broad arc, in an effort to spread them as widely and evenly as possible. The radio spectrum offered the perfect technology for translating this model of husbandry into the cultural realm since broadcast signals were readily available to anyone who lived within range of a transmitter and had a working aerial and receiving set.

There had been earlier experiments with wired connections, most notably in Hungary, where the Budapest Messenger, launched in 1895, used dedicated telephone lines to distribute a daily service of news, music and talk to 6,000 subscribers around the city. It was still operating in 1918, albeit in a much-reduced state (see Briggs, 1977). Talk of radio telephones continued into the broadcast age, but the idea was never seriously pursued and broadcasting came to be understood as a wire-less system of mass communication using networks of land-based transmitters. Cable connections were developed in Britain and elsewhere, but, as the name of one of Britain's leading operators, Rediffusion, indicates, they were confined to relaying broadcast signals to homes where clear off-air reception was difficult or impossible due to the surrounding terrain. They were not permitted to sell additional services. This technological settlement had important consequences for the political economy of broadcast services.

First, it required the radio spectrum to be centrally managed in order to stop signals interfering with one another. The interruption of military communication by transmissions from enthusiastic radio amateurs had created problems that were of particular concern to governments. Since spectrum space could not be owned it had to be assigned through some form of state intervention. In Britain, this took the form of a public monopoly, the British Broadcasting Corporation (BBC), launched in 1926. In the United States, the main alternative point of reference in debates over broadcasting, the Communications Act of 1934, established a central regulatory body, the Federal Communications Commission, to allocate frequencies and oversee the performance of franchise holders, almost all of whom were commercial companies. They were expected to fulfill some public service requirements but these were minimal.

Secondly, unlike most popular cultural products, broadcasting was not a commodity sold for a price for personal use but a public good in the technical sense used by economists. Whereas a cinema seat could be used only by one person at a time, and patrons might find themselves with a large head or hat or a talkative couple in the seat in front spoiling their pleasure, broadcast programming could be received by everyone at the same time without interfering with anyone else's enjoyment. As John (later, Lord) Reith, the BBC's first Director General, noted in his 1924 book, *Broadcast Over Britain*, "It does not matter how many thousands there may be listening; there is always enough for others, when they wish to join in" (Reith, 1924, p. 217). In his view this potential for universality, when coupled with broadcasting's removal from the price system, made it a uniquely democratic form of popular communication since "There is nothing in it which is exclusive to those who pay more" (Reith, 1924, pp. 217–18). The absence of direct customer payments, however, left only two main ways to finance broadcasting services. They could be funded by public subsidies either in the form of a direct grant from general taxation or an earmarked payment from a dedicated tax, usually raised by requiring set owners to pay a yearly license fee. Alternatively, they could seek payments from advertisers wanting to reach mass audiences in their homes. The choice between these two forms of funding or, in many instances, the balance struck between them, had fundamental implications for the imagined communities audiences were beckoned to join and the way they were encouraged to picture themselves as social agents capable both of remaking themselves and contributing to the common good.

In complex modern societies, everyone is a communard. They belong to multiple imagined communities offering identities anchored in dedicated rituals, narratives, and networks of support. These may be communities of religious belief, locality, political conviction, occupation, gender, age, or style. From the outset, broadcasters sought to use the medium's potential for universality to transcend these particularistic loyalties. The emergence of mass production and mass democracy had generated three master identities that cut across boundaries: worker, citizen, and consumer. The first of these presented difficulties for broadcasters since it was central to the political rhetorics of socialist and communist movements and spoke to forms of collective organization and militancy that many Western governments, witnessing the Bolshevik seizure of power in Russia and widespread labor and social unrest at home in the aftermath of World War I, saw as deeply threatening to democratic order. Even after the immediate crisis had been weathered, the coverage of left-wing parties and trade unions and their access to the airwaves presented continuing problems. The struggle to address mass audiences, therefore, centered on the two other master identities of modernity: consumer and citizen.

The culture of consumerism was most advanced in the United States. In 1913 Henry Ford introduced the assembly line process and started to mass-produce his Model T motor car. With the arrival of the washing machine in 1916 and the refrigerator in 1918, a new domestic landscape opened up promising that everyone could be born again, leaving behind the old struggle to maintain basic living standards and entering the domain of lifestyles in which every market choice was an act of self-expression. The rule of necessity would yield to open horizons of choice. The early marketers were convinced that women, as custodians of the household budget, were in the vanguard of this movement. This made the domesticity of broadcasting a particularly enticing arena for promotion. For most households in Europe, however, and many in the United States hit by the Great Depression, this new consumer landscape remained an unvisited country until the 1950s. By creating an imaginary landscape in which advertising and promotion were integrated into a continuous flow of program pleasures, commercial broadcasting set out to make it the destination of choice when real incomes finally caught up with aspirations.

In Europe, a landscape still recovering from the damage and devastation of World War I, there were other priorities. With the arrival of universal suffrage for adult males and the extension of the vote to women in some (but not all) major countries, it was possible for the first time to talk of mass popular participation in the political process. People were no longer loyal subjects of kings and princes, subjected to forms of rule they had no say in determining. They were citizens with "the right to participate fully in social life with dignity and without fear and help formulate the forms it might take in the future" (Murdock, 1999, p. 8), coupled with the obligation to extend these same rights to everyone else. However, the Bolshevik seizure of power in Russia and the defeat of the armed opposition (supported by substantial number of troops from both Britain and the United States) in the ensuing Civil War, had rekindled the dark images of mob rule. For many commentators of the time, the only sure way to avoid the triumph of the crowd and unreason was to construct a new culture of responsible citizenship.

Model citizens were the exact opposite of members of a crowd. They acted individually rather than collectively. They were open to rational argument rather than swayed by emotion, dedicated to finding non-violent resolutions to conflicts of interest, and prepared to welcome difference and accept dissent. To this end, they were exhorted to seek out information on key social issues, to listen attentively to contending positions, to consider what options for action would enhance the general public good as well as their own individual interests, and to register their decisions in the silence and secrecy of the voting booth. Constructing a culture of democracy that would nourish these habits was seen as essential to managing mass political participation and countering Bolshevism. Once again, broadcasting's potential universality gave it a pivotal role in this project.

Cultivating Democracy

For many commentators, what separated democracy from both the old forms of autocratic rule and the emerging systems of dictatorship, was the fact that decisions were never imposed by fiat but always grounded in a continuing process of open deliberation. Within broadcasting studies, Jürgen Habermas's key concept of the public sphere has been the most influential variant of this ideal (see, for example, Dahlgren, 1995).

For Habermas, democracy is truly deliberative only when it establishes the validity of the norms and values that govern decisions by allowing everyone affected by their application to enter into debate about their validity with the aim of arriving at a provisional agreement. This requires two basic conditions to be met. First, every speaker and position must be accorded an equal opportunity to be heard even if their claims contradict the beliefs of other participants. Secondly, deliberation must be governed by the expectations that speakers will support their claims by speaking truthfully, sincerely, openly, and without coercion. Wherever such discussions concern issues "connected with the practice of the state," whether in casual conversation, a specially convened meeting, or a medium of mass communication, like television, they constitute, for Habermas, "a political public sphere" (Habermas, 1989, p. 231). He sees this space of argument originating in the coffee houses and newspapers of eighteenth-century

London, but admits that it was largely restricted to men who could vote and read fluently and excluded both women and the poor and rapidly eroded by the commercialization of public communications. PSB offered the chance to universalize the mediated political public sphere by providing open access to three essential cultural resources for full citizenship: comprehensive and accurate information about contemporary events and the actions of power holders; access to the contextual frameworks that convert raw information into usable knowledge by suggesting interpretations and explanations; and access to arenas of debate where contending accounts, aspirations, and positions can be subjected to sustained scrutiny. Faced with populations that they saw as essentially ignorant and "untutored," however, public service broadcasters took it upon themselves to teach the skills of deliberation by staging demonstrations. As Charles Lewis, the BBC's first Organiser of Programmes, noted, broadcasting offered the public "an opportunity they have never had before of hearing both sides of a question expounded by experts" (quoted in Smith, 1974, p. 43). Buried in this statement are two assumptions that came to govern much of the actuality programming produced by public broadcasters: that there were always only two major positions on any issues with professional broadcasters acting as a neutral chair; and that the role of the audience was to listen and learn not to speak. This avowedly paternalistic stance generated continuing conflicts around representation in both the senses that term carries in English. There were disputes over who had the right to speak about other people's lives and articulate their views and arguments about the value and relevance of particular program forms as ways of organizing expression and debate. In this process, top-down practices for managing mass political participation have been continually challenged by communities of interest claiming to be neglected, misrepresented, and excluded from the mainstream of programming.

In his more recent reformulation, written partly in response to the rise of new social movements, Habermas re-presents the political public sphere as a space where the issues generated by the myriad interest groups and grassroots movements that mobilize people as communards can secure a hearing. He sees it acting as an early "warning system" picking up the initial tremors of possible social eruptions and thematizing and dramatizing them "in such a way that they are taken up and dealt with by parliamentary complexes" (Habermas, 1996, p. 359). As we will see presently, public service broadcasting has responded to the same mounting pressures from below by developing new forms of representation. The emphasis on "thematization" and "dramatization" in this formulation, however, does nothing to address another underlying problem with Habermas's model of the political public sphere – its emphasis on information and argument.

Again, we see this replicated within the ethos of public service broadcasting with the privileged emphasis given to news, current affairs, and documentary programming as the essential supports for active citizenship. The problem is that the majority of what most people watch on television is not actuality programming but fiction, entertainment, and comedy. In his original work on the public sphere, Habermas points to a second cultural arena, the literary public sphere, centered on the novel. He sees this as offering a space in which people can explore what it means to be human, to be in love, to become ill, to die, and can imagine what it might be like to walk in someone else's shoes. For him, this is a parallel space that has little or nothing to do with the political public sphere. But if we accept that a culture of democracy requires citizens to grasp the links between the good life and the good society and to see their own life chances as inextricably tied to the general quality of communal life, then the habits of sympathy and projection required by fiction and the capacity of comedy and art to decenter established ways of looking, are essential resources. These "affective, aesthetic and emotional modes of communication" constitute a cultural public sphere alongside the political public sphere (McGuigan, 2004).

In Habermas's model of deliberative democracy, however, people participate as discrete, autonomous subjects who know their own intentions, desires, and preferences and set out to realize them by bartering with others. This certainly involves debate in which already formed positions are advanced and defended but deliberation is more open-ended. It presupposes citizens who acknowledge that their understanding is incomplete and who enter into dialog to discover other "partial perspectives that can be woven into a new whole" (McAfee, 2000, p. 182). This politics of "inclining toward and welcoming the other" (op. cit., p. 125) is an essential precondition for the dismantling of stereotypes required by open deliberation. Consequently, developing a democratic culture then entails mobilizing the expressive resources provided across the whole range of programming. For much of the history of public service broadcasting, however, this ideal of a creating a cultural commons skeptical of all entrenched assumptions and open to the exploration of difference and dissent has been in conflict with strategies for managing mass participation by constructing the nation as a unified imagined community.

Securing the Nation

In the immediate aftermath of World War I, the ties binding the imagined community of the nation to the administrative ensembles commanded by states were under pressure in a number of European countries, from both the tenacity of regional identities and the resilience of class solidarities. It was against this background that public broadcasting came to be seen as the key to cementing the primacy of the nation as a source of social solidarity that took priority over localized or sectional loyalties. Under Reith, the BBC introduced a series of invented traditions designed to knit the nation together. They included broadcasting the chimes of Big Ben and the clock on the Houses of Parliament bulletins as well as covering key national sporting events such as the Cup Final and the Oxford and Cambridge Boat Race. The monarchy played a central role in these rituals of unity. As Reith argued in 1925 in his submission to the committee that established the BBC as a public service organization, by bringing the king's voice into every home with a radio set, broadcasting George V's opening speech at the British Empire Exhibition, had the effect of "making the nation as one man" (quoted in Scannell, 2000, p. 48). Later the king was persuaded to give an annual broadcast address to the nation and the empire on Christmas Day, a tradition that has continued down to the present. These broadcasts also helped to support Britain's international position by linking the disparate territories of the empire, particularly the white settler communities, to the imagined homeland, with some success. In their annual report for 1935–36, for example, the Canadian Radio Broadcasting Commission, which had been set up three years before to oversee the development of a national public broadcasting system, hailed the King's Christmas message as "the chief event broadcast in Canada" that year (Vipond, 1994, p. 164).

These periodic celebrations of national (and imperial) social unity were supported by the general promotion of a version of national culture based on the selections already made by established public institutions – museums, libraries, concert halls, and, above all, schools and universities. It was designed to demonstrate how the distinctive qualities of the nation, and by extension of the Western Christian tradition, found their highest expression in works that had entered the official canon. Reith was adamant that one of public broadcasting's central missions was to ensure that "the wisdom of the wise and the amenities of culture are available without discrimination" (Reith, 1924, p. 218), but he took it for granted that what constituted "wisdom" and "culture" would be defined by intellectual and creative elites.

This exercise in cementing distinctions had contradictory effects. On the one hand, it was an openly paternalistic project, which justified the devaluation of vernacular forms of creativity and expression, thereby further compounding problems of representation. On the other, by equalizing access to the cultural capital required for success within the formal education system, it offered working-class children a route to sponsored mobility. For many of its practitioners, however, public service broadcasting was also "educational" in the original Latin sense of "leading out," opening up new horizons and experiences for those who would otherwise be denied them. They saw it as a classroom, museum, library, and concert hall without walls. They envisaged culture as a ladder which people would steadily climb, moving from the lowest rungs of packaged commercial entertainment to the highest rungs of consecrated cultural artefacts. Mixed programming schedules, in which light entertainment or comedy would be followed by a classic music concert or a dramatization of a great play, would convert them by stealth, using their existing tastes as a point of entry to something more "elevated."

The selective celebration of national culture and character was given added impetus by the growing cultural domination of the United States. The global ascendancy of Hollywood and the increasing popularity of jazz in the years following World War I led many observers in Europe and elsewhere to view

American-style commercial broadcasting as one more agent of imaginative annexation. By 1928 the BBC was concerned enough to commission an internal report on the growing reach of the American entertainment industry. Entitled "The Octopus," it floated the idea "that the national outlook and, with it, character, is gradually becoming Americanised" (quoted in Frith, 1983, p. 103). The permeable border between domestic and American culture was even more an issue in Canada where the struggle for public service broadcasting was fought under the slogan, "The State or the United States" (see Raboy, 1998, p. 163) and advertising supported services were presented as suitable only for those "who believe that Canada has no spirit of her own, no character and soul to express and cultivate" (quoted in McChesney, 1999, p. 30). This concern with the vitality of national cultures and languages has generated continuing efforts to defend domestic production by imposing quotas on imported programming.

Institutionalizing an Ideal

Institutionally, public service broadcasting was founded on two core organizing principles: keeping commercial and market pressures at arm's length and ensuring that editorial and creative decisions remained independent from state or government intervention. In practice, both these ideals often proved difficult to maintain in full.

As Charles Lewis argued in 1924, when the BBC was pressing to become a public corporation, PSB "is not Governmental, it would be fatal for it to become the cat's paw of any political policy. It must establish itself as an independent public body" (quoted in Smith, 1974). Supporters of this argument saw PSB as constitutionally like the universities, an institution whose existence can only be guaranteed by the state but one which remains resolutely independent of state direction in determining what cultural activities it will undertake and how. This conception was informed by a deep conviction that institutions in the public domain should be administered by independent professionals in whom "pride in a job well done or a sense of civic duty or a mixture of both" replaces the search for profits (Marquand, 2004, p. 1). Animated by a proper sense of vocation they would ensure that the public interest takes precedence over private interests and "citizenship rights trump both market power and the ties of neighbourhood and connection" (Marquand, 2004, p. 135). It is this philosophy that adds the keyword "service" to "public service broadcasting."

Faced with civil unrest or the need to mobilize popular support behind military action abroad, however, democratic governments could rarely resist pressurizing broadcasters to toe the prevailing political line and speak for the "national interest" as they defined it. As a consequence, the history of the BBC is punctuated with collisions between broadcasters and governments of the day, stretching from the General Strike of 1926, through the invasion of Suez, the Falklands/ Malvinas War and the "Troubles" in Northern Ireland down to the recent bitter dispute over the doctoring of the key intelligence briefing on Saddam Hussein's command of "weapons of mass destruction" that was used to justify Britain's support for the invasion of Iraq. In all these instances the application of pressure stopped short of assuming direct control. Elsewhere, however, governments had no such qualms. In France in the 1930s, news content was orchestrated by a cabinet minister through phone links to publicly owned stations and in 1939 the prime minister placed the whole public network under his own direct management (Smith, 1998, p. 41). With Hitler's seizure of power in Germany, broadcasting became an arm of the Nazi state, and with Franco's victory in the Spanish Civil War, stations were placed in the safe hands of friends and supporters. Keeping commerce at arm's length also proved difficult in many places.

In the years immediately following 1918, most experiments with broadcasting were initiated by amateurs, educational groups, and entrepreneurs exploring the commercial potential of the new medium. The manufacturers of radio sets were particularly active since they realized that people were more likely to invest in their products if they had access to regular program services. In 1922, the British Post Office, which oversaw the use of the radio spectrum, granted the monopoly right to develop program services to a consortium of radio manufacturers, the British Broadcasting Company. It was a temporary arrangement, which owed more to administrative convenience than to conviction. But the tireless proselytizing undertaken by Reith, then the Company's managing director, coupled with the widespread support (stemming from wartime experience) for using public bodies to manage scarce resources, persuaded the 1925 Government Committee that was appointed to decide the future shape of British broadcasting to support the creation of a monopoly public service broadcaster funded by a compulsory license fee levied on set ownership. In 1926 the BBC was converted from a private company to a public corporation and Reith was appointed its first director general.

Across the Atlantic, however, the future of broadcasting was far from settled. Throughout the 1920s a variety of educational, religious, community, and labor groups had experimented with non-profit broadcasting, creating a mounting demand for spectrum space. In an effort to manage this situation, the Radio Act of 1927 established a temporary regulatory agency, the Federal Radio Commission, to allocate frequencies. Their decisions successively marginalized non-commercial initiatives and consolidated control in the hands of the two major commercial players, NBC and CBS. By 1931 their networks of owned and affiliated stations, supported by advertising, accounted for nearly 70 percent of American broadcasting (McChesney, 1993, p. 29). This arrangement was formally endorsed by the Communications Act of 1934 and from that point on "it was clear, and forever the case, until this day, that commercial interests dominate American broadcasting. They have first claim to it, and any public service interests will come only after the needs of the commercial broadcasters have been satisfied" (McChesney, 2003, p. 11). The case for a national publicly funded broadcasting system was also weakened by the near impossibility of ensuring equity of service across such a large land mass. Consequently, "the USA would have found it difficult to impose a European-style receiver licence fee upon viewers even if it had wanted to, simply because of the difficulty (before the era of the satellite) in providing many states with reception of the same programme" (Smith, 1998, p. 40).

By the mid-1930s, then, two major solutions to managing broadcasting had been arrived at, each a mirror image of the other. In the United States, broadcasting was overwhelmingly a market-driven enterprise, run by privately owned companies and dedicated to assembling mass audiences for sale to advertisers. Public service initiatives, where they existed, were pushed to the outer reaches of the system. Conversely, in Britain broadcasting was a public monopoly, established under Royal Charter financed by a compulsory license fee and charged with providing the cultural resources required for full citizenship. American-style commercial programming was available in the south of England but it was very much peripheral, coming from stations located in Continental Europe beaming signals across the English Channel.

These clear alternatives, produced by the leading world powers of the time, influenced thinking on broadcasting across the world. In the countries where the two rival powers exercised special influence, their preferred options were vigorously promoted and exported. Hence many Latin American countries adopted the US solution, while the former British colonies, such as India, were persuaded to follow the British model. Elsewhere, however, necessity and national politics were often as influential as ideology in determining the choice of systems.

The European nations mostly opted for the public service monopoly model, the majority following the BBC in nationalizing an originally commercial system, as in France in 1933, and in Finland in 1934, when the state bought the majority of shares in Oy Ylesradio AB, a company which had been privately owned until then. The single exception was Luxembourg, which developed a purely private system that came to play a crucial role in breaking public service monopolies by broadcasting commercial programming across borders. Elsewhere, however, ideological preferences bumped up against practical constraints. In New Zealand, for example, the relatively small and scattered population meant that no government was prepared to fund the full costs of sustaining broadcast services out of the public purse. The result was a dual system in which public stations competed with commercial operators and hybrid financing with public funding topped up advertising revenues. Elsewhere, public broadcasting came to an accommodation with private investors, particularly newspaper interests, enlisting them as collaborators rather than competitors. This strategy was most fully developed in Sweden where public broadcasting was operated by a not-for-profit organization in which 60 percent of the shares were allocated to popular movements, 20 percent to general business interest and 20 percent to the press. Other departures from the BBC model were shaped by more overtly political considerations.

Although the BBC had operating divisions in the major regions, it was a strongly centralized system directed from London, an arrangement that led to constant disputes over control. In contrast, in Germany, where the process of national unification was still relatively recent, although public programming was partly funded by the compulsory license fee introduced by the German Post Office, station performance was overseen by committees based in each province or Lande. In the BBC system, this responsibility was vested in a Board of Governors, appointed by government from a secret list of the "great and the good." Members of the Board who were responsible for appointing the Corporation's senior managers, including the director general, were expected to leave their particular interests at the door when they hung up their coats and to act in the general public interest. As a way of organizing social representation and popular participation in broadcast governance, this was less than ideal. Other nations tackled the problem of representing interest groups within civil society in other ways, though there was a marked tendency to favor the more established groupings around the main churches and political parties. In the Netherlands, for example, the Catholics, Protestants, Social Democrats, and Liberals were designated as the four main "pillars" of civil society, each being allocated their own stations and a percentage of the total output.

Visions of Plenty: Television and Reconstruction

Television broadcasting had begun in the mid-1930s. In August 1936, the Olympic Games in Berlin were covered by television and three months later a regular service was launched in Britain, but the first transmissions in the United States had to wait until April 1939. Television services were discontinued in both Britain and the United States for the duration of World War II with services resuming in the late 1940s. In this initial period there were opportunities to rethink the organizational arrangement set in place in the radio age. They were not pursued and the earlier structures developed in the radio era were simply transferred across the new medium. The BBC retained its national monopoly. The major American networks maintained their dominance. In Germany, prompted by the desire to avoid the centralization and politicization imposed by the Nazi regime, the Lande were confirmed in their role as the sole sites of broadcast regulation and "pillarization" continued in the Netherlands.

Although television services were up and running in the major European countries (Britain, France, and West Germany) by the early 1950s, the medium was not finally installed across the whole continent until 1960, when broadcasts finally began in Norway. As the medium gained momentum, however, so pressures from private investors wishing to gain access to the growing audiences commanded by public service broadcasting monopolies began to intensify.

Television's expansion in the 1950s coincided with a concerted push towards reconstruction, but once again this movement was imagined in two ways. In Europe, it was understood primarily as a collective project. After the damage inflicted by saturation bombing and occupation, governments promised that living conditions and life chances would be incrementally improved by new welfare states that expanded educational entitlements, socialized housing and health care, and redistributed income. Running alongside this collective project and its imagined landscape of active citizenship and equal entitlement, however, was an alternative vision of emancipation based on expanded personal consumption. This vision traded on images of personal mobility, domestic plenty, and individual style. Commercial programming, either imported from the United States or modeled on its major formats, was the perfect medium for bringing this vision home. As the camera moved seamlessly between advertisements and shots of the stylish kitchens and domestic interiors featured in situation comedies and soap operas, the effect was a "visceral dazzle, an absorbing sense of pleasure in the act of perusal. Costumes. Things. Things to look at. New things. The latest things" (Marling, 1994, p. 5). In market rhetorics, the contest between these competing utopias of the welfare state and the supermarket was mapped onto the Cold War between the United States and the Soviet Union. Personal choice and state management, individualism and collectivism, were presented as implacably opposed.

Visions of Independence: Broadcasting and Nation Building

The struggle between these competing conceptions was most bitterly contested in the "developing" countries that were making the transition from former colonies to independent nation-states. They appeared to many observers as a "Third World," situated uneasily between the First World created by the major capitalist powers and the Second World being developed by the communist powers of the Soviet Union and China. Persuading them to enlist in the legions of the "free world" became a major aim of the Cold War. In regions situated within the US sphere of influence, particularly in Central and Latin America, this aim was largely secured by installing advertising supported channels at the heart of the system. In Mexico in 1947, for example, having considered the relative merits of the British and American systems a government-appointed commission opted for a US-style solution of privately owned stations regulated by federal agencies. The result was structurally national, with foreign ownership prohibited, but culturally strongly oriented to America with extensive program imports. An advertising-free public educational channel, operated by the National Polytechnic Institute, was introduced in 1958, but, as in the US system, it remained very much on the margins.

In contrast, India, which had gained independence from Britain in 1947, saw raising educational standards and cementing national unity as central to the core policy of pursuing economic self-sufficiency. To this end, in 1959 the government established a state-owned television monopoly, Doordarshan, along the lines of the BBC, devoted to nation building. Initially confined to New Delhi, it gradually expanded as the transmitter system was extended, reaching Bombay in 1972 and Calcutta in 1975. That same year, the Indian government became one of the first to mobilize the educational potential of relatively new technology of satellite broadcasting, launching the Satellite Instructional Television Experiment (SITE) in 1975 to beam programmes on topics such as birth control and improved agricultural techniques to more than two thousand villages across the continent.

Commercial Competition and Market Failure

In Europe, too, most countries continued to see PSB as the cornerstone of cultural welfare providing imaginative resources for citizenship alongside the material resources distributed by the emerging welfare states. As commercial pressure for access to television audiences increased, however, cracks began to appear in this ideological wall. In Finland, in a unique accommodation, public service programming was partly financed by selling a section of the broadcast time commanded by the public broadcaster, YLE, to a privately owned company, Mainos Television (MTV). By the mid-1980s, MTV was providing around a quarter of total programming on the two YLE-owned channels. But it was Britain, hitherto the bastion of PSB, that open up broadcasting to commercial interests most comprehensively.

With rising real incomes producing a steady expansion of mass consumerism, business interests exerted mounting pressure on governments to introduce commercial television services. They succeeded and on July 30, 1954, the Independent Television Act formally ended the BBC's monopoly over broadcasting services – "independent" in this usage meaning free from the pressures imposed by state funding. Within this new dual system, however, competition was carefully controlled. The ITV companies were given monopoly rights to broadcast and collect advertising revenues within their franchise areas, while the BBC retained its sole entitlement to the license fee. There was competition for audiences, but not for funding. Moreover, the regulatory body established to oversee the new sector imposed substantial public service conditions on franchise holders, requiring them to produce a range of educational, minority, and other programs that would not have met purely commercially calculations. These commitments imposed a double cost, the outlay on production and the opportunity costs of filling slots with material likely to command lower audiences, and hence less advertising revenue than more immediately popular programs.

Before 1980, the only other European country to develop a fully dual television system was Italy, where Silvio Berlusconi exploited a loophole in the law banning national commercial channels to launch a network of local stations whose near synchronicity of transmissions offered de facto nationwide coverage. On the basis of this initial breach, Berlusconi developed three major national channels – Canale 5, Rete 4, and Italia 1. Ranged against them were the three channels operated by the public broadcasting organization, Radiotelevisione Italiana (RAI). By 1987 the entrenched control formerly exercised by the Christian Democrats had been broken and the channels parceled out among the three major parties, with the first channel going to the Christian Democrats, the second to the Socialists, and the third to the Communists. None of these channels provided objective reporting or open debate and comment, but their biases were at least known. Following a succession of political corruption scandals, however, popular support for this system of *lottizzazione* declined sharply, providing Berlusconi with an opportunity to create a new political party, Forza Italia, in 1994 and to mobilize his channels behind his successful bid to be elected prime minister on a "clean hands" ticket.

Critics of public service broadcasting's paternalism argue that privately run stations have often "fulfilled a great number of the aims seen as unique to PSB" and have "an excellent record in the performance of public interest tasks" (Jacka, 2003, p. 180). This assessment cannot, however, be applied to Berlusconi's channels which became a by-word throughout Europe for their commercial excesses, the poor quality of their programming, and their subservience to their owner's political ambitions. In contrast, the experience of the British ITV companies during the years of regulated duopoly lends this argument strong support. They were populist, but populism has two dimensions. On the one hand, it assembles the broadly based coalitions of interests required by mass advertising by appealing to cultural preferences and worldviews that are already well entrenched and widely shared. It celebrates popular tastes and common sense and derides the "eggheads," know-alls, and busy-bodies set on telling "us" what to do. On the other hand, this distrust of authority also has a radical edge. ITV was populist in both senses. Its best programming was both more rooted in vernacular cultures and everyday lived experience than most of the BBC's output in the years of monopoly and simultaneously more disrespectful and questioning of power holders. This competition revivified the Corporation and opened programming to a wider range of voices and perspectives. This outcome, however, was based on a unique set of regulatory and financial arrangements, which critics wanting to introduce more stringent competition denounced as an "all too 'comfortable duopoly.'" It cannot serve as a general defense of commercial broadcasting's ability to provide comprehensive resources for citizenship.

This is because even under the most favorable conditions, it remains reliant on advertising. It is not simply that the need for mass audiences works against minority representation by mainstreaming programming and pushing it toward the already familiar, accepted, and successful, or that advertisers seek to influence programming in the interests of securing a positive selling environment. Dependency on advertising undermines the core project of providing full resources for citizenship in more fundamental ways. First, by setting aside a fixed amount of time in every broadcast hour for product promotion, commercial broadcasting privileges the rights of commercial speech and constricts the space available to other voices. Secondly, democratic deliberation requires people to trust other participants to be speaking truthfully and sincerely. "Commercial communication promotes the idea that there are no truths, only strategies and claims" (quoted in Wintour 2004, p. 12). Thirdly, advertising invites viewers to see themselves primarily as consumers with a sovereign right to realize their aspirations through personal acts of purchase. It offers them the chance to buy their way out of the social contract. Why support improved public transport if you can afford a car? The practical democracy of the mail order catalog and the supermarket seems to offer more immediate benefits than the politics of the ballot box. Prices appear as the gateways to freedom and taxes as the denial of choice. Private interests take precedence over the public good.

It was precisely these features that led Newton Minow, President Kennedy's appointee as the Chair of the Federal Communications Commission, to give a speech in the spring of 1961, in which he denounced American television as a "vast wasteland" dominated by "game shows . . . audience participation shows, formula comedies. And endlessly commercials – many screaming, cajoling and offending" (Minow, 1964, p. 55). His criticisms provided welcome ammunition for the growing campaign to combat the fragmentation of not-for-profit initiatives and put public broadcasting in the United States on a more secure footing. This pressure eventually produced the Public Broadcasting Act of 1967 establishing a Corporation for Public Broadcasting supported by congressional funds. It was explicitly conceived as a way of addressing market failure by producing resources for citizenship that the major networks did not or would not provide, an ambition later encapsulated in the slogan: "If PBS Doesn't Do It, Who Will?" In particular, it was charged with providing a more open "forum for controversy and debate" and helping "us see America whole, in all its diversity" (quoted in Hoynes, 2003a, p. 122). From the outset, however, these aims were undercut by the funding arrangements.

PBS's money came from three main sources: federal and state grants, corporate sponsorship, and donations from viewers. Each exerted pressures on programming. Because public funding came in the form of discretionary grants, it was continually open to political horse-trading and cuts by unsympathetic political incumbents. Corporate sponsors had vested interests in programming that fitted snugly with their promotional strategies. And "the focus on viewers as direct contributors" encouraged PBS stations "to develop programming aimed at potential *donors* instead of a *public* that is more broadly conceived" (Hoynes, 1993b, p. 43; emphasis in the original), and since regular donors were likely to be better educated and to earn more than average, production was constantly pulled towards their tastes and preferences. This made it difficult for PBS to fulfill its aim of providing "a voice for groups in the community that might otherwise be unheard." The question of how best to represent marginalized groups was not confined to PBS, however. It was becoming a problem for public service broadcasting as a whole.

Representation at Issue

The second half of the 1960s saw important shifts in the composition of civil society. The established organizations, particularly the trade unions and political parties, were losing their purchase on popular loyalties. New communities of interest and points of mobilization were emerging. Some were organized around a revivified sense of place and the rights of minority-language speakers. Others grew out of social movements pressing for environmental protection, extended rights for women, and full recognition for gays. Others again were rooted in the postcolonial experience of migration and diaspora and the struggle to dismantle imperial mentalities and develop multi-ethnic cultures based on equality of recognition and respect. There were two basic responses to these emerging constituencies of communards: new forms of programming, such as the BBC's *Open Door* and *Video Diaries* series, which mobilized these new communards as collaborators rather than simply subjects; and, more ambitiously, new public broadcasting organizations.

In 1977, Australia responded to the gathering influx of immigrants from southern and eastern Europe, Asia and the Middle East by establishing a new public service body called the Special Broadcasting Service. Originally restricted to radio and intended as a multilingual service to the new migrant communities, it established a television service in 1980 and later broadened its mission. As its corporate plan for 1990–3 explained, its aim was to provide "an innovative and quality multilingual and multicultural . . . service which depicts the diverse reality of Australia's multicultural society and meets the needs of Australians of all origins and backgrounds" (quoted in Debrett, 1996, p. 66). But arguably the most ambitious response to the changing landscape of civil society was the launch of Channel 4 in the United Kingdom.

In 1977 a government report on the future of British television had launched a strong attack on the BBC–ITV duopoly for failing to properly represent emerging constituencies of interest. This critique breathed new life into the lobby pressing for the vacant fourth national channel to be allocated to a new, independent force operating as a publisher-broadcaster, commissioning its programs from a wide range of freelance producers, many of whom would have close ties with the new communards. After a heated debate, this vision prevailed and in 1979 Margaret Thatcher's incoming Conservative government gave its backing. It also endorsed nationalist claims for Channel 4 in Wales to carry a substantial amount of programming in the Welsh language.

At first sight, it appears odd that an avowedly neo-liberal administration dedicated to extending the reach of market forces should lend its support to an initiative welcomed so enthusiastically by radical program makers and movement activists. With the benefit of hindsight, however, we can see that two core features of the channel's organization fitted perfectly with the government's general promotion of marketization. First, the reliance on independent production broke the BBC–ITV duopoly on program making and opened up space for just the kind of small and medium-sized businesses that were central to the Conservative Party's vision of enterprise. Secondly, the channel was funded entirely by advertising. Initially the right to sell advertising space was granted to the ITV companies in return for an agreed annual levy paid to the channel. By placing programming at one remove from direct advertising pressure, this arrangement gave commissioning editors considerable freedom to pursue the channel's statutory goals of providing for "tastes and interests not generally catered for by ITV" and encouraging "innovation and experiment in the forms and content of programmes." At the same time, it extended the ITV companies' monopoly control of advertising, and in 1990, in the interests of introducing greater competition, the cross-subsidy system was discontinued and the channel allowed to sell its own advertising. For many commentators, this marked the end of its experimental period and the beginning of its incorporation into the mainstream of commercial services, a transition marked by the virtual disappearance of "diversity," the lynch-pin of its original project, from its corporate lexicon and its replacement by the standard vocabulary of business – "let the viewers decide," "risk-taking," "product quality," and "commerce" (see Born, 2003, p. 782). This was not an isolated instance, but part of a concerted and much more broadly based movement to marketize public service broadcasting.

Marketizing Public Service

In 1980, only two of the broadcasting systems in the 17 major Western European countries, the UK and Italy, were dual systems in which public service channels competed with terrestrial commercial services. The remaining 15 were still public service monopolies, although only four (Norway, Sweden, Denmark, and Belgium) were supported entirely out of public funds. The rest relied on a mixture of public money and advertising revenues. By 1997, ten more countries had introduced dual systems and three more were actively discussing introducing them. Fully funded public service monopolies had completely disappeared (see Siune and Hulten, 1998, p. 27). This dramatic structural shift was a measure of the growing momentum of marketizing across Western Europe, propelled by policies which aimed to introduce more competition into the television marketplace whilst simultaneously relaxing the public service requirements imposed on commercial providers and promoting market criteria of judgement as the yardsticks against which the performance of all organizations, including those still formally in the public sector, would be evaluated (see Murdock and Golding, 2001).

With the collapse of the Soviet Union, the process of marketization was extended to the countries of Central and Eastern Europe. Although idealists saw an opportunity to develop a variant of public service broadcasting by turning "state" into "social" broadcasting directly managed and controlled by society, the deep distrust of public intervention, the weakness of civil society institutions, and the belief that free expression was best ensured by free markets comprehensively undermined this project. The result, depending on the relative strength of competitive democracy, was either the capture of broadcasting by dominant political interests or the installation of dual systems in which the logic of commercial enterprise determined the rules of engagement (see Jakubowicz, 2004).

Public service broadcasting was also under pressure in its major stronghold in the "developing world," India. In the spring of 1990, the AsiaSat 1 satellite was launched with a "footprint" covering 38 countries in the Middle East and Asia. In 1991 a Hong Kong-based consortium took advantage of this new "platform" to launch the Star TV service beaming five channels into India. Its availability was accelerated by the mushrooming growth of improvised cable connections linking communal satellite dishes to city apartment blocks. In 1993 Rupert Murdoch bought a controlling interest in Star and a year later it was estimated that its programming, dominated by American imports, could be received in almost a quarter of all television homes. From the outset, however, it faced stiff competition from Zee TV, launched in 1992, broadcasting in Hindi rather than English, and offering a schedule based around Bollywood films and local adaptations of game show and chat show formats.

The emerging consumer landscape promoted by the new satellite channels ran directly counter to the Doordarshan's historic commitment to nation building. The channel had been permitted to take advertising since 1976, but in a major move the government responded to the changed competitive environment by requiring the channel to generate 80 percent of its operating costs from advertising. Meeting this demand required the introduction of a new programming strategy which sought to maximize its appeal to the new consumer-oriented middle class by concentrating on soap operas, game shows and other populist program forms. In the process, less commercially viable programming was pushed to the margins or eliminated entirely.

This instance of corporatization is part of a more general process of marketization which proceeds along three other broad fronts: privatization, liberalization, the re-gearing of regulation.

Privatization, in which assets and resources previously held in common are sold to private investors, is a direct continuation of the enclosure movements that incorporated common land, open for use by all, into private landed estates (see Murdock, 2001). Comparatively few major publicly owned television stations have been privatized thus far, although the sale of the most popular French public channel, TF1, is a notable exception. However, the two decades between 1980 and 2000 saw all the European PTT's (Post, Telegraph and Telephone organizations) move from being publicly owned utilities to profit-oriented public companies freeing them up to invest in commercial television services. They have been particularly active in new services delivered by cable and satellite, both of which are dual technologies used for both telecommunications and television.

By constructing a third sector based on direct customer subscriptions, commercial cable and satellite services have played a leading role in liberalizing broadcast markets and intensifying competition across Europe. In France, the satellite pay-TV service, Canal +, was launched in 1984, three years before the sale of TF1. This period, 1984–5, also saw the public service monopoly in West Germany broken by the rise of commercial cable and satellite services. In both Norway and Sweden, it was the launch, in 1988, of the satellite channel TV3, operated by the Swedish company Kinnevik, that initially breached the historic national public service monopolies held by SVT and NRK, with national commercial channels only being introduced in both countries four years later in 1992. In the United Kingdom, the major competition to the BBC and ITV companies has come from Rupert Murdoch's Sky TV satellite system. Originally launched in 1988, it achieved a monopoly position when it took over its failing rival, British Satellite Broadcasting, in 1990. Since BSB had been selected to operate the British national satellite service partly on the basis of its commitment to domestic production, the government could have insisted on re-advertising the franchise. However, the consortium's well-publicized failure to shore up its finances by establishing an extended shareholder base among other major players in the country's commercial broadcasting and leisure sectors suggested that credible national bidders were unlikely to come forward and this option was never pursued. Nor was the European Union inclined to intervene. On the contrary, in line with its general policy of opening hitherto protected markets to competition, in 1989 it issued the Television Without Frontiers Directive with the aim of creating a single market for television programs in the European Community. This specified that only the state where the broadcast originated from, not the receiving state, had the right to control programming according its national laws. Uplinking TV3 from London, for example, meant that it did not have to abide by the stricter rules governing television advertising in the Scandinavian countries it broadcast to.

In Europe, the intensified competition for viewers led the established terrestrial commercial channels to pressure national regulators for a relaxation in the rules governing their operations. In Britain, they met with considerable success with both major parties favoring a "light touch" approach that enlarged the space for commercial maneuver in the key areas of ownership and advertising. A succession of mergers and acquisitions progressively reduced the number of separate companies in the ITV system, eventually producing a single consolidated company covering all the major English markets. The rules governing the amount and types of advertising permitted were relaxed and program sponsorship extended. At the same time, arguments for diluting or removing public interest programming requirements gathered momentum. Again, they met with an increasingly sympathetic hearing from regulators. As the Office of Communication, the body now responsible for overseeing both telecommunications and broadcasting in Britain, recently noted, "increased competition for audiences and revenues will continue to place pressure on the profitability of the commer-

Public Broadcasting and Democratic Culture

cial terrestrial broadcasters . . . [and] affect their ability to meet their regulatory obligations in future" (Ofcom, 2004, p. 8), strengthening the case for confining their public service remit to news, regional news, and original UK production (op. cit., p. 10). This pattern has been repeated in the United States with the networks securing an unprecedented relaxation of the rule governing ownership and the cancellation of crucial public service obligations.

At the same time as lobbying for greater freedom of action in the marketplace, however, the commercial television companies have also pressed for more access to the resources historically commanded by public service broadcasters. One outcome has been the requirement that PSB channels commission an increasing amount of their total programming from independent producers rather than making it in-house. Another has been the increasingly vocal argument that public monies for programming should no longer be monopolized by public service organizations.

The rationale for this rests on a sharp distinction between those programs that people in their role as consumers of television most enjoy and those that, as citizens, they value for their contribution to the overall quality of public life. This distinction is underpinned by radically different definitions of what constitutes the "public interest." Whereas the market demand model equates it with what the public is interested in as evidenced by audience size, the social demand model identified it with providing the cultural resources required to define and develop the public good through the active exercise of citizenship (see Raboy, Proulx, and Dahlgren, 2003). As Ofcom noted: "Even if the TV market provided all the programming that consumers desired and were willing to buy, it would probably not offer sufficient programmes that are valued by society as a whole" (op. cit., p. 9). This argument was first outlined in a key report on the future of the BBC commissioned by Margaret Thatcher and published in 1986. Known as the Peacock Report, after its convinced neo-liberal chair, it advocated separating production and distribution. The BBC's monopoly right to a license fee would end and a new body would dispense the public money invested in broadcasting to program proposals that were "supported by people in their capacity as citizens and voters but [were] unlikely to be commercially self supporting in the view of broadcast entrepreneurs" (Home Office, 1986, para. 133). The "programmes of a public service kind" the Committee had in mind included, "critical and controversial programmes," "high quality programmes on the Arts," "avowedly educational programmes," and programs that experimented with new styles and presentational forms (op. cit.). Providing they met these criteria, any producer could make a submission for funds. This idea made little headway in Britain at the time, but it was tried out in New Zealand where the government gave over the license fee to a new body, the Broadcasting Commission (later renamed New Zealand on Air), with a remit "to reflect and develop New Zealand identity and culture by enhancing the range of New Zealand-made programming which reaches our screens" (New Zealand on Air, 1997, p. 1). Although NZOA was "New Zealand's first single-minded advocate for local content" (Horrocks, 2004,

193

p. 31), it was faced with two major problems in fulfilling its remit. First, despite a rise in the Public Broadcasting Fee, the amount of money it commanded was too small and too thinly spread to support the full diversity of domestic public service programming. Secondly, decisions over whether or not a proposal that secured funding would find a slot in the schedules still lay with the stations, and in a fiercely competitive environment, screening a minority interest program entailed substantial opportunity costs in the form of lost advertising revenues that could have been earned from a more immediately popular program placed in the same slot (see Murdock, 1997). Nor could the public service broadcaster, Television New Zealand, ignore this raw economic logic since the same package of reforms that had created New Zealand on Air had converted it to a state-owned enterprise charged with generating as much income as possible for the Treasury.

The New Zealand initiative was an extreme variant of corporatization, the process of requiring or cajoling public enterprises to act as though they were private companies. In more modified forms, this key dimension of marketization was pursued in a number of countries. As the BBC's case shows, however, success in the marketplace generates new pressures. Through its operating subsidiary, BBC World, for example, the Corporation has been increasingly successful in generating income from commercial activities. These include sales of merchandise (magazines, books, records, and toys) spun off from its programs; joint ventures with major US-based commercial partners; program and format sales in overseas markets; and the launch of new advertising-supported "offshore" channels. On the one hand, these initiatives have been welcomed by market-oriented governments as evidence that the Corporation is operating more efficiently, maximizing the returns on its creative investments and generating additional monies that can be used to enhance and extend its national services. At the same time, the more successful they are, the more they erode claims to sole access to public subsidy. As a Parliamentary Committee report on the BBC's future noted, "should the BBC find a new, profitable commercial role . . . it might be very difficult, if not impossible, to justify the existence of a licence fee at all" (National Heritage Committee, 1993, para. 105). This was written in 1993. In the decade since then, the situation has been further complicated by the continuing switch-over from analog to digital systems of broadcasting.

Digital Deliberations

Digitalization, coupled with the growing integration of television and tele-communications, accelerates three major shifts in the broadcasting environment. First, by compressing broadcast signals, it massively increases the potential number of channels available and intensifies competition. Secondly, by enabling programming to be delivered through a range of other devices, from home computers to mobile phones, it breaks the television set's historic control of viewing. Thirdly, it allows broadcasters to develop opportunities for viewers to interact with what appears on the screen, by voting in an instant poll, ordering goods displayed in programs and advertisements, or following the program onto the Internet, interacting with program makers and other viewers or delving deeper into the issues and arguments presented.

For commercial companies, the potentially negative impacts that flow from the intensification of competition and the dissolution of the mass audiences are off-set by the installation of customer payments as the dominant form of channel financing, by the opportunities to create new niche channels that cater for leisure and personal interests with a strong articulation to consumption, and by the chance to develop forms of promotion that combine purchasing and pleasure in new ways by mobilizing the full range of interactive possibilities. Enthusiasts of these developments celebrate the unprecedented extension of freedom of customer choice they see emerging. Their capacity to erode democratic culture is less well publicized. There are three obvious problems. First, funding new channels primarily by customer subscriptions breaks with broadcasting's historic universality and establishes precisely the differences between "first" and "second" class services, or "premium" and "basic" packages, as they are now called, that Reith spoke against so passionately. Access to a full range of cultural resources for citizenship becomes inextricably tied to ability to pay. Secondly, by integrating commercial promotion even more thoroughly into program forms and flows, managed interactivity squeezes still further the space available for other voices and perspectives. Thirdly, while building niche channels around already existing consumption and interest communities makes sound commercial sense, it reduces the possibility that people will be encouraged to enter unfamiliar cultural landscapes and lean toward others in ways that are essential for building the understanding of difference and openness to deliberation that democratic culture depends on.

For many commentators, digitalization abolishes the case for public service broadcasting. They see PSB as an idea whose time is over, overtaken by innovations in technology that render it obsolete. As Adam Singer, a manager with considerable experience of commercial cable services, put it: "The traditional model, using scarce publicly owned air-waves for the benefit of society, does not hold up, once all scarcity is removed" (Singer, 2004, p. 17). This is a convenient misreading of history. As we noted earlier, the organization and ethos of public service broadcasting was always the product of cultural strategies and political requirements as well as technical considerations. This argument still holds. Multiplicity of channels is no justification for consigning PSB to the museum of cultural curiosities. On the contrary, there is a convincing case to be made that for the first time the possibilities opened up by digital technologies allow public service broadcasting to realize its full potential as a thick cultural resource for citizenship and an open space for continuing deliberation across social boundaries. Finding ways to anchor this potential in workable organizational arrangements and new forms of audience engagement is the major challenge facing not simply public broadcasters but anyone interested in developing and deepening democratic culture. It requires two conditions to be met. First, all services offered must remain free at the point of use and as far as possible, free from advertising. Secondly, broadcasters must follow their audiences onto the Internet by establishing themselves as the portal of first resort and by providing access to the widest possible range of resources for information and interpretation, together with spaces for deliberation hospitable to the widest possible range of voices and positions. Public service broadcasters are currently grappling with these challenges. They are launching new digital channels in an increasingly competitive multi-channel environment, often with considerable success. In the autumn of 2004, the first official review of the BBC's moves in this direction concluded that after only two years on-air, its service for pre-school children, CBeebies, was "a triumph" and an "exemplary PSB service" and that its new arts and documentary channel, BBC4, had "successfully established itself as 'a place to think'" (Barwise 2004, pp. 82–3). They are also developing comprehensive websites offering free access to additional resources related to particular programs, issues of the day, and links to sites produced by a variety of social movements and interest groups. These initiatives create, for the first time, an extended and dynamic space of encounters between citizens and communards in which sectional claims can be assessed against the general good and the meaning of the public interest can be re-negotiated. Building on the possibilities offered by this emerging digital cultural commons and using them to reconstruct the relations between television and democratic culture is arguably the central task facing public service broadcasting in the coming years. If it is successful, the present moment of transition will come to mark the end of the beginning rather than the beginning of the end.

References

Barwise, P. (2004) *Independent Review of the BBC's Digital Television Services*, London: Department of Culture, Media, and Sport.

BBC (2004) *The Year Ahead: BBC Statement of Programme Policy 2004/2005*, London: British Broadcasting Corporation.

Born, G. (2003) "Strategy, Positioning and Projection in Digital Television: Channel Four and the Commercialization of Public Service Television in the UK," *Media,Culture and Society*, 25(6), 773–99.

Briggs, A. (1977) "The Pleasure Telephone: A Chapter in the Prehistory of the Media," in I. de Sola Pool (ed.), *The Social Impact of the Telephone*, Cambridge, MA: MIT Press, pp. 40–65.

Dahlgren, P. (1995) *Television and the Public Sphere: Citizenship, Democracy and the Media*, London: Sage Publications.

Debrett, M. (1996) *Public Service Television – Constraints and Possibilities: A Study of Four Systems*, unpublished Masters thesis, Department of Political Studies, University of Auckland.

Frith, S. (1983) "The Pleasures of the Hearth: The Making of BBC Light Entertainment," in T. Bennett et al. (eds.), *Formations of Pleasure*, London: Routledge, pp. 101–23.

Habermas, J. (1989) *The Structural Transformation of the Public Sphere: An Inquiry into a Category of Bourgeois Society*, Cambridge, MA: MIT Press.

Habermas, J. (1996) *Between Facts and Norms: Contributions to a Discourse Theory of Law and Democracy*, Cambridge, MA: MIT Press.

Home Office (1986) *Report of the Committee on Financing the BBC*, London: HMSO, Cmnd 9824.

Horrocks, R. (2004) "The History of New Zealand Television: 'An Expensive Medium for a Small Country,'" in R. Horrocks and N. Perry (eds.), *Television in New Zealand: Programming the Nation*, Melbourne: Oxford University Press, pp. 20–43.

Hoynes, W. (2003a) "Branding Public Service: The 'New PBS' and the Privatization of Public Television," *Television and New Media*, 4(2), 117–30.

Hoynes, W. (2003b) "The PBS Brand and the Merchandising of Public Service," in M. P. McCauley et al. (eds.), *Public Broadcasting and the Public Interest*, Armonk, NY: M. E. Sharpe, pp. 41–51.

Jacka, E. (2003) "'Democracy as Defeat': The Impotence of Arguments for Public Broadcasting," *Television and New Media*, 4(2), 177–91.

Jakubowicz, K. (2004) "Ideas in our Heads: Introduction of PSB as Part of Media System Change in Central and Eastern Europe," *European Journal of Communication*, 19(1), 53–74.

McAfee, N. (2000) *Habermas, Kristeva, and Citizenship*, Ithaca, NY: Cornell University Press.

McChesney, R. W. (1993) *Telecommunications, Mass Media, and Democracy: The Battle for Control of US Broadcasting, 1928–1935*, Oxford: Oxford University Press.

McChesney, R. W. (1999) "Graham Spry and the Future of Public Broadcasting," *Canadian Journal of Communication*, 24(1), 25–47.

McChesney, R. W. (2003) "Public Broadcasting: Past, Present and Future," in M. P. McCauley et al. (eds.), op. cit., pp. 10–24.

McGuigan, J. (2004) "The Cultural Public Sphere," Loughborough University, Department of Social Sciences, Inaugural Lecture as Professor of Cultural Policy.

Marling, K. A. (1994) *As Seen on TV: The Visual Culture of Everyday Life in the 1950s*, Cambridge, MA: Harvard University Press.

Marquand, D. (2004) *Decline of the Public: The Hollowing-out of Citizenship*, Oxford: Polity Press.

Minow, N. N. (1964) *Equal Time: The Private Broadcaster and the Public Interest*, New York: Atheneum.

Murdock, G. (1997) "Public Broadcasting in Privatised Times: Rethinking the New Zealand Experiment," in P. Norris and J. Farnsworth (eds.), *Keeping it Ours: Issues of Television Broadcasting in New Zealand*, Canterbury: Christchurch Polytechnic, pp. 9–33.

Murdock, G. (1999) "Rights and Representations: Public Discourse and Cultural Citizenship," in J. Gripsrud (ed.), *Television and Common Knowledge*, London: Routledge, pp. 7–17.

Murdock, G. (2001) "Against Enclosure: Rethinking the Cultural Common," in D. Morley and K. Robins (eds.), *British Cultural Studies*, Oxford: Oxford University Press, pp. 443–60.

Murdock, G. and Golding, P. (2001) "Digital Possibilities, Market Realities: The Contradictions of Communications Convergence," in L. Panitch and C. Leys (eds.), *A World of Contradictions*, London: Merlin Press, pp. 111–29.

New Zealand on Air (1997) *Local Content Research: New Zealand Television 1996*, Wellington: New Zealand on Air.

National Heritage Committee (1993) *The Future of the BBC Volume 1: Report and Minutes of Proceedings*, House of Commons Session 1993–4, London: HMSO.

Ofcom (Office of Communications) (2004) *Ofcom Review of Public Service Television Broadcasting. Phase 1 – Is Television Special?*, London: Office of Communications.

Raboy, M. (1998) "Canada," in A. Smith and R. Patterson (eds.), *Television: An International History*, Oxford: Oxford University Press, pp. 162–8.

Raboy, M., Proulx, S., and Dahlgren, P. (2003) "The Dilemma of Social Demand: Shaping Media Policy in New Civic Contexts," *Gazette*, 65(4–5), 323–9.

Reith, J. C. W. (1924) *Broadcast Over Britain*, London: Hodder and Stoughton.

Scannell, P. (2000) "Public Service Broadcasting: The History of a Concept," in E. Buscombe (ed.), *British Television: A Reader*, Oxford: Clarendon Press, pp. 45–62.

Singer, A. (2004) "Television in the Digital Age," *The Guardian*, May 3, p. 17.

Siune, K. and Hulten, O. (1998) "Does Public Broadcasting Have a Future?," in D. McQuail and K. Siune (eds.), *Media Policy: Convergence, Concentration and Commerce*, London: Sage Publications, pp. 23–37.

Smith, A. (1974) *British Broadcasting*, Newton Abbot: David and Charles.

Smith, A. (1998) "Television as a Public Service Medium," in A. Smith (ed.), *Television: An International History*, Oxford: Oxford University Press, pp. 38–54.

Vipond, M. (1994) "The Beginnings of Public Broadcasting in Canada: The CRBC, 1932–36," *Canadian Journal of Communication*, *19*(2), 151–71.

Wintour, P. (2004) "Media Blamed for Loss of Trust in Government," *The Guardian*, May 6, p. 12.

CHAPTER TEN

Culture, Services, Knowledge: Television between Policy Regimes

Stuart Cunningham

Despite the claims of electronic gaming – console, video, computer and online – and the explosion in Internet use, television retains its claim to be "overwhelmingly the most pervasive contemporary mass medium" (Collins, 1990, p. 22). Studying television is important because it is both a vastly pervasive popular entertainment medium and also perceived as a key to influence and commercial success in the information age. In their systematic study of the ownership strategies of the biggest players in the ECI (entertainment–communications–information) industries in the 1990s, Herman and McChesney (1997) show how virtually all have moved to acquire or consolidate holdings in television. Strategically, television bridges, partakes in, or provides a major platform for significant elements across the continuum of entertainment (cinema, music, computer gaming), information (journalism, news) and communications (carriage of signals, satellites, broadband cable, Internet) and thus stands at the center of the convergent ECI complex, the most dynamic growth sector of the information age. It remains at, or close to, the center of powerful players' corporate and political strategies.

However, television is also a mature industry across the developed world, and it bears the hallmarks of stasis, decline (in some respects, especially free-to-air), budding-off, parallel growth and reformulation. The shape, scope, and style of television-in-the-future is contestable. In this chapter, I treat television as an object of policy regimes that have been and may become influential on its shape and possible futures.

I should stress that this is not an exercise in historical periodization, laying out quasi-organic stages of the long-term "business cycle" of television as industry and culture, a cycle that moves from the innovation and diffusion of a new technology, to its establishment and system growth as a communications industry, then to a period of maturity and popularity followed by indicators of specialization, diversification and decline. (For a treatment of US television from this perspective, see Comstock, 1991; for Australian television, see Cunningham, 2000.) Recent discussions of what shape US television in particular may assume beyond the standard "broadcast" and "cable" age periodization, such as Rogers, Epstein, and Reeves's (2002) "mass," "niche," and "brand" marketing ages, bring us closer to a sense of overlapping analytical frameworks by which to attempt to understand shifts in the nature of television.

However, this chapter goes further than ex post facto analytical frameworks. It takes an explicit policy-oriented focus and tracks television as it has been or could be deliberated across policy "regimes." These three grids of understanding – "culture," "services," and "knowledge" – also serve as historical and/or possible rationales for state intervention in television, as well as the industry's own under-standings of its nature and role. To emphasize the dynamic, overlapping and in part contesting nature of these regimes, I use Raymond Williams's (1981, p. 204) distinction between residual, dominant, and emergent cultural forces. The first, the *residual*, regime of cultural policy, is of well-established vintage for television but is under siege. The second, the *dominant*, the service industry model, is the most widespread regime. The third, *emergent* regime, the place of television in the knowledge economy, is embryonic at this stage of its development.

My argument is that there was a cultural industries and policy "heyday" around the 1980s and 1990s, as the domain of culture expanded, which benefited television through its being seen as a central cultural industry. Cultural policy fundamentals are, however, being squeezed by the combined effects of the "big three" – convergence, globalization and digitization – which are underpinning a services industries model of industry development and regulation. This model, despite clear dangers, carries advantages in that it can mainstream cultural in-dustries like television as economic actors and lead to possible rejuvenation of hitherto marginalized types of content production.

But new developments around the knowledge-based economy point to the limitations for wealth creation of focusing solely on microeconomic efficiency gains and liberalization strategies, the classic services industries strategies. Recognizing that such strategies won't "push up" the value chain to innovative, knowledge-based sectors, governments are now accepting a renewed interven-tionary role for the state in setting twenty-first-century industry policies.

But the content and entertainment industries, such as television, *don't as a rule figure* in R&D and innovation strategies. The task is, first, to establish that these industries indeed engage in what would be recognizable as R&D and exhibit value chains that integrate R&D into them; and, secondly, to evaluate whether the state has an appropriate role to support such R&D in the same way and for the same reasons as it supports science and technology R&D.

Residual Regime: Cultural Policy

Culture is very much the home patch of us content proselytizers – where many of us grew up intellectually and where we feel most comfortable. Further, it has been around as a fundamental rationale for government's interest in regulation and subsidy for decades. The "cultural industries" was a term invented to embrace the commercial industry sectors – principally film, television, book publishing, and music – which also delivered fundamental, popular culture to a national population. This led to a cultural industries policy "heyday" around the 1980s and 1990s, as the domain of culture expanded. (In some places it is still expanding, but is not carrying much heft in the way of public dollars with it, and this expansion has elements trending towards the – perfectly reasonable – social policy end of the policy space, with its emphasis on culture for community development ends.)

Meanwhile, cultural policy fundamentals are being squeezed, since they are nation-state specific during a period dominated by the WTO and globalization. Cultural nationalism is no longer in the ascendancy socially and culturally. Policy rationales for the defense of national culture are less effective in the convergence space of new media. Marion Jacka's (2001) recent study shows that broadband content needs industry development strategies, not so much cultural strategies, as broadband content is not the sort of higher-end content that has typically attracted regulatory or subsidy support (see Cunningham, 2002a). The sheer size of the content industries and the relatively minute size of the arts, crafts, and performing arts sub-sectors within them underline the need for clarity about the strategic direction of cultural policy (John Howkins in *The Creative Economy* (2001) estimates the total at $US2.2 trillion in 1999, with the arts at 2 percent of this). Perhaps most interestingly, and ironically, cultural industries policy was a "victim of its own success": cultural industry arguments have indeed been taken seriously, often leading to the agenda being taken over by other, more powerful, industry and innovation departments (see O'Regan, 2001 and Cunningham, 2002b).

Where does this leave television? Television content – as national cultural output and expression – has been regulated for specifically cultural outcomes in Europe, Canada, Australia, and several other jurisdictions over decades. A study by Goldsmith, Thomas, O'Regan, and Cunningham (2001, 2002) of contemporary state action in broadcasting systems around the world concluded that:

> Notwithstanding globalizing ideology that asserts that the nation is going out of business, national governments continue to develop models and working policy frameworks for asserting cultural and social principles in converging media systems. They do this, however, knowing that the environment for such activity is changing very rapidly and in very complex, uncharted ways. (Goldsmith et al., 2002, p. 106)

201

The overall trend, though, is for most new forms or refinements of state action around television broadcasting to occur with more of a social and informational remit (a services industry model for regulation) than an explicitly cultural remit.

This may be due to the fact that much cultural rationale for television content regulation is based as much on an industry development or protection basis as on a sophisticated cultural rationale. General transmission quota regulation (such as the European Union's direction to its member states "Television without Frontiers," and rules for Canadian or Australian content across most of the transmission time of broadcasters) is based on a broad, generic cultural remit (national culture is represented by whatever content is found on television; anthropological account of culture). But regulation for specific forms of national television, such as high-end fictional drama and social documentary, is based on cultural exceptionalism. The official argument goes that high-end fictional drama is an exceptionally key genre of the national culture because it heightens, dramatizes, and narrativizes national stories while also providing crucial alternatives to the US hegemony in audiovisual fictional drama. The official argument has to be of this more intense (culture as art) nature because such content may not be produced without state intervention. The rationale then becomes one of market failure to provide such high-end genres, because of their cost relative to imported hegemonic content (usually US but also UK and even some major European-language programming).

However, if there is emerging evidence that there is a decline in audience demand for high-budget series and one-off TV drama, the market failure argument is weakened, and specifically cultural policy for television is no longer based on a popular cultural mandate but is pushed back to more of an "arts and audience development" strategy. Is the trend away from nation-defining drama to reality TV, from authored texts to branded experiences, a cultural, generational shift; or how much is it the corporate strategies of the television industry driving unit costs of content creation inexorably down in the face of the exploding multichannel marketplace and the fragmentation of the audience base? While the latter is undoubtedly true, it is too early to say whether the former – the cultural shift – has irrevocably taken place.

The "squeeze" on national cultural policy for television has produced a displacement of cultural policy focus to the regional (intra-country) and the supranational. Charles Leadbeater and Kate Oakley's study of *The Independents: Britain's New Cultural Entrepreneurs* (1999), for example, gives a concise account of the crucial role that Britain's regionalized television production capacity plays in sustainable cultural development in provincial cities. They stress the added value that local broadcasting production capacity brings to provincial cities in the UK, such as Glasgow and Cardiff, as compared to cities without such capacity, such as Sheffield. In Cardiff, for example, television broadcasters play an important role in stimulating employment beyond their immediate workforce and locale. Where broadcasting distribution opportunities exist, independent producers and post-production houses also exist and contribute to creative sector development. The digital broadcast and broadband future will need to pay more and more attention to regions, cities and districts as much as nations as cultural milieux.

At the supranational level, a key development has arisen from Canada's response to the adverse findings of the WTO in its periodicals dispute with the United States. As Goldsmith et al. (2003, p. 97) point out, it has moved to "forum shift" whereby discussion and the potential for multilateral or a series of bilateral deals moves to a new decision-making forum (Braithwaite and Drahos, 2000, pp. 28–9) and facilitate fora such as the International Network on Cultural Policy, a network of over 40 national cultural ministers which was established in 1998 following the Stockholm UNESCO *Intergovernmental Conference on Cultural Policies for Development*; and the International Network for Cultural Diversity, a network of hundreds of non-governmental organizations from over 50 countries "dedicated to countering the homogenizing effects of globalization on culture" (incd.net). These fora are important sites at which international coalitions premised on the preservation and maintenance of cultural sovereignty mechanisms such as content regulations may be established. Their lobbying efforts have seen the Commission IV (culture) of the UNESCO General Conference commit, in 2003, to a Draft Convention on Cultural Diversity titled the *Convention for the Safeguarding of the Intangible Cultural Heritage*. The Draft Convention is designed to act as a legal instrument to protect and enable governments to enact domestic policy measures to support a diverse range of cultural expression.

Dominant Regime: Services Industry Model

This doesn't get talked about much in the cultural/audiovisual industries "family," but it's *sine qua non* in telecommunications and in, well really, pretty much the rest of the economy. We can begin to see how this "services" conception of content and entertainment industries might work by considering television – a, if not *the*, major content and entertainment industry – as also a central service industry.

Many of the content and entertainment industries – especially the bigger ones such as publishing, broadcasting, and music – can be and are classified as service industries. But the broader and larger service industries, such as health, telecommunications, finance, education, and government services, require increased levels of creativity through increased intermediate inputs, and it is here that much of the growth opportunities for content creation is occurring. Just as it has been received wisdom for two decades that society and economy are becoming more information-intensive through ICT uptake and embedding, so it is now increasingly clear that the trend is toward "creativity-intensive" industry sectors. This is what Lash and Urry (1994) refer to as the "culturalization of everyday life" and why Venturelli (2002) calls for "moving culture to the center of international public policy."

It is not surprising that this is where the growth opportunities are, as all OECD countries display service sectors which are by far the biggest sectors of their respective economies (the services sector is in the 60–80 percent range for total businesses; total gross value added; and employment across all OECD economies), and that relative size has generally been growing steadily for decades.

Much convergence talk has it that a potent but as yet unknown combination of digital television and broadband will become a, if not *the*, prime vehicle for the delivery or carriage of services. Education, banking, home management, e-commerce and medical services are some of the everyday services which types of interactive television and broadband might deliver.

But for television to be considered *as* a central service industry takes the convergence tendency to a new level. For most of its history, media content, and the conditions under which it is produced and disseminated, have typically been treated as issues for cultural and social policy in a predominantly nation-building policy framework. They have been treated as "not just another business" in terms of their carriage of content critical to citizenship, the information base necessary for a functioning democracy and as the primary vehicles for cultural expression within the nation.

In the emerging services industries policy and regulatory model (which some – for example, Damien Tambini (2002) in talking about recent UK communications reforms – might dub "new" public interest), media content could be treated less as an exception ("not just another business") but as a fundamental, yet everyday, part of the social fabric. Rather than television's traditional sectoral bedfellows cinema, the performing arts, literature and multimedia, it is seen as more related to telecommunications, e-commerce, banking and financial services, and education.

This entails a rethinking of television's place in the public sphere. The public sphere, in its classic sense advanced in the work of Jürgen Habermas ([1962] 1989), is a space of open debate standing over against the state as a special subset of civil society in which the logic of "democratic equivalence" is cultivated. The concept has been regularly used in the fields of media, cultural, and communications studies to theorize the media's articulation between the state and civil society. Indeed, Nicholas Garnham (1995) claimed in the mid-1990s that the public sphere had replaced the concept of hegemony as the central motivating idea in media and cultural studies. This is certainly an overstatement, but it is equally certain that, almost 40 years since Jürgen Habermas first published his public sphere argument, and almost 30 since it was first published in outline in English (Habermas, 1974), the debate over how progressive elements of civil societies are constructed and how media support, inhibit or, indeed, are coterminous with, such self-determining public communication, continues strongly.

The debate is marked out at either end of the spectrum, on the one hand, by those for whom the contemporary Western public sphere has been tarnished or even fatally compromised by the encroachment of particularly commercial media and communications (for example, Schiller, 1989). On the other, there are those for whom the media have become the main if not the only vehicle for whatever can be held to exist of the public sphere in such societies. "Media-centric" theorists such as John Hartley can hold that the media actually *envelop* the public sphere:

> The "mediasphere" is the whole universe of media . . . in all languages in all countries. It therefore completely encloses and contains as a differentiated part of itself the (Habermasian) public sphere (or the many pubic spheres), and it is itself contained by the much larger semiosphere . . . which is the whole universe of sense-making by whatever means, including speech . . . it is clear that television is a crucial site of the mediasphere and a crucial mediator between general cultural sense-making systems (the semiosphere) and specialist components of social sense-making like the public sphere. Hence the public sphere can be rethought not as a category binarily contrasted with its implied opposite, the private sphere, but as a "Russian doll" enclosed within a larger mediasphere, itself enclosed within the semiosphere. And within "the" public sphere, there may equally be found, Russian-doll style, further counter-cultural, oppositional or minoritarian public spheres. (Hartley, 1999, pp. 217–18)

We can think of Hartley's *Uses of Television* (1999) as a key theoretical argument for a services industries model of media, or, in other words, as a provider of educational services. For Hartley, the media, but especially television, have a "permanent" and "general," rather specific and formal, educational role (1999, p. 140) in the manners, attitudes, and assumptions necessary for citizenly participation in communities. "[C]ontemporary popular media as guides to choice, or guides to the attitudes that inform choices" (1999, p. 143) underpin Hartley's allied claim for the media's role in promoting "Do-it-yourself" (DIY) citizenship.

Hartley claims that television is a "transmodern" teacher or informal pedagogue on a vast scale. That is, it has a bardic function, employing or embodying *pre-modern*, oral, forms of communication based on family and a domestic setting. It is classically modern:

> it is no respecter of differences among its audiences; it gathers populations which may otherwise display few connections among themselves and positions them as its audience "indifferently", according to all viewers the same "rights" and promoting among them a sense of common identity as television audiences. (1999, p. 158)

It is also a major postmodern form, embodying disjunctive aesthetics, clashes of the superficial and the serious, and pervasive image construction that "threatens" to become constitutive of reality rather than merely reflective of it. Television in its traditional broadcast form achieves effective cross-demographic communication because of its continuing modernist role. Within the accepted understanding of television's role as a provider of information and entertainment, is its postmodern provision of contemporary cultural and DIY citizenship – the construction of selves, semiotic self-determination and functional media literacy: "citizens of media remain citizens of modernity, and the rights struggled for since the Enlightenment are not threatened but further extended in the so-called 'postmodern' environments of media, virtuality and semiotic self-determination" (1999, p. 158).

This, then, is a way of thinking of television as a cultural technology in a services industries model – a major arena within a community through which processes of informal citizenship, public and cross-demographic communication and plebiscitary "democratainment" takes place. As a supranational policy regime, however, the services industries model carries dangers. As the concerns about the WTO expressed through UNESCO's Global Alliance for Cultural Diversity or the International Network for Cultural Diversity show, it subjects all television systems to a normative, globalizing perspective and thus weakens the specifics of a cultural case for national regulation and financial support. Its widespread adoption would see the triumph of what might be called the US regulatory model, where competition is the main policy lever and consumer protection rather than cultural development is the social dividend. The application of this model across the board is not a universal panacea for all industry regulatory problems, as most mid-level and smaller countries need to, or do, acknowledge.

However, there are also possible advantages. As mentioned in discussing the cultural policy regime, a range of initiatives are being taken around television broadcasting to strengthen its social and informational remit within a services industry model. Hitherto marginal programming could be significantly upgraded in a services industries model. Programming produced for and by regional interests might be regarded as fundamental as the guarantee of a basic telephone connection to all regardless of location. The need for programming inclusive of demographics such as young people and children might be as crucial as free and compulsory schooling. Moves in various jurisdictions, including the EU and Canada, to give greater weighting to regional, infotainment, youth and children's programming signal a shift in the priority of content regulation to include these alongside a continuing emphasis upon drama and social documentary (see Goldsmith et al., 2002). While the latter advance core cultural objectives such as quality, innovation, and cultural expression, the former warrant greater consideration in a services industries model of media content regulation in terms of their contribution to diversity, representation, access, and equity.

Emergent Regime: The Knowledge-Based Economy

Content and entertainment industries are beginning to be seen as an element of high-value-added, knowledge- and innovation-based industries. This is an emergent regime, but it is the one most likely to advance new positionings of the high-growth, cutting-edge content and entertainment industries into the future.

To make this argument, it is necessary to consider how and why content and entertainment industries might qualify as high-value-added, knowledge-based industry sectors. From where has this new macro-focus emerged? In part, it's been around for some time, with notional subdivisions of the service or tertiary industry sector into quaternary and quinary sectors based on information management (4th sector) and knowledge generation (5th sector). But the shorter-term influence is traceable to new growth theory in economics which has pointed to the limitations for wealth creation of focusing solely on microeconomic efficiency gains and liberalization strategies (Arthur, 1997; Romer, 1994, 1995). These have been the classic services industries strategies.

Governments are now attempting to advance knowledge-based economy models, which imply a renewed interventionary role for the state in setting twenty-first-century industry policies, prioritization of innovation and R&D-driven industries, intensive reskilling and education of the population, and a focus on universalizing the benefits of connectivity through mass ICT literacy upgrades. Every OECD economy, large or small, or even emerging economies (e.g. Malaysia) can try to play this game, because a knowledge-based economy is not based on old-style comparative factor advantages, but on competitive advantage – namely, what can be constructed out of an integrated labor force, education, technology and investment strategy.

As noted previously, the content and entertainment industries *don't as a rule figure* in R&D and innovation strategies, dominated as they are by science, engineering, and technology. But they should. Creative production and cultural consumption are an integral part of most contemporary economies, and the structures of those economies are being challenged by new paradigms that creativity and culture bring to them.

Worldwide, the content and entertainment industries have been among the fastest-growing sectors of the global economy, with growth rates better than twice those of advanced economies as a whole. In the United States, entertainment has displaced defense as the driver of new technology take-up, and it has also overtaken defense and aerospace as the biggest sector of the Southern Californian economy (Rifkin, 2000, p. 161). Rifkin (2000, p. 167) claims that cultural production will soon ascend to the first tier of economic life, with information and services moving to the second tier, manufacturing to the third tier and agriculture to the fourth tier.

Most R&D priorities reflect a science and technology-led agenda at the expense of new economy imperatives for R&D in the content industries, broadly defined. But the broad content industries sector – derived from the applied social and creative disciplines (business, education, leisure and entertainment, media and communications) – represents 25 percent of the US economy, whilst the new science sector (agricultural biotech, fiber, construction materials, energy, and pharmaceuticals) for example, accounts for only 15 percent (Rifkin, 2000, p. 52).

In fact, all modern economies are consumption driven and the social technologies that manage consumption all derive from the social and creative disciplines. We can no longer afford to understand the social and creative disciplines as commercially irrelevant, merely "civilizing" activities. Instead they must be recognized as one of the vanguards of the new economy. R&D strategies must work to catch the emerging wave of innovation needed to meet demand for content creation in entertainment, education, and health information, and to build and exploit universal networked broadband architectures in strategic partnerships with industry.

Political economy and critical cultural studies (for example, see the *International Journal of Cultural Studies*, vol. 7(1) March 2004) might view these kinds of claims for creativity in the new economy as reductionist economism, and a "cheerleading" boosterism fatally deflated by the dotcom bust. However, I would argue that the creative and informational economy poses a serious challenge to traditional "scale and scarcity" economic orthodoxy as well as heritage notions of culture. Also, that the trends toward the "culturization" of the economy are more longer-term than the hothouse events of the late 1990s and early 2000s. As Venturelli argues (2002, p. 10), "the environmental conditions most conducive to originality and synthesis as well as the breadth of participation in forming new ideas comprise the true tests of cultural vigor and the only valid basis for public policy." There is enough in new growth theory, and evolutionary and institutional economics, to suggest progressive new takes on traditional political economy. Creativity, once considered as marginal, has had to be brought toward the heartland of economic thought, and with it its values. What was once considered as the only model for innovation (science and technology) has had to make some way for creative content and process.

Despite the difficulties in shoehorning content and entertainment industries into innovation frameworks – designed as they are fundamentally for the manufacturing sector – it is beginning to happen, as innovation and R&D policies evolve. Lengrand et al. (2002) talk of "third generation" innovation policy, while Rothwell (1994) contemplates five generations of innovation. The trend is the same, however. Earlier models are based on the idea of a linear process for the development of innovations. This process begins with basic knowledge breakthroughs courtesy of laboratory science and the public funding of pure/basic research and moves through successive stages – seeding, pre-commercial, testing, prototyping – until the new knowledge is built into commercial applications that diffuse through widespread consumer and business adoption. Contemporary models take account of the complex, iterative, and often non-linear nature of innovation, with many feedback loops, and seek to bolster the process by emphasizing the importance of the systems and infrastructures that support innovation.

What, then, is R&D in content and entertainment industries? Major international content growth areas, such as online education, interactive television, multi-platform entertainment, multi-player online games, web design for business-to-consumer applications, or virtual tourism and heritage, need *research* that seeks to understand the interrelation of complex systems involving entertainment, information, education, technological literacy, integrated marketing, lifestyle and aspirational psychographics, and cultural capital. They also need *development* through trialing and prototyping supported by testbeds and infrastructure provision in R&D-style laboratories. They need these in the context of ever-shortening innovation cycles and greater competition in rapidly expanding global markets.

What Applications Does This Regime Have in Television?

A policy report, "Research and Innovation Systems in the Production of Digital Content," exploring the application of contemporary innovation system approaches to the content industries (QUT CIRAC and Cutler&Company, 2003) identified the importance of television, particularly as it migrates to digital platforms, as the gateway between established and emergent *content creation* (major popular entertainment and informational formats transmigration to interactivity and mass customization) and *industry structure* (highly centralized distributional models to more networked and distributed models). Understanding the interaction between the potent legacy of broadcasting and the potential of convergent broadband media is the key to positioning innovative opportunities in content creation if they are to remain close to the mainstream of popular cultural consumption rather than being siphoned off into science or art alone.

One of the recommendations of "Research and Innovation Systems in the Production of Digital Content" was to strengthen broadcasting's role in the innovation system and ensure an active digital community broadcasting sector.

Public broadcasters often have or could develop a role as an R&D or innovation "laboratory" for their respective national television systems and may be able to set standards of televisual innovation that become internationally recognized and emulated. The charters or legislative frameworks under which public broadcasters operate might be strengthened to address innovation through community engagement and content diversity. They could be encouraged or required to make available television windows for innovative digital content. Their content-sourcing policies could be aligned with an innovation agenda through a mandated or voluntary quota of independently and diversely sourced digital content. The emerging digital television environment represents an innovation incubator for the carriage and distribution of digital content production. The world's foremost public broadcaster, the BBC, has led the way in establishing an explicit R&D institutional milieu. R&D for the BBC is an integral element in an ongoing process of developing new cutting-edge programming, understanding consumption with the advent of new technology and integrating content production and distribution with new technologies (see the BBC R&D website at www.bbc.co.uk/rd).

The two primary engines are BBC R&D and its sister group Creative R&D (a subunit of BBC New Media Technology). The primary goal of BBC R&D is to "innovate in technology in support of the BBC's public service purpose – delivering both immediate value to current projects, and helping the long-term strategic objectives of the BBC and the nation" (BBC, 2003, p. 4). R&D activity includes delivery, production, spectrum planning, and technology in the home development. Creative R&D undertakes research into new forms of content for future programming with the aim of complementing technical work undertaken by BBC R&D:

> Creative Research and Development are particularly interested in the intersection where audience behaviour impacts on the consumption of digital media and where trends in technology development impact on audiences. CR&D formed a visual Navigation Group with the Navigation Group at BBC R&D to link our understanding of audiences, with our understanding of technology . . . More can be done to create navigational tools and interfaces to contribute to consumers' enjoyment of the experience of finding content, and to keep them with our content services longer. (Hooberman et al., 2003, pp. 3–5)

Beyond the specificities of public broadcasting, there is scope to leverage digital television licensing arrangements to establish R&D testbeds for trialing interactive TV possibilities in partnership with advertisers, television companies, and other stakeholders in the provision of interactive services. Such "testbeds" could address the current slow and uneven uptake of digital TV receivers by facilitating the uptake of digital set top boxes in schools and other centers where the trialing of digital content can be carried out.

Another focus for television's role in innovation recommended in "Research and Innovation Systems in the Production of Digital Content" is the promotion of open content repositories to fuel creative content production for broadcasting and other outlets. Open content repositories, or public domain digital content, is the content and entertainment industries equivalent of open source software. It *selectively* addresses barriers to production and unintended cultural outcomes of prevailing copyright and IP regimes through an alternative *opt-in* model, which can operate in parallel with existing regimes. As such, it can be a powerful structural mechanism to support a rich "digital sand pit" for creative content producers. The measure would facilitate the active re-purposing and re-use of digital content assets. Misuse of this public domain material would be protected under the provisions of a General Non-Exclusive Public Licence scheme.

The proposal would provide clear policy direction for public agencies. Cultural agencies would be given the mission to act as such repositories, and be required to make, hold, and administer their content collection assets on this open content basis. Because of the scale of the public sector assets involved, scale and impact is achieved through this initiative. It also encourages moves to integrate a wide range of producers, curators, and administrators of creative content more fully within the content and entertainment industries innovation system through seeking its placement on broadcasting and broadband outlets.

A specific broadcasting policy initiative which could arise from this development of the innovation system would be to encourage or mandate retransmission and an open access channel regime for subscription TV and broadband channels covering third-party content and open content repository material. This proposal is one measure to address the major issue about distribution bottlenecks and builds on the existing (if limited) policy provisions in broadcasting policy regimes.

This proposal not only provides alternative distribution channels for independent content producers, but also promotes content diversity. Within an interactive media environment, the issue of public access channels to public domain content repositories requires freely available access channels. In many countries, the most rapid digitization of television is taking place in the subscription TV domain. Developing access regimes on the basis that they are to be used as testbeds for content innovation as much as a community "safety valve" is important.

So much for policies about television in an innovation system; what about content? The high-end of content innovation is exemplified in the BBC's Creative R&D unit (and its predecessor, the Imagineering workshop), with its innovative series such as *Walking with Dinosaurs*, *Walking with Beasts*, and *Walking with Humans* which have explored the potential of various modes of interactivity on digital platforms.

At another, more demotic level, the potential of multi-platform television has been explored effectively by programs such as *Big Brother* and other reality "event" television. *Big Brother* has been a series of multi-platform, cross-promotional world media events centered on television, with claims to technical, cultural, broadcasting, Internet, advertising, marketing, and event management innovation. It was accessible in the traditional way on free-to-air, via the official *Big Brother* website with discussion forums, on unofficial fan sites. It was catchable via radio updates. There was telephone voting, SMS updates to mobiles, and in the UK, there was live coverage and unedited rushes on a digital channel in Britain up to 18 hours a day.

It is possible to argue that there is a *Big Brother* innovation "system." It is an international "learning" system, which achieves technology transfer and format and style upgrades around the world very rapidly. Format franchising and the associated IPR issues have been handled with aplomb. The package assists in solving problems for major services sectors like advertisers and marketing, which benefit from integrated marketing solutions. There are technological and process/project management innovations in the successful trialing of a large-scale convergent, multi-platform delivery system. There is industry innovation due to successful trialing of diverse and, in many cases, marginalized regional capacity around the world for such large-scale real-time, prime-time TV production.

Big Brother and its ilk are, of course, not television fare for everyone and there is a thoroughgoing critique that could be made of their "bread-and-circuses" exploitation of the attractions of celebrity (and also, of course, of their perform-ance of contemporary demotic rituals of self-formation and empowerment-from-below). However, what if the *content* of *Big Brother* was to be substituted for, say, a similarly resourced experiment in the convergent, multi-platform, delivery of crucial government or community services to a client base similar to that which tuned in or accessed the website or bought the products marketed through the program? In that case, a clear advance in contemporary strategies of community engagement and innovative service delivery would be claimable. And the poten-tial for just such advances is increased by the commercial innovation of multi-platform reality event television.

Acknowledgments

Thanks to Terry Cutler, Greg Hearn, Mark Ryan, and Michael Keane, co-authors with me of *Research and Innovation Systems in the Production of Digital Content* (QUT CIRAC [Creative Industries Research and Applications Center] and Cutler & Company, 2003) and Ben Goldsmith, Julian Thomas and Tom O'Regan, co-authors with me of "Asserting Cultural and Social Regula-tory Principles in Converging Media Systems," in M. Raboy (ed.), *Global Media Policy in the New Millennium* (London: University of Luton Press, 2002). This chapter has incorporated several passages from these studies.

References

Arthur, B. (1997) "Increasing Returns and the New World of Business," in J. Seely Brown (ed.), *Seeing Differently: Insights on Innovation*, Boston: Harvard Business Review Books, pp. 3–18.

Braithwaite, J. and Drahos, P. (2000) *Global Business Regulation*, Cambridge: Cambridge Univer-sity Press.

British Broadcasting Corporation (2003) *BBC R&D Annual Review 2002–2003*, British Broadcast-ing Corporation, April 2002–March 2003: http://www.bbc.co.uk/rd/pubs/rev_03/bbcrd-ar-2002-3-complete.pdf.

Collins, R. (1990) *Television: Policy and Culture*, London: Unwin Hyman.

Comstock, G. (1991) *Television in America*, Newbury Park, CA: Sage.

Cunningham, S. (2000) "History, Context, Politics, Policy," in G. Turner and S. Cunningham (eds.), *The Australian TV Book*, St Leonards: Allen & Unwin, pp. 13–32.

Cunningham, S. (2002a) "Policies and Strategies," in K. Harley (ed.), *Australian Content in New Media: Seminar Proceedings*, Network Insight, Sydney: RMIT, pp. 39–42.

Cunningham, S. (2002b) "From Cultural to Creative Industries: Theory, Industry, and Policy Implications," *Media Information Australia Incorporating Culture & Policy*, *102* (February), 54–65.

Garnham, N. (1995) "The Media and Narratives of the Intellectual," *Media, Culture & Society*, *17*, 359–84.

Goldsmith, B., Thomas, J., O'Regan, T., and Cunningham, S. (2001) *The Future for Local Content? Options for Emerging Technologies*, Queen Victoria Building, NSW: Australian Broad-casting Authority, June.

Goldsmith, B., Thomas, J., O'Regan, T., and Cunningham, S. (2002) "Asserting Cultural and Social Regulatory Principles in Converging Media Systems," in M. Raboy (ed.), *Global Media Policy in the New Millennium*, London: University of Luton Press, pp. 93–109.

Habermas, J. ([1962] 1989) *The Structural Transformation of the Public Sphere: An Inquiry in a Category of Bourgeois Society*, Cambridge: Polity Press.

Habermas, J. (1974) "The Public Sphere," *New German Critique* 1 : 3, 49–55.

Hartley, J. (1999) *Uses of Television*, London: Routledge.

Herman, E. S. and McChesney, R. W. (1997) *The Global Media: The New Missionaries of Corporate Capitalism*, London and Washington: Cassell.

Hooberman, L., Winter, G., Glancy, M., and Seljeflot, S. (2003) *The Potential of Visual Navigation: BBC R&D White Paper*, WHP 075, September. http://www.bbc.co.uk/rd/pubs/whp/whp-pdf-files/whp075.pdf.

Howkins, J. (2001) *The Creative Economy: How People Make Money From Ideas*, London: Allen Lane.

Jacka, M. (2001) *Broadband Media in Australia: Tales from the Frontier*, Sydney: Australian Film Commission.

Lash, S. and Urry, J. (1994) *Economies of Signs and Space*, London: Sage.

Leadbeater, C. and Oakley, K. (1999) *The Independents: Britain's New Cultural Entrepreneurs*, London: Demos.

Louis Lengrand et Associes, PREST and ANRT (2002) *Innovation Tomorrow. Innovation Policy and the Regulatory Framework: Making Innovation an Integral Part of the Broader Structural Agenda*. Innovation papers No. 28, Directorate-General for Enterprise, Innovation Directorate, EUR report no. 17052, European Community.

O'Regan, T. (2001) *Cultural Policy: Rejuvenate or Wither?*, Griffith University Professorial Lecture, http://www.gu.edu.au/centre/cmp/mcr1publications.html#tom.

QUT CIRAC (Creative Industries Research and Applications Center) and Cutler & Company (2003) *Research and Innovation Systems in the Production of Digital Content*, Report for the National Office for the Information Economy, September, http://www.noie.gov.au/projects/environment/ClusterStudy/Research%20and%20innovation%20systems%20in%20production%20of%20digital%20content.pdf; or http://www.cultureandrecreation.gov.au/cics/Research_and_innovation_systems_in_production_of_digital_content.pdf.

Rifkin, J. (2000) *The Age of Access: How the Shift from Ownership to Access is Transforming Modern Life*, London: Penguin.

Rogers, M. C., Epstein, M., and Reeves, J. L. (2002) "*The Sopranos* as HBO Brand Equity: The Art of Commerce in the Age of Digital Reproduction," in D. Lavery (ed.), *This Thing of Ours: Investigating The Sopranos*, New York: Columbia University Press, pp. 42–57.

Romer, P. (1994) "The Origins of Endogenous Growth," *Journal of Economic Perspectives*, 8(1) (Winter), 3–22.

Romer, P. (1995) "Interview with Peter Robinson," *Forbes*, 155(12), 66–70.

Rothwell, R. (1994) "Towards the Fifth-generation Innovation Process," *International Marketing Review*, 11(1), 7–31.

Schiller, H. (1989) *Culture Inc: The Corporate Takeover of Public Expression*, New York: Oxford University Press.

Tambini, D. (2002) "The New Public Interest," Australian Broadcasting Authority Conference, Canberra, May.

Venturelli, S. (2002) "From the Information Economy to the Creative Economy: Moving Culture to the Center of International Public Policy," Center for Arts and Culture, Washington, www.culturalpolicy.org.

Williams, R. (1981) *Culture*, Glasgow: Fontana.

PART FIVE

Television/ Commerce

CHAPTER ELEVEN

Television Advertising as Textual and Economic Systems

Matthew P. McAllister

When I was a young child, my mother routinely took me with her while grocery shopping, as most mothers would. She was amazed when, as we strolled up each aisle, my two-year-old self would see different products on the shelf and begin singing the commercial jingles that I had heard on television associated with those products (e.g. "Snap, Crackle, Pop, Rice Krispies!"). She tells this story as one of the first times that she saw me as a being who was developing an identity and intellect separate from her: in this case, a commercial identity.

Now it is not like I was some sort of freakish advertising savant; in the United States as early as 1955 five-year-olds were reported singing beer commercial jingles (Samuel, 2001, p. 72). Many adults, too, express their fondness for the occasional commercial. The percentage of US viewers who claim to watch the Super Bowl *only* for the commercials has gradually risen over the years (McAllister, 2003). People watch 30-minute-long commercials (infomercials) or TV channels that are essentially nothing but commercials, like home-shopping channels (Cook, 2000). On the other hand, most of us do not watch TV for the commercials. The obnoxiousness of much of television advertising is a major pain for millions of viewers, often prompting them to channel surf or mute the television. As will be discussed at the end of this chapter, the proclivity of viewers to escape ads, and technological advances that facilitate this escape, may greatly influence the future role of promotional messages on television.

The television commercial is perhaps the most consistent and pervasive genre of television content – maybe even of *all* modern culture. The influence of the TV ad goes way beyond sticking a stupid song in our head, irritating us, or even the selling of specific products. Television advertising's power comes from its presence and visibility both as a textual, symbolic system and as an economic system. As a textual system, commercials pervade television and therefore our lives as viewers. Although designed to sell products, television commercials have a high degree of symbolic complexity and have unintended effects beyond the

217

selling goal. However, the ads themselves may be not be the most significant aspect of advertising. Economically, American television from the very beginning was driven by dollars from advertising. Television advertisers are the source for television's funding. This is especially true of broadcast television, but also significantly so for most cable and satellite television networks. As the main funding source for television, advertising influences the nature of television programming in profound ways.

This chapter will review many of the key ideas of advertising as textual/ symbolic and economic systems of television, focusing especially on points raised by the extensive critical literature on television commercials. Although examples from many countries will be discussed, the focus in this chapter will be on research about US television advertising, in many ways the "archetype" broadcast advertising model that set the bar – a gaudily decorated and corrosive bar, admittedly – for the rest of the world's commercial television systems. As Magder (2004, p. 142) writes, "TV almost everywhere relies heavily on advertising dollars. Even so, the US TV system stands out – not any longer because it is commercial but because of the scale of the money in the system."

Beginnings of Television Advertising

American television was, as Samuel (2001, p. xiv) notes, the "first exclusively commercial medium in history." By this he means that other media – radio, newspapers, magazines – had some roots in non-advertising revenue streams, most commonly where audience members would foot at least part of the bill. TV in the United States, though, was *designed* from the very beginning to be advertising-supported.

Adopting the sponsorship system from radio, advertisers and their agencies had a greater degree of operational control over US television programming than in later years. Advertising agencies often served as program producers as one sponsor funded the production of a program and comprise the majority of commercial time. One implication of this was that in addition to the then-typical 60-second spot advertisements between program segments, early television was dominated by product selling *during* the programming. Such techniques as "integrated commercials" (promotional messages integrated into variety skits) and "host selling" would blur the distinctions between programming and advertisement (Alexander, Benjamin, Hoerrner, and Roe, 1998; Samuel, 2001). When factoring in host selling, product placement, on-set sponsor signage, and the spot advertisements themselves, some early programs may have been as much as 70 percent promotional (Samuel, 2001).

Eventually, sponsorship left television for the more "magazine-style" spot advertisement system, with many advertisers buying commercial time during a program, rather than one exclusive sponsor. There are several reasons for the decline of exclusive sponsorship in television, including the high cost of producing an entire program, the increased revenue potential to the networks of the magazine system, and the increased control over programming decisions and scheduling desired by the networks. In addition, as Barnouw notes, television was sufficiently commercialized by the 1960s, with commercial logic completely dominating the medium, that the direct control offered by sponsorship was no longer needed. Barnouw (1978, p. 4) argues that "A vast industry has grown up around the needs and wishes of sponsors. Its program formulas, business practices, ratings, demographic surveys have all evolved in ways to satisfy sponsor requirements."

Television's embrace of advertising may be seen in the amount of time devoted to commercials. In the United States, roughly one-fourth of all time on advertising-supported broadcast and cable television is devoted to commercial and promotional messages, a figure that is rising (Chunovic, 2003). According to data from the 1990s (and, therefore, likely to be conservative), the typical child viewer is exposed to 40,000 television commercials a year, nearly double the figure from 20 years earlier (cited in Kunkel and McIlrath, 2003). Such "clutter" may be even more prominent on television systems in developing countries (Mueller, 1996). From a fiscal point of view, by 2003 spending on US television advertising across all categories (broadcast network, local, cable, and syndicated programming) reached over $57 billion, comprising approximately 23 percent of all advertising revenues generated in the United States and making television by far the biggest advertising medium (Coen, 2003). The dominance of television as an advertising medium is more pronounced in many other countries. In the top 20 countries in terms of advertising growth from 2002 to 2003 (a list which includes Russia, China, India, and Brazil), television accounts for over 50 percent of all advertising revenues (Global Advertising, 2003).

Given their obvious and pervasive roles as symbols, what are a few of the key points that critical scholars make about television commercials as cultural messages? Given their economic import, how might the commercial nature of television affect its programming? The next two sections consider these questions.

Television Commercials as Cultural Texts: Commodity Fetishism and Representation

Television commercials are designed, first and foremost, to sell a product. Yet commercials often have effects beyond the purchase of a product. With often little material difference between products to promote, commercials must grab our attention and enhance the image of the product. They use a variety of sounds, dialog, visuals, and motion and editing techniques to communicate. They must appeal to particular demographic groups and not others. As a form of storytelling, they represent people, institutions, and practices in particularly self-serving ways. And they have a short time frame in which to communicate (typically 15 or 30 seconds). For these reasons, television commercials often cultivate effects beyond their immediate purpose that may have profound implications. In this section, two themes raised by television advertising critics will be explored: the degree to which television commercials fetishize the commodity form, and the ideological implications of representation in ads.

Television commercials as commodity fetishism

Macintosh's "1984" commercial, aired during the (appropriately enough) 1984 Super Bowl, is an important text in the history of television commercialism (Berger, 2004; McAllister, 1999; Stein, 2002). This ad, which announced the debut of the Macintosh personal computer, is credited with being the first ever "event commercial," legitimizing advertising campaigns as newsworthy and as forms of entertainment in and of themselves, and establishing the Super Bowl as a vehicle for high visibility campaigns. Stein (2002) also argues that this advertisement was especially ideological; it presents persuasive images in a way that is embedded in social power. By constructing IBM as an oppressive, "Big-Brotherish" force in our lives, the ad positions the Macintosh, and the consumer's decision to buy Macintosh, as an emancipatory solution to the industrialized alienation. Through a complex use of cinematic techniques (composed by the film director Ridley Scott) and cultural symbolism, the ad constitutes – creates – the consumer as someone who can act upon their frustratingly limited social environment simply by buying the product. To not buy Macintosh (itself, of course, a product of industrialization), the ad implies, is to be a powerless lemming with no social agency.

The 1984 ad is an archetypal example of one important textual characteristic of most television commercials. TV ads routinely "fetishize commodities" and the act of purchasing commodities (for a discussion of this Marxian concept as applied to advertising, see Jhally, 1987). Raymond Williams noted about advertising that the reason modern advertising is effective is not because humans in a modern society are hyper-materialistic. If we were purely materialistic, then ads themselves would not even be necessary or, at most, would just need to show the product, perhaps in especially flattering camera angles. We would simply want the product for the product's sake. The reason ads look the way they do, though, is that "our society quite evidently is not materialistic enough" (Williams, 1980, p. 185). Because humans have other needs besides the material, in order to be persuaded to purchase products, advertisers have to symbolically link other personal, social, and cultural values to the product. In the ads, the material product must be elevated beyond the material.

Advertising in all media can do this, but television advertising seems especially suited to the linkage of products to social values. Print advertising has more space for words, so factual information and logical argument about the product is easier to incorporate in a one-page ad than in a 30-second commercial. Besides using sound in creative ways, television has an image advantage over radio in that it can use visual symbolism to link products to desirable images and emotions. Samuel (2001) argues that television commercials are so effective at linking products to happiness that this form may have revitalized the "American Dream" and the imperative of consumer culture in the United States in the 1950s, an era that followed times of economic scarcity (the Depression) and material scarcity (World War II).

Butler (2002) argues that at least eight different "meaning categories," or social values, are found in modern US television commercials. Depending upon the specific product positioning, different visual techniques link products to one or more of these meaning categories: luxury & leisure; individualism; nature; folk culture; progress & novelty; romance & sexuality; easing/elimination of pain, fear & guilt; and utopia & freedom from dystopia (p. 289). SUV commercials, for example, link those vehicles to the idea of entering nature and gaining security over other drivers (Andersen, 2000); Cascade dishwashing detergent becomes a way to keep peace in a marriage and make men more domestically competent (Budd, Craig, and Steinman, 1999); Saturn ads connect the car to warm community-based images of the past (Goldman and Papson, 1996); and children's commercials present toys and cereals as a gateway to a utopian world where "kids rule" and adults are clueless (Seiter, 1993). Such symbolic linkages of product to social value imbue products with a "magical" quality (Williams, 1980), elevating the status beyond the material. The commodity is thus "fetishized" in a way that increases its cultural value beyond its purely material function.

What are some implications of how advertising, especially television advertising, fetishizes its commodities? Advertising's celebration of consumption may mask troublesome production practices. The Nike corporation airs TV campaigns celebrating consumer choice and empowerment in the ads while simultaneously promoting exploitative industrial practices in underdeveloped countries (Andersen, 1995, ch. 2; Stabile, 2000). Nor can advertisements deliver what they promise. As Andersen points out, alienation and dissolution may result when consumers realize that buying a particular product does not lead to magical solutions that improve their lives (1995, p. 89). Television advertising may also be one factor that leads to the over-consumption of commodity goods, encouraging overwork, a decrease in true leisure time, and an increase in consumer debt and ultimately life dissatisfaction for many (Schor, 1999). Scholars have noted the destructive impact upon the environment that overconsumption encourages (Budd, Craig and Steinman, 1999; Schor, 1999), so much so that Jhally (2000) wonders about the role of "advertising at the edge of the apocalypse" (p. 27). If true, such problems become more globally salient as television commercials in Eastern countries such as Hong Kong (Wong, 2000), China (Zhang and Harwood, 2004) and Malaysia (Holden, 2001), although perhaps still reflecting indigenous value systems, increasingly incorporate Western consumerist values and the celebration of acquiring goods. Indonesia banned television advertising for many years in large part because of concerns over consumerism in the form (Mueller, 1996, p. 147).

Representation in television commercials

Another central issue of television commercials as textual systems is representation: to what extent are certain social groups systematically portrayed in ways that may enhance or undermine their social authority and agency? Gender in television advertisements, for example, has been much studied by scholars (for reviews, see Furnham and Mak, 1999; Shields, 1997). There are several incentives for the use of sexism in television commercials: sex as a way to grab attention; the fit of advertising stereotypes in gender-based demographic segmentation (such as male-oriented programming); the use of stereotypes to shortcut the storytelling process (i.e. the "dumb blonde type"); the linkage of sexual success to the product as a way to (appropriately enough) fetishize it; and, again, the representational tools for enhancing emotional communication – including eroticism – that television provides such as film-style visuals (i.e. slow motion, camera angles), editing and music.

Shields with Heinecken (2002, p. 19) argues that although portrayals of women in ads have become more diverse, the stereotypical representations found in earlier television still exist. Research from a variety of countries highlights that although the overall ratio of male to female actors in commercials may be fairly equal, there are still consistent trends in presenting women in sexualized ways, in domestic settings, and in less authoritarian roles (such as voice-over narrators) than males (Bresnahan, Inoue, Liu, and Nishida, 2001; Furnham and Mak, 1999). Television commercials also tend to use thin models at least as often as standard-sized women and much more than heavier-than-average women, often linking physical attractiveness to slenderness (Peterson and Byus, 1999).

In children's advertising, studies indicate that, similar to adult ads, the percentage of portrayals of boys and girls in television commercials have evened out, but boys tend to be portrayed as more active and aggressive (Kunkel and McIlrath, 2003). Even the language in children's advertising tends to reinforce gender stereotypes, as commercials for girl-oriented products tend to use more passive and "feeling-based" verbs, whereas ads centered on boys use more action- and control-oriented words (Johnson and Young, 2002). Men also do not escape gender stereotyping in television commercials, as in, for example, the linkage of masculine initiation and control with alcohol in television beer commercials (Strate, 2000).

How may gendered portrayals affect viewers? There is evidence that viewing television commercials with excessively thin models may encourage adolescent girls to be dissatisfied with their own bodies (Hargreaves and Tiggemann, 2003), and that gender stereotyping of such notions as women's aversion to math in commercials may reinforce that stereotype in some women viewers' attitudes and behaviors (Davies, Spencer, Quinn, and Gerhardstein, 2002).

Racial representation has also been studied extensively by advertising critics and researchers (for a review, see Wilson and Gutiérrez, 2003), with the majority of work focusing on portrayals of African Americans. Although African

222

Americans appear in about one-third of all prime-time television commercials (cited in Entman and Book, 2000), the specific nature of these representations are often racially biased. In commercials during children's television, for example, African Americans are essentialized as musicians and athletic (Seiter, 1993). Similarly, others have noted the still-circumscribed roles that African Americans play in prime-time ads, such as athletes and low-wage workers (Bristor, Lee, and Hunt, 1995).

In terms of behavior and lifestyle, African-American males tend to be portrayed as aggressive and African-American females are often excluded from affect-oriented roles where romantic or domestic life is shown as pleasurable (Coltrane and Messineo, 2000). Shades of skin color may also be ideologically slanted in commercials, with more light-skinned actors appearing in commercials than dark skinned, especially among females, and with dark-skinned performers associated with athleticism (Entman and Book, 2000). When African Americans are presented prominently and sympathetically, the context of this portrayal in a television commercial may introduce problematic elements. In a study of the Budweiser "Whassup!" ad campaign, Watts and Orbe (2002) argue that although the ad portrayed African-American male friendship in a positive way, much of the humor and popularity of the commercial series, at least for white audiences, is the "otherness" of black male culture and the clownish "hyperbolic black acting" that characterizes the ads (p. 10).

Even though commercials are something we see every day and the symbols they employ are formidable and calculatingly refined by market research, perhaps a greater influence of television advertising upon our lives is less obvious: the influences advertisers and the advertising system exert upon other elements of television beyond the 30-second spot. Besides their symbolic import, commercials are also the main revenue source for television. The next section reviews many of the critical claims made about advertising's economic impact upon television programming.

Television Commercials as Economic Source: Seven Effects

The economic logic of a simple financial transaction like buying a glass of lemonade from a child's stand is easy to see: you are the buyer, the child is the seller, and the lemonade is the product. Television is a bit more complicated to map in this way, and this complication is due in large part to the role of advertising. As Smythe (1977) noted, because US commercial television generates virtually 100 percent of its revenues from advertising, in this system the advertisers are the "buyers," the television networks (for national advertising) and stations/cable systems (for local) are the sellers, and we, the viewers, are the product. The cost of placing a 30-second ad during a program, after all, depends upon the ratings for that program, the measure of the size and demographics of the viewership. Advertisers will purchase time on a program when this audience measure promises a sufficient audience, both in terms of size and kind. Ideally, advertisers would also like this product – the viewership – to be in an appropriate mental/emotional state to view ads.

If advertisers purchase audiences from television organizations, then what economic function do the programs serve? They are a "bribe" or "free lunch" or bait to grab the audience (Smythe, 1977, p. 5). In the United States, given this subordinate economic role that programming plays, the dominance of advertising as the main revenue generator for commercial television, and the 50-plus-year entrenchment of this system, it should not be surprising that its television programming is affected by all of these factors. As other countries adopt a more privatized and "Americanized" television system, it is logical to assume that the economic logic of advertising will continue to shape programming worldwide as well. In fact, the advertising industry in many countries is more dependent upon television than in the United States. As noted earlier, other countries, such as Brazil, Greece, Hong Kong, Mexico and Spain, spend more than 50 percent of their advertising expenditures on television (Mueller, 1996).

Television advertising's economic role has many effects upon society generally, including its role in decreasing competition (Bagdikian, 2000, ch. 8), and its financial marginalization of print media (Frith and Mueller, 2003). It also has many effects upon television programming, both blatant and subtle. This section will focus on seven of the more significant programming effects. Of course, these effects are not absolute: exceptions are often found as creative personnel bend rules and contradictory industrial factors collide (for example, the need to attract young, hip audiences through "edgy" programming versus the need to avoid alienating audiences). But, given the structural incentives built into television's economic logic, it is difficult for television personnel to continually and fundamentally counter the effects discussed below.

The "don't bite the hand" effect

Simply put, for content creators it is easier to avoid fundamental criticism of advertising on television than it is to include such criticism given the potential economic fallout. Advertisers feel, logically enough, that programming critical of them will not put audiences in the mood to buy their product. All advertising-supported media feel such pressure. But given that some forms of television are completely supported by advertising and that there is a very high level of competition for television advertising dollars at both the national and local level (the latter unlike US newspapers, which typically have a monopoly in their market), television is especially vulnerable to advertising pressures (Soley, 2002).

The removal or avoidance of advertising criticism can be seen throughout the history of television. Sometimes this can reach absurd levels: during the single-sponsorship days of TV, Miles Labs, a drug company and the sponsor of *The Flintstones* during its original prime-time manifestation, forbad any character on the show from having a stomachache or headache (Samuel, 2001, p. 139). Sometimes the issue is not so trivial: consumer reporter David Horowitz was fired from WCBS in New York, perhaps due to the repeated complaints from car advertisers about his stories on auto safety (Soley, 2002, pp. 12–13). More systematically, one local TV station executive noted that "you never see an investigative report on [local] car dealers. It's an unspoken rule" (quoted in Andersen, 1995, p. 23; for other examples, see Baker, 1994). In one survey of local television reporters and editors, nearly 75 percent reported that advertisers had attempted in the previous three years to influence a story, 40 percent stated that advertisers had succeeded at their station (Soley, 2002). One survey of US network news correspondents, though, found significantly less pressure from advertisers, although CNN correspondents reported more pressure than other networks (Price, 2003). At least at the local level, self-censorship becomes a concern as media personnel anticipate advertiser response and reporters learn to avoid altogether the kind of stories that cause organizational headaches.

The "plugola" effect

The flip side to the suppression of ideas critical to specific advertisers is the insertion of ideas that favor specific advertisers. Compared to other media, television seems somewhat more willing to offer such promotional incentives. It is commonplace on American television, for example, to have "co-sponsored" segments of a program: "This ABC Sports exclusive presentation brought to you by . . . Tostitos Gold . . . and Circuit City." Television often goes beyond this. In news, for example, local television reporters are sometimes pressured to do stories that feature advertisers (Soley, 2002). News organizations routinely receive public relations materials from corporations, including their advertisers, that make it easy to use such stories. On the morning I wrote this sentence, for example, there was a story on the local NBC affiliate about the new lime flavor of Diet Coke, a story that was no doubt PR-originated and advertiser-flattering in effect. One also sees news stories that feature or mention advertisers on national news. The night of the final episode of *Seinfeld* on NBC also saw on that network a news segment on the *NBC Nightly News* about Garden Burger, an advertiser during the program (McAllister, 2002).

Product placement and sponsorship are increasingly commonplace on television. In some countries product placement has been a standard practice for many years. In Brazil, for example, corporations may sign annual contracts to integrate products in popular *telenovelas* (Frith and Mueller, 2003). With American television, product placement has not always been the easiest fit. While marketers spend $360 million a year to place products in film (James, 2003), the practice was problematic for US television due to the medium's need to meet FCC rules of revenue disclosure and to avoid treating certain advertisers more favorably than others. Competition for advertising dollars sparked by cable and the Internet and the looming threat of the digital personal recorder have changed this. Some forms, such as sports, daytime television, and reality-based shows, have been more accepting of product placement than prime-time "scripted" shows. In reality-based programs, for example, advertisers will provide visible, even heroic, products for abused contestants, such as the stranded players of *Survivor* (Deery, 2004). However, deals like the Volkswagen Bug in *Smallville* and Ragu in *Everybody Loves Raymond* may signal a wider pervasiveness, as will be discussed at the end of this chapter (Steinberg and Vranica, 2004).

Sometimes entire programs will be built around one advertiser, especially in sponsorship deals reminiscent of early television. Besides the end-of-season US college football games (like the Tostitos Fiesta Bowl), sponsored specials include MTV-Japan's *Super Dry Live* (sponsored by the beer company Asahi) and CBS's *Victoria's Secret Fashion Show*. The 2003 installment of this latter program aired several commercials for the lingerie retailer – between segments of the program that were also essentially commercials for the store – and both the national and local news aired promotions for tie-in news stories *about* the program *during* the program. Regular series such as *EA Sports NFL Matchup* on ESPN used images from the EA Sports videogame "Madden NFL Football" to illustrate strategy, blurring the program with intertwined commercials for EA videogames with similar images.

As with product placement in film, the intrusion of specific advertiser images in television programming raises issues of programming autonomy and creative integrity. It also raises concerns about the degree to which television creators can air non-consumption-friendly programming ideas given the increasingly close relationship of advertising to production, a concern raised by the following section.

The "don't rock the boat" effect

Another related effect of advertising upon media content is the "blanding" of television programming (Bagdikian, 2000; Baker, 1994). All things being equal, advertising discourages controversy. As one marketing professional who screens controversial media content for advertisers explained, "Basically, we look for the Big Six: sex, violence, profanity, drugs, alcohol and religion" (quoted in Richards and Murphy, 1996, p. 22). Controversy, it is believed, will cause some audience members to be in a non-buying mood and others to leave altogether, thereby eroding ratings. This effect is true of all advertising-supported media, but especially true of television. In a newspaper, if a story angers a reader, that reader may very well just turn the page. In this case, the reader stays with the same economic entity (the newspaper). With television, though, if viewers are angered, they will turn the channel, eroding its ratings. And with sound and visuals being a part of the medium, television is more intrusive than print, since again the latter just requires looking at another part of the page to selectively ignore something offensive. The intrusiveness of television, in addition to the total reliance upon advertising by television, may also make the medium more vulnerable to advertising boycotts by upset protesters of television. These factors indicate that television advertisers may be especially touchy about airing a commercial during controversial content.

Richards and Murphy (1996, p. 24) provides a long list of national and local programming that felt heat from advertisers before or after airing controversial ideas. An example from 2003 may be telling as well. Although CBS president Leslie Moonves denied that threats from conservative groups to engage in an advertising boycott was a factor, the television film *The Reagans* was moved from the advertising-supported CBS to the advertising-free (and much less accessible) pay-cable network Showtime amidst such threats (for a review of this incident, see Carter, 2003). This decision was made before advertisers even had a chance to screen the film. Regardless of CBS's intention, the image of such apparent public acquiescence to a possible advertiser boycott may send a chilling message to future creative personnel in TV.

On the other hand, other industrial factors may occasionally encourage controversy on television, with advertisers accepting, even welcoming, it. To preview a latter effect, controversy seems more welcome on certain narrowly targeted cable networks that can deliver the right audience. Thus, controversial animated programs such as *Beavis and Butthead* in the early 1990s on MTV and *South Park* on Comedy Central bring in young, hip viewers to advertisers attempting to sell products – like movies and soft drinks – with a similar lifestyle image.

The "conspicuous consumption" effect

If one cannot have controversial or potentially disturbing programming, then what may take its place? In fact, another effect that advertising may have upon television programming is a greater emphasis upon consumer values in the programming itself, even beyond individual advertiser influence. Advertisers desire television to put viewers in a "buying mood" (Baker, 1994, pp. 62–6). As television has matured, this advertiser-friendly orientation has seeped into the medium to such a degree that consumerist values are implemented without the influence of individual advertising pressure. So even beyond the touting of individual advertisers and the removal of criticism of individual advertisers, television is consumption oriented both in some of the prominent genres found on television and in the portrayal of lifestyles in other genres that are not inherently consumerist.

Consumption-oriented programming is found throughout the television schedule. During the daytime, for example, US game shows test consumer knowledge – such as the Pax network's *Shop 'Til You Drop* and *Supermarket Sweep* and CBS's *The Price Is Right* (the latter a hit among college students) – or blatantly celebrate consumer overspending, such as the now-defunct *Debt* (Merskin, 1998). Some specialized cable channels are "niched" with consumption-oriented lifestyles in mind, such as Home and Garden Television and the Food Network in the United States, creating very conducive programming environments for advertisements. On the Food Network, although the network does air programs about how to cook frugally without brand names prominently shilled, it also includes programs like *Unwrapped*, which celebrates mass-produced (and mass-marketed) foods such as branded candies and pastries.

Similarly, while most press discussions of the Bravo/NBC program *Queer Eye for the Straight Guy* focused on its risky advancement of gay visibility on television, in another sense the consumption orientation of the program was perfectly conventional for commercially supported television. The morale of this program is the same as most commercials: "consumption will make your life better." Episodes prominently feature/tout particular brands of skin care products, wines, fashions, and furnishings; these specific products are then also found listed on the program's website. The program's hosts were not shy about voicing the transformational power of such commodities. "We're taking you to a relationship counselor," the fashion guru Carson says in one episode to one commodity-challenged couple, "I mean, no, we're taking you to Crate and Barrel."

Other programs may not center around commodity competence or knowledge, but, rather, assume levels of "normal" class and lifestyle that raise expectations about the quality and quantity of personal possessions. As television looks to attract affluent audiences and create the proper upscale environment for advertisers, characters on television tend to display well-to-do lifestyles much more commonly than in real life. *Friends* and *Frasier* live in spacious urban dwellings, and soap operas feature the lifestyles of the rich and powerful. Schor (1999) argues that, more than commercials, the display of conspicuous consumption on television programs may promote overspending and indebtedness among television viewers. The rich may hide from us in gated mansions in real life, but they are on display on television. Indeed, the more one watches television, the more that one associates possessions/products with affluence, although this relationship is mitigated by income and education level (O'Guinn and Shrum, 1997). In terms of young audiences, there is evidence that exposure to television generally (and television advertising specifically) leads to an acceptance of more materialist values (Smith and Atkin, 2003).

The "that's entertainment" effect

Television is an entertainment-oriented medium; it often is fun to watch. Funny sitcoms, absorbing dramas, and perversely fascinating reality-based shows dominate the broadcast schedule. And everyone likes to be entertained. But can there be too much of a good thing, and, if so, is television advertising responsible for this in some way? Postman argues that television's entertainment orientation is largely due to its characteristics as a communication medium: the audio and visual nature of television delivers information in such a way that is non-linear and stimulating to the senses. The content of the medium, then, has increasingly exploited the characteristics of this image and sound delivery system as television has matured.

However, advertising's dominance as a funding system and the resulting emphasis on ratings also play a significant role in television's entertainment orientation. Just as advertisers abhor controversial ideas that may offend viewers and cause them to leave the channel, they also do not embrace ideas that are too complex or even boring for some viewers that may trigger channel surfing or perhaps even critical thinking not conducive to the reception of often-stupid advertising messages.

We see this effect not just in the dominance of certain titillating genres like sitcoms and reality-based shows, but also in trends in television news. For example, in terms of audiovisual elements, the "pace" of television news is significantly faster now than in the early days of television, with increasingly heavier editing and elements such as the incredibly shrinking video-quote soundbite (Hallin, 1992). In a trend that also illustrated the influence of the Internet on television news, the cluttered screens of cable and broadcast news coverage of the 2003 Iraq War shows the pressure to keep interest high. Split screens, multiple scrolls, and labels placed throughout the screen do not just mimic websites, but also attempt to hold viewer attention to the screen through over-stimulation. Perhaps more disturbing are news topic trends that may be influenced by ratings pressure. While television news coverage of celebrity and scandal increases, coverage of less exciting topics such as international news decreases (McChesney, 1999; McAllister, 2002, 2003).

The "pardon the interruption" effect

This next effect is painfully obvious: commercials interrupt the TV programs we are watching. Although the US model of constant interruptions is normalized for those viewers, such a model does not necessarily follow the implementation of a commercial television system. In the United Kingdom, commercials were circumscribed to the start and finish of programs or during "natural breaks" (Samuel, 2001, p. 69). Even more confining, early in Italy's commercial system ads were compartmentalized into three pre-established time periods (Samuel, 2001, p. 185). When interrupting programs, as is now the norm in most countries, television also has a qualitatively different interruption effect than print, since in that medium ads can be easily ignored while reading a story.

In the United States, there are no legal limits on the amount of time that television commercials may take up or the interruption pattern that commercials may use, except with regard to children's television programming (Kunkel and McIlrath, 2003). Indeed, if such limits existed then some types of commercial forms, such as the infomercial or home-shopping channels, may be illegal. Even with regular programming, the amount of time devoted to intrusive spot advertising is significant. One concern is that the amount of promotional time that viewers are exposed to – sometimes called "clutter" – is increasing. In 1991, for example, ABC averaged 12 minutes, 50 seconds per hour of non-programming content in prime time (mostly commercials and program promos); in 2002, the

229

same network averaged 15 minutes, 16 seconds (Fleming, 1997; Chunovic, 2003).

The growth and interruptability of commercial messages on television have several implications. One such implication is that, if programs are the bait or reward for viewers, then we are now receiving less and less reward from television. Television programs have gotten shorter as commercial and promotional clutter has increased. This is frustrating not just for the audience, but also for the creative personnel: "It's getting worse and worse," said one executive producer about the decrease in creative time, "we have less time to tell a story, and it's brutal for the writers" (Flint, 1996/1997, p. 32).

Postman (1986) wonders if commercials, along with the increased pace of television (that is also commercially triggered), decrease attention spans and make us less tolerant of extended argument. A similar effect is the trivialization that may occur when important ideas are interrupted by often-silly and certainly irrelevant-to-the-program selling messages. As Postman and Powers argue, the implicit message given when commercials interrupt serious content is "You needn't grieve or worry about what you are seeing. In a minute or so, we will make you happy with some good news about how to make your teeth whiter" (1992, p. 126). Thus, when the United States began the Persian Gulf War by launching missiles on Iraq on the evening of January 17, 1991, during the same time as the Evening News in the Eastern United States, Tom Brokaw on the *NBC Nightly News* announced that, "It does appear that an attack of some kind, what dimensions we cannot say, is underway tonight on Baghdad itself and perhaps other military targets in Iraq. As soon as we can learn more, we'll be to you with additional details." Fade to black. Then, a Brokaw-esque man informs us, "One day there was just too much gray. And I thought, should I or shouldn't I? . . . I wanted natural looking color . . . I figured, if it was from Clairol, I could trust it." This was followed by another message for prunes, another for Vaseline Intensive Care lotion, and then Brokaw returning by saying, "We're back, we want to tell you that there's an attack underway on Iraq . . ." Of course, television executives know that advertising may delegitimize or trivialize serious discussion, hence their promotion of especially prestigious programming (NBC's 1997 broadcast of *Schindler's List*, for example) as "commercial free" or "with limited interruption." Normal programming, clearly, is not so lucky.

The "youth (and other advertising-friendly groups) will be served" effect

The seventh effect upon programming is the fragmentation of audiences into different target markets, a fragmentation that takes many forms, but often privileges youth. Advertisers do not want just large audiences, they want audiences that are most likely to purchase their products. This means, then, that advertisers are looking to target their marketing efforts for the best possible efficiency: niche audiences that match in demographics, lifestyle and interests the profile of

a niche market. Following other advertising-supported media that have segmented audiences (such as radio and magazines), the technology of cable and
satellites allow the creation of hundreds of channels which are created to cultivate the complex demographic combinations for efficient delivery of specialized
audience product to potential advertisers (Turow, 1997). So while the Discovery
Channel, with its straight-up nature and science emphasis, is the most "generic"
of Discovery Communications Inc. holdings, Discovery Kids delivers a younger
demographic with the same interests. Other cable channel holdings like Discovery Wings, Discovery Times, and Discovery Health similarly drill down to
deeper demographic layers.

One implication of the segmentation of television is that society may be
"broken up" with commercial criteria as marketers define social groupings
according to market imperatives. Less commonalities may be evident to society-
at-large as cable stations (and other advertising-driven media) look to subdivide
people into demographically defined markets. Marketers do not just react to
existing consumer groups; they help define and solidify such groups in their
selective advertising purchasing and subsequent formations by media outlets
(Turow, 1997).

Another implication of this is that some audiences are more desirable as
products than others. Young people, for example, have disposable income and a
propensity to spend it, are key markets for leisure activities (like movies), and
are willing to try new brands. Older people, who may have as much potential
spending power but not as likely to be persuaded to switch brands by advertising, are less desirable as audience-product. This means fewer television choices
for these less advertiser-coveted audiences. One television executive used this
criterion in justifying the cancellation of the well-viewed but older-skewing
CBS program *Dr. Quinn, Medicine Woman*: "Unfortunately we get paid zero –
not a nickel, but zero – for anybody over 55. Despite the show's very loyal and
devoted audience, the amount of money we were getting per advertising spot was
disastrous" (quoted in Budd, Clay, and Steinman, 1999, p. 74).

Conclusion: Three Future Trends

This chapter explored some of the major themes about advertising as a symbolic
and economic system. In the critical scholarship on television commercialism
there is often a separation between these two commercial elements. Rhetorical
critics, cultural studies scholars, and content analysts concentrate on the messages in television commercials, while political economists focus on advertising's
economic impact. When one looks at the future of television advertising, one sees
trends that will affect both arenas. Advertisers have been frustrated with the
viewers' ability to avoid television commercials since at least the 1980s, when
remote controls, VCRs, and advertising clutter became more prevalent (Andersen,
1995). With newer concerns over the ease with which personal digital recorders

allow viewers to skip commercials, advertisers' anxiety – and attempts at further control – are heightened. Recent trends in fact may further combine the symbolic machinations of a television commercial with the financial impact of the "ad dollar" that infuses the television industry.

Digital manipulation

One trend that may become increasingly salient in commercials is the use of digitally created imagery to enhance the effectiveness of selling symbols in both commercials and programs. We of course see this now. Digital manipulation makes products shimmer more brightly, models sexier, and production values more spectacular. Digital technology, as it does in film, creates very realistic-looking phenomena that do not exist but that may enhance branding strategies, such as long-dead celebrities interacting with the latest products (DeSalvo, 1998). We have only begun to see the potential for symbolic persuasion and manipulation that visual-digital technology will bring to advertisements. But such technology also enhances the commercial presence in television programs. "Virtual advertising" includes the increasingly accepted practice of digitally inserting product signage into televised events, such as rotating ads behind the home plate of a baseball game that only the television viewers can see and targeting virtual billboards for different regions of the viewership. Such technology would also allow "retroactive product placement" in which new products can be inserted into the syndicated or DVD versions of older programs and movies; such insertion occurred on a limited basis in 2003 (Kafka, 2003).

Advertiser-controlled television programs

Another way to counter such factors as clutter and personal recording technology is for advertisers to demand more operational control of program production. This would return TV to the days of early sponsorship by creating and shaping programs to make them more conducive to the seamless insertion of commercial messages into programs. For example, in December 2003 ABC and a subsidiary of the mega-advertising conglomerate the WPP Group announced plans for a project to develop programs together. An ABC executive said, "This gives us a chance to have a collaborative relationship with our major clients [advertisers] early in the process" (Reuters, 2003, p. C2). In describing the blurring effect between commercial interests and programming that such a "collaboration" may have, an executive for Coca-Cola noted, "We're headed to ideas . . . Ideas that bring entertainment value to our brands and ideas that integrate our brand into entertainment" (Lewis, 2003, p. 20). In late 2003, the Nielsen Company announced that it would begin charting product placements and calculating audience exposure, perhaps signaling the further institutionalization of the practice (James, 2003). Clearly, this direction would enhance not just the visibility of products in programs, but also the degree to which programs should be commercially slanted. The "branded program" *Nike Training Camp*, shown on the cable station College Sports Television, typically "will feature only teams that have endorsement deals with Nike" (Ives, 2003, p. C8).

Database marketing and interactive television

A final trend that advertisers are increasingly embracing is to co-opt interactive, digital systems like TiVO rather than seeing it as the enemy. One advantage of such interactive systems is the collection of viewing information that such systems may offer. If interactive systems record, even predict, our preferences, there is concern that this data may be collected by marketers to enhance their consumer profiles about viewers, and create hyper-targeted marketing campaigns based upon this information. Further, the integration of increased home shopping opportunities on television, even in routine programming (i.e. "Pause here to buy a tie like Regis's!"), may enable highly targeted selling messages to be placed near these buying moments. Such techniques would not only enhance the degree to which these selling messages may fetishize commodities – creating an especially targeted way of celebrating the product – but would also make it easier for viewers to acquiesce to the buying impulse the targeted ads encourage. The "moment of pause" between seeing an ad and buying the product, a characteristic of the physical distance between home televisions and retail outlets, would disappear. Scholars have noted that viewers may disregard the privacy concerns of such systems given their promise of viewing flexibility (Andrejevic, 2002; Campbell and Carlson, 2002).

A Final Word: Commercials and Resistance

Commercials are fully integrated into television and much of our everyday culture. As argued above, this integration may increase in the new digital television era. Although commercials may give us product information and entertain us, there are associated costs. The symbols in commercials are often problematic in terms of their portrayal of social groups and their framing of commodities as a solution to various social and personal problems. Economically, commercials shape the media messages we receive, or don't receive, in ways that may be unintended, but that are nevertheless troublesome for a democratic society that thrives on a diversity of ideas.

But the integration will not take place without the protests of those who resist the further commercialization of television as well as the larger society. As writers such as Naomi Klein (1999) have argued, not everyone is willing to be an obsessed-with-buying consumer or even to buy into the messages of commercials. Organizations like Adbusters (www.adbusters.com) provide alternatives to consumer culture and Media Watch (www.mediawatch.com) highlight racism and sexism in advertisements. Many of these resistant activities target television specifically. In 2003, the group Commercial Alert (www.commercialalert.org) filed complaints to both the Federal Communication Commission and the Federal Trade Commission about the increased but unlabeled practice of product placement on US television, a practice that this group argues violates some of the earliest policies about the full disclosure of commercial influences in broadcasting (Bauder, 2003). Such groups and the resistance strategies they advocate, although standing against a massive commercial force that knows how to publicize and promote self-serving ideas, can nevertheless keep vibrant ideas that may go against the televised commercial grain.

References

Alexander, A., Benjamin, L. M., Hoerrner, K., and Roe, D. (1998) "'We'll Be Back in a Moment': A Content Analysis of Advertisements in Children's Television in the 1950s," *Journal of Advertising*, *27*(3), 1–9.

Andersen, R. (1995) *Consumer Culture and TV Programming*, Boulder, CO: Westview.

Andersen, R. (2000) "Road to Ruin: The Cultural Mythology of SUVs," in R. Andersen and L. Strate (eds.), *Critical Studies in Media Commercialism*, Oxford: Oxford University Press, pp. 158–72.

Andrejevic, M. (2002) "The Work of Being Watched: Interactive Media and the Exploitation of Self-disclosure," *Critical Studies in Media Communication*, *19*(2), 230–48.

Bagdikian, B. (2000) *The Media Monopoly*, 6th edn., Boston: Beacon Press.

Baker, C. E. (1994) *Advertising and a Democratic Press*, Princeton, NJ: Princeton University Press.

Bauder, D. (2003) "FCC Complaint Decries Hidden Ads on TV Shows," *The Pittsburgh Post-Gazette*, October 4, p. C12.

Berger, A. A. (2004) *Ads, Fads, and Consumer Culture: Advertising's Impact on American Character and Society*, 2nd edn., Lanham, MD: Rowman & Littlefield.

Barnouw, E. (1978) *The Sponsor: Notes on a Modern Potentate*, New York: Oxford University Press.

Bresnahan, M. J., Inoue, Y., Liu, W. Y., and Nishida, T. (2001) "Changing Gender Roles in Prime-time Commercials in Malaysia, Japan, Taiwan, and the United States," *Sex Roles*, *45*(1/2), 117–31.

Bristor, J. M., Lee, R. G., and Hunt, M. R. (1995) "Race and Ideology: African-American Images in Television Advertising," *Journal of Public Policy & Marketing*, *14*(1), 48–59.

Budd, M., Craig, S., and Steinman, C. (1999) *Consuming Environments: Television and Commercial Culture*, New Brunswick, NJ: Rutgers University Press.

Butler, J. G. (2002) *Television: Critical Methods and Applications*, 2nd edn., Mahwah, NJ: Lawrence Erlbaum Associates.

Campbell, J. E. and Carlson, M. (2002) "Panopticon.com: Online Surveillance and the Commodification of Privacy," *Journal of Broadcasting & Electronic Media*, *46*(4), 586–606.

Carter, B. (2003) "Showtime to Present 'Reagans' This Month," *The New York Times*, November 18, p. E4.

Chunovic, L. (2003) "Clutter Reaches All-time High," *Television Week*, May 12, p. 19.

Coen, R. (2003) *Insider's Report: Robert Coen Presentation on Advertising Expenditures*, December, New York: Universal McCann. Retrieved January 12, 2004 at http://www.universalmccann.com/ourview.html.

Coltrane, S. and Messineo, M. (2000) "The Perpetuation of Subtle Prejudice: Race and Gender Imagery in 1990s Television Advertising," *Sex Roles*, *42*(5/6), 363–89.

Cook, J. P. (2000) "Consumer Culture and Television Home Shopping Programming: An Examination of the Sales Discourse," *Mass Communication & Society*, 3(4), 373–91.

Davies, P. G., Spencer, S. J., Quinn, D. M., and Gerhardstein, R. (2002) "Consuming Images: How Television Commercials that Elicit Stereotype Threat can Restrain Women Academically and Professionally," *Personality and Social Psychology Bulletin*, 28(12), 1615–28.

Deery, J. (2004) "Reality TV as Advertainment," *Popular Communication*, 2(1), 1–20.

DeSalvo, K. (1998) "Spot World's Dawn of the Living Dead," *SHOOT*, April 17, pp. 1–3.

Entman, R. M. and Book, C. L. (2000) "Light Makes Right: Skin Colour and Racial Hierarchy in Television Advertising," in R. Andersen and L. Strate (eds.), *Critical Studies in Media Commercialism*, Oxford: Oxford University Press, pp. 214–24.

Fleming, H. (1997) "PSA Slice Shrinks as Commercial Pie Grows," *Broadcasting & Cable*, March 31, pp. 19, 21.

Flint, J. (1996/1997) "Intro Course Focus: How to Stop Surfing," *Variety*, December 23–January 5, pp. 31, 32.

Frith, K. T. and Mueller, B. (2003) *Advertising and Societies: Global Issues*, New York: Peter Lang.

Furnham, A. and Mak, T. (1999) "Sex-role Stereotyping in Television Commercials: A Review and Comparison of Fourteen Studies Done on Five Continents over 25 Years," *Sex Roles*, 41(5/6), 413–37.

Global Advertising (2003) "World Media on the Move," *Campaign*, May 9, p. 10.

Goldman, R. and Papson, S. (1996) *Sign Wars: The Cluttered Landscape of Advertising*, New York: Guilford.

Hargreaves, D. and Tiggemann, M. (2003) "The Effect of 'Thin Ideal' Television Commercials on Body Dissatisfaction and Schema Activation during Early Adolescence," *Journal of Youth and Adolescence*, 32(5), 367–73.

Hallin, D. C. (1992) "Sound-bite News: Television Coverage of Elections, 1968–1988," *Journal of Communication*, 42(2), 5–24.

Holden, T. J. M. (2001) "The Malaysian Dilemma: Advertising's Catalytic and Cataclysmic Role in Social Development," *Media, Culture & Society*, 23(3), 275–97.

Ives, N. (2003) "Product Placement Goes to College," *The New York Times*, October 27, p. C8.

James, M. (2003) "Nielsen to Follow Popularity of Product Placement on Prime-time Television," *The Los Angeles Times*, December 5, p. C1.

Jhally, S. (1987) *The Codes of Advertising: Fetishism and the Political Economy of Meaning in the Consumer Society*, New York: Routledge.

Jhally, S. (2000) "Advertising at the Edge of the Apocalypse," in R. Andersen and L. Strate (eds.), *Critical Studies in Media Commercialism*, Oxford: Oxford University Press, pp. 27–39.

Johnson, F. L. and Young, K. (2002) "Gendered Voices in Children's Television Advertising," *Critical Studies in Media Communication*, 19(4), 461–80.

Kafka, P. (2003) "Spot the Spot," *Forbes*, November 10, p. 68.

Klein, N. (1999) *No Logo: Taking Aim at the Brand Bullies*, New York: Picador.

Kunkel, D. and McIlrath, M. (2003) "Message Content in Advertising to Children," in E. L. Palmer and B. M. Young (eds.), *The Faces of Televisual Media: Teaching, Violence, Selling to Children*, 2nd edn., Mahwah, NJ: Lawrence Erlbaum Associates, pp. 287–300.

Lewis, N. (2003) "The Future of Soap Operas," *Brand Strategy*, September, pp. 20–2.

Magder, T. (2004) "The End of TV 101: Reality Programs, Formats and the New Business of Television," in S. Murray and L. Ouellette (eds.), *Reality TV: Remaking Television Culture*, New York: New York University Press, pp. 137–54.

McAllister, M. P. (1999) "Super Bowl Advertising as Commercial Celebration," *The Communication Review*, 3(4), 403–28.

McAllister, M. P. (2002) "Television News Plugola and the Last Episode of Seinfeld," *Journal of Communication*, 52(2), 383–401.

McAllister, M. P. (2003) "Is Commercial Culture Popular Culture?: A Question for Popular Communication Scholars," *Popular Communication*, 1(1), 41–9.

McChesney, R. W. (1999) *Rich Media, Poor Democracy: Communication Politics in Dubious Times*, Urbana, IL: University of Illinois Press.

Merskin, D. (1998) "The Show for Those Who Owe: Normalization of Credit on Lifetime's *Debt*," *Journal of Communication Inquiry*, 22(1), 10–26.

Mueller, B. (1996) *International Advertising: Communicating Across Cultures*, Belmont, CA: Wadsworth.

O'Guinn, T. and Shrum, L. J. (1997) "The Role of Television in the Construction of Consumer Reality," *Journal of Consumer Research*, 23(4), 278–97.

Peterson, R. T. and Byus, K. (1999) "An Analysis of the Portrayal of Female Models in Television Commercials by Degree of Slenderness," *Journal of Family and Consumer Sciences*, 91(3), 83–91.

Postman, N. (1986) *Amusing Ourselves to Death: Public Discourse in the Age of Show Business*, New York: Penguin.

Postman, N. and Powers, S. (1992) *How to Watch Television News*, New York: Penguin Books.

Price, C. J. (2003) "Interfering Owners or Meddling Advertisers: How Network Television News Correspondents Feel about Ownership and Advertisers Influence on News Stories," *The Journal of Media Economics*, 16(3), 175–88.

Reuters (2003) "ABC, Ad Agency to Develop TV shows," *Los Angeles Times*, December 2, p. C2.

Richards, J. I. and Murphy, J. H., II (1996) "Economic Censorship and Free Speech: The Circle of Communication between Advertisers, Media, and Consumers," *Journal of Current Issues and Research in Advertising*, 18(1), 21–34.

Samuel, L. R. (2001) *Brought to You by: Postwar Television Advertising and the American Dream*, Austin, TX: University of Texas Press.

Schor, J. B. (1999) *The Overspent American: Why We Want What We Don't Need*, New York: HarperPerennial.

Seiter, E. (1993) *Sold Separately: Children and Parents in Consumer Culture*, New Brunswick, NJ: Rutgers University Press.

Shields, V. R. (1997) "Selling the Sex that Sells: Mapping the Evolution of Gender Advertising Research across Three Decades," in B. R. Burleson (ed.), *Communication Yearbook 20*, Thousand Oaks, CA: Sage, pp. 71–109.

Shields, V. R., with Heinecken, D. (2002) *Measuring Up: How Advertising Affects Self-image*, Philadelphia, PA: University of Pennsylvania Press.

Smith, S. L. and Atkin, C. (2003) "Television Advertising and Children: Examining the Intended and Unintended Effects," in E. L. Palmer and B. M. Young (eds.), *The Faces of Televisual Media: Teaching, Violence, Selling to Children*, 2nd edn., Mahwah, NJ: Lawrence Erlbaum Associates, pp. 301–25.

Smythe, D. W. (1977) "Communications: Blindspot of Western Marxism," *Canadian Journal of Political and Social Theory*, 1(3), 1–27.

Soley, L. (2002) *Advertising Censorship*, Milwaukee, WI: Southshore Press.

Stabile, C. A. (2000) "Nike, Social Responsibility, and the Hidden Mode of Production," *Critical Studies in Media Communication*, 17(2), 186–204.

Stein, S. R. (2002) "The '1984' Macintosh Ad: Cinematic Icons and Constitutive Rhetoric in the Launch of a New Machine," *Quarterly Journal of Speech*, 88(2), 169–92.

Steinberg, B. and Vranica, S. (2004) "Prime-time TV's New Guest Stars: Products," *The Wall Street Journal*, January 12, p. B1.

Strate, L. (2000) "Intoxicating Consumptions: The Case of Beer Commercials," in R. Andersen and L. Strate (eds.), *Critical Studies in Media Commercialism*, Oxford: Oxford University Press, pp. 145–57.

Turow, J. (1997) *Breaking Up America: Advertisers and the New Media World*, Chicago: University of Chicago Press.

Watts, E. K. and Orbe, M. P. (2002) "The Spectacular Consumption of 'True' African American Culture: 'Whassup' with the Budweiser Guys?," *Critical Studies in Media Communication*, 19(1), 1–20.

Williams, R. (1980) *Problems in Materialism and Culture: Selected Essays*, London: Verso.

Wilson, C. C. and Gutiérrez, F. (2003) "Advertising and People of Color," in G. Dines and J. M. Humez (eds.), *Gender, Race and Class in Media: A Text Reader*, Thousand Oaks, CA: Sage, pp. 283–92.

Wong, W. S. (2000) "The Rise of Consumer Culture in a Chinese Society: A Reading of Banking Television Commercials in Hong Kong during the 1970s," *Mass Communication & Society*, 3(4), 393–413.

Zhang, Y. B. and Harwood, J. (2004) "Modernization and Tradition in an Age of Globalization: Cultural Values in Chinese Television Commercials," *Journal of Communication*, 54(1), 156–72.

CHAPTER TWELVE

Watching Television: A Political Economic Approach

Eileen R. Meehan

Introduction

Do you watch television? For most Americans, this query means: do you have a television set turned on? But for political economists, watching television is a rather different matter. We watch television by tracing political and economic forces, entities, and structures that foster the production, distribution, and continuation of some kinds of expression – shows, plots, characters, assumptions, and visions – rather than other kinds of expression. That task requires varied approaches to television.

One such approach analyzes markets to determine how media corporations construct the television industry. Another involves tracing relationships between companies, as well as between companies and governmental entities, to see how these markets are negotiated. Yet another examines corporate structures in order to uncover the internal relationships and elements that shape a company's goals, actions, and interactions. Understanding the systems that comprise the television industry helps us understand why we get what we get on television. In this way, we study television in terms of its political economy, by examining "the social relations that mutually constitute the production, distribution, and consumption of resources" (Mosco, 1996, p. 25).

In this chapter, I will first discuss some of the ways in which political economists have studied television and then discuss two foci for research on the political economy of programming. The initial focus examines the three interlinked markets that comprise television: the markets in which programs, audiences, and television ratings are sold. The point here is to test that truism that television gives viewers what they want and thus reflects us and our beliefs. The subsequent focus examines how corporate structure and the governmental policies supporting those configurations can integrate industries. This tests the truism that the separate industries of network television, cable television, and satellite television compete fiercely to persuade us to patronize their particular technology.

My analysis will focus entirely on American television because of the historical dominance of American programming in national markets and the increasing legitimacy accorded to the strictly commercial form of television articulated in the United States. In the 1950s and 1960s, a combination of favorable export laws and foreign aid policy positioned American television producers to control emerging markets for programming (Barnouw, 1990). Further, within those emerging markets, American producers could price their wares below the cost of local production because they had already covered those costs and earned their profits in the United States. These early advantages translated into long-term dominance in the country-by-country markets for television programs.

Since the 1980s, that dominance has intensified as neo-conservative governments attacked noncommercial television in the United States and Western Europe as the demise of Communist governments privatized and marketized state-run television systems in Eastern Europe. While national variations still exist among television systems, overall, these systems have come to focus increasingly on commercialism in the American style: that is, on markets for advertising, for viewers that advertisers want, and for programs that attract the viewers that advertisers want to reach. That style has itself been intensified in the United States as neo-conservative administrations have sought to delegitimize service to the public interest and to eradicate or relax limitations on ownership, restrictions on mergers, regulation of commercial speech, etc. (Wittebols, 2004). Over time, then, as political economists have watched television in the United States and elsewhere, television has become increasingly commercialized, privatized, and marketized – thereby emphasizing and globalizing the peculiarities of the American focus on markets constrained by advertisers' preferences. With these reflections in mind, let us turn to a brief review of the research on the political economy of television.

Tracking Television

The research literature addresses the political construction of American television as an industry, the economic structures comprising that industry, and the programming delivered via television technologies. Whether addressed individually or together, these three areas intrigue political economists of television, allowing us to ask questions about the impact of governmental and corporate policies on television's universal curriculum (Gerbner, 1996) delivered to us in our homes as well as in "sites of commerce, bureaucracy, and community . . ." (McCarthy, 2001). Television's ubiquity is particularly noteworthy given its consistent mission to deliver "eyeballs" to advertisers over the last 50-plus years. The decisions and structures that undergird an omnipresent system dedicated to fostering the ideology of consumerism are naturally of interest to scholars who want to understand how consumerist ideology rises to dominate the larger cultures that comprise a nation's way of living and being in the world (Andersen, 1995; McAllister, 1996). Because the chapter deals mainly with economic structures, I will concentrate on scholarship addressing the politics and ideology of television.

Political Supports for the American Television Industry

Throughout its history, from the inception of radio to the present day, broadcasting has depended upon political supports (Douglas, 1987; Barnouw, 1990). At the end of World War I, when the Department of the Navy could not persuade Congress to give it a monopoly over radio, the Navy asked its long-term contractor, General Electric (GE), to take radio in hand. GE proposed the creation of the Radio Corporation of America to AT&T and Westinghouse, its partners in the wartime Patent Pool. The result was RCA and its NBC networks, with a representative of the Navy on RCA's Board of Directors.

Besides its long-term connections to the military, the broadcasting industry has relied on legislation to define its operating rules and on regulatory agencies to enforce those rules in its favor (McChesney, 1993; Streeter, 1996). The industry has depended on Constitutional precedents to grant it freedom of speech, shelter its operations from local control as a form of interstate commerce, copyright its programming, and protect its technologies via patents (Bettig, 1996; Tillinghast, 2000). Political economists have examined these supports as well as the networks' news coverage of these topics (Brown, 1998) and the shared interests that connect the primary companies within the industry to politicians, political parties, and ideological factions (Gitlin, 1980). In tracing these relationships, political economists have illuminated the institutions, precedents, and structures that fold radio, television, cable, and satellite distribution of signals into the American political system. More importantly, they demonstrate how these media are constructed as industries through the active involvement of corporations and the state.

From this perspective, the state sets in place economic rules, incentives, and protections that foster the privatization and commercialization of technologies of mass communication. Analytically, this creates a picture of television in which the foreground is filled with individuals, corporations, trade associations, non-profit organizations, and governmental entities pursuing their particular interests against a background of institutional structures, rules, policy processes, and political agendas that together constitute the state. A focus on the foreground reveals the dynamic nature of these debates, discourses, and outcomes: individuals rise and fall in prominence; debates among corporations, trade associations, and non-profit organizations shift grounds; administrations and Congresses argue over and alter policies. A focus on the background reveals the stability of the political system's commitment to a privately owned, for-profit industry that uses public property for strictly private ends – to wit, earning the highest revenues possible by embedding advertisements in programming properties whose raison d'être is to deliver consumers to advertisers.

The extent of that commitment is noteworthy. Since 1980, Republican and Democratic Administrations, Congresses, and politicians have joined with lobbyists to restructure the American economy generally and the media industries particularly. Variously called Reaganism, monetarism, neo-conservatism or neo-liberalism, the political deregulation of the economy encouraged companies in disparate media industries to indulge in "merger mania" (Bettig and Hall, 2003). Where regulation had allowed corporations to constitute oligopolies on an industry-by-industry basis, deregulation encouraged corporations to diversify and intensify their media holdings. This has restructured media industries as a series of overlapping oligopolies comprised by media conglomerates (Wasko, 1994).

The impact of transindustrial oligopolies on mediated expression has been notable indeed. For television programming, this has meant an increasing number of advertisements between program segments, as well as an increasing intrusion of product placements and other commercial content into programming (Andersen, 1995). Transindustrial conglomerates used deregulation to erase the traditional separations between advertising, news, and entertainment on television just as they used deregulation to erase traditional separations between ownership of television networks, production units, cable channels, and cable systems. Before discussing media oligopolies spanning broadcast, cable, and satellite television, let's examine the television industry's interlinked markets.

Television's Interlinked Markets

Three interlinked markets comprise broadcast television in the United States: the market in which networks commission and select programs; the market in which advertisers demand and buy audiences; and the market in which the A. C. Neilsen Company (ACN) sells ratings to advertisers and networks as proof that networks' programs deliver the demanded audiences in acceptable numbers. I will refer to these three markets as, respectively, the markets for programs, audiences, and ratings. Much discussion of these markets can be found in trade papers like *Variety* or *Advertising Age* as well as in media outlets like Cable News Network's "The Hollywood Minute" (owned by AOL Time Warner), *Entertainment Tonight* television series (Viacom), *Entertainment Weekly* magazine (AOL Time Warner), and E! Entertainment cable channel (Disney 40 percent, Comcast 40 percent, AT&T 10 percent, Liberty Media 10 percent). These sources can be useful, but typically report stories in terms of unique personalities, one-of-a-kind events, singular relationships, and unparalleled outcomes. This emphasis obscures the corporate, market, and legal structures that provide the parameters within which individuals interact to create personae, events, relationships, and outcomes (Gitlin, 1985; Meehan, 1986; Meehan, Wasko and Mosco, 1994). While individuals do exercise agency, they do so within the confines of institutional structures.

To see how this works, we examine the market for ratings at its inception, analyzing that initial structure to identify six elements that set the parameters for nationally distributed, commercial programming. These six elements shape the market for ratings, audiences, and programs – regardless of the technology delivering ads and programs. The ratings interlink these markets as they define a program's success and set the price for advertisers' gaining access to audiences.

Creating a Market

The market for broadcast ratings was created in November 1929 when the Association of American Advertising Agencies and the Association of National Advertisers founded the Cooperative Analysis of Broadcasting (CAB) to measure radio audiences. The Associations did not allow CAB to sell ratings to radio networks or radio stations, thus cutting broadcasters out of the market. CAB's measurements were based on telephone interviews, which defined radio's audience in terms of telephone subscription at a time – the beginning of the Great Depression – when telephones were a luxury. In this way, CAB separated bona fide consumers – listeners with the necessary desire, disposable income, and access to the retail system – from the rest of the audience.

Further, CAB's questions appear designed to deflate the size of radio's audience among bona fide consumers. The first question asked if respondents had listened to radio on the previous day. Those who said "yes" were asked to name every program heard during the previous day. With most programs running 15 minutes, CAB was asking for a phenomenal feat of memory. Unsurprisingly, CAB found that radio attracted a small quantity of high-quality listeners. Advertisers argued that prices for programming slots should be modest given radio's modest number of listeners. RCA and CBS each countered with its own numbers, demonstrating large quantities of high-quality listeners, implying that prices should be higher. This market structure was obviously unstable, but it illuminates some central elements of the market for ratings.

Analytically, this sketch indicates an important continuity in the demand for audiences and for audience measures: both advertisers and networks share an interest in bona fide consumers. The sketch also illuminates an important discontinuity: because price depends upon quantity and quality, advertisers are interested in measures that underestimate the quantity of that high-quality audience because such measures depress prices. With networks interested in earning the highest revenues possible, networks have the contrary interest: measures that overestimate quantity raise prices. This mixture of continuity and discontinuity in demand ensured that the CAB's days were numbered: it created an opportunity for an independent company to manipulate discontinuities while serving continuities in a manner that would appeal to both advertisers and broadcasters. In the ratings market, then, the structures of demand invite a would-be ratings producer to rebalance continuity and discontinuity in order to gain a foothold in the market.

Such a manipulation put an end to CAB within two years. The C. E. Hooper Company (CEH) rebalanced continuities and discontinuities to include broadcasters as buyers. This served broadcasters' demands for ratings and increased the number of CEH's revenue sources, allowing it to lower costs overall, thereby decreasing costs for advertisers or agencies. CEH also used telephone interviews, thus preserving quality. By changing the interview, CEH got better information about listening, effectively increasing quantity. Respondents were asked if they had listened to radio in the last 15 minutes; if so, to what. Listeners were then asked about the previous 15 minutes. Suddenly, the ratings showed that radio attracted a large quantity of high-quality listeners. This was achieved at less cost per ratings report to advertisers and agencies because broadcasters also paid for the improved information about consumers' listening.

Here I have somewhat simplified the history of the CAB and CEH in order to clarify six of the rudimentary economic relations in the markets for audiences and ratings. First, advertisers want bona fide consumers – people with sufficient disposable income, desire, and access to goods and services that they can afford luxuries like telephones in 1929 or cable subscriptions in 1989. Second, networks want to produce what advertisers want to buy – thus, creating continuity in demand regarding who ought to be counted. Third, conflict over prices between networks and advertisers introduces discontinuity in demand, which opens space for companies to struggle over industrial definitions. Historical examples of successful campaigns to change such definitions include NBC's campaign to make 18- to 34-year-old consumers the premium demographic in the 1960s and Turner Broadcasting System's campaign to get Nielsen to measure cable networks in the late 1970s. Currently, CBS is waging a campaign to redefine the premium audience as upscale consumers in their 40s and 50s who have access to the Internet. Discontinuity allows companies to agitate for their vested interests.

Fourth, that structural wiggle room allows the ratings producer to be more than a slave to demand. Like its clients, the ratings producer can creatively manipulate discontinuities and rivalries for its own benefit.

Just as CEH replaced CAB by rebalancing demand, so CEH was replaced by ACN. Familiar to advertisers because of reports on displays of product and promotional materials in retail outlets, ACN used meters to monitor radio tuning. Using a clever mixture of appeals to technological snobbery, manipulation of rivalries between NBC and CBS, and long-term relations with advertisers, ACN took over the monopoly of radio ratings. It parlayed this into an eventual monopoly over television ratings and is currently one of two operations rating Internet sites. By understanding the ability of individual companies to utilize such structural wiggle room, we recognize both firms' abilities to exercise agency and institutional limits on that agency.

The example of CAB and CEH also indicates a fifth element: markets for ratings, audiences, and programs are interlocked. Neither the ratings producer nor the networks nor program producers have any economic interest in producing commodities that are unresponsive to advertisers' demand for bona fide consumers. No network can afford to program for an audience that is undemanded and unmeasured. No producer can afford to pitch a show that targets an audience unwanted by advertisers. Because high ratings mean success for programs, producers often model this season's programs on last season's ratings hits. This encourages producers to generate creative imitations of last year's hits by recombining elements in new ways. The market for ratings, then, sets the parameters for the market in which programs are conceptualized, pitched, and sometimes scheduled.

Thus, at any particular point in time, depending upon the ratings monopolist's measurement practices and advertisers' demands, some sections of the audience will be ignored. The monopolist simply has no interest in measuring viewers whose measurements are not wanted. This is one reason for ignoring institutionalized viewers: people in dormitories, barracks, prisons, nursing homes, etc. are not demanded and not measured. The design of measurement methods, then, is very important. For example, from about 1934 to 1963, ACN used the same group of metered households to calculate ratings first for radio, then for radio and television, and finally for television. This was useful for RCA and CBS, which were trying to persuade advertisers to move from radio to television by simultaneously broadcasting programs on radio and television. Regardless of medium used, RCA and CBS were selling the same commodity audience to advertisers. This was the audience in which advertisers were interested and which was encapsulated in ACN's operations. Thus, ACN had no incentive to construct a new sample comprised by early adapters of television. Instead, ACN continued to report ratings from its radio households, adding more information whenever a household bought a television set.

This "transitioned" old radio programs to the new medium, but it also set the mold for creating new television shows. Radio's genres, character types, and formats became the bases for television's programming because television ratings depended on ACN's prewar radio sample. That reliance reinforced creative imitation as a programming strategy, subjugating a new medium to the old. When RCA, CBS, and advertisers agreed that ACN radio households were the measure of television's success, they effectively agreed that television programming would not fully exploit television's technological potential. This decision set the television industry on a particular road – one that ignored technological potential in favor of incorporating new technologies into the old economic structures of the markets for programming, audiences, and ratings.

Over the decades, that effect has persisted so that new technologies were defined in terms of the pre-existing ACN households and their programming preferences expressed through the ratings. Such redefinition limits innovation in programming as well as the full exploitation of technological capabilities. In the 1960s and 1970s, cable's technological capabilities were recognized as profoundly different from broadcast television by policy makers, telecommunications and media corporations, non-profit foundations, and media activists. Cable promised not merely to bring us more television channels but also to facilitate interaction between individuals, social groups, community organizations, non-profit asso-ciations, unions, ad hoc movements, public servants, governmental entities, corporations, trade associations, etc. (de Sola Pool, 1973). Aside from AOL Time Warner's QUBE cable, which defines interactivity as responses to polling and commercial questions, owners of multiple cable systems (MSOs) rejected interactivity as too expensive and lobbied against it. MSOs preferred to define cable in terms of broadcasting and, with Turner Broadcasting System's success in persuading ACN to rate TBS, that was achieved. Cable was integrated into the interlocked markets.

That integration reflected cable's primary functions in the 1970s: distributing sports events and Hollywood movies, rerunning old network programming, and purchasing programming that followed broadcast genres, character types, and formats. In the 1980s, as cable channels originated more programming, innova-tion was defined as network-style programs with harsher language, more nudity, more explicit violence, and more adult themes. Because cable channels and broadcast networks are treated as competitors in the ACN sample and by adver-tisers, networks then incorporated more naughty language, etc. into broadcast programming. Thus, interlinked markets can exert effects far beyond their im-mediate focus at any particular time.

Sixth and finally, the smooth and inexpensive operations of these three markets require a single source of ratings. In this way, advertising slots can be bought and sold as a matter of routine, using low-paid employees who quickly transact their business. This encourages advertisers and broadcasters to support a monopoly in ratings production. For a new firm to enter that ratings market as a producer, that company must put together some faction comprised by disgruntled networks, advertisers, and agencies seeking advantages not available from the current monopolist. Despite such challenges, ACN remains the ratings monopolist.

Stability and Dynamism

The markets for audiences, programs, and ratings have been remarkably stable in terms of the structure of demand. Advertisers want measurements of high-quality consumers and they want to buy access to those consumers via programs that earn high ratings, i.e. that attract the targeted consumers. Networks and channels want to earn revenues from advertisers and thus also demand ratings that measure advertisers' targeted audiences as well as programs that will earn high ratings. This fosters the practice of limited creativity in the market for programs. Buyers seek programs that will attract households in the ratings sample and sellers creatively twist the elements of ratings hits. There is no indication that this structure of demand will change in the near future.

But while these relationships are stable, relationships based on discontinuities in demand provide the basis for the television industry's apparent dynamism. Discontinuities open up spaces for strategic behavior, ranging from Hooper's challenge to the CAB for the monopoly over ratings in the 1930s to CBS's current campaign to make baby boomers the most demanded demographic. Strategic behavior encourages rivalries between networks as each strives to be number one in the ratings, to earn the most revenues by producing the most demanded demographic. Networks craft their schedules, using the tricks of their trade to snatch the best demographics in the Neilsen sample away from their rivals. These visible rivalries are easily personified and widely reported in the trade and popular press.

This reportage hides the fact that this dynamism is rooted in the structures of demand that construct the interlinked markets for ratings, audiences, and programs. These stable structures set the constraints within which rivals create strategies and pursue their shared goals of beating the others, of being "number one." By understanding the structural stability that marks the television industry, we can also understand the dynamism within each market as individuals and corporations seek to use those market structures for their own best advantage. These interlinked markets have not been destabilized by deregulation, regardless of deregulators' promises to introduce the inherently destabilizing condition of competition. In the next section, we examine the effects of deregulation on corporate structure and on the television industry.

Deregulation, Transindustrialism, Limited Rivalries

So far, we have been treating broadcast networks and cable channels in terms of their positions in the markets for ratings, audiences, and programs – that is, as active entities expressing and reacting to demand, and seeking ways to manipulate discontinuities in demand in order to serve their particular interests. Yet, as we know, networks and channels are merely subsidiaries of larger corporations. This has been the case since General Electric, Westinghouse, and AT&T created RCA and through RCA the National Broadcasting System to run two radio networks. Deregulation allows the networks' owners to reorganize themselves as transindustrial conglomerates integrating ostensible competitors like network, cable, and satellite television. This collapses multiple television industries into a single industry. In this section, I will define deregulation and transindustrial conglomeration, and then discuss some of the outcomes of deregulation and their dampening effect on competition.

Under the Reagan Administration in the 1980s, Federal officials and industry representatives began rewriting the rules that had undergirded the American economy. Among other things, these rules had been set in place to stabilize industries by softening "boom-and-bust" cycles and eliminating cutthroat competition leading to monopoly. Often, stabilization meant allowing groups of companies to control an industry and then regulating those oligopolists (Baran and Sweezy, 1966). Operating as rivals, oligopolists attempted to get the best share of available revenues while serving the public interest, convenience, and necessity. In this way, regulated industries accepted governmental protection and promised public service in return for stability. Under regulation, neither the ideal of competition nor the evil of monopoly was achievable. Hopefully, public and private interests would be balanced and markets would be fair.

This logic undergirded both the Radio Act of 1927 and the Communications Act of 1934. The rules articulated under regulation tended to shelter the big companies (Kellner, 1990) but also ensured that listeners got radio services. At times, public interest in decreased concentration of ownership was manifested in regulations that required network owners to reorganize their operations. For example, RCA divested one of its two radio networks due to concerns about horizontal integration, thereby creating ABC. Subsequent concerns over vertical integration led to RCA, CBS, and ABC divesting subsidiaries that produced television programming. In exchange, network interests were generally protected as in the FCC's early regulation of cable, which was designed to contain any threat that the new technology might pose. The FCC's price for minimizing that threat was the networks staying out of cable. All that changed with deregulation.

Under the new rules, companies can own multiple television networks, production units, cable channels, cable systems, and infrastructural companies. The new rules allowed mergers that could bring every type of media operation under the same corporate umbrellas. This merger mania (Bettig and Hall, 2003) fostered the creation of transindustrial media conglomerates integrating operations in production, distribution, and delivery/exhibition in film, broadcast television, and cable television – and also integrating film, broadcast television, and cable television. At this point, each of the five conglomerates that own the six major broadcast networks – Time Warner, Disney, General Electric, News Corporations, and Viacom – have operations in cable television.[1] All except GE produce and distribute theatrical films under multiple labels; but GE is attempting to buy Vivendi Universal Entertainment, which includes operations in film production, distribution, and exhibition as well as recorded music and cable.[2] Let's turn to the Big Five's television holdings.

The Big Five and Broadcast Television in the United States

In television, deregulation means that conglomerates can own a production company, distribution company, and network (vertical integration). But they are also allowed to own multiple production companies or multiple distribution companies or multiple networks (examples of horizontal integration). The results can be seen in the current structure of broadcast television operations of the five major companies.

In terms of networks, three conglomerates still own one network each: Time Warner (WB), Disney (ABC), and News Corporation (Fox). Viacom owns both UPN and CBS; General Electric owns NBC, Telemundo, and 32 percent of Paxson Communications' generic Christian network PAX.[3] Each network has operations in production and distribution. In entertainment programming, four conglomerates have multiple production units; GE is trying to add Universal Television to its NBC Studios. Disney produces television series under four labels: ABC Studios, Disney Productions, Touchstone Television, and Buena Vista Productions. Viacom produces under eight labels: CBS International, CBS Enterprises, King World Productions, Paramount Television, Big Ticket Television, Spelling Television, Viacom Enterprises, and NickToons. News Corporation uses three labels – Twentieth Century Fox Television, Fox Television Studios, and Twentieth Television – and owns a 20 percent stake in Regency Television.[4] Time Warner produces television entertainment under seven labels: Castle Rock, HBO Downtown Productions, HBO Independent Productions, New Line, Fine Line, Telepictures Productions, and Warner Bros. Television.

Because of deregulation, the conglomerates that own networks are now allowed to produce programming for their own networks and for their rivals' networks. While the former can be seen as a return to the status quo of the 1950s and 1960s, the latter is a radical departure that undercuts rivalry between networks. Any company that produces a show has a vested interest in the success of that show because producers generally earn most of their revenues from a program when it goes into syndication, rerunning on cable channels or on local television stations. When a television conglomerate produces a program for another television conglomerate, both companies share an interest in the program's success. The producing conglomerate wants the program to earn high ratings and run for a long time in order to be able to syndicate the show at the highest price possible for the longest run possible. The conglomerate that buys the rights for the show's first run also wants the program to earn high ratings and run as long as possible in order to earn the highest revenues possible from advertisers. This shared interest encourages the producing conglomerate not to aggressively counter-program its network(s) when its programs are running on rival networks.

Let's consider some of these deals for the 2002 television season. When dealing with co-productions featuring small companies, I will refer only to the dominant partner in the production. This reflects the Big Five's policies regarding programming: for an independent producer to get a show on a major network, that small company has to cede ownership rights to the conglomerate (Gomery, 2000). First, we turn to deals between conglomerates.

For prime time in 2002, GE's NBC ran one program produced by Viacom (*Frasier*) and five programs produced by Time Warner (*ER*, *Friends*, *Good Morn-*

ing Miami, *Third Watch*, *West Wing*). On CBS, Viacom ran two programs produced by Disney (*CSI* and *CSI: Miami*) and two by Time Warner (*Without a Trace* and *Presidio Med*). On UPN, Viacom licensed *Buffy the Vampire Slayer* from News Corporation. News Corporation's Fox ran Time Warner's *Fastland*. Time Warner's WB ran two programs from News Corporation (*Angel* and *Reba*). Disney continued running *NYPD Blue* and *The Practice*, both produced by News Corporation. From Time Warner, Disney licensed *The Drew Carey Show*, *George Lopez*, and *Whose Line Is It Anyway?* Disney contracted with Time Warner and News Corporation for a "first look" at television programs being developed for broadcast.

Licensing programs from rivals is but one aspect of programming entanglements, since companies also co-produce programs. GE and Viacom co-produced *Ed* and *Inlaws*; Viacom and News Corporation did *Standing Still*. Co-production intensifies each firm's interest in a program doing well on one of their networks. Viacom also co-produces internally using its production units operated under the CBS, UPN, and Paramount brands. In the 2002 season, CBS provided UPN with the series *Half and Half* and joined with UPN to co-produce *Haunted*, which ran on UPN. Joining with Paramount Television, UPN co-produced *Bram and Alice*, which aired on CBS. Paramount continued providing CBS with the series *JAG*. This internal trade in programming suggests that Viacom uses its subsidiaries to support each other's operations. As the networks cooperate with each other, their rivalry diminishes.

One final connection between two of the Big Five should be noted. Broadcast networks are constructed using a combination of stations that are owned-and-operated by the conglomerate and stations (or companies owning groups of stations) that sign contracts to become affiliated with the conglomerate's network. News Corporation owns 37 stations, of which 10 are affiliated with Viacom's UPN. In seven cities, News Corporation owns the Fox station and the UPN station. While UPN and Fox may appear to be rivals in the market for audiences, News Corporation has a vested interest in UPN flourishing and in Viacom profiting from its UPN operation.

As conglomerates' interests intertwine through these contracts, competition between them stalls. When firms share vested interests in a particular program, for example, both seek to serve their shared interests so that each may profit accordingly. When News Corporation produces a television series for Disney, Fox has little impetus to aggressively compete for the audience demographic watching News Corporation's show on Disney's ABC. If Fox succeeded in pushing the rival program off the air, it would be a loss for both News Corporation and Disney. Better to seek a different demographic by scheduling shows from a different genre and thereby achieve a "win–win" situation for both News Corporation and Disney, both Fox and ABC. Deals that entangle companies and give them shared interests in each others' profits make it hard for oligopolists to act as rivals.

While the trade press and entertainment-oriented news coverage often note that News Corporation has a contract with Disney or that Time Warner provides programming to General Electric, the presses rarely draw the obvious conclusion: that broadcast television is dominated by an oligopoly whose members cooperate in joint ventures, share vested interests in the success of licensed programs, and have other financial dealings with each other.

Although deregulation has expanded the network oligopoly from three to five, it has allowed these five companies to intertwine their programming operations and, in the case of News Corporation and Viacom, has allowed the former to own stations affiliated with the latter. Deregulation has fostered vertical integration of production, distribution, and networking within each of the Big Five conglomerate. It has encouraged horizontal integration achieved via the ownership of multiple production subsidiaries in four cases and multiple networks in two cases.

Deregulation has knit the Big Five together as they provide programming to each other, co-produce programming with each other, and promise "first looks" at developing programs. News Corporation's decision to buy stations affiliated with Viacom's UPN and to keep those stations in the UPN network indicates the degree to which competition has become an illusion: in seven cities News Corporation's Fox stations "compete" with News Corporation's UPN stations. This is far from Adam Smith's notion of full, free competition in which each firm strives for the triumphal moment: proving its worth by putting the competition out of business.

Transindustrialism and Overlapping Oligopolies

For our purposes, a final effect of deregulation that changes how we watch television is the decision to allow companies to own as many media outlets and to operate across as many media industries as possible. Where separate oligopolies once dominated distinct media industries in film and television, deregulation has fostered the integration of these oligopolies. Film production and distribution is oligopolized by Disney, Viacom, Vivendi, News Corporation, Sony, and Time Warner. Television production and distribution for broadcast and for cablecast is oligopolized by Disney, General Electric, Viacom, Vivendi, News Corporation, Sony, and Time Warner. If the GE–Vivendi deal goes through, the oligopoly in television production and distribution will shrink from seven to six and GE will replace Vivendi in the oligopoly in film production. The oligopoly in television networks overlaps the oligopoly in television production with Disney, GE, Viacom, News Corporation, and Time Warner serving as major forces in both oligopolies. These overlapping oligopolies effectively integrate these portions of the film and television industries into a single industry.

The picture is more complex in terms of cable channels. The extent of ownership in cable channels varies among the Big Five, but all are involved to some extent. General Electric owns Bravo, MUN2, and Telemundo Internacional; and in addition it co-owns CNBC and MSNBC. Disney owns the Disney Channel, SoapNet, Toon Disney, Disney Worldwide (12 channels), ESPN International, and Fox Kids Latin America. Disney co-owns E! Entertainment with Comcast and Liberty Media (MSOs) and with AT&T, which had owned Liberty Media. Disney also co-owns Style, ESPN (4 channels), and Lifetime (3 channels). Viacom now owns Black Entertainment Television (BET, 4 channels) and Comedy Central. Viacom holds 50 percent of the Sundance Channel with Robert Redford and Vivendi as co-owners and joins with Liberty Media in the BET/STARZ channel. News Corporation owns FX, Fox News, Fox Movie, and Fox Sports Networks. But the owner of the most channels is undoubtedly Time Warner with its CNN Newsgroup (5 channels), HBO and Cinemax pay channels (14 channels), Turner Channels (Boomerang, Cartoon Network, TBS, Turner Classic Movies, TNT, and Turner South), and its 50 percent stake in Court TV. These channels alone would make the Big Five in network television into major stakeholders in cable.

Mirroring entanglements in network programming are entanglements among the Big Five in cable channels and programming. Disney, GE, and Viacom co-own History and History International. GE and Disney co-own A&E and A&E International. GE and News Corporation are major stakeholders in National Geographic and National Geographic International. Disney and News Corporation co-own Fox Kids Europe. If the GE–Vivendi deal is consummated, Viacom and GE may share Sundance with Redford. In programming, entanglements abound, but here are a few that strike me as interesting. Time Warner and GE share the rights to NASCAR races, with Time Warner running the races on cable and GE broadcasting them on NBC. Disney has granted Viacom's pay channel Showtime exclusive rights to "first runs" of Disney films released under the Hollywood, Touchstone, and Miramax labels. These and other contractual relationships intertwine the broadcasting's Big Five in cable television.

Vested Interests, Infrastructural Interests

While three of the Big Five also have infrastructural interests in cable and satellite television, all of these firms gained a vested interest in satellite television with the passage of the Telecommunications Act of 1996. This law redefined satellite television as just another way to distribute cable television channels. Given the Big Five's ownership of cable channels, the 1996 Act gave them a new way to distribute those channels. This connected the vested interests of the Big Five with DirecTV, which effectively monopolizes satellite television in the United States (Gomery, 2000). In 2003, News Corporation acquired 34 percent of DirecTV, adding it to News Corporation's worldwide satellite operations and reinforcing News Corporation's infrastructural interest in a technology once seen as cable's competitor.

251

As a long-term owner of multiple cable systems, Time Warner had a considerable interest in cable infrastructure before founding WB. In the late 1990s, that interest entangled Time Warner with various other MSOs, including TCI, AT&T, and Comcast. Under pressure to simplify these relationships, Time Warner holds a 79 percent stake in Time Warner Cable, partnering with Comcast. Time Warner also shares two ventures with rival publisher Advance Publications/ S.I. Newhouse. These ventures provide broad-band communications and long-distance telephony to corporate and governmental clients. While reconfigurations simplified relationships, they reaffirmed companies' intertwining interests.

Finally, GE's subsidiary American Communications (Americom) owned multiple satellites and tracking stations that served governments and corporations including those with broadcast or cable operations. In 2001, GE sold Americom to SES Global for a combination of cash and stock. This gave GE a 31 percent stake in the world's largest provider of satellite services – services upon which broadcast networking and cable channels depend.

Joint ventures and contractual arrangements between the Big Five and cable MSOs also deserve mention because they demonstrate once again the degree to which cable and broadcast television are not competitors. This is clearest in the case of the cable MSO Liberty Media. Liberty owns 4 percent of Time Warner and 18 percent of News Corporation. Liberty also owns 3 percent of Vivendi Universal, but it is not clear if that would become a stake in GE if Vivendi and GE consummate their deal. However, GE is already connected to Liberty by a contractual relationship. GE leases three hours of its Saturday morning schedule to cable's Discovery Channel, which is co-owned by Liberty and Discovery Communications. Besides joining Viacom in co-ownership of Court TV and BET/STARZ, Liberty joins its previous owner AT&T, fellow MSO Comcast, and Disney in co-ownership of E! Entertainment. When companies active in networking, cable channels, and cable systems have intertwined their interests and profits, the claim that cable competes with broadcasting becomes untenable.

Watching Television

From this political economic perspective, we can dismiss both of the truisms noted at the beginning of this chapter. Our analysis of the network broadcasting's interlinked markets for ratings, audiences, and programs demonstrates that these markets are organized around satisfying advertisers' demand and that ratings are taken as the sole measure of satisfaction. In the ratings market, the monopolist crafts its sample to capture bona fide consumers, using methods to balance continuities and discontinuities in demand. In the market for programs, networks seek new versions of last year's programs that attracted the commodity audience as defined by the rater's methodology. In the market for audiences, each network deployed programming in order to attract more of the ratings sample than its rivals. These interlinked markets exclude the bulk of television's audience. Thus, the operations of our most ubiquitous medium are driven by dynamics between advertisers, networks, and the ratings monopolist as each firm pursues its private interest in markets closed to the people whom television is supposed to serve. The analysis of interlinked markets illuminates that disconnection, falsifying the truism that television gives us what we want.

By focusing on demands, that analysis implies that competition exists within the markets for programs and audiences: networks compete against each other for advertisers' dollars and program producers compete against each other for commissions from networks. This truism becomes untenable under deregulation. Production subsidiaries do not try to push each other out of business. Conglomerates connected by programming contracts and co-production are not engaged in full, free competition. Conglomerates that own multiple networks or stations affiliated with another firm's network do not pit their subsidiaries against each other to see which will survive. To the degree that the Big Five share an interest in each other's profitability as part of their own profitability, their oligopoly in network television fails to produce rivalry and begins to approximate monopoly.

Further entanglements among the Big Five in cable and satellite television serves to collapse all three television technologies into a single industry. The Big Five's ownership and co-ownership of cable channels refutes the claim that cable channels are deadly rivals to the six networks. Channel ownership and co-ownership positions the Big Five so that they have a vested interest in the success of satellite television, regardless of their varied infrastructural interests in cable systems, broadband services, and satellites.

In sum, the Five all profit from the redefinition of cable and satellite as just another, more expensive, slightly naughtier version of network television. The three interlinked markets in broadcasting networking now encompass cable as well and the traditional separation of broadcasting, cable, and satellite television has been erased both in law and practice.

Broadcast, cable, and satellite television are fully integrated. As such, television's production and distribution systems are organized as series of overlapping oligopolies in which Disney, GE, News Corporation, Viacom, and Time Warner are both major players and frequent partners. Thus, the second truism is proven false: network television, cable television, and satellite television are not separate industries and not fierce competitors.

This picture of the television industry differs startlingly from that drawn in the popular and trade presses. While rivalry over ratings evokes fractious rhetoric among networks, that rivalry is softened through contracts and joint ventures that marry the interests of the networks' owners. When we watch television as political economists, we uncover relationships and structures that define the field upon which the game of television is played. As long as that game is played by these rules and these rulers, we can only expect more of the same, regardless of the number of channels or available technologies. Combining political economy with the common experience of ". . . fifty-seven channels and nothin' on . . ." (Springsteen, 1992) gives us the tools necessary to explain why that is so.

Notes

1 All five companies are publicly traded. However, Sumner Redstone's National Amusements owns a controlling stake in Viacom.
2 This includes the cable channel USA and Music Corporation of America (MCA).
3 At this writing, GE is extricating itself from Paxson, which may cost Paxson as much as $6 million.
4 Monarchy Enterprises, a holding company with its headquarters in the Netherlands.

References

Andersen, R. (1995) *Consumer Culture and TV Programming*, Boulder, CO: Westview Press.

Baran, P. A. and Sweezy, P. M. (1966) *Monopoly Capital: An Essay on the American Economic and Social Order*, New York: Monthly Review Press.

Barnouw, E. (1990) *Tube of Plenty: The Evolution of American Television*, second rev. edn., New York: Oxford University Press.

Bettig, R. V. (1996) *Copyrighting Culture: The Political Economy of Intellectual Property*, Boulder, CO: Westview Press.

Bettig, R. V. and Hall, J. (2003) *Big Media, Big Money: Cultural Texts and Political Economics*, Lanham, MD: Rowman & Littlefield.

Brown, D. (1998) "Dealing with a Conflict of Interest: How ABC, CBS, and NBC Covered the Telecommunications Act of 1996," Paper presented at the Broadcast Education Association conference, Las Vegas, Nevada, April 6–9.

De Sola Pool, I. (ed.) (1973) *Talking Back: Citizen Feedback and Cable Technology*, Boston, MA: MIT Press.

Douglas, S. J. (1987) *Inventing American Broadcasting 1899–1922*, Baltimore: Johns Hopkins University Press.

Gerbner, G. (1996) "The Hidden Side of Television Violence," in G. Gerbner, H. Mowlana, and H. I. Schiller (eds.), *Invisible Crises: What Conglomerate Control of the Media Means for America and the World*, Boulder, CO: Westview Press, pp. 27–34.

Gitlin, T. (1980) *The Whole World Is Watching: Mass Media in the Making and Unmaking of the New Left*, Berkeley and Los Angeles: University of California Press.

Gitlin, T. (1985) *Inside Prime Time*, New York: Pantheon Books.

Gomery, D. (2000) "The Television Industries: Broadcast Cable, and Satellite," in B. M. Compaine and D. Gomery, *Who Owns the Media?: Concentration of Ownership in the Mass Communications Industry*, Mahwah, NJ: Lawrence Erlbaum Associates, pp. 193–284.

Kellner, D. (1990) *Television and the Crisis of Democracy*, Boulder, CO: Westview Press.

McAllister, M. P. (1996) *The Commercialization of American Culture*, Thousand Oaks, CA: Sage.

McCarthy, A. (2001) *Ambient Television: Visual Culture and Public Space*, Durham, NC: Duke University Press.

McChesney, R. W. (1993) *Telecommunications, Mass Media, and Democracy: The Battle for the Control of US Broadcasting, 1928–1935*, New York: Oxford University Press.

Meehan, E. R. (1986) "Conceptualizing Culture as Commodity," *Critical Studies in Mass Communication*, *3*(4) (December), 448–57.

Meehan, E. R., Wasko, J., and Mosco, V. (1994) "Rethinking Political Economy," in. M. R. Levy and M. Gurevitch (eds.), *Defining Media Studies: Reflections on the Future of the Field*, New York: Oxford University Press, pp. 347–58.

Mosco, V. (1996) *The Political Economy of Communication*, London: Sage.

Springsteen, B. (1992) "57 Channels (And Nothin' On)," *Human Touch*, New York: Columbia.

Streeter, T. (1996) *Selling the Air: A Critique of Commercial Broadcasting in the United States*, Chicago: University of Chicago Press.

Tillinghast, C. H. (2000) *American Broadcast Regulation and the First Amendment: Another Look*, Ames, IA: Iowa State University Press.

Wasko, J. (1994) *Hollywood in the Information Age*, Austin, TX: University of Texas Press.

Wittebols, J. H. (2004) *The Soap Opera Paradigm*, Lanham, MD: Rowman & Littlefield.

CHAPTER THIRTEEN

Keeping "Abreast" of MTV and Viacom: The Growing Power of a Media Conglomerate

Jack Banks

It was the breast seen round the country. During the half-time show for Super Bowl XXXVIII televised on CBS on February 1, 2004, Janet Jackson and Justin Timberlake performed a duet that ended with Timberlake ripping off part of Jackson's costume to expose her breast. What was termed a "costume malfunction" immediately led to a fierce condemnation of CBS, the National Football League, and MTV, which produced the show. Even though both performers and all of these organizations apologized for the incident, many government officials were furious and demanded swift action. House and Senate panels chastised representatives from MTV's parent company Viacom and the NFL, and asked Federal Communication Commission (FCC) officials how something like this could happen. Michael Powell, the FCC Chairman, put the occurrence in a broader context of a growing problem with the media: "The now infamous display during the Super Bowl half-time show, which represented a new low in prime-time television, is just the latest example in a growing list of deplorable incidents over the nation's airwaves" (Geewax, 2004, p. 1E). As a result of the fiasco, Powell said that the FCC would increase the punishment for violating its indecency rules, by fining a media outlet for each incident rather than for each program. Members of Congress also planned to introduce legislation to increase the fines for indecency from $27,500 to ten times that amount.

Yet while members of Congress and the FCC focused on this issue, almost completely ignored was the underlying issue of media ownership that prompted this bare breast in the first place. MTV, which produced the half-time show, is owned by the media conglomerate Viacom, which also owns the CBS television network that broadcast the Super Bowl. Viacom, like all of today's major media conglomerates, encourages its subsidiaries in various media to cooperate with each other in many ways to enhance the profits of the corporation as a whole.

This coordination between MTV and CBS on the Super Bowl broadcast is just one example of this practice. Even with this one instance involving the Super Bowl, there were other less publicized collaborations among Viacom units. The half-time show was promoted by the Viacom website MTV.com with an item headlined "Janet Gets Nasty!" describing a "kinky finale that rocked the Super Bowl to its core . . . Armchair quarterbacks, fair weather fanatics and fans of Janet Jackson and her pasties were definitely in the right place" (Rosenthal, 2004, p. 12). Viacom's cable program services, Nickelodeon, TV Land and Country Music Television (CMT), all promoted the upcoming Super Bowl on CBS. *Broadcasting and Cable* noted that "at least a half dozen" of Viacom's program services "were concocting some kind of special programming tied to the Super Bowl" (McClennan, 2003, p. 16). Besides the scandalous half-time show, CBS originally commissioned MTV to produce a one-hour show modeled after its popular live program Total Request Live (TRL) to be shown prior to the Super Bowl (McClellan, 2003).

This intricate web of arrangements involving MTV and other units of Viacom suggest that Congress and the FCC are asking the wrong kinds of questions and proposing ill-advised policies when it comes to dealing with today's media. Rather than focusing obsessively on indecency, advocacy groups like The Free Press Team and mediareform.net argue that the half-time show illustrates the danger of media concentration:

> The Super Bowl half-time show is just the latest example of synergy gone wild – CBS (which aired the game) and MTV (which produced the half-time show) are corporate cousins in the Viacom media empire. And last summer, the FCC passed rules that will allow media giants like Viacom to get even bigger, narrowing the range of debate and stifling minority and independent voice. (Free Press Team, 2004)

So, even though the FCC is consumed with increasing penalties for indecency, it has moved toward relaxing ownership restrictions to allow Viacom and other conglomerates like Disney and Time Warner to further dominate the ideas, views, and creative expression presented in mainstream media. Frank Rich, a columnist for the *New York Times*, suggests that Powell is so loudly complaining about this incident involving Viacom as a way to foster the illusion of his independence from such media conglomerates:

> Mr. Powell's real agenda here is to conduct a show trial that might counter his well-earned reputation as a wholly owned subsidiary of our media giants. Viacom has been a particularly happy beneficiary of the deregulatory push of his reign, buying up every slice of the media pie that's not nailed down. (Rich, 2004, p. 1)

A survey of MTV and its role within Viacom provides a potent example of how media concentration centralizes control over culture in the hands of a few, especially in the field of popular music.

MTV's Place within Viacom

As a result of a series of mergers and acquisitions, Viacom has grown into a vast conglomerate with interests in a range of media. MTV Networks is the Viacom subsidiary that operates a range of program services, including those with music-oriented programming such as MTV, the 24-hour service that targets a young audience. MTV2 plays non-stop music videos to compensate for MTV's shift away from videos toward non-music shows like *The Real World* and *The Osbournes*. MTV X plays hard and alternative rock videos, MTV Hits highlights current popular songs, and MTV S presents Spanish-language videos. VH1 is a music-oriented channel for an older audience from 25 to 44 years old that features a mix of music videos, movies, series like *Behind the Music* and *I Love the 80s*, and special events like *VH1 Divas Live*. MTV Networks has also created several variations of VH1 that play mostly videos, including VH1 Classic with rock from the 1960s and 1970s, VH1 Soul with classic R&B and urban music and VH1 Country. Country music is also featured in the Viacom service Country Music Television (CMT), while jazz is highlighted on BET Jazz. Two other services devote a significant amount of their programming to music: BET, which offers programs for an African American audience, and College Television Network, which is offered on college and university campuses (Viacom, 2002, 2004).

Apart from its music services, Viacom operates Nickelodeon for children, Noggin for preschoolers and The N for teens and pre-teens. Nick at Nite and TV Land focus on old television series like *Bewitched*, *The Mary Tyler Moore Show* and *Happy Days*. Spike TV is a new reincarnation of The National Network (formerly The Nashville Network) that targets a male audience, and Comedy Central presents shows like *South Park* and the faux news program, *The Daily Show*. In addition, Viacom owns several premium movie channels, operated under the Showtime Networks division that encompasses many variations of Showtime, The Movie Channel and Flix. Viacom is also part-owner of the Sundance Channel, a station dedicated to broadcasting independent films.

In broadcast television, Viacom operates two television networks, CBS and UPN (United Paramount Network), and production companies like CBS Productions and Paramount Television that make shows for its own networks as well as others. The Viacom Television Stations group owns 39 stations in the United States, 20 affiliated with CBS, 18 with UPN and one independent, with duopolies in eight cities including Dallas, Miami, Philadelphia and San Francisco. King World distributes and syndicates shows such as *The Wheel of Fortune*, *Jeopardy* and *The Oprah Winfrey Show*. The subsidiary in radio broadcasting, Infinity owns more than 180 radio stations, the vast majority of them in the largest 50 radio markets. In the area of film, Viacom owns Paramount Pictures film studio and in the home video market it operates the chain of Blockbuster Video stores, although it has announced plans to spin-off the chain due to its poor performance (Fabrikant, 2003b).

In addition, Viacom owns several book publishers including Simon & Schuster, Free Press and Pocket Books, as well as Famous Music, a music publisher with over 100,000 copyrights. The Viacom Outdoor Group sells advertising space on over one million display faces, including billboards, buses, trains and terminal stations. A series of Paramount Amusement Parks like Great America include attractions and rides featuring other Viacom media properties. Finally, Viacom operates Internet services associated with many of its media outlets, such as MTV.com, VH-1.com, Nickelodeon.com and CBS.com (Viacom, 2002, 2004).

Programming Strategies of Viacom and the MTV Networks

One of Viacom's main strategies for its media subsidiaries is to strictly limit expenses for productions, maintaining tight control over budgets. MTV Networks has embraced this objective by focusing on low-cost programming. The music videos played on its many music-oriented program services are a very cheap source of material that is supplied mainly by record labels. Viacom has signed contracts with the labels to allow all of its services to present these music clips. In recent years, MTV Networks has moved toward "reality shows" like *The Real World* and *Road Rules* that have limited costs because the company does not have to pay for writers, actors or other creative personnel associated with traditional television shows such as sitcoms or dramas. MTV's large payment for a second season of its hit show *The Osbournes*, featuring the dysfunctional Osbourne family, was a notable exception to this practice. Expenses are also limited by MTV's practice of creating its own on-air personalities to host its shows rather than hiring established celebrities for a much higher price. "We are known in the industry as a place where presenters are made," says James Dearlove, an executive in charge of MTV's non-music talent (Asthana, 2003, p. 10). This policy of inexpensive programming is very lucrative for MTV, giving it one of the largest profit margins in the media business.

This practice of cutting costs is extended to Viacom's subsidiaries in the film business, including Paramount Pictures and MTV's film studio division. Among the major studios, Paramount has a reputation for aggressively limiting budgets for film projects and exerting firm creative control over the films. One top agent laments, "Once you make a deal and arrive at a budget you have to start reducing your terms to accommodate their budget. It seems to happen there more than at most places" (Fabrikant, 2003a, p. 1). However, this fiscally conservative stance may have contributed to recent problems Paramount has had at the box office, having few hits and many films that did not do well. Some Hollywood executives suggest that Paramount's disappointing performance may be due to the studio's unwillingness to invest in films and give creative control to talented directors.

The need to have commercially successful services has led MTV to stress certain themes in its programs. In order to attract audiences, MTV and its associated channels highlight sexually oriented content in its shows. Music videos often feature scantily dressed beautiful women, and fairly explicit sexual activity in bars, bedrooms, and other locations. MTV's series like *The Real World* and *Undressed* highlight the sexual activities of young people, voyeuristically depicting an ever-changing arrangement of sexual partners. Special programs like its Spring Break coverage and Beach House shows also focus on the bare, tan bodies of young people in sexually suggestive situations. This commercial imperative also influences the kinds of music featured on MTV. In order to reach the largest audience of young people, MTV leans more toward mainstream, conservative tastes in music. In the mid-1990s, MTV pandered to a pre-teen audience by focusing on bland acts like Britney Spears and 'n Sync, while more alternative performers were shuffled over to MTV2.

During the 2003 war in Iraq, this conservatism in musical tastes carried over into political issues. MTV Europe refused to show videos that depicted the war in Iraq, and more generally banned videos that dealt with war in visual images or song lyrics. While MTV in the United States said that the policy only applied to its European services, it did say that its American counterpart was "being responsive to the heightened sensitivities of its audience" (Strauss, 2003, p. 1). Moreover, during the war, MTV in the United States said that it would not accept antiwar public service commercials because "MTV does not accept advocacy ads" (Strauss, 2003, p. 1). Yet MTV has always been willing to air advertising from the branches of the US armed forces that depicts military combat in a very favorable manner. These policies are quite consistent with those of Clear Channel Communications, owner of over 1,000 radio stations in the US, which discouraged its stations from playing anti-war songs and banned acts like the Dixie Chicks which disagreed with Bush Administration policies toward the war.

MTV Networks' Dominance over Music on Cable

The range of services owned by Viacom and MTV Networks that focus on music programming in whole or in part give Viacom an almost complete dominance over popular music presented on cable television systems. Even as US cable systems have greatly expanded the number of channels offered to subscribers by converting to digital systems, Viacom's tight control over music-oriented programming remains in place. A review of the channel line-up on a cable system in Connecticut operated by Comcast illustrates Viacom's reach in this area. Like most cable systems, Comcast separates its channels into two groups: channels that are included in basic cable for a general monthly fee, and those channels in a premium tier available for an additional charge (which Comcast identifies as "digital cable" services). In the basic cable category, Comcast offers three music-oriented services, all owned by Viacom: MTV, VH1, and BET. So, for those subscribers that have only basic cable service, they will only have access to Viacom music programming.

In the premium "digital" group, there are eight additional music services, six of which are owned by MTV Networks: MTV2, VH UNO, MTV Espanol, VH1 Country, MTV Hits, VH1 Classic, VH1 Soul, and BET on Jazz Digital. Comcast offers every national music service provided by Viacom except CMT. The only two services not owned by Viacom are Fuse and HTV Musica. Each of these two channels must be purchased by the subscriber in different packages, further limiting their availability. Digital cable and satellite television services also often offer many audio-only channels featuring various genres of music. However, this service is generally provided by one service like Music Choice or DMX that controls and selects the music presented on all of these channels.

The extent of Viacom's control over music content on cable is expanded when one takes into account the conglomerate's program services that present other kinds of programming. The N airs music videos between its regular programs like DeGrasse that reach a preteen audience. The many variations of the premium channel Showtime regularly present concerts featuring popular musical acts. Viacom also owns two commercial television networks, CBS and UPN, that also feature musical concerts and various awards shows like the Grammys that spotlight hot acts. There is also regular cooperation between these various program services so that the music-oriented show on one Viacom outlet is promoted on another service. For instance, both MTV and VH1 have promoted the Grammy Awards shows that have aired on CBS since 1974 (Friedman, 2004). This range of services owned by Viacom and the collusion among them give the conglomerate great control over the range of music that is available on US cable television systems, and perhaps significant cultural influence to shape current trends in popular music.

As noted above, all but one of Viacom's program services are on the Comcast system. The broad range of its cable and broadcast program services give Viacom significant clout in its relationship with cable and satellite television systems. Viacom can use its ownership of these entities to pressure a company to include all or most of its services on its system, and discourage it from dropping any particular channel. For instance, Viacom may tell a cable system that in order to have access to MTV, it must include VH1 as well. Alternately, Viacom could dissuade any system from dropping MTV by threatening to deny access to the CBS television network. The conglomerate can use its group of services as a club to dissuade cable operators from adding new services that directly compete with the format of Viacom channels, putting any new fledgling independent service at an immediate disadvantage. The range of services also protects MTV and other Viacom services from any pressure from cable and satellite television providers to reduce the price that they are charged for these services.

MTV's Power in a Concentrated Music Industry

The dominance of MTV Networks and Viacom over music on cable and satellite television is complemented by the tight oligopoly of global corporations that control the production and distribution of music. A large majority of the music sold throughout the world is controlled by just five major record labels, each a subsidiary of a larger media conglomerate: Universal Music Group, Warner Music Group, Sony, Bertelsmann Music Group (BMG) and EMI Group. In February 2004, two of these companies, Sony and BMG, announced plans to merge, which would reduce the oligopoly to four companies. The new company Sony BMG would be jointly owned by the parent companies of both music labels and would combine Sony, now the second largest company with artists like Beyonce and Bruce Springsteen, with BMG, currently the fifth largest company with a catalog of artists including Britney Spears and Elvis Presley. The proposed deal was attacked by a group of independent record labels under the trade group IMPALA and was also investigated by the European Commission because of concerns that the four major companies would control 80 percent of the record market in Europe and other major countries (*Los Angeles Times*, 2004; Timmons and Sorkin, 2003). These companies have well-established marketing divisions for promoting their artists, and a vast distribution network for getting their artists' music to retail outlets.

MTV Networks has a symbiotic relationship with these major record labels where MTV promotes the majors' acts and the labels provide MTV with access to their artists for music videos, interviews, and performances for appealing programming. MTV Networks' dominance of the market for music television reinforces the centralized control of the music business by these five, soon to be four, companies. MTV has always sought to be the main outlet on television for major labels to promote their acts. Most of the music videos presented on MTV's music services have always been for artists signed to these major labels, while acts associated with independent labels have been consigned to the margins.

MTV has also long used its collusion with the major labels as a way to undermine any serious competition to it in the area of televised music programming. Since the 1980s, MTV had signed contracts with the major labels providing MTV with access to their artists' music videos, and exclusive rights to certain videos by popular acts for long periods of time. This latter provision for exclusive videos contributed to the failure of potential competitors to MTV like the Cable Music Channel because they were denied access to the most desirable music clips (Banks, 1996). Because of the emergence of a new music service, Fuse, MTV has in the past few years aggressively pressured record labels to agree to these exclusive arrangements as a way to undercut Fuse and prevent it from directly challenging MTV's dominance. MTV has received exclusive rights to recent videos by popular acts, including Radiohead's "There There," Limp Bizkit's "Eat You Alive," and Beyonce Knowles' "Crazy in Love" (Leeds, 2003a). Viacom's description of these arrangements with the record labels in its 10-K report to the SEC is rather vague:

> MTVN, in exchange for cash and advertising time or for promotional consideration only, license from record companies' music videos for exhibition on MTV,

MTV2, VH1, CMT and other MTVN program services. MTVN has entered into multi-year global music video licensing agreements with the major record companies. These agreements generally cover a three- to five-year period and contain provisions regarding video exhibition. (Viacom, 2002, pp. 1–7)

More specific press accounts of these agreements say that the contracts give MTV exclusive rights to up to 20 percent of a record label's music videos, and allow MTV to have sole access for up to six months (Leeds, 2003a). In return, the record labels receive free advertising time from MTV.

The collusion between MTV Networks and the major labels is intended to maintain the dominance of these companies in their respective fields, while undercutting competition from other music networks or record labels. However, there are persistent conflicts between MTV and labels over their financial arrangements. The record labels often complain about paying for expensive music videos in addition to the costs associated with their acts appearing and performing on MTV, like travel expenses and stage sets. MTV pays about $5 million each year to the larger record labels for licensing fees, but the labels charge that this does not cover their costs (Leeds, 2003b). Recently, some labels have been refusing to pay for some of their acts' appearances on MTV, and have been making their own product placement deals with companies like General Motors and Verizon without MTV's approval (Leeds, 2003a). The major record labels would ideally like to take direct control of the marketing of their artists on television rather than relying on MTV as a middleman with its own agenda. In 1994, these companies tried to launch their own music video channel, but they dropped the idea after the Justice Department launched an inquiry into the project.

MTV's Competitor: Fuse

Today, MTV's only real competition comes from Fuse, a program service that aspires to focus on music videos in much the same way that MTV did in its early years. Fuse was originally MuchMusic, a Canadian music channel that was bought by Cablevision in 2000 and has been a fledgling service in the United States over the last few years. In 2003, MuchMusic was relaunched as Fuse with a new format and identity as a corporate rebel. Such a characterization is odd since the service is owned by an established corporation, Cablevision Systems, which also provides the channel with a measure of vertical integration to ensure that Fuse is included on cable systems owned by the parent firm. As of December 2003, Fuse was available to about 34 million households through cable and satellite television, an audience that can be compared to the reach of MTV2 (50 million households) and the original MTV (86 million) (Leeds, 2003a). Music industry observers see MTV's exclusivity contracts with the labels as an attempt to undercut Fuse's future growth. "Whether anyone wants to admit it or not, the

Fuse is becoming a player," said one anonymous artist representative. MTV "is seeing a spark, and they want to keep them from playing" (Leeds, 2003a, p. 1).

Fuse distinguishes itself by playing music videos all the time, while mocking the original MTV for moving away from music. A billboard for Fuse that was put up across from MTV's Times Square headquarters, asks "Where's the M in Emptee-vee?" (Dreisinger, 2004). Other billboards feature Sally Struthers saying, "Right now, a music video is being neglected," and thanking Fuse for "saving the music video," a parody of her "Save the Children" commercials (Press, 2003). While Fuse tries to differentiate itself from MTV, some observers have noticed more similarities among the services. Dreisinger (2004) says that Fuse adopts MTV's current practice of playing blocks of videos in a similar genre, rather than playing the eclectic mix of videos that MTV did in its early years. Fuse also has the same kind of hosts as MTV, and also seems to be gradually increasing the kind of non-music programming that it supposedly derides. While both Fuse and MTV2 have a similar format of non-stop music videos, Fuse differentiates itself from MTV2 through an emphasis on interaction with its viewers through such means as instant messaging, games and Internet connections with the television service (Press, 2003).

Synergy at MTV Networks and Viacom

Tom Freston, the president of MTV Networks, publicly dismisses the corporate strategy of "synergy," where various subsidiaries of a conglomerate cooperate in various ways for the mutual benefit of these units and the parent company. "In most corporations synergy has been an excuse for one division to rip off another division," say Freston (in Burt and Larsen, 2004, p. 12). "It just doesn't work like that in real life." However, Viacom has strongly encouraged its units to collaborate, and MTV is often at the center of such joint ventures. In contrast to Freston's derision of synergy, the Viacom website strongly emphasizes MTV's movement into other media, implicitly acknowledging cooperation with units in these other media:

> MTV is a multidimensional youth brand that extends across virtually all media. The network has launched home videos, consumer products, and books featuring MTV programming and personalities. Its affiliated Websites provide innovative online music experiences to music lovers around the world. In addition, MTV pursues broadcast network and first-run syndication television opportunities through MTV Productions. (Viacom, 2004)

Viacom also continually examines where their different units can cooperate to promote the products of other units. This issue is addressed at the Viacom Marketing Council, which meets roughly every six weeks and is attended by the marketing chiefs of MTV Networks, CBS, Paramount, Infinity Radio, Blockbuster, Viacom Outdoor, Simon & Schuster and Comedy Central (McClintock). There are frequent collaborations between the CBS television network and various MTV Networks services, such as the examples mentioned earlier about the Super Bowl and the Grammy Awards, both televised on CBS. CBS's new shows for the 2003–04 fall season were promoted on VH1 and TV Land, and the CBS show *Survivor: Pearl Islands* was featured on a prime-time special on VH1.

MTV Networks has also cooperated with Paramount Pictures, which is responsible for releasing MTV Films. MTV Films and Paramount have collaborated to produce films that reach a young audience such as the Paramount film *Jackass*, based on the MTV series of the same name. Viacom's Blockbuster video unit plays a pivotal role in promoting DVDs and videos of MTV and Paramount Productions in its chain of stores. These products receive special attention in video monitors or displays in the stores. Paramount Parks, Viacom's amusement parks, and National Amusements, its chain of movie theaters, also promote new products from MTV Networks by putting up lots of posters and promotional material about these media, as was done with *Rugrats Go Wild*, a Paramount film featuring characters from two cartoons on Nickelodeon, a MTV Networks service for children (Beard, 2003). Sometimes this cooperation between units fails, as was the case when the autobiography of Viacom's Chairman Sumner Redstone, published by Simon & Schuster and heavily promoted in Blockbuster stores, did not sell. However, more generally, this strategy allows Viacom to channel the public's attention and interest toward the media content that it is offering in its range of subsidiaries.

As part of a larger media conglomerate, MTV Networks is also in an advantageous position in its dealings with advertisers. MTVN has adopted a strategy of offering advertisers special package deals where they can advertise on all of the MTV Networks program services simultaneously. These arrangements let an advertiser show commercials for the same product on MTV, VH1, MTV2, CMT, Nickelodeon, Nick at Nite, TV Land, Spike and the College Television Network. In May 2004, MTV Networks made an elaborate sales presentation to advertisers in order to encourage them to buy ads from all of its services as a group in the "upfront market" where firms purchase commercial time before the new fall season begins. MTV's Freston described these program services as "one big shopping cart" for advertisers to reach a much larger audience than they would with any one service (Romano, 2004, p. 23). These package deals that are also practiced at other media conglomerates put smaller companies with only a single program service at a disadvantage because they are unable to offer such arrangements.

MTV and Globalization

MTV Networks continues its international expansion with variations of its original US services like MTV, VH1 and Nickelodeon, operating 96 channels around the world that reach more than 340 million households in 140 countries (Burt and Larsen, 2003; Viacom, 2004). As of October 2003, there were 38 versions of MTV and 11 of VH1. MTVN says that its programming philosophy is to "think globally, act locally" where each of its international services "adheres to the overall style, programming philosophy and integrity of the MTV network while promoting local cultural tastes and musical talent" (Viacom, 2004). There remain questions about how these competing objectives are balanced, since the promotion of a certain style of music, fashion, and culture necessarily have consequences for the local and regional culture of a country. MTV's programs, music videos and advertisements promote a highly consumerist, materialist value system that often clashes with local cultures. Moreover, MTV tends to feature music acts that are consistent with its brand and image, excluding artists with idiosyncratic local music that do not conform to this style.

Much of the original local programming for these MTV services also tends to be copies of program formats that originated in US services. For instance, there are 10 locally produced versions of the *MTV Video Music Awards*, and copies of the US MTV series *Unplugged*, such as *La Ley MTV Unplugged* in MTV Latin America. The MTV services also cut programming costs by repeating programs originally presented on other MTV channels. After the *MTV Music Awards Latin America* was broadcast on that MTV channel, MTV Networks planned to air this on other MTV channels at later dates.

The list of MTV Networks' international program services in Viacom's 10-K SEC report (I-5, 6, 7) reveals three kinds of financial arrangements for these services. Some of the services like MTV Latin America and MTV Asia are wholly owned by Viacom. Other services are joint ventures between MTV Networks and other companies that are often based in those regions. For instance, MTV Italia is jointly owned with Holding Media e Communicazione, and MTV Russia is a joint venture with Russia Partners Company and others. In other areas, MTV has a licensing arrangement with another company which owns and operates the service that is licensed by MTV to use its name and format. Viacom has such licensing contracts with Optus Vision Pty Limited for MTV Australia and Transglobal Media SRL for MTV Romania. MTV Networks' cooperation with other companies allows it to aggressively expand to other regions, while sharing the financial risk of such ventures and taking advantage of local companies' knowledge of media systems in the area.

A cursory review of MTV's services demonstrates the company's global sweep in music programming. MTV's services in Asia include MTV Mandarin, MTV Southeast Asian, and MTV India, and two channels that were launched in 2001, MTV Japan and MTV Korea. Viacom points out that its Asian channels play a heavy proportion of regional music: it says MTV Mandarin's play list comprises 60 percent Mandarin music videos, and MTV India's programming is 70 percent devoted to Indian film and popular music. The rest of the time for both services is made up of "international music videos" (Viacom, 2004). MTV

Europe's service is divided into five separate 24-hour services: MTV in the UK and Ireland, MTV Central, MTV European and MTV Southern, and MTV Nordic. As it has in the United States, MTV is also creating new specialized digital program services, such as MTV Base, MTV Extra and VH1 Classic, which are shown in the UK. MTV Russia differs from most of the MTV channels in that it is broadcast as a "free over the air" service that can be seen in 18 million homes. MTV Latin America has three services – North, South, and Central – that are tailored to viewers in each region. This service reaches 11.5 million households in 21 territories in Latin America and some areas in the United States with large Latino populations. MTV Brasil is a separate Latin-themed MTV program service for that country.

Viacom's 10-K report (2002, pp. 1–25) notes that MTV's international services do have competition from other music program services, but for the most part these are services that are partially owned by major international record labels. For example, Channel V, a music service available in Asia and Australia, is owned by Star TV, a satellite television provider, and four major record labels. Similarly, Viva and Viva 2 are music channels distributed in Germany that are owned primarily by four large record companies. This leads to a peculiar form of "competition" since MTV cooperates extensively with the big international record labels and promotes their acts in its global program services. Moreover, the label-owned services like Channel V and Viva primarily market the artists signed to these major record companies. Thus, viewers in many countries are left with a choice between watching MTV Networks music channels that focus more on major label acts and those owned by these labels outright that do the same, albeit more brazenly. Moreover, MTV Networks and the major international record companies that operate these music channels are all subsidiaries of larger global media conglomerates that use these services in order to promote the full spectrum of their media products in their various subsidiaries.

MTV Networks and Viacom have extensive and growing control over large areas of media and popular culture, both in the United States and globally. Viacom's services from MTV to CMT dominate the range of music presented on cable and satellite television systems, giving it great influence to favor certain artists and music genres. This influence over music is enhanced through its collaboration with major record labels that serves to promote major label acts and marginalize independent labels and alternative sources for music television programming. MTV Networks also becomes a conduit for promoting the media properties of a range of Viacom subsidiaries, pushing the audience to pay attention to Viacom properties rather than other independent media outlets. Finally, Viacom's international operations extend these trends globally, providing MTV Networks with more influence over music and popular culture in Europe, Asia and Latin America. This survey suggests that when it comes to the activities of MTV and its parent firm Viacom, an exposed breast at a football game is the least of our worries.

References

Ali, L. and Gordon, D. (2001) "We Still Want Our MTV," *Newsweek*, July 23, p. 50.

Associated Press (2003) "MTV to Produce 2004 Super Bowl Half-time Show," Associated Press, December 8.

Asthana, A. (2003) "MTV's Fame Academy for Small-Screen Wannabes," *The Observer*, May 25, p. 10.

Banks, J. (1996) *Monopoly Television: MTV's Quest to Control the Music*, Boulder, CO: Westview Press.

Beard, A. (2003) "Rugrats Help Viacom Make a Splash: Characters' Return to the Cinema Shows How Cross-Promotion of Products Has Worked for Some Media Groups," *Financial Times*, June 14, p. 10.

Burt, T. and Larsen, P. T. (2003) "To Sell the World on a Song," *Financial Times*, October 10, p. 12.

Carter, B. (2003) "MTV Goes All Out to Grab Business from the Broadcast Networks," *New York Times*, May 6, sec. C, p. 4.

Columbia Journalism Review (2004a) Internet website page titled "Who Owns What: Viacom Corporate Timeline," retrieved February at www.cjr.org/tools/owners/viacom-timeline.

Columbia Journalism Review (2004b) Internet website page titled "Who Owns What: Viacom," retrieved February at www.cjr/org/tools/owners/Viacom.

Comcast (2004) Listing of channels on Comcast cable system for Bloomfield, CT, retrieved February at www.comcast.com.

Dreisinger, B. (2004) "I Want My Fuse TV," *Los Angeles Times*, January 18, pt. E, p. 54.

Fabrikant, G. (2003a) "Troubles at Paramount: Is It Just the Money?," *New York Times*, December 6, sec. C, p. 1, 4.

Fabrikant, G. (2003b) "Viacom Will Pursue Spinoff of Blockbuster," *New York Times*, February 11, sec. C, p. 12.

Free Press Team (2004) Email message sent from the Free Speech Team, retrieved February at www.mediareform.net.

Friedman, W. (2004) "Super Bowl, Grammys Give CBS Promo Clout; Back-to-Back High Profile Events Create Cross-Promotional Opportunities," *Television Week*, January 12, p. 46.

Furman, P. (2004) "Viacom Set to Shed Blockbuster Chain," *New York Daily News*, February 11, p. 71.

Geewax, M. (2004) "Lawmakers, FCC Vow to Push for Cleaner Airwaves," *Atlanta Journal-Constitution*, February 12, p. 1E.

Goldsmith, J. (2003) "Viacom Merger Set Trend," *Daily Variety*, October 30, p. A6.

Larson, M. (2003) "Cable TV: Viacom Buys a Laugh," *Mediaweek*, April 28.

Leeds, J. (2003a) "As Rival Gains, MTV Locks Up New Videos," *Los Angeles Times*, December 17, part C, p. 1.

Leeds, J. (2003b) "Fuse Music Channel Crosses Frequency of Omnipresent MTV," posted on Kansas.com, December 28.

Lieberman, D. (2003) "Viacom to Buy Rest of Comedy Central," *USA Today*, April 23, p. 1B.

Los Angeles Times (2004) "EU to Examine Deal to Link Sony Music, BMG," *Los Angeles Times*, February 13, part C, p. 5.

McClennan, S. (2003) "Super Bowl XXXVIII: CBS Eyes a $170 Million Sunday; That's Not Counting Viacom Cable-Net Tie-In Programs," *Broadcasting and Cable*, November 3, p. 16.

McClintock, P. (2003) "Viacom Synergy Builds Up Brand," *Daily Variety*, October 30, p. A28.

PR Newswire (2003) "MTV Latin America's Second Annual 'MTV Video Music Awards Latin America 2003' to Air Via Broadcast in Latin America," *PR Newswire*, October 15.

Press, J. (2003) "The Music TV Wars: Reality Killed the Video Star," retrieved July 21 at www.villagevoice.com.

Rich, F. (2004) "My Hero, Janet Jackson," *New York Times*, February 15, sec. 2, p. 1.

Romano, A. (2004) "MTVN Is a Little of a Lot of Things; Freston Brandishes His Array of Channels at Broadcast," *Broadcasting and Cable*, May 12, p. 23.

Rosenthal, P. (2004) "Cover Story So Bad, Even FCC Sees Through It," *Chicago Sun-Times*, February 3, p. 12.

Strauss, N. (2003) "The Pop Life: MTV is Wary of Videos on War," *New York Times*, March 26, sec. E, p. 1.

Timmons, H. and Sorkin, A. R. (2003) "BMG-Sony Music is Said to Be Near," *New York Times*, December 12, sec. C, p. 6.

Viacom (2002) Form 10-K, filed with the Securities and Exchange Commission, December 31.

Viacom (2004) Information about businesses, retrieved February at website www.viacom.com.

CHAPTER FOURTEEN

The Trade in Television News

Andrew Calabrese

NewsProNet is launching a TV news service containing highly promotable news stories to enable stations to expand their special reporting efforts and fuel topical news promotion every week of the year. SweepsFeed™ client stations receive monthly market-exclusive, audience-tested story component packages – video, sound, scripts, background, source contacts and creative promo treatments – as well as audience demographic research results for each story as determined by our exclusive monthly Impact Tracker national audience survey.

Press Release, August 22, 1997

SweepsFeed™ delivers 8 in-depth special reports every month. 96 proven, promotable special reports every year. The stories can be run as delivered with the addition of local talent narration, or localized and inter-cut with local sound bites and stand-ups to maximize the local impact. You can have a local, high impact, promotable story produced and on the air in less time, saving your budget without scrimping on quality. Each story also comes with scripts, complete with localization tips to help you make the story your own, source contact information, and in-depth, interactive companion content. This service is designed to assist you in maximizing local content creation and promotion while delivering audiences you seek both on-air and online.

From NewsProNet's Website, June 8, 2004 (www.newspronet.com)

Marketing research and strategic business consulting and planning have an ever-growing presence in the making of television news in the United States. Whether one chooses to respond to the creative new forms of influence the profit motive has on deciding what is news with moralist outrage or amoral irony, I intend no cynicism in suggesting that the outcome may be about the same. *Irony* is cool, and widely recognized as salable, but *outrage* also has market appeal within the cottage industry of the virtuecrats.[1] The major media have shown no lack of capacity to absorb and profit from what passes as high-minded self-criticism. In the late 1990s, the subject of tabloid radio and television became a news story. *Newsweek* (18 January 1999) concluded that "sizzle trumps substance" in American media: "News, public affairs and history are morphing into entertainment" (Alter, 1999, p. 24). In a well-calibrated mix of scandal, wit, irreverence, and admiration, including a cover story about how "shock jock" talk radio host Don Imus "turns politics into entertainment" (cover), and another about how "Year by year the anchors of news give way to the ringmasters of talk" (Alter, 1999, p. 25), *Newsweek*'s cover series endeavored to show how "the old media mold" was being broken by the new "titans of tude" (p. 32).

Concerns over the tabloidization of mainstream media have tended to focus on the personalization of news story selection and framing. However, as the growing emphasis on the star-power of television reporters, anchors, and talk show hosts illustrates, the cult of personality is just as fundamental to the formula of tabloid television's success. Given the fame and high earnings associated with such celebrity, it is no wonder that one of the most desired occupations among students of journalism in the United States is to become on-air television "talent." It is reasonable to wonder whether such hyper-commodified aspirations are at odds with ideals about the potential contributions of television news workers to democratic discourse. With this in mind, my aim in this chapter is to illustrate the tension that exists between commodity culture and the scope of what passes as civic discourse in the realm of commercial news.

Sweeping Success: Industry Restructuring and the Sovereign Consumer

Over the past 25 years, the preoccupation with marketing, promotion, and market share in American television news production and distribution has intensified. This chapter outlines some of the structural changes and resulting new approaches to survival and competition in this environment, paying particular attention to the pervasive impact of industry restructuring and accompanying new marketing practices. The fact that the US television industry was a commercial one from the start (notwithstanding the very small role played by public television since the late 1960s), gives the clearest indication that commercial values have always been central to determining what is "news" in the United States and how news will be presented. Indeed, to the extent that the idea of "public service" has been of concern in television history at all in the United States, it has been predominantly with the understanding that the needs, interests, and desires of the public can and should be met within a commercial system.

Whether the subject is Michael Jackson, Prince William, O.J., J-Lo, Monica Lewinsky, or any of hundreds of other less celebrated but no less sensationalistic stories, "real-life" dramas result in countless hours of apolitical "infotainment" in a variety of new formats. In the late 1980s, tabloid television shows like *Inside Edition* and *Entertainment Tonight* began competing head-to-head with local and network news. *Entertainment Tonight* became a major rival to network news in some markets in those years. In response, network news programs devoted twice as many minutes to arts and entertainment coverage in 1990 as in they did in

1989 or 1988 (Robins, 1990, p. 3). Not surprisingly, the major television networks – Fox, ABC, NBC, CBS (with Fox pushing the pace) – have moved progressively into the tabloid genre, with its emphasis on human drama, celebrity, scandal, and crime. Prime-time network shows such as NBC's *Dateline* and ABC's *20/20* can afford to spend much more per episode than syndicated tabloid shows, which has led to fears among producers in the latter group that their genre has been "co-opted" or mainstreamed by the networks (Mifflin, 1999). For example, the size of the audience of ABC's *20/20* is difficult for syndicated "news" magazine shows to match. In response to criticism that the network news is becoming "infotainment," executive producer of *NBC Nightly News* Steve Friedman claimed "there isn't anything geopolitical left" to report (Cook, Gomery, and Lichty, 1992, p. 4).

From a commercial standpoint, personality topics are of unquestionable good value, but coverage of them has also heightened ambivalence about the capacity of popular television journalism to offer a meaningful political space for democratic discourse (Dahlgren and Sparks, 1991; Dahlgren, 1995). Today, there is widespread concern that tabloid television is not just a niche-market phenomenon, but rather that its principles and practices are infecting the mainstream, so to speak, in a race to the bottom of public taste. What follows is a brief overview of the industrial restructuring of television that is pertinent to understanding the mainstreaming of the tabloid genre.

The germ of what became widespread interest in deregulation in the United States began in the late 1960s among largely politically marginalized think-tanks interpreting the work of Chicago School economists. Within ten years, market liberalization began taking effect, first during the presidential administration of Jimmy Carter, and more fully blown under Ronald Reagan, as well as during the Clinton years, and of course during the presidency of George W. Bush. In 1981, Reagan administration Federal Communications Commission (FCC) Chair Mark Fowler proselytized faith in the free market and set about dismantling the Commission's interference in the market's ability to devise and serve the "public interest." Fowler and his aide Daniel Brenner provided a neo-liberal manifesto for the public interest with respect to the means of communication:

> Communications policy should be directed toward maximizing the services the public desires. Instead of defining public demand and specifying categories of programming to serve this demand, the [Federal Communications] Commission should rely on the broadcasters' ability to determine the wants of their audience through the normal mechanisms of the marketplace. The public's interest, then, defines the public interest. (Fowler and Brenner, 1982)

Amid the climate of deregulation during the Reagan era, Wall Street became much more predatory. Stocks increasingly were traded by money managers who were evaluated based on short-term performance. Investment banks ceased to take on their traditional roles as cautious counselors and more aggressively vied to lend money to speculative investors, while hostile takeovers raged. In the television industry, reasons for media mergers and acquisitions have often included, but are not limited to, the desire for the vertical integration of programming and distribution. Mergers and acquisitions are also a way to enter emerging media, such as cable, Internet and DBS, by the process of simply buying companies with significant market share in these media.

Based on recent efforts by the FCC to further deregulate the television industry in the name of "synergy," ownership concentration will continue to intensify (Calabrese, 2004). For a glimpse into the future of television ownership concentration in the United States, we need only take a brief look at the recent history of radio deregulation. Since 1996, when Congress removed restrictions on the number of stations that one parent corporation could own, three companies have come to control 60 percent of the stations in the top 10 US markets. The number of radio station owners has declined by more than a third, and advertising rates have skyrocketed (and are passed on to consumers). Texas-based Clear Channel Communications, the country's largest radio company, expanded from 40 to 1,240 stations between 1996 and 2003. Clear Channel distributes national playlists for music, and it provides centralized news services for simulated "local" reporting. Clear Channel's "local news" networking operation is a "hub and spoke" system that has reporters reading news copy from a hub that is distant (even out of state) from the location where the reported news has taken place. As one study notes, in such a system reporters have very little idea of what news is important to the community on which they are reporting (Hood, 2001). The company exerts considerable leverage on artists, and on the market for music by imposing its conservative political views on its airwaves. Its political views are a show of mutuality toward a federal administration that rewards it through deregulation and the opportunity to expand its holdings, and it is a pattern that will likely be reproduced by the television networks as their holdings and power grow. Rupert Murdoch's Fox TV network already resembles Clear Channel both in its tendency to build its brand identity around "brand America," and in its strongly centralized control over programming and editorial viewpoint. Due to flagging ratings, the other major networks have demonstrated complete willingness to follow Fox's lead (Calabrese, in press).

The process of network integration, both vertical and horizontal, has occurred in a climate of intensified competition among news organizations, manifest partly by the injection of entertainment strategies into previously more understated news programming and promotion practices, and even by the consolidation of network management of news and entertainment divisions. According to NBC's Gulf War correspondent Arthur Kent (who was fired by NBC after criticizing the network), entertainment and tabloid coverage started immediately after GE bought the network in 1986. GE demanded that all its divisions show higher profits and consolidated the management of its news and entertainment divisions. NBC news operated at a loss from 1979 to 1988. The staff was downsized and began using checkbook journalism (paying for interviews and "exclusives") and private detectives to comply with demands for greater efficiency and flashier news stories, according to Kent. Michael Gartner, the man GE hired to turn NBC News around, resigned over the 1992 scandal in which the *Dateline NBC* magazine program admitted to rigging explosions of GM trucks in a story about faulty gas tanks. Three *Dateline* producers also lost their jobs in connection with the scandal. Kent cites the case as an indication of the prevalence of unprofessional newsgathering in the face of revenue pressure and management by MBAs rather than journalists (Connor, 1997). It is a lament that also can be heard from newspaper critics (Underwood, 1995).

There is no clearer indicator of the dictatorship of the market over political communication than that of the application of television ratings to the evaluation of news and public affairs programming. In an environment of constant ratings pressure, failure to deliver an audience on one "important" news story can cause heads to roll. Following CBS News' poor ratings on the night that Princess Diana died, CBS reorganized its news management team with promotions and demotions, leaving the man in charge of "hard news," Lane Venardos, to be held as the "fall guy" ("More Viewers Spurn the Nets," 1997). The preoccupation with quarterly ratings sweeps and short-term profitability is a major determinant in news production and program decision making. Expensive news specials, which are produced with the intent to "hype" the ratings during sweeps periods, are the rule rather than the exception, as *Chicago Tribune* media critic Steve Daley notes: "Traditionally, sweeps stories run the gamut from the cynical to the downright moronic. Sex and horrific illness are the hallmarks of any true sweeps period, as are blatant and unvarnished plugs for a station's own network entertainment programming" (quoted in Moritz, 1989, p. 123).

The most significant change in television news in recent years has been the shrinking viewership for the three main broadcast networks. Since the mid-1980s, ABC, CBS, and NBC have been besieged by declining market share and increased competition from multi-channel cable programming, Murdoch's Fox network, independent stations, pay-per-view channels, videocassette rentals, home satellite dishes, and other sources of alternative entertainment and information. Between 1978 and 1988, the prime-time network audience fell from 92 to 68 percent. According to Nielsen's November 1996 ratings sweeps, the big three networks collectively held a 50 percent share of the television market (Flint, 1996). A year later, the big three each saw an average decline of about one million viewers from the previous year ("More Viewers Spurn the Nets," 1997).

Follow the Money: New Techniques in Local News

A 1997 survey of local broadcast news profitability, sponsored by the Radio and Television News Directors Association (RTNDA), found that 62 percent of local TV news operations are profitable, and that on average 35 percent of a station's revenue is generated by news programs. Given these figures, it is not surprising that the emphasis on technological innovation and on-site reporting are on the rise ("Northeast is Biggest News Money Maker," 1997). This section provides a brief review of recent technological innovations in local news production that are aimed at enhancing stations' competitive programming strategies.

On-Site Reporting: Since the 1980s, it has been commonplace for local stations to send their reporters off to other parts of the United States and other countries in order to give a local flavor to distant news. The infatuation with images, coupled with cutthroat competition, has led to a global race for on-site footage. An international presence upstages local competition and steals thunder from the network news, which previously had a virtual monopoly on international coverage. Increasingly, local reporters are sent great distances, and at great expense to their stations, to illustrate just how "in the thick of it" their news operations can be. Typically, the coverage lacks any depth of analysis and perspective, and suffers from the worst symptoms of "parachute" journalism. Regardless, the widespread availability of high-quality field equipment, along with microwave and satellite uplink capabilities, has made the on-site report an essential part of broadcast news operations.

Many network affiliates supplement their broadcasts with CNN and Conus newsfeeds of on-site and live video, thus often "scooping" the network broadcast with footage. Since 1990, CNN's Newsource has grown from 159 to about 600 affiliates. The greatest demand is for live shots. ABC's "News One" service, which feeds 16 hours of news stories to local affiliates each day, is increasing the number of correspondents it uses around the country. ABC's News One carried out 300 live shots and the figure had risen to 700 by 1993. News One vice-president Don Dunphy acknowledges the "need" to do more live shots, particularly for morning and midday news, as the business becomes more competitive. NBC's "News Channel" and CBS's "Newspath" are similarly anxious about the threat from CNN (McAvoy, 1997).

The speed with which on-site reporting occurs following the outbreak of conflict at military flash points is accelerating. During the Persian Gulf War in 1990–1, home viewers got a bird's-eye view of the decisive power and superior weaponry of the US military. Since that time, up-close images of many other military conflicts and natural disasters have been delivered live via portable satellite uplink. During the recent Iraq war, "embedded" journalists were able to use much lighter and more powerful mobile equipment to produce news from the field (Calabrese, in press).

Fast Pace, New Venues: Market forces have led to a speeded-up style of reporting. Stories are told with a faster visual pace, ever-shorter sound bites, and increased use of computer graphics. Because so little information can be transmitted in shorter clips and bites, reporters must be more interpretive and more emotive than they once were. Gone are the two-minute excerpts of politicians' speeches in favor of eight-second sound bites that are meant to exemplify the reporter's thesis about the meaning and significance of the situation at hand. By

275

reducing the average sound bite from over 43 seconds in 1968 to under nine seconds in 1988, and all but replacing campaign messages with reporter commentary, contemporary news has driven political candidates to seek formats that allow more direct access to audiences (Hallin, 1994, p. 134). During the 1996 Presidential election campaign, candidates Bill Clinton and Bob Dole appeared on hitherto mostly un-presidential forums, such as talk shows, MTV, CNN's *Larry King Live*, and other popular entertainment shows. Clinton and running mate Al Gore had begun to explore such venues in the 1992 election, and similar appearances were commonplace by the 2000 election.

Amateur Video: Another technological trend in news reporting is the growing use of amateur video. In January 1987, CNN started its "News Hound" hotline, which encouraged viewers to call in with scoops and send in amateur camcorder footage. By 1992, CNN was using about four of these scoops per month (Cook, Gomery, and Lichty, 1992, p. 15). Many local news stations have similar hotlines or routinely use such footage. Particularly bizarre, shocking, or legitimately newsworthy footage may be shown on local stations across the country. The most famous use of such footage was the video recording of the beating of Rodney King by Los Angeles police. Using amateur footage is often touted as "community involvement," but the content of such footage is rarely newsworthy in the traditional sense, and sometimes it is staged and fraudulent.

Network–Local Tie-Ins: All the networks now focus on "tie-ins" between late night local newscasts of news stories directly related to preceding prime-time entertainment programs. For example, on April 25, 1991, NBC was broadcasting an episode of *L.A. Law* in which a transsexual model character sues a cosmetics company. NBC affiliates were previously sent a memo urging them to follow up the program with a late-night news report on "transsexuals fitting into the community." Three days later, NBC was broadcasting a made-for-TV movie, "White Hot," about the mysterious death in 1935 of screen star Thelma Todd. A memo from NBC suggested running a segment on the late news on "how the rich and famous live in your area" or conducting a "satellite interview with Loni Anderson," the star of the movie. In all, the NBC "Sweeps Taskforce" issued a ten-page memo listing 31 news tie-ins for 25 different entertainment programs.

In sum, the uses of new technologies, particularly portable electronic news-gathering (ENG) equipment combined with satellite uplinks, enhances the power, speed, and reach of local news producers and poses significant threats to old forms of network dominance over the availability of remote, on-site coverage of late-breaking events. Just as in the world of computers the model of storage and processing has shifted from centralized production and distribution to decentralized modes, the same might be said of the television industry. The erosion of market share in television news competition by the major networks is the clearest evidence of this. This is not to suggest that in a more distributed system of multi-channel news sources and outlets there are no massive profits to be made through diversified ownership. The profits come increasingly from owning stakes in multiple sources and outlets, as the strategies of the world's top media groups recognize. The game is not simply to own a good (or profitable) operation, but also to hedge bets by owning stakes in as many of one's competitors as is possible. TCI's John Malone embodies the textbook illustration of this pattern of behavior. In industry jargon, it is a "win–win" situation.

Dressing for Success: The Cult of Personality in TV News

In the 1960s, research about network affiliate stations found that the popularity of local news programs was a major determinant of the affiliates' market share during prime-time network programming. As a result, affiliates hired audience researchers to help develop strategies for keeping audience attention. Paul Klein, an analyst for NBC, discovered that people did not really "watch a program" as much as they "watch television" (Rapping, 1986, p. 44). So Klein developed the "Least Objectionable Programming" concept, which provided the market rationale for "happy talk" and a shift toward a more entertainment-oriented emphasis in news programming, a strategy that functions to manage the "flow" of audience attention across programs.

In 1974, *News Center 4*, an NBC affiliate program in New York, captured only 333,000 viewers compared to ABC's 697,000 and CBS's 937,000. That same year, they hired consultants who used standard marketing techniques to gauge audience tastes and attitudes toward different styles and features of the newscast. By following the marketers' recommendations, the affiliate more than doubled its viewership and went from last to first place in only 17 months (Rapping, 1986, p. 45). Before the end of the decade, marketing researchers and consultants were permanent fixtures at the networks as well.

The consulting industry that shapes the image of the evening news is very much preoccupied with the cult of personality and personal image. NewsPronet, a television news research and consulting company, provides a variety of services aimed at enhancing the market share of network and local television news organizations (www.newspronet.com). Among NewsPronet's services are "Talent Performance Tips," which are guidelines for on-air news talent needing advice on appearance and delivery, "Hot Tease Tips," strategies for getting viewers to stay tuned through the commercials, especially between programs, and a range of special strategies for enhancing viewership during ratings sweeps. In 1997, NewsPronet conducted a survey comparing the three major commercial network news anchors – Peter Jennings of ABC, Tom Brokaw of NBC, and Dan Rather of CBS – noting the differences in demographic appeal ("Jennings is at his strongest among women, 25–64 years old, middle-class viewers, and whites"; "Brokaw is at his strongest among men, those 65+, lower- and upper-class viewers and whites"; "Rather enjoys surprisingly strong support among 18–24 year olds, many of them minorities"), as well as reasons why viewers prefer each of the three anchors ("'American Voice' Poll of Network Television News,"

1997). Given the enormous salaries paid to these stars, it should not be surprising that their market value is fundamentally important to the success of the news organizations for which they work, and that market research is conducted to gauge the competitiveness of such assets.

The types of suggestions made by marketing consultants about news programs and talent include more soft-feature stories, more emotive delivery, more use of graphics, and close attention to the youthful and attractive appearance of (female) on-air talent. Based on focus groups, marketers can pinpoint specific clothing fashions, hairstyles, body types, voice qualities, and personality traits that are most appealing to the audience (Friedman, 1988, p. 4). These observations have been codified by consultants and their influence on the recruiting, selection, evaluation, and promotion of newsworkers (Carstens, 1993). Resistance to pressures to conform to consultants' prescriptions has included lawsuits based on gender and age discrimination. In one of the most prominent cases, in 1983 a federal jury awarded former television anchor Christine Craft $500,000 in damages. Craft alleged that she was demoted because she was "too old, too unattractive, and not deferential enough to men." Craft also was offered $20,000 less in salary than her male counterparts who held similar positions. Craft eventually lost the case when in 1985 a federal appeals court overturned lower juries' judgments in her favor (Friedman, 1988). Every year since, several similar lawsuits have been filed. Recently, a 48-year-old veteran female anchor at a CBS affiliate in Connecticut lost her $250,000-a-year job after focus groups found her unattractive and a shopping mall survey found her ranking last in its "net enjoyment score" (Randolph, 1999). In essence, the legal system functions to enhance the property interests of the TV stations. Within the logic of the system, on-air talent is a particularly important and volatile commodity that stations "need" to be able to trade at will, just as professional sports team owners trade players.

During the Persian Gulf War, NBC correspondent Arthur Kent, known fondly by many as the "scud stud" because of his good looks and on-air charisma, did much to glamorize televised war reporting. Of course, good looks are not the only factor. The perceived bravery of CNN's Peter Arnett, the only network correspondent who was allowed to report from Baghdad during the Gulf War, made him a role model for many young adventure- and glory-seeking aspiring television journalists.[2]

The emphasis on personality has been the same for many years, as one broadcast producer observed in 1986:

> The average local newscast, almost anywhere in the country, is a kind of succotash served in dollops and seasoned by bantering between anchorpersons, sportspersons, weatherpersons, and person-persons. And these people had better be good-looking, sparkling or cute – weathermen with party charm, anchorladies with good teeth and smart coiffures, sportscasters with macho charisma. It doesn't matter if they have a news background or not. (Corwin, 1986, p. 33)

In conclusion, the economic value of the physical appearance and overall image of television news personalities is a rational part of the system that is in place, and there is nothing to suggest that there is any sort of crisis or moral revulsion at present or on the horizon that will counter this trend.

Search for Civic Reform

Before the present era of hyper-commodification in television news, in 1961 FCC Chairman Newton Minow delivered his famous "Vast Wasteland" speech to the National Association of Broadcasters, the industry trade organization. In his speech, Minow tried to argue that the market cannot and should not be the sole arbiter of the meaning of "quality" television. He concluded that "It is not enough to cater to the nation's whims – you must also serve the nation's needs" (Minow, 1991, pp. 26–7). Minow's speech epitomized the more general contradictions of the Great Society era of policy making in the United States. The ideal broadcasting system he envisioned should seek and achieve a balance between the roles of the consumer, who presumably is driven by whim and desire, and the citizen, who is driven by rational thought and needs. It is a theme which American intellectuals who have attempted to strike what is viewed as a proper balance between media profitability and responsibility have visited before and after Minow's speech.

American television journalism has never been governed exclusively by professional ideals, or rather such ideals have always been deeply influenced by a commercial logic that includes the commodification of politics. Such was the complaint of Edward R. Murrow in his 1958 speech to the Radio and Television News Directors Association (RTNDA) (Murrow, 1958). There is always navel-gazing going on among television industry leaders, one relatively recent and high profile example being Dan Rather's 1993 speech to the RTNDA, titled "Call It Courage," in which he pays homage to Murrow and criticizes the increasing tabloid orientation and audience pandering in television news (Rather, 1993).[3] A harsh response to Rather by TV critic Walter Goodman was published three weeks later in the *New York Times*. Goodman wrote:

> The ratings may not be all that scientific, but the bottom-liners have learned that they are more reliable guides to the nation's taste than high-minded journalists. Corporate executives are not by and large suicidal. If they were persuaded by the figures that news from other countries, economic news and serious substantive news of any kind would bring in more money than game shows or crime shows, America would have an hour's worth of such nourishment every night . . . So good luck to Mr. Rather in his campaign to stand up to the bad guys and rouse others in his trade to do likewise. But he had better recognize that the fat cats he is fighting have nothing more devious in mind than catering to the enormous audience he wants to serve, and that bold talk notwithstanding, he is more a beneficiary of the show-biz system than a victim. (Goodman, 1993)

What has changed over the nearly four decades between the speeches of Murrow and Rather is the extent to which public affairs broadcasting has become more and more entangled with and driven by the mechanisms of market priorities, gradually displacing what existed of a tradition of professional culture. In addition, public expectations have adapted. It would be delusional to think that the "fat cats" mentioned by Goodman care a whole lot about devising alternatives to what Rather calls the "fuzz and wuzz" of contemporary TV news. While Goodman's apparent low estimation of the television audience seems fairly undemocratic, if not misanthropic, it does not gainsay the importance of his point. As marketing researchers and consultants in any field of production clearly understand, for better and for worse, it is not simply the quality of a product that determines its economic success, and indeed its survival in the market. More important to market survival and success is the ability of the producers to position their commodity effectively. Distasteful as that reality seems from a public service standpoint, it is one that is faced by all broadcasters.

As attempts to stem the overshadowing of responsible journalism by market vulgarity, voluntary codes of media responsibility, such as advocated by Minow, do not work. A poignant illustration is the miserable failure of the US Hutchins Commission of 1947. The "blue ribbon" Hutchins Commission was established under a $200,000 grant by magazine baron Henry Luce because he wished to find ideological support for his own profitable enterprises, but instead the Commission disappointed Luce with a scathing report, published under the title *A Free and Responsible Press* (Leigh, 1947). The report accused the press of failing to live up to its responsibility to provide truthful and intelligent coverage of current events, or to provide a lively and representative medium of democratic discourse. In essence, the report found that the press had failed to make profitability secondary to responsibility.

Based on an assessment of the Hutchins Commission's efforts and ultimate failure to bring about any change in the practices of commercial news reporting, Stephen Bates argues that voluntary codes of professional responsibility do not work. While the Hutchins Commission was addressing the print media, Bates's observations are no less applicable in the television news environment (Bates, 1995). In the United States, the standard response by media organizations to any pressure to change their practices is to cloak themselves in the mantle of the First Amendment and claim that the right of free speech is at stake. More accurately, what they perceive to be at stake is their unbridled freedom to pursue profit in a destabilized and intensifying competitive system.

Among the more recent attempts to re-calibrate the balance between media responsibility and the quest for profitability in favor of the former has been the so-called "civic journalism" movement in the United States (also known as "public journalism"). Civic journalism represents an attempt to improve the present poor image and reality of newspapers, and to move them away from their entertainment and tabloid style and more toward being a central node for facilitating public debate about issues of arguably broad concern. According to civic journalism's two chief intellectual proponents, Davis Merritt and Jay Rosen, the proper role for journalism in a democratic society is to stimulate and enhance the quality of political participation and public debate as a matter of the "public trust":

> Today, the only way for journalists to protect that trust is to strengthen, through journalism, America's civic culture, by which we mean the forces that bind people to their communities, draw them into politics and public affairs, and cause them to see "the system" as theirs – public property rather than the playground of insiders or political professionals . . . We are far from believing that journalists or journalism can cure what ails politics and public life in America. That would ascribe too much power to the press . . . Our claim is a more modest one: if changes are necessary for America to meets its problems and strengthen its democracy, then journalism is one of the agencies that must change. That is the conviction on which public journalism stands. (Merritt and Rosen, 1994, pp. 4–5)

In affirming an ideal role for the press that is not unlike that of the Hutchins Commission, the proponents of civic journalism similarly align themselves against the evils of tabloidization, trivialization, sensationalism, and the like. Echoing those like Newton Minow who have argued for striking a balance in favor of press responsibility above profitability, civic journalism's advocates also evoke the distinction between consumer and citizen: "If the first form of understanding – seeing people as *consumers* – is typical of 'the media' as a business, the second – regarding people as *citizens* – characterizes 'the press' as an American institution" (Rosen, 1994, p. 16; emphasis added).

While not typically explicit in their condemnation of the trends in television news discussed above, the implicit message of civic journalism's advocates regarding television is quite clear: The reliance on image and fast pace, and the short-circuiting of an ideal of rational discourse among a literate public, leaves television increasingly out of the picture, so to speak. Civic journalism resembles the spirit of the Hutchins Commission in that it is opposed to the prevailing, if lamentable, understanding of what is "news." But what is striking is the fact that television is so obviously and deliberately overlooked in a movement that professes to speak to democratic ideals. While literacy and reading are fundamental to democratic processes in the broadest sense, how can a meaningful and relevant definition of civic journalism be articulated without also recognizing the many ways in which political power is exercised through the uses of the medium of television? Perhaps television journalism is a hopelessly anti-democratic medium of communication, as many of its critics contend, but the case that print journalism is more democratic has not yet been clearly made. As Harvey Graff (1987) notes in his excellent study, literacy has functioned throughout its history as a means of creating and maintaining social and cultural hegemony (11–12). To

281

assume that the promotion of print literacy was unambiguously aimed at the promotion of a democratic culture of critical discourse is probably as erroneous as the assumption that the promotion of the consumption of commodified television news is unambiguously antithetical to such an aim.

Regardless of whether it is applied to television or to newspapers, perhaps civic journalism's greatest point of vulnerability is that it inadequately confronts or explains the tension defining the production of news in a commercial media environment, namely, the tension between ideals of journalistic integrity and demands for commercial success. Advertising dollars, media conglomeration, and interlocking media industry directorates all place an increasingly serious set of overlapping limitations on the willingness and ability of news organizations in any medium to bite the hands that feed them. Attacks on decadent journalism, combined with new proposals for responsible journalism that focus on the profession while paying no attention to the political-economic environment in which journalists operate, are myopic to say the least.

The public service ideology of journalism in the United States strains against the realities of the commercial media environment, as is the pattern increasingly in other affluent countries. Today, the cult of personality is likely to continue dominating television news, along with trends toward blurring historically recognized distinctions between "news" and "entertainment." For example, the use of tie-in strategies to link news to entertainment programming is a tried and true practice of managing audience flow. Advertising shapes television (and print) news story selection and framing more than most self-respecting broadcast journalists are prepared to acknowledge publicly. If it were not the case that the two are so intertwined, then it would be unimportant for networks and stations to be increasingly obsessed with ratings sweeps for their news shows, which will determine the prices they can charge advertisers for thirty seconds of air-time. Despite this reality, it is neither surprising nor unreasonable (nor unwelcome) to find widespread concern about the quality of a cultural environment in which the focus and legitimacy of political discourse is heavily shaped by the sale of cars, pet food, pharmaceutical products, and deodorant.

Given this reality, it is not surprising that some have attempted to find ways to reform the production of news while accommodating the structure of the commercial system, one highly visible model being the civic journalism movement. For others, a source of optimism about how a competitive market environment might improve the quality of television news in the future lies in the hope that the degree of destabilization of traditional commercial outlets, combined with the declining cost of high-quality video production and means for distribution (particularly the Internet), will continue to open up new niches for independent news sources to emerge. Of course, there is no reason to assume that sensible and heavily capitalized opportunists in the existing market will fail to continue emerging there as dominant players, or that commercial imperatives will not otherwise pervade the entrepreneurial news publishing environment. In essence, the insidious challenge that liberal reformers typically ignore remains, namely, market censorship. Unfortunately, liberal intellectuals and media policy makers generally fail to acknowledge and address in any depth how commercial enterprise shapes political communication.

Conclusion

It is reasonable to doubt that if the quality of civic discourse is truly in decline, it can be reduced to the problem of the progressive commodification of news. Furthermore, it is also reasonable to question whether such a problem can be resolved by small surgical adjustments to media industry practices. While there are good reasons to expect more from television news organizations than they currently offer, it seems myopic and technologically deterministic to think that commercial television has fundamentally transformed citizenship rather than reproducing and, in essence, modernizing the pattern of its historical conditions. My point is not that there is nothing new under the sun, nor is it to apologize for market-driven journalism, but rather it is that beyond recent polemics about the commodification of news, this category of concern – generally about the vulgarization of civic discourse – has deep roots in a range of competing theoretical perspectives about the cultural origins of social decay (Brantlinger, 1983). Indeed, if we are to take seriously a familiar theme in critical theory about how a "culture-consuming public" has been displacing a "culture-debating public" (Habermas, 1989, pp. 159–75), we must recognize that the pattern of transformation has not been a sudden one, as ahistorical panic about the encroachment of commodity culture upon civic culture tends to suggest.[4]

In Peter Riesenberg's (1992) account of the history of Western citizenship, he highlights two general concepts of the citizen – "active" and "passive" – the former prevailing from the time of the Greek city-state up through the French Revolution, and the latter having lasted since that time. "Passive citizenship safeguarded everyone's person, property, and liberty. Active citizenship was reserved for the adult male who would contribute to the welfare of the state with his body and property." Riesenberg argues that while there has been a progressive expansion of equality in the enjoyment of basic human rights and dignity, "politics remained largely in the hands of traditional elites." In that respect, he concludes that neither the French nor the American revolutions broke with the past, and that the model of active citizenship advanced by Rousseau, based on his image of the Geneva of his childhood, "proved an attractive, but illusory goal throughout modern history" (pp. 271–2; see also p. xviii).

We can see how, in modern terms, active and passive citizenship find a rough analogy in the modern dichotomy between the *active citizen* and the *passive consumer*. Today, the idea that citizenship can find meaningful expression in the reception of one-way, mass-mediated messages meets with widespread objection. Dissatisfaction with the mass media has led some to hope that the "interactive" mode of ostensibly de-commodified Internet communication will provide citizens with the greater capacity to resist passivity and become more actively engaged in political discourse and action, a view that warrants healthy skepticism (Calabrese and Borchert, 1996). For others, the characterization of *mass* communication is wrong in the first place because it underestimates the selective and critical capacities of the active audience.[5] Yet another view, and one that I think has more reasonable and less polarizing potential, is to think of the system of mass communication as a system of representation (e.g. Keane, 1991, p. 44; Dahlgren, 1995, pp. 15–17). As Peter Dahlgren writes, "We can safely assume that the mass media are not about to fade from the scene – and in fact they continue to grow – and that just as representation in democracy is unavoidable, so is representation in communication. Neither by itself necessarily means the demise of civilization, even though both generate special problems" (Dahlgren, 1995, p. 16). With this view in mind, I believe it is fair to say that the idea of passive citizenship, as defined by Riesenberg, is not something we can manage without. Nor is the idea of active citizenship, and more generally *civic competence*, something we should reject as unattainable in all areas of our lives (Calabrese, 2001a, 2001b).[6]

It is reasonable to conclude that commercial television is widely viewed as one of the most lethal carcinogens of public life in late capitalist society. It is a view that sees the quality of citizenship as being undermined not only by the constant bombardment of product promotions but, more importantly, by the second-order commodification of civic discourse itself by virtue of the commercial pressures and constraints imposed on public discussion. Furthermore, commercial television news is often seen as setting (indeed lowering) the standards of commercial print news as well. While there are good grounds for holding these views, as the problems raised throughout this chapter illustrate, too little research, theory, and political practice has attempted to develop an articulate, systematic, and affirmative view of the relationship between the means of communication and our ideals of what it means to be a competent citizen. Perhaps the first step in such an effort would be to develop a more accurate and deeper understanding of what the idea of the citizen has meant historically, as suggested above, which would be a first step toward developing realistic expectations of citizens in complex modern society.

Clearly, civic journalism has touched a nerve in the body politic by self-consciously attempting to reform journalistic practices in such ways as to promote greater press responsibility in gauging and helping to shape the agenda of political communication. While this movement has been nearly blind to the question of how the framework of market-driven journalism shapes professional ideology among journalists, it is at least an example of effort to move toward a more reflexive relationship between the media and the citizen. But can we hope for more fundamental transformations? Is it indeed possible any longer to effectively sustain a decommodified, or less commodified, relationship between the means of communication – particularly television – and the citizen in a hyper-commodified cultural environment? Certainly public service television's declining audience and growing tendency to mimic its commercial competitors has shaken any bed of confident certainty on this matter. If in fact our expectations have been unalterably lowered for us, then one response must be: how can the tide be stopped or stemmed? On the other hand, in order to fight the temptation to recover an idealized and nonexistent past, it is reasonable to pursue and realize a historically grounded normative concept of citizenship and, more importantly, of civic competence. Such efforts must lie at the heart of any meaningful critique of market-driven journalism, or of any search for more democratic alternatives.

Notes

1   The recovery of "virtue" in the contemporary lexicon of American politics has been nothing short of explosive. William Bennett, Ronald Reagan's Secretary of Education and George Bush's "drug czar," has become the most widely acclaimed expert on virtue, whether or not such a distinction holds any validity by definition. Aimed ostensibly at restoring the moral compass of American society, and deriving much of its inspiration from principles advanced in communitarian philosophy, the rhetoric of virtue found its most popular, if not distorted and instrumentalized, expression in the 1992 "Republican Revolution" in the US Congress, and it has more generally underlain much of the contemporary assault on the US welfare state. Although Bennett is not alone among popular authors on the subject, his bestselling work has gained the most notoriety (Bennett, 1995, 1996, 1998). For useful commentaries on Bennett and the industry of virtue, see Goodgame (1996) and Katz (1996).

2   Arnett's career took a downturn during the 2003 Iraq war when, after he spoke on Iraqi television in favor of the US anti-war movement, he was recalled from the field by CNN. Another well-known television personality, Geraldo Rivera, sought glory in Iraq, although his blunder in revealing on television the general direction of US troop movements led to his being recalled from assignment (Addis, 2003; Stanley, 2003).

3   Ironically, in a BBC interview, Rather included himself in implicating American journalists for being cowed by politicians who might accuse them of being unpatriotic for persisting with questions about justifications for war and the conduct of war: "None of us in journalism have asked the questions strongly enough or long enough about this business of limiting access and information for reasons other than national security" (Rather, 2002; see also Holt, 2002).

4   It should be noted that Habermas has modified his view on this matter by acknowledging that the dichotomy between a "culture-debating" and a "culture-consuming" public is simplistic in its pessimistic denial of any sort of individual agency. Although there is no doubt that Habermas remains concerned about the progressive encroachment of instrumental reason and commodity culture upon the prospects for critical political deliberation, he also has argued that the contemporary relationship between politics and media culture "is more complex than a mere assimilation of information to entertainment" (Habermas, 1992, p. 439).

5   In this chapter, I have tried to avoid reinforcing the familiar – and arguably false – dichotomy between citizen and consumer, which has parallels in the binary opposition of information and entertainment. Despite the temptation to readily accept this dichotomy, I believe that the tenableness of these antinomies is worthy of greater scrutiny, perhaps of the sort that Derrida has applied to the binary opposition set up between speaking and writing in Plato's *The Phaedrus*, in which Socrates treats writing as an unnatural and inferior "supplement" to what is characterized as the more authentic experience of face-to-face communication (Derrida, 1981). But for the present, I wish simply to acknowledge that while the distinction may have some usefulness in terms of describing ideal types for analytical purposes, it makes no practical sense from an ethnographic viewpoint to normatively conceive of citizens as rational monads seeking

only to be "informed," as if such an experience would negate the possibility of emotional engagement, including the possible experience of pleasure. As to whether such a view puts us on a slippery slope to finding "the political" nowhere in cultural practices but in so-called "oppositional" forms of consumption, I would disagree. However, this is not the place in which to add further to polemics about media reception that tend to converge on a straw-man whose embodiment is often seen to reside in the work of John Fiske.

6   I wish to acknowledge a problem that was beyond the scope of this contribution, but that is worthy of the increasing attention it has been getting, which is the mounting set of challenges in social and political thought to the very idea of citizenship as a category of social, cultural, political, and economic inclusion and exclusion. I would at least assert here that no such discussion of this problem would be complete without considering the role of the means of communication, new and old, in both sustaining and challenging the concepts and practices of citizenship.

References

Addis, D. (2003) "Rivera and Arnett both Scripted Their Own Demise," *Virginian-Pilot*, 2 April, p. B.1.

Alter, J. (1999) "The New Powers That Be," *Newsweek*, 18 January, pp. 24–5.

"American Voice" Poll of Network Television News (1997) NewsPronet Website (accessed on 24 January 1999), Available at <www.newspronet.com>.

Bates, S. (1995) *Realigning Journalism with Democracy: The Hutchins Commission, Its Times, and Ours*, Washington, DC: The Annenberg Washington Program in Communications Policy Studies.

Bennett, W. J. (1995) *The Children's Book of Virtues*, New York: Simon & Schuster.

Bennett, W. J. (1996) *The Book of Virtues: A Treasury Of Great Moral Stories*, New York: Simon & Schuster.

Bennett, W. J. (1998) *The Death of Outrage: Bill Clinton and the Assault on American Ideals*, New York: The Free Press.

Bradford, K. (1993) "The Big Sleaze," *Rolling Stone*, 18 February, at <www.tvmuse.com/tabloid_tv.html>.

Brantlinger, P. (1983) *Bread & Circuses: Theories of Mass Culture as Social Decay*, Ithaca, NY: Cornell University Press.

Calabrese, A. (2001a) "Justifying Civic Competence in the Information Society," in S. Splichal (ed.), *Vox Populi, Vox Dei?*, Cresskill, NJ: Hampton Press, pp. 147–64.

Calabrese, A. (2001b) "Why Localism? Communication Technology and the Shifting Scale of Political Community," in G. Shepherd and E. Rothenbuhler (eds.), *Communication and Community*, Mahwah, NJ: Lawrence Erlbaum Publishers, pp. 251–70.

Calabrese, A. (2004) "Stealth Regulation: Moral Meltdown and Political Radicalism at the Federal Communications Commission," *New Media and Society*, 6(1), 18–25.

Calabrese, A. (in press) "Casus Belli: US Media and the Justification of the Iraq War," *Television and New Media*.

Calabrese, A. and Borchert, M. (1996) "Prospects for Electronic Democracy in the United States: Re-thinking Communication and Social Policy," *Media, Culture and Society*, 18, 249–68.

Carstens, P. (1993) "Selling the Show: A Study of the Impact of Market Pressures on a Local Television News Department and its Newsworkers," PhD thesis, University of Iowa.

Connor, L. (1997) "Caught in the Network News Wars," *The Indianapolis Star*, August 23, p. A13.

Cook, P. S., Gomery, D., and Lichty, L. W. (eds.) (1992) *The Future of the News*, Washington, DC: Woodrow Wilson Center Press.

Corwin, N. (1986) *Trivializing America: The Triumph of Mediocrity*, Seacaucus, NJ: Lyle Stuart.

Dahlgren, P. (1995) *Television and the Public Sphere*, London: Sage.

Dahlgren, P. and Sparks, C. (eds.) (1991) *Communication and Citizenship: Journalism and the Public Sphere in the New Media Age*, London: Routledge.

Derrida, J. (1981) "Plato's Pharmacy," in J. Derrida, *Dissemination* (B. Johnson, trans.), Chicago, IL: University of Chicago Press, pp. 61–171.

Flint, J. (1996) "NBC Claims Crown: Web Declares a Win in 8th Straight Sweeps," *Variety*, November 27, p. 1.

Fowler, M. and Brenner, D. (1982) "A Marketplace Approach to Broadcast Regulation," *Texas Law Review, 60*, 207–57.

Friedman, B. (1988) "Women in Prime Time: Does Age Matter?," *Woman Inc.*, July, p. 4.

Goodgame, D. (1996) "The Chairman of Virtue Inc.," *Time*, September 16, pp. 46–9.

Goodman, W. (1993) "What Parson Rather Left Out of His Sermon," *New York Times*, October 17, p. H33.

Graff, H. J. (1987) *The Labyrinth of Literacy*, London: Falmer Press.

Habermas, J. (1989) *The Structural Transformation of the Public Sphere* (T. Burger, trans.), Cambridge, MA: MIT Press. (Original work published 1962.)

Habermas, J. (1992) "Further Reflections on the Public Sphere," in Craig Calhoun (ed.), *Habermas and the Public Sphere*, Cambridge, MA: MIT Press, pp. 421–61.

Hallin, D. (1994) *We Keep America On Top of the World*, New York: Routledge.

Hansell, S. and Harmon, A. (1999) "Caveat Emptor on the Web: Ad and Editorial Lines Blur," *New York Times on the Web*, February 26, at: <www.nytimes.com>.

Holt, M. (2002) "Is Truth a Victim?," BBC Newsnight, May 16, at: http://news.bbc.co.uk/1/hi/audiovideo/programmes/newsnight/1991885.stm.

Hood, L. J. (2001) "The Local News Audience and Sense of Place: A Home in the Global Village," PhD dissertation, Boulder, CO: University of Colorado.

Katz, J. (1996) "The Crook of Virtues," *GQ*, March, 268–76.

Keane, J. (1991) *The Media and Democracy*, Cambridge, UK: Polity Press.

Leigh, R. D. (ed.) (1947) *A Free and Responsible Press*, Chicago: University of Chicago Press.

McAvoy, K. (1997) "Live News: The Competitive Edge," *Broadcasting and Cable*, July 21.

Merritt, D. and Rosen, J. (1994) "Introduction," in J. Rosen and D. Merritt, *Public Journalism: Theory and Practice*, Occasional Paper of the Kettering Foundation, New York: Kettering Foundation, pp. 3–5.

Mifflin, L. (1999) "Tabloid Television Era on the Verge of Dying Out," *New York Times on the Web*, January 18, at <www.nytimes.com>.

Minow, N. (1991) "The Vast Wasteland," Address to the National Association of Broadcasters, May 9 1961, Washington, DC, in N. Minow, *How Vast the Wasteland Now?*, New York: Gannett Foundation, pp. 21–33.

"More Viewers Spurn the Nets," *Studio Briefing* (1997) [online media industry news service list], 8 October, at http://newshare.com/sb/.

Moritz, M. (1989) "The Ratings Sweeps and How They Make News," in G. Burns and R. J. Thompson (eds.), *Television Studies: Textual Analysis*, New York: Praeger, pp. 121–36.

Murrow, E. R. (1958) "Lights and Wires in a Box," Speech at the annual convention of the Radio-Television News Directors Association (RTNDA). Chicago, Illinois. October 15.

"Northeast is Biggest News Money Maker," *Broadcasting and Cable* (1997) April 21, p. 35.

Randolph, E. (1999) "When an Anchor's Face is Not Her Fortune," *New York Times on the Web*, January 26, at <www.nytimes.com>.

Rapping, E. (1986) *The Looking Glass World of Nonfiction TV*, Boston: South End Press.

Rather, D. (1993) "Call It Courage," Remarks to the Radio and Television News Directors Association Annual Convention. Miami, FL, September 29.

Rather, D. (2002) "Interview with Madeleine Holt," *BBC Newsnight*, 16 May, at http://news.bbc.co.uk/1/hi/audiovideo/programmes/newsnight/1991885.stm.

Riesenberg, P. (1992) *Citizenship in the Western Tradition: Plato to Rousseau*, Chapel Hill, NC: University of North Carolina Press.

Robins, J. M. (1990) *Variety*, July 18, p. 3.

Rosen, J. (1994) "Public Journalism: First Principles," in J. Rosen and D. Merritt, *Public Journalism: Theory and Practice*, Occasional Paper of the Kettering Foundation, New York: Kettering Foundation, pp. 6–18.

Stanley, A. (2003) "Two Correspondents, One Predictable Outcome," *New York Times*, 1 April, p. B.14.

"Titans of Tude," *Newsweek* (1999) 18 January, pp. 32–4.

Underwood, D. (1995) *When MBAs Rule the Newsroom*, New York: Columbia University Press.

PART SIX

Television/ Programming, Content, and Genre

CHAPTER FIFTEEN

Configurations of the New Television Landscape

Albert Moran

Introduction

The first years of the new millennium seems like an appropriate moment to suggest that a more nuanced engagement with the term "globalization" – as a means of understanding recent developments in the field of television – is increasingly called for. Hence, for example, as against those who would claim the irresistible and inevitable onset of worldwide interconnection in such areas as economics and culture, recent attention paid by television researchers to geolinguistic regions, "cultural continents," as significant mediators of such linkages gives pause to a term such as "globalization" as an accurate description of recent developments (cf. Mattelart, Delcourt, and Mattelart, 1984; Sinclair, Jacka, and Cunningham, 1996; Straubhaar, 1997; Sinclair, 2000). Similarly, we would argue that an understanding of the specific historical trajectory of the institution that is television along the lines developed here will also give pause to easy generalizations about planetary developments in the present.

Thus, we would suggest that over the course of the past 10 to 15 years or so, television in many parts of the world finds itself in a new period or era that is marked off from earlier configurations of the institution (cf. Moran, 1989; Curtin, 1996; Saenz, 1997; Rogers, Epstein, and Reeves, 2002; Dell, 2003). In contrast with the present moment, earlier stages of television might be usefully designated as Live Television, Filmed-series Television and "Quality" Television. On the other hand, the characterization of the contemporary moment as that of New Television is suggested because of a unique intersection of new technologies of transmission and reception, new forms of financing, and new forms of content that have come together in recent years. What is novel and original about New Television is the fact that, by the 1990s, the centralized broadcasting arrangements in country after country across the world, mostly in place with minor changes since the beginning of television broadcasting in that place, have been increasingly transformed and reconstituted. The oligopolistic model is currently undergoing a profound transformation. Television is rapidly becoming something else characterized by new patterns, agendas, and structures. Hence, analogous to the adoption of the title of New Hollywood to designate the post-studio era in US motion picture history, so we espouse the term New Television to characterize the institution of television in the present epoch.

What, then, is "new" about New Television? The short answer is: the rapid multiplication of services of every kind. Television has indeed become the tube of plenty and it is to this phenomenon that we turn.

The Multi-Channel Landscape

Beginning in the last years of the previous century and quickening since 2000, television systems in many parts of the world have been distinguished by an ongoing reconfiguration of the institutional field. Television is currently undergoing a sustained shift, away from an oligopolistic-based scarcity associated with broadcasting toward a more differentiated abundance or saturation associated with the proliferation of new and old television services, technologies, and providers.

In the multi-channel environment of the present and the near future, television is and will be delivered via existing and new technological arrangements (McChesney, 1999; Flew, 2002). Meanwhile, a transformed system also comes to provide additional services to viewers, now increasingly referred to as consumers. These data services are complementary to the information and entertainment provisions of broadcasters and are increasingly more interactive than the older services.

Television may once have been defined by an oligopoly of broadcast channels, frequently as few as one or two in any center of population. Today, however, it seems more and more likely to be defined by licensed or "free-to-air" providers together with others as the system becomes more differentiated. The new institutional players come from both within and without the sector. Thus, for example, in different European markets, there has been a significant increase in the number of television broadcasting networks on the air, public service and, especially, commercial (Blumler and Nossiter, 1991; Noam, 1992; Wieten, Murdock and Dahlgren, 2000). Meanwhile, the past 20 years have also seen the onset of satellite, cable, and pay TV services (Gross, 1997; Paterson, 1997). New players have entered the distribution arena, including companies based in the telecommunications and computer sectors and newspapers (Constantakis-Valdes, 1997; Strover, 1997). In addition, new trade agreements seem likely to encourage other groups, both local and international, to enter the television arena.

In turn, the new multi-channel environment is served and stimulated by new distribution technologies such as satellite, cable, and microwave and also by new computer software including the Internet. Television is also characterized by a multiplying non-exclusivity of content that is now becoming available through other modes, including marketing and the worldwide web (McChesney, 1999; Flew, 2002). The convergence with computers and mobile phones yields new forms of interactivity, including electronic commerce, online education and tele-working. Meanwhile, digital TV, Web TV, and personal video recorders (PVRs) may further strengthen a tendency toward niche and specialized programming.

At the reception end of the reconfigured system, the television set now em-braces many functions, including television broadcast program reception, off-air taping and replay of videotape, engaging in computer games, playing of DVDs, surfing across channels, telecommunicating including accessing the Internet and email, using dedicated information services and an engagement in home shop-ping. In other words, "content" has ceased to be synonymous with the television program and programming. Instead, it has also come to include the creation of new sequences of image and sound, availing and engaging in interactive services and the accessing of dedicated data and information (Saenz, 1997).

Of course, one must tread carefully here and avoid premature proclamations of apocalyptic change. For indeed, the near future is likely to have much in common with the recent past. Hence, for example, despite a 10 percent fall in audience and a 25 percent increase in pay-TV households in the decade since 1993, prime-time television on the five Australian free-to-air broadcast networks seems likely, at least for the foreseeable future, to continue to define the indus-trial, financial, and aesthetic norms of local television. At the same time, it has also to be recognized that the future may yet see them yield economic and other ground to dominant global interests.

Whatever the case of ultimate ownership and control, the salient underlying feature of the new television landscape of the early twenty-first century remains the same. This is the multi-channel environment of television that highlights the passing of the broadcast oligopoly in favor of new financial and program arrangements.

One major consequence of these changes is likely to be a falling audience for any particular television show, no matter how popular it seems to be. With so many channels and technologies of distribution and circulation, it has been increasingly impossible for any hit show, no matter how successful, to register the kinds of ratings achievable in earlier phases of television.

In turn, several responses to this situation are now evident. One of these is stagnation, if not a drop, in the system's demand for more expensive forms of prime-time programming. In the United Kingdom, for example, there has been a decline in demand for both drama and current affairs programming in prime time, a trend that has its parallels elsewhere such as Australia (Brunsdon et al., 2001; Moran, forthcoming; Lawson, 2002; Meade and Wilson, 2001; Perkin, 2001; Mapplebeck, 1998). In other words, in characterizing the present era of New Television as one of abundance, it has to be borne in mind that this tend-ency only occurs with certain programming genres, indeed occurs at the expense of other types of content.

What, then, is the motor or source of this differentiated abundance that is already a central feature of the new landscape? How does it register as a phenomenon and how does it come about?

Adaptation

The most significant dynamic, then, seems to be one of adaptation, transfer, and recycling of narrative and other kinds of content (Bellamy, McDonald, and Walker, 1990; Pearson and Urrichio, 1999; Thompson, 1999, 2003; Brenton and Cohen, 2003). Not surprisingly, this tendency is not limited to television; rather, it is characteristic of many media and related areas of culture industries. Nor is it unique to the present epoch but is familiar in the past. However, in the present age of international media conglomerates, recycling, and adaptation of content across different media platforms is rapidly multiplying to the point of marginalizing if not extinguishing other economic and cultural practices (cf. Bellamy, McDonald and Walker, 1990; Cooper-Chen, 1994; Pearson and Urrichio, 1999; Horn, 2002). Many of these different kinds of adaptations are familiar. So, for example, films become television series just as television series trigger feature films. Remakes are equally common, although these are sometimes referred to by other names – the sequel, the spin-off or even the prequel. Clearly, a more encompassing name for these various phenomena is that of the serial or even the saga (Eco, 1987; Thompson, 2003). Nor does this general phenomenon of a content-genealogy end there. Instead, narratives can span several media – theatrical film, television, video, DVD re-release, videogames, CD soundtrack, radio, comics, novels, stage shows, musicals, public concerts, posters, merchandising, theme parks and so on. Fanzines and Internet websites further spin out these contents. Individually and collectively, this universe of narrative and content constitutes a loosening of the notion of closure and the self-contained work of art (Thompson, 1999, 2003).

Behind this proliferation of transfers, this ever-expanding recycling of content, is a set of new economic arrangements designed to secure a degree of financial and cultural insurance not easily available in the multi-channel environment of the present. Adapting already successful materials and content offers some chance of reproducing past and existing successes. Media producers, including those operating in the field of television, attempt to take out financial and cultural insurance by using material that is in some way familiar to the audience (Fiddy, 1997; Moran, 1997). Having invested in the brand, it makes good business sense to derive further value from it in these different ways (Todreas, 1999; Bellamy and Trott, 2000; Rogers, Epstein, and Reeves, 2002). And, of course, in turn, this tendency of recycling is further facilitated by the fact of owning the copyright on the property in the first place.

In the age of New Television, then, there is a clearly identified need to derive as much financial mileage out of an ownership as possible – hence the importance of the idea of intellectual property (IP). This move to safeguard and control content-related ideas formalizes ownership under the protection of property laws such as those of trademark, brand name and registered design as well as those of copyright law (Lane, 1992; van Manen, 1994; Moran, 1997; Freeman, 2002). Indeed, the era of New Television may come to be characterized as one of a heightened awareness and emphasis on program rights.

However, for all the recent rhetoric about IP proclaimed by industry associations and lobbyists as well as individual companies, with attendant discourses concerning "piracy," "plagiarism," "theft," and so on, we are dealing here with the transformations facing international and national television industries in changing market conditions. Despite rhetoric to the contrary, rights are not innate or inherent. Rather, they are constructed aspects of the competition between different program producers, local and international, and between different users of program content and "brands," broadcasters, cable, radio, telephone and Internet.

The interests in rights held by television companies – both producers and broadcasters – who have joined the newly formed, Cologne-based Format Registration and Protection Association (FRAPA) are not defined abstractly, but change according to commercial circumstances. Thus, for example, the income generated from the licensing of a TV program into public usage has to be measured against its use as a means of promotion. As Frith (1987) has pointed out, copyright is generally used to make money rather than to control use.

Nevertheless, this emphasis on rights helps secure the general conditions for the process of selling the same content over and over again across a series of different media that has already been mentioned as a key feature of the present epoch. In the particular case of New Television, the process of worldwide geographic dispersal and recycling of existing content goes under the specific name of TV format adaptation and it is to this phenomenon that we now turn.

The TV Program Format

As already suggested, there is no single agreed-upon industry or critical term governing the ever-expanding system of adaptation of narrative and other content across different media. However, within the purview of television, just such a term – the format – exists although critical researchers have been slow to recognize such a development. Until 1998, for example, the only book dealing with the subject of the television format was a legal handbook published in Dutch (van Manen, 1994). There is no entry concerning television formats in the *New York Times Encyclopedia of Television* (Brown, 1977), while only a short note appears in the recent *Museum of Broadcasting Communication Encyclopedia of Television* (Fiddy, 1997). At the same time, however, it is also worth noting that there have been several discussions of specific program recyclings, even if the researchers involved have not paid any conceptual attention to the general phenomenon of format adaptation (Heinderyckx, 1993; Cooper-Chen, 1994; Gillespie, 1995; Pearson and Urrichio, 1999).

On the other hand, inside the international television industry, broadcasters and producers have been quicker to embrace the term and the concept. Indeed, the TV format is now a crucial mechanism in regulating the recycling of program content across different television systems in the world (cf. Heller, 2003; Moran and Keane, 2003). Thus, in contradistinction to general use of the term and even its specific application in radio (cf. Johnson and Jones, 1978), a format in television is understood as that set of invariable elements in a program out of which the variable elements of an individual episode are produced. Or, as a more homely recent explanation for would-be format devisors would have it:

> A format sale is a product sale. The product . . . is a recipe for re-producing a successful television program, in another territory, as a local program. The recipe comes with all the necessary ingredients and is offered as a product along with a consultant who can be thought of as an expert chef. (Bodycombe, 2002)

However, although international television industries talk confidently of the format as a single object, it is in fact a complex, abstract and multiple entity that is, typically, manifested in a series of overlapping but separate forms. At the point of programming and distribution, a format takes the cultural form of different episodes of the same program. Meanwhile, at the production end, these different industrial manifestations include the paper format, the program Bible, a dossier of demographic and ratings information, program scripts, off-air video-tapes of broadcast programs, insertable film or video footage, computer software and graphics, and production consultancy services.

These various manifestations underline the point that a format is not a single or a simple entity. Nevertheless, for good pragmatic reasons the industry ignores this complexity, because the TV format has become one of the most important means of functioning industrially in the era of New Television. As an economic and cultural technology of exchange inside the institution of television, the format has meaning not because of a principle but because of a function or effect. In other words, the important point about the format is not what it is, but rather what it permits or facilitates (Moran, 1997).

Of course, the adaptation from one territory to another of TV program ideas as a means of producing a new series is not novel. Replicating experiences in radio, television broadcasters and producers in many parts of the world have for many years looked to both the United States and the United Kingdom for program templates that would guarantee the production viability of new content ideas and help ensure the success of the finished program (cf. Wikipedia, 2003). And, of course, the fact that these ideas had been "indigenized" or "vernacularized" in the local culture was one important step toward that success. In the past, especially in the Anglophone countries, this borrowing mostly occurred in a series of lower-cost genres, including news and current affairs, game shows, children's programs, and variety programs, such as *Tonight*.

Historically, these adaptations were usually not paid for (and, often, too, not even explicitly acknowledged), a practice shared by producers from Latin America, Europe, Asia, and the Pacific (van Manen, 1994; Moran, 1997; Staubhaar, 1997).

It is equally worth noting in passing that the practice of unauthorized adaptation is far from dead. It continues as a matter of course in very large television markets such as those of India and PR China (Thomas and Kumar, 2003; Alford, 1995; Keane, Donald and Yin, 2002; Keane, 2003). Additionally, as legal disputes and even court cases attest, it also arises from time to time in the West.

When payments were made at all in the past, these were in the form of a compliment or a gesture of goodwill rather than as a tariff or fee. At least three factors were at work in this situation. First, the original producer was frequently unaware of a recycling taking place elsewhere. Much more significant was the fact that most program devisors and owners lacked the international reach that would have enabled them to pursue legal action against format appropriation in other territories. Under the Berne Convention, legal action against perceived infringement must take place in the jurisdiction of the territory where the infringement occurred. Finally, too, these borrowings, no matter how frequent, tended to be *ad hoc* and one-offs so that devisors and owners did not organize long-term international legal protection. The net result of these elements was that international TV program idea transfer occurred in a milieu of apparent benign ignorance and indifference.

Historically, this pattern of transfer was one means among several that helped to make television production in many different parts of the world economically viable. Adapting successful overseas programs (mostly from the United States or the United Kingdom) meant that local broadcasters and producers saved the expense of the relatively costly research and development (R&D) work involved in the first production of a successful TV program. Thus, adapting a popular program that had been on-air in the US or in the UK meant that a local producer or broadcaster was accessing a template that had already withstood two rounds of R&D, first to survive development and trialing before broadcasting executives and, secondly, to survive further testing before viewing audiences. Thus, recycling notable TV program ideas took much of the guesswork out of local television production in many lower-cost genres. After all, these were not endowed with the more costly, quality production values that would make them highly desirable and exportable. Thus, in an English-speaking market such as that of Australia, television programs in other genres, most especially fiction, were not – for the most part – imitated or adapted precisely because the original US or UK programs were imported. However, in other markets shielded by language or other cultural barriers – such as, for example, those of Russia and India – adaptation certainly takes place in fiction (Thomas and Kumar, 2003; Heller, 2003; Iwabuchi, 2003).

In the recent present, this process of international adaptation of TV program ideas has been consciously routinized and formalized through a series of related measures. These include a deliberate generation of value-adding elements under the name of the format (such as the format Bible), format marketing arrangements (industry festivals and markets), licensing processes and a form of self-regulation within the industry administered by a new industry association (FRAPA).

TV program ideas are now claimed as intellectual property (with a constant, consequent industry rhetoric about "piracy"). Meanwhile, there has been a concerted international attempt to formalize and commodify program ideas under the label of format. In turn, this has led to a degree of regularity in the licensed adaptation of program ideas to other producers and to other places. Additionally, as already indicated, the format commodity can also circulate within any particular multi-channel system, generating income from a variety of cross-platform sources.

Clearly, this new situation and arrangement formalizes what was once casual and spontaneous as a means of deriving financial benefit, most especially from overseas adaptations. Now, TV programs are not simply devised and produced for local buyers with the, often faint, hope that they might sell elsewhere in the world. Instead, they are consciously created with the deliberate intention of achieving near simultaneous international adaptation. Additionally, increased communication and company linkages around the world have meant that the unauthorized appropriation of TV program formats is less and less likely to go unchallenged.

This reconfiguration of the international circulation of TV program ideas has facilitated the emergence of new national sources of TV formats – a development that has affected Australia as well as many other territories, even including the United States and the United Kingdom. Thus, in the recent present, formats have come to Australian television not only from the traditional sources of imported programs and formats – namely Hollywood, the US networks, the BBC, ITV, and Granada – but also from producers in other countries as diverse as the Netherlands, Japan, and New Zealand. Of course, the United States and the United Kingdom continue to be the major sources of formats. Nevertheless, because formats can be linguistically neutral, templates derived from more culturally and linguistically "foreign" sources are, nonetheless, proving to be capable of being "indigenized" inside particular national television systems.

In other words, the recent era has witnessed the emergence of a new type of television, namely the globally-local program (Lie and Servaes, 2000; Moran and Keane, 2003). Of course, since 1969 and the televised Moon landing, one kind of global television program has existed, a type of television public event whose other recent forms have included major international sporting encounters, ceremonial events and the outbreak and pursuit of wars (Dayan, 1997). This kind of program is constituted by the simultaneous transmission of the same live event to different audiences in different parts of the world.

Meanwhile, however, the international formalization of the TV format trade has created a second kind of global television program. Here, there is not a single live event being transmitted simultaneously. Instead, there are a series of parallel events being transmitted in the same proximate time to a series of audiences who might, cumulatively, be said to have seen the "same" program. Thus, for example, although two billion people worldwide may be said to have watched *Big Brother*, in fact this figure is made up of a series of smaller audiences who have seen a succession of adaptations (*Big Brother US* Series 1, *Big Brother UK* Series 2, *Big Brother Australia* Series 3 and so on) based on the *Big Brother* format. In other words, with the television format, we encounter a program that is abstract and international in type while simultaneously local and concrete in its particular manifestations. Hence, formats generate regional, national, or even pan-national series even while the program itself is international or global.

Finally, before turning to a specific example of the format program in the new landscape, it is also worth noting an attendant shift in the preferred forms of prime-time genres. For just as different eras of television have promoted various genres and program types over others, so the same is true in the era of New Television. As already suggested, the format program is a sign of differentiated abundance and it is worth asking about the contours of this plenty. Hence, it should be noted that one principal genre type not favored for format adaptation is that of television fiction. The term refers not only to drama series and serials, but also to situation comedy and children's drama. Television fiction of these different types is marked by narrative, itself supported at the production end of television by the use of scripts and professional performers in various on-camera roles. What is striking about the area of television fiction is the very limited extent of format adaptation that is occurring or has occurred in the recent present. The few exceptions that can be cited – such as the remaking of the Australian soaps *The Restless Years, Sons and Daughters* and *Prisoner: Cell Block H* in Germany, the Netherlands and elsewhere in the 1990s – highlights the extent to which format adaptation is not occurring in fiction (Moran, 1997; O'Donnell, 1999). Instead, adaptation is occurring elsewhere, including makeover or lifestyle programs, competitions and game shows, talent shows and "reality" or docudramas (Brenton and Cohen, 2003). A useful term to cover this range might be "live" programs. Van Manen (1994) has distinguished this type from that of fiction on the basis of the absence of episode scripts and the heavy, although not total, use of non-professional performers in on-camera roles.

Of course, none of these types are new to television. Each has its precedent. The point is, though, that earlier incarnations of these subgenres were, usually, scheduled either in daytime slots or else programmed on minority channels whereas now, they have often become prime-time programming. Additionally, as part of this shift, the forms of these programs have also been reconfigured, most especially in terms of their *mise en scène* and their narratives. Not only are these newcomers interesting to watch, but they are frequently more pleasurable and engaging than were their predecessors.

More generally, in the process of this reconstitution, these subgenres have also undergone considerable transformation. This is particularly true in the case of what is sometimes called the docudrama, or "live" soap, which characteristically involves a narrative with a beginning, a middle and an end, elongated to span many weeks on-air, and generically expanded into several different modes, including that of observation, confession, talk, and entertainment-spectacular (Johnson-Woods, 2002; Hill, 2002; Roscoe, 2002; Brenton and Cohen, 2003). At the same time, the general budget of this kind of program has increased to help endow it with some of the textual features and values appropriate to prime-time rather than daytime scheduling. However, that said, it has to be realized that the production expenditures of a formatted program, such as that of live soap, must be balanced against several other factors including the savings generally achieved through development and trialing occurring in other television markets, through general spin-off and franchising arrangements and through network savings secured through reduced demand for fiction.

Finally, to complete this account, we can look briefly at one particular program as a paradigmatic instance of the program format in the era of New Television.

Live Soap: *Big Brother* (*Australia*) as Paradigm

Background

By now, the general history of this program format is well known so that only a brief summary need be provided (Ritchie, 2000). Established in the Netherlands in 1993, floated on the stock exchange in 1995 and bought by Spanish telecommunications giant Teléfonica in 2001, Endemol is the largest television format devisor, owner, and producer in the world. The development of a format catalog containing over 400 different items has been an integral part of the company's growth. In terms of format origination, the company has shown a marked interest and capacity in devising hybrid formats, such as those combining human interest with entertainment. Faced with the recent inflexion of the "reality" or docusoap form, most notably with the international success with *Survivor*, Endemol devised *Big Brother* (Brenton and Cohen, 2003; Johnson-Woods, 2002; Roscoe, 2002). The latter program first went to air in the Netherlands in 1999 and its outstanding success there immediately led to other national versions across Europe, the United Kingdom, the United States and elsewhere (Johnson-Woods, 2002).

The Australian-based film and television distributor and producer Southern Star signed a joint venture agreement with Endemol in 1999, so it was inevitable that the Australian commercial television networks would be offered an opportunity to broadcast a local version of the series. In April 2001, the first local series of *Big Brother* went to air. Altogether, the series ran for 13 weeks and was broadcast across Australia and in New Zealand. When it ended on July 16, it had

achieved the same outstanding success that it had attained in many other territories with the notable exception of the United States. Nationally, it chalked up
an audience of just on three million for the last night of the series when the
winner was announced. Additionally, the commissioning channel, Network Ten,
witnessed a big jump in its viewing public not only for the program but for
others in its schedule. It also took the opportunity to launch a new weekly soap
opera, *The Secret Life of Us*, immediately after the last episode of the program.
Subsequently, further series of Australian *Big Brother* made annual reappear-
ances – Series 2 in 2002, Series 3 in 2003 and Series 4 in 2004. Although ratings
slowly declined, nevertheless, Network Ten was still pleased with audience sizes
both for the program and for its effect on the network's yearly performance.

Commodity

Discussion of the different national versions and national series of *Big Brother* in
the previous paragraph serves as a reminder of the point made above that a TV
format is a complex, dispersed entity. Indeed, as an industrial commodity, it also
warrants such names as franchise or brand name, a useful reminder that, from a
business perspective, the product in question is the total package of commercial
rights together with the supporting television vehicle that was *Big Brother* (Dicke,
1992; Todreas, 1999; Bellamy and Trott, 2000). In other words, *Big Brother*
circulates not only through broadcast TV, but also through various other platforms, including radio, advertising, telephone, billboards, merchandising and
personal management.

Thus, the franchising associated with the format involved the purchase of a
bundle of services that helped guarantee profitability for the broadcasting network. Managing a package rather than an individual right ensured the generation
of incomes from a series of ancillary operations that are not normally licensed but
rather exist in the area of publicity. At the same time, in exploiting these rights,
Endemol also needed to exercise some commercial prudence. Like any franchising
parent, it was disposed to build its product, to maintain and strengthen the overall, international commercial reputation of *Big Brother* and to secure an Australian
licensing fee that was compatible with this objective.

Given this general disposition, the guises under which the format appeared in
Australia as it did elsewhere are obvious enough (Roscoe, 2001; Johnson-Woods,
2002; Hill, 2002). Merchandising included such standard elements as video,
music CD, magazines, T-shirts, bunny ears and masks. The format also conferrred
celebrityhood (both temporal and more durable) on former housemates, who in
turn made "star" appearances not only in subsequent series of the program, but
in other programs, including talk and game shows. One managed to move from
the live soap of *Big Brother* to the fictional soap of *Neighbours*. Even the Big
Brother house basked in this glory, being opened to the public as an attraction at
its Dream World theme park site on the Gold Coast (Johnson-Woods, 2002;
Roscoe, 2002).

While on-air, Australian *Big Brother* also briefly pervaded the world of local commerce where it coupled with brand names for other commercial products and services. Thus, for example, as occurred elsewhere, various TV commercials such as those for Pizza Hut and McDonalds were consciously molded to "fit" with the program's content (Moran, 2001). This kind of linkage was consolidated and extended by other marketing strategies such as on-screen product placement and phone line voting facilities.

Generic family

Earlier, I suggested that TV formats are multiple, complex entities and this same point is again underlined in our study of *Big Brother*. For this format signifies not a single program but rather a constellation of programs, a saga of dispersed, reconfigured, perhaps even perpetually open narratives and narrative effects including imagery, *mise en scène* and celebrities (Johnson-Woods, 2002; Thompson, 2003). Not surprisingly, the general macro feature of differentiated abundance and saturation, characteristic of the era of New Television, has its more micro parallel in the multiple, complex circulations – known variously as format, franchise, brand name and so on – of *Big Brother*.

Thus, rather than thinking of the format as giving rise to a single program in any territory, it makes more sense to speak of *Big Brother* as a kind of television genealogy, a constellation of different programs that collectively constitute a clan or extended family of programs. Indeed, the notion of genre is also highly relevant to grasping the phenomenon of the format program (Moran and Keane, 2003). Thus, we can identify a series of different features of programs and program effects. Inside the immediate family of the format, there were those programs that appeared in the schedule under the name of *Big Brother*, made up of the different program contents scheduled differently during any particular series (Johnson-Woods, 2002). These included the five-evenings-a-week half-hour live soap, a weekly Thursday night "talk show" version (*Big Brother Uncut*) and a talk and variety *Saturday Show*. But the premier of these was a *Live Eviction Show* on Sunday night, "specials" that combined the spectacle of the variety show with the intimacy of the talk show.

Similarly, in any particular territory, including Australia, repeated seasons of the program put to air in successive years can be seen to form sequels to the original national series. But the *Big Brother* format exists in time in a more contiguous way so that we can also talk of a diachronic chain of adaptations. Thus, in Australia, forerunners to the main program or group of programs included *Big Brother Is Coming* and *Big Brother Revealed*, while spin-offs included *Rove Big Brother Special, Big Brother Beauty and the Beast, Big Brother – Where Are They Now?* and *Celebrity Big Brother*.

In the meantime, synchronic versions of the format were appearing or had appeared in more than 20 different territories across the world, from Argentina to the United States. And, finally, too, various other adaptations of *Big Brother* also exist. Some of the latter, including *Big Diet, The Flat, The Bus, The Bar* (aka *Bar Wars*) and *Jailbreak*, bear obvious generic resemblances. Unauthorized and pornographic spin-offs included *Sex Survivor 2000* and *Pornstar Survivor*. And indeed, the reference to both *Pop Stars* and *Survivor* in the titles of these last two, again highlight general broad resemblances between this format and those of both *Pop Stars* and *Survivor*. The similarity between *Big Brother* and the latter was not lost on the owners of the *Survivor* format who took legal action against Endemol alleging copyright infringement (van Manen, 2002).

Live soap

Reference to the notion of genre is one step further toward a greater cultural understanding of *Big Brother* and of program format adaptation. In particular, the format of this program involves a radical renegotiation of two specific forms of television – the drama serial, or soap opera, and the game show.

Taking the latter first, it is apparent that any series of *Big Brother*, like either a single episode or a series of episodes, is formally structured around a competition. The gradual elimination of contestants means that there will be a finite number of episodes. Thus, unlike the fictional serial or soap opera (although like both the Latin American *telenovela* and the mini-series), the live soap that is *Big Brother* is not indefinitely open-ended so far as its narrative structure is concerned (Tulloch and Alvarado, 1982; Eco, 1987). Instead, it is finite with a beginning, a middle and an end.

Meanwhile, any weeknight episode is structured according to the narrative rhythms of the drama serial or soap opera (Roscoe, 2002). Related to this is the ontological fact that in this genre no on-screen figure is guaranteed narrative permanence. Instead, in both fictional soap and in the live soap that is *Big Brother*, the on-screen figures always suffer from being "guests," rather than "permanent" members of the cast, perpetually facing the possibility of disappearing from the program (Tulloch and Moran, 1987). In turn, this relationship with fictional soap opera provides a clue to some of the social meanings being generated in this format adaptation.

Self-identity and family

Culturally, then, *Big Brother* foregrounds and celebrates expressive individualism (Allen, n.d.; Brenton and Cohen, 2003). For what is at stake in the regular rituals of internal nomination and public voting is identity. Based on the performative behavior of the housemates, this individualism is assessed first by the group itself and secondly by the voting audience. As so often in fictional soap, the notion of "family" is central to this construction of self in *Big Brother*, functioning both as context and as organizing principle. Indeed, the program becomes a set of ongoing lessons in how to perform family. For, unlike more masculine adventure reality formats such as *Survivor* or *Boot Camp*, the emphasis in this format is domestic and communal. The setting is interior, the Big Brother house, a private space that is central and all-encompassing, a world where the individual figures are brought together. They live in this space that becomes home just as they become family.

However, this is by no means a nuclear or a traditional family. There are no parents or elders, although there is authority vested in the voice of Big Brother. But that authority, although absolute so far as ensuring obedience on the part of the housemates, is, at the same time, frequently a theatrical, performative one, a caricature or parody of actual familial authority in terms of the tasks and requests that he assigns. Equally, when it comes to outcomes of these tasks and assignments, there is a kind of playing at family processes. For while Big Brother rewards, he does so not with love and affection but with consumer goods of one kind or another.

Despite its title, the series is centrally concerned with the *ad hoc* "blended" family members who are the housemates in the house. Here, there is an obvious performance of family. The on-screen figures talk, touch, hug, move around, eat, and sleep together but also spat, argue, and fight. For this modern family, presence is arbitrary – they can be voted out at any time. Thus, the relations between these family members are contingent rather than enduring. With a void of biological, ethical, and legal ties, the self and the family in *Big Brother* are a result of unending processes of self-assembly. Family is not a given but is, rather, performed through family-type roles and rituals.

This is important not just for the on-screen figures involved in *Big Brother* but also for the viewer. For here, too, the series is reality television. It is about the modern family. What *Big Brother* offers its viewer is dramatically educative of processes of family life in the present. The hierarchies and attendant lines of authority associated with parents and offspring, the old and the young, male and female, have gone. What continues is a looser grouping of post-teens engaged in the ongoing process of relating to each other as a family of sorts (Allen, n.d.).

Increasingly, families are the product of human choice, but now these choices are made in the absence of authorized sources of approval or arbitration. Instead, both the self and the family are the result of unending processes of self-assembly, the guidelines for which have been radically privatized. Now a family is defined through the performance of family-type roles and rituals rather than through biological or legal elements. In addition, it is thoroughly penetrated by consumer culture. As Allen has put it:

> This culture now steps in to police consumption as a mark, as a set of mutually reinforcing set of consumer behaviours, as a set of overlapping generational imperatives and as an imprimatur to be placed on a set of suitable/appropriate products and services. (n.d., p. 22)

Thus, the formatted program *Big Brother* filmed "live" using "real" people secures itself not only as entertainment but more especially as education and information. And what it educates and informs its audience about are the various modes, styles, languages, and so on – the expressive individualism of living in this post-modern family.

Conclusion

This chapter has argued that the present era of television exhibits a decisive break with the recent past. With its reconfiguration of new technologies of delivery, reception, and storage, new agencies and players, new contents and new financial arrangements, the medium has changed markedly from what it has been, justifying the name of New Television. In the present landscape, a new global type of the television program has emerged in the form of the format adaptation. Drawing upon but transforming older practices of transnational adaptation, the format is simultaneously international in its dispersal and local and concrete in its manifestation, a practice underlined and considerably strengthened by developments in the area of intellectual property. In turn, these changes lead to a situation of differentiated abundance in the contemporary era of television. Finally, by way of concretizing and further focusing the argument, I have offered a short case study concerning the program *Big Brother*, canvassing various economic and cultural implications.

References

Alford, P. (1995) *To Steal a Book is an Elegant Offence: Intellectual Property Law in Chinese Civilisation*, Palo Alto, CA: Stanford University Press.

Allen, R. C. (n.d.) "The Movie on the Lunch Box: Demographics, Technology, and the Transformation of Hollywood Cinema," Unpublished paper.

Bellamy, R. V., McDonald, D. G., and Walker, J. R. (1990) "The Spin-off as Television Program: Form and Strategy," *Journal of Broadcasting and Electronic Media*, 34, 283–97.

Bellamy, R. V. and Trott, P. L. (2000) "Television Branding as Promotion," in S. Eastman (ed.), *Research in Media Promotion*, Mahwah, NJ: Lawrence Erlbaum Associates, pp. 127–59.

Blumler, J. and Nossiter, T. (eds.) (1991) *Broadcasting Finance in Transition: A Comparative Handbook*, New York: Oxford University Press.

Bodycombe, D. (2002) "Format Creation," www.tvformats.com./formatsexplained.hum ref re chef/recipe.

Brenton, S. and Cohen, R. (2003) *Shooting People: Adventures in Reality TV*, London and New York: Verso.

Brown, L. (1977) *The New York Times Encyclopedia of Television*, New York: New York Times Press.

Brunsdon, C., Johnson, C., Moseley, R., and Wheatley, H. (2001) "Factual Entertainment on British Television: The Midlands Research Groups 8–9 Project," *European Journal of Cultural Studies*, 4(1), 29–63.

Constantakis-Valdes, P. (1997) "Computers in Television," in H. Newcomb (ed.), *Encyclopedia of Television*, Chicago and London: Fitzroy Dearborn Publishers, pp. 412–13.

Cooper-Chen, A. (1994) *Games in the Global Village: A 50-Nation Study*, Bowling Green, OH: Bowling Green University Popular Press.

Curtin, M. (1996) "On Edge: US Culture in the Neo Network Era," in R. Ohmann (ed.), *Marketing and Selling Culture*, Case Western Reserve: Wesleyan University Press, pp. 181–203.

Dayan, D. (1997) "Public Events Television," in H. Newcomb (ed.), *Encyclopedia of Television*, Chicago and London: Fitzroy Dearborn Publishers, pp. 1028–30.

Dicke, T. S. (1992) *Franchising in America, 1840–1980*, Chapel Hill, NC, and London: University of North Carolina Press.

Dell, C. E. (2003) "The History of 'Travellers': Recycling in Prime Time American Network Programming," *Journal of Broadcasting and Electronic Media*, *47*(2), 260–79.

Eco, U. (1987) *The Role of the Reader*, Bloomington, IN: Indiana University Press.

Fiddy, D. (1997) "Format Sales, International," in H. Newcomb (ed.), *Encyclopedia of Television*, Chicago and London: Fitzroy Dearborn Publishers, pp. 623–4.

Flew, T. (2002) *New Media: An Introduction*, Melbourne: Oxford University Press.

Freeman, M. (2002) "Forging A Model for Profitability," *Electronic Media*, *21*, 13–15.

Frith, S. (1987) "Copyright and the Music Industry," *Popular Music*, *71*, 57–75.

Gillespie, M. (1995) *Television, Ethnicity and Cultural Change*, London and New York: Routledge.

Gross, L. S. (1997) "Cable Network," in H. Newcomb (ed.), *Encyclopedia of Television*, Chicago and London: Fitzroy Dearborn Publishers, pp. 265–71.

Heinderyckx, F. (1993) "Television News Programmes in Western Europe," *European Journal of Communication*, *8*(4), December, 425–50.

Heller, D. (2003) "Russian 'Sitcom' Adaptation: The Pushkin Effect," *Journal of Popular Film and Television*, *31*(2), Summer, 60–86.

Hill, A. (2002) "Big Brother: The Real Audience," *Television and New Media*, *3*(3) August, 323–40.

Horn, J. (2002) "Franchi$e Fever," *Newsweek*, April 22, p. 50.

Iwabuchi, K. (2003) "Feeling Glocal: Japan in the Global Television Format Business," in A. Moran and M. Keane (eds.), *Television Across Asia: TV Industries, Program Formats and Globalisation*, London: RoutledgeCurzon, pp. 21–35.

Johnson, J. C. and Jones, K. (1978) *Modern Radio Station Practices*, Belmont, CA: Wadsworth.

Johnson-Woods, T. (2002) *Big Bother: Why Did That Reality TV Show Become Such a Phenomenon?*, Brisbane: University of Queensland Press.

Keane, M. (2003) "A Revolution in Television and a Great Leap Forward for Innovation? China in the Global Television Format Business," in A. Moran and M. Keane (eds.), *Television Across Asia: TV Industries, Program Formats and Globalisation*, London: RoutledgeCurzon, pp. 88–104.

Keane, M., Donald, S., and Yin, K. (2002) *Media in China: Consumption, Content and Crisis*, London: Curzon Press.

Lane, S. (1992) "Format Rights in Television Shows: Law and the Legislative Process," *Statute Law Review*, 24–49.

Lawson, A. (2002) "Independent Television Does It Tough," *The Australian*, October 17, p. 17.

Lie, R. and Servaes, J. (2000) "Globalization: Consumption and Identity: Towards Researching Nodal Points," in G. Wang, J. Servaes and A. Goonasekera (eds.), *The New Communications Landscape: Demystifying Media Globalization*, London and New York: Routledge, pp. 307–32.

Mapplebeck, V. (1998) "The Mad, the Bad and the Sad," *DOX*, *16*, pp. 8–9.

Mattelart, A., Delacour, X., and Mattelart, M. (1984) *International Image Markets: In Search of Alternative Perspectives*, London: Comedia Publishing.

McChesney, R. W. (1999) "The New Global Media," *The Nation*, *269*(18), November 29, pp. 11–15.

Meade, A. and Wilson, A. (2001) "Oh Brother, What's Next?," *The Australian*, 19 July, p. 33.

Moran, A. (1989) "Three Stages of Australian Television," in G. Turner and J. Tulloch (eds.), *Australian Television: Programs, Pleasures and Politics*, Sydney: Allen and Unwin, pp. 1–14.

Moran, A. (1997) *Copycat TV: Globalisation, Program Formats and Cultural Identity*, Luton: University of Luton Press.

Moran, A. (forthcoming) "The Culture of Television Markets," in *Working Papers in Communications*, No. 3/4, Brisbane: School of Arts, Media and Culture, Griffith University.

Moran, A. and Keane, M. (2003) "Joining the Circle," in A. Moran and M. Keane (eds.), *Television Across Asia: TV Industries, Program Formats and Globalisation*, London: RoutledgeCurzon, pp. 197–204.

Noam, E. (1992) *Television in Europe*, New York: Oxford University Press.

O'Donnell, H. (1999) *Good Times, Bad Times: Soap Operas and Society in Western Europe*, Leicester: Leicester University Press.

Paterson, C. (1997) "Satellite," in H. Newcomb (ed.), *Encyclopedia of Television*, Chicago and London: Fitzroy Dearborn Publishers, pp. 1438–9.

Pearson, R. A. and Urrichio, W. (eds.) (1999) *The Many Lives Of The Batman*, London and New York: Routledge.

Perkin, C. (2001) "It May Be Real Life But It's Death For Drama," *Sydney Morning Herald*, July 15, p. 49.

Ritchie, J. (2000) *Big Brother: The Unseen Story*, London: Channel 4 Books.

Rogers, M. C., Epstein, M., and Reeves, J. S. (2002) "*The Sopranos* as HBO Brand Equity: The Art of Commerce in the Age of Digital Reproduction," in D. Lavery (ed.), *This Thing Of Ours: Investigating The Sopranos*, New York: Columbia University Press; London: Wallflower Press, pp. 42–57.

Roscoe, J. (2000) "Big Brother Australia: Performing the Real Twenty-Four-Seven," *International Journal of Cultural Studies*, 4(4), 473–88.

Saenz, M. (1997) "Programming," in H. Newcomb (ed.), *Encyclopedia of Television*, Chicago and London: Fitzroy Dearborn Publishers, pp. 1301–8.

Sinclair, J. (2000) "Geolinguistic Region as Global Space: The Case of Latin America," in G. Wang, J. Servaes and A. Goonasekera (eds.), *The New Communications Landscape: Demystifying Media Globalization*, London and New York: Routledge.

Sinclair, J., Jacka, E., and Cunningham, S. (1996) *New Patterns in Global Television: Peripheral Visions*, New York: Oxford University Press.

Staubhaar, J. D. (1997) "Distinguishing the Global, Regional and National Levels of World Television," in A. Sreberny-Mohaammadi, D. Winseck, J. McKenna and O. Boyd-Barrett (eds.), *Media in Global Context*, London: Edward Arnold Thompson.

Strover, S. (1997) "Telcos," in H. Newcomb (ed.), *Encyclopedia of Television*, Chicago and London: Fitzroy Dearborn Publishers, pp. 1630–1.

Thomas, A. O. and Kumar, K. J. (2003) "Copied from Without and Cloned from Within: India in the Global Television Format Business," in A. Moran and M. Keane (eds.), *Television Across Asia: TV Industries, Program Formats and Globalisation*, London: RoutledgeCurzon, pp. 122–37.

Thompson, K. (1999) *Storytelling in the New Hollywood: Understanding Classical Narrative Technique*, Cambridge, MA: Harvard University Press.

Thompson, K. (2003) *Storytelling in Film and Television*, Cambridge, MA: Harvard University Press.

Todreas, T. M. (1999) *Value Creation and Branding in Television's Digital Age*, London: Quorum Books.

Tulloch, J. and Alvarado, M. (1982) *Doctor Who: The Unfolding Text*, London: Macmillan.

Tulloch, J. and Moran, A. (1987) *"Quality Soap": A Country Practice*, Sydney: Currency Press.

Van Manen, J. R. (1994) *Televiseformats: en-iden naar Netherlands recht*, Amsterdam: Otto Cramwinckle Uitgever.

Van Manen, J. R. (2002) "Interview with Albert Moran."

Wieten, J., Murdock, G., and Dahlgren, P. (2000) *Television Across Europe: A Comparative Introduction*, London and New York: Sage.

Wikipedia (2003) http://www.wikipedia.org/wiki/List-of-British-TV-shows-remade-for-the-American-market.

CHAPTER SIXTEEN

The Study of Soap Opera

Christine Geraghty

It is no surprise that television genres have become central to popular discussion about television and to academic research, nor that soap opera should feature extensively in such debates. Defining genres, marking out the boundaries and then crossing them with glee is a practice engaged with by producers and viewers of television alike while genre definition, in a relatively new discipline like television studies, is a crucial way of mapping the field, and of identifying precisely what it is that is there to be studied. The study of soap opera has been particularly influential in the discussion of genres and debates about television as a whole. First, defining soap opera was one way of separating the characteristics of television drama from drama in theater or cinema and of assessing distinctions within television drama itself by setting soap opera against other forms such as the series or serial. More recently, the fictional elements in cross-generic programs have been described by comparisons with soaps in the development of docusoaps, for instance, and of the various formats of reality TV. Secondly, how soap opera has been studied and defined has affected the development of television studies itself and continues to shape the way in which we consider certain kinds of issues. As we shall see, work on soap opera has allowed an entrée for feminist work on television; it has also provided the basis for cross-cultural explorations of considerable richness. Finally, in debates about the mass media, soap opera continues to brand television as a whole as a mass medium that produces particular kinds of products. That the term "soap opera" is often used as a metaphor for rather tacky activity in other spheres – politics, sport, business – tells us something about how the pleasures and possibilities of popular television are defined.

Essays like this tend to function as summaries of the body of work that has led to this point. The emphasis is on a smooth account of the origins of the genre and the various by-ways taken which at some point merge together to give an accepted definition and allow the topic to be sorted out and slotted into the television studies curriculum. With media studies now a globally established area of study, the proliferation of readers and textbooks encourages these readily transmittable versions of work that was often originally more tentative and certainly more provocative than now seems to be the case. I will explore this issue further in the first section of this chapter, but here I want to acknowledge that these types of accounts can only be partial. Work on television crosses disciplines and the programs under discussion here are made worldwide so it is impossible to provide an all-embracing account. I should also acknowledge that I am writing as what might be called an "observant participant," reversing the terms of the anthropological "participant observer." This chapter, although it is not autobiographical, reflects my particular familiarities with textual and feminist work and with British programming. But it also draws on a personal history of working on soap opera during the time when it was being established as worthy of study. Charlotte Brunsdon, one of the most influential writers on soap, has produced a body of work marked by a complex account of the way in which issues of feminism, femininity, and identity are not only the subject of soap opera but also layered into writing about it. This study will not replicate that but will, I hope, share some of Brunsdon's sensitiveness to the position from which theory is/can be spoken.

The first section of this chapter considers the beginnings not so much of soap opera but of its study within a television studies (rather than a mass communications) tradition. I have not tried to replicate the historical accounts available elsewhere; rather, I draw attention to how the object of study and the historical context have been constructed in particular ways. The remaining three sections are organized around three questions: to what extent can soap opera as a genre be defined by its narrative structure; to what extent can soap opera be described as women's fiction; and what kind of intervention can soaps make in the public sphere? Discussing answers to these questions will allow me to reflect on how the field has been explored and draw attention to some newer work that indicates further possibilities.

Beginnings and Definitions

We can find a recent account of the kind referred to above in the British Film Institute's *The Television Genre Book* (Creeber, 2001). "Studying Soap Opera," an exemplary summary essay by Anna McCarthy, describes the development of soap opera as a form. It is followed by her second essay, "Realism and Soap Opera," which describes some of the key academic texts in the study of the genre. This is a smooth, clear account that starts with a point of origin in 1930s radio and concludes with a recognition that contemporary soap opera cannot be studied within closed generic boundaries. I do not want to dispute this as a legitimate historical account, but I do want to suggest that such an approach has particular consequences. Undoubtedly, it privileges a particular national version of soap opera, a fact indicated by the way in which the statement that "the format emerged from the radio sponsorship by detergent companies in 1930s radio" apparently requires no indication that it is US radio and the development of US soaps which is being referred to. Of course, British programs and academic work on them can be fitted into the schema developed from this originating point, partly because British soaps were developed in deliberate counterpoint to the US programs and John Tulloch's subsequent account of "Soap Operas and Their Audiences" is similarly based on anglophone work. It is not surprising therefore that it is difficult to incorporate work whose development is different – it is significant that Thomas Tufte's section on "The Telenovela" comes last and does not relate directly to material in the earlier sections. Unwittingly, a hierarchical relationship is implied in which the US versions retain the dominance ascribed at their point of origin.

Equally contentious is the identification of a starting point for academic work on soap opera in television studies. Here I want to refer not to origin but to a proclaimed break. In *Speaking of Soap Operas* in 1985, Robert Allen marked a break with what he calls the empiricist work on soap opera being undertaken in mass communications research. Allen traces the development of this work from classic mass-media work in social sciences beginning in the late 1930s to the content analysis prevalent at the time he was writing. Such work is condemned for its narrow focus – which ruled out, for instance, the aesthetic experience of the audience – and for its emphasis on counting standardized responses to limited questions rather than examining the complexity of soap opera's production, textual organization and relationship with its predominantly female audience. Although Allen looks at the institutional history of the daytime soaps and production practices of *The Guiding Light* in particular, it was chapter 4 on "A Reader-Orientated Poetics" which was the most influential, the title indicating that he was using an approach developed from semiotics and reader-response theory which emphasized the range of responses made possible by a complex and extended text. The nature of the break Allen was making can be seen by the way that it was reviewed by sociologist Muriel Cantor (1986), herself a soap opera specialist. She identifies Allen as "part of a new teaching subdiscipline called television criticism" (p. 386), criticizes chapter 4 as "obscure" (p. 387) and suggests that he would have done better to build on earlier work rather than dismissing it. However, Allen's approach prevailed as the new discipline of television studies developed and it is only recently that connections between earlier work and that associated with Allen's cultural and textual approach have been made. Brunsdon (2000) finds in the 1940s work on soap opera audiences by Arnheim and Herzog "tropes, themes, concerns and characters that recognizably return in [later] feminist work" (p. 51), while Tamar Liebes (2003) (whose own work has made a major contribution to the field) has suggested that Hella Herzog's work might be connected in "a matrilineal line" (p. 44) with that of Janice Radway and Ien Ang, scholars much more strongly associated with the cultural studies traditions Allen was moving toward.

Whatever the differences between US scholars, however, they were at least referring to the same format – an unending, daytime, fictional program shown five times a week. Stempel Mumford emphasized the importance of "dailiness," though she recognized the difficulties this might cause in systems in which soaps were shown in prime time perhaps two or three times a week. An example of a more confused approach to the object of study can be found in British work of the 1970s and 1980s. Here there was much less certainty about whether the rather foreign term "soap opera" could be usefully applied. My own 1981 article on *Coronation Street* referred to the "continuous serial" partly in deference to the refusal of the production company to call its program a soap and partly in acknowledgment that the analysis was based on textual work of definition rather than production work on origins. The term is used in some, but not all, of the other essays, and perhaps the term is most clearly used by Terry Lovell when she discusses the pleasures that *Coronation Street* might offer women viewers. By the early 1990s, the term had become more widely used both in British television production and in the daily press while feminist work by Ang, Modleski, and others had made it central in academic debates. The title of my *Women and Soap Opera* therefore named the genre and simultaneously made the US association with a female audience. But the definition I offered in that book's introduction struggled to maintain the broader approach necessary for the study of British television – soaps here are defined "not purely by daytime scheduling or even by a clear appeal to a female audience, but by the presence of stories which engage an audience in such a way that they become the subject for public interest and interrogation" (p. 4).

By comparison, the *telenovelas* maintained a different name and the possibility of a different identity. Those who have studied them trace a distinctive form with its own version of antecedents in newspapers and radio, a particular relationship with melodrama and links to the storytelling, songs and verses of oral culture. Despite this, the pressures of genre connection mean that *telenovelas* are regularly subsumed under the term soap opera as was the case in Allen's 1995 collection *To Be Continued*, which is subtitled "soap operas around the world," making the US term the uniting concept. That tradition continues in this essay but as more indigenous examples are explored, the originating point in US radio seems to become less and less appropriate, blocking off more than it reveals. Other sources are being explored in literary serials, written and film melodrama, realist novels with their emphasis on everyday detail, folktales and histories, religious sagas and ancient legends. In this essay, I have situated US soaps alongside rather than above all of the other sources. Serial dramas of all kinds come out of a tradition of human storytelling which is highly valuable to modern broadcasting systems since it draws regular audiences to repeated and repeatable scenarios. As we shall see, soaps speak both to television's capacity for intimacy and to its role as public educator.

Narrative

The double emphasis on the formal qualities of narrative and their usefulness to audience-seeking television companies was noted by Raymond Williams in a 1969 television review when he observed how narrative was used to organize viewers' relationship with a new medium; a series, he remarked, "is a sort of late version of character training: encouraging regular habits in the viewers; directing them into the right channels at certain decisive moments in their evening lives" (p. 81). Much early work in television studies carried out in the late 1970s and early 1980s considered how narrative worked on television, mapping out the key features which distinguished a series from a serial, a mini-series from a classic serial, a situation comedy from a soap.

Soaps were of central interest in this debate because they seemed to be the clearest example of television's difference from other narrative-dominated media. For critics coming from literature and cinema, the defining feature of soap opera was the way in which its narratives operated differently from two common (though often implicit) poles of comparison being used – the American feature film and the bourgeois realist novel. It needs to be remembered that work on soaps developed in a situation in which theoretical debate was much concerned with the interaction between formal and ideological properties of particular forms. Endings were strongly associated with the ability of dominant ideology to close down and overdetermine progressive or radical possibilities that might have been raised in the ongoing narrative. Endings resolved the problems, giving the reader an illusory sense of control and power. The never-ending nature of soaps and their "sense of an unwritten future" (Geraghty, 1981, p. 12) was a key feature, setting up a different relationship for the audience and apparently refusing the ideological closure of other texts. For Allen (1985), it meant that the soap opera text, as it developed over the years, is "ungraspable as a whole at any one moment" (p. 76) and can never be understood from the position of closure when the meaning of action is clear and ambiguities and contradictions have been ironed out. While "our desire for narrative closure" was seen as fundamental to the fictions of Hollywood cinema, soap opera with its lack of closure "has openness, multiplicity and plurality as its aims" (Flitterman-Lewis, 1992, p. 217).

For a period, then, when theoretical work was being developed through US and British examples, the lack of an ending (even when a program ceased) was the defining quality of a soap. The association of unending seriality with a more open text has been challenged in recent work. Jostein Gripsrud (1995), for instance, argues that open endings cannot be equated with ideological openness while Stempel Mumford (1995) suggests that US daytime audiences expect stories to be resolved so that the narrative can move on and that these closures strongly reassert capitalist patriarchy. More importantly, the emphasis on endlessness proved too rigid to take on other forms of serials that needed to be incorporated into the soap opera debate. *Telenovelas*, for instance, work toward closure in a manner that is often the subject of extensive controversy and popular debate. Allen recognized this in *To Be Continued* (1995), which emphasized the worldwide attraction of the serial form. In his introduction, the distinctive emphasis on endlessness is less prominent though he does preserve a distinction between open and closed serials and still associates "the absence of a final moment of narrative closure" in the former with the indefinite postponement of "final ideological or moral closure" (p. 21). As a further refinement, we might note that *telenovelas* also have their open and closed versions; with closed versions, the ending is already decided when the program starts screening while open versions have not been completed at this point and their endings are therefore more subject to audience reactions and preferences (O'Donnell, 1999, p. 5).

As work beyond anglophone television narratives continues, the generic definitions based on narrative have loosened. Popular narratives considered under the term soap opera, but which are not defined by the lack of an ending, include different forms of Latin American *telenovelas*; European versions of the form produced in Italy and Spain, for example; serializations of Hindi sacred texts; the *shinei ju* or "indoor drama" from China and the *hsiang-tu-chu* or rural soaps of Taiwanese television. A prime-time slot is now the norm for many viewers worldwide. Soap opera narrative work now places less emphasis upon the lack of an ending and instead defines the form by its extended, complex, and interweaving stories; a wide range of characters, allowing for different kinds of identification; the delineation of an identifiable community, paying attention to domestic and familial relationships; and an emphasis, often expressed melodramatically, on the working through of good and evil forces within a family or community. In some appropriations of the term, it is clear that the dominance even of an extensive and interrupted narrative as a defining feature is becoming weak. One critic discussing soap opera in China describes *Kewang* and *Bejingren sai Niuyue* as marking "the maturity of 'soap opera' as a full-blown Chinese genre" (Lu, 2000, p. 26), though they had only 50 and 21 episodes respectively. Therefore it is worth bearing in mind when reading the literature that programs are treated as soap operas in some critical contexts that would not be in others and that US writers are more likely to retain the original model of the daytime, endless serial.

A number of points emerge from this shift in concept. First, it may be that the "pure" soap opera like *The Guiding Light* and *Coronation Street* is no longer the most characteristic form of soap. If so, the implications of that need further discussion – in particular, attention needs to be paid to the formal narrative processes which mark forms that are not united by their lack of closure. Yean Tsai (2000) returns to the earlier work of Propp to analyze Taiwanese soap operas, but other models are needed to dig below the familiar assertion that soaps are characterized by interweaving stories. This might lead us to explore either Nelson's proposal that television fiction is dominated by "flexi-narratives" or, alternatively, the "Scene Function Model" proposed by US researchers as an analytic tool for television narrative (Porter et al., 2002). The implication here

313

Figure 16.1   Self Portrait With Television. Photograph © 1986 by Diane Pansen

would be that, at the level of the narrative unit, soaps operate in the same way as other forms of television fiction and that it is misleading to see the soap opera genre as distinctive. Alternatively, it might be that there is merit in clearly delineating differences between narrative formats within the soap opera genre and being more careful about applying soap opera as a blanket category to different forms. A final possibility is that soap operas are distinctive, but that their defining features lie elsewhere, which leads us on to our second question.

"Should We Still Classify Soap Operas as Women's Programs'?"

It could be argued that the notion that soap opera is fiction for women is largely a product of a particular contingency. Work on soap opera was developed by theorists with a common background in feminist film theory in relation to a very particular mass media product (US daytime soaps) at a time when feminism was having some impact on the academic world and beyond. Asking to what extent soap operas are women's fiction enables us to look at the various ways this question has been understood and to trace a shift from posing the relationship as an aesthetic one to an emphasis on women as audience.

314

Tania Modleski's early intervention in this area was premised on a knowledge of and commitment to avant-garde aesthetics which we would be unlikely to find today. Understanding her chapter on US daytime soaps in *Loving with a Vengeance* requires a knowledge of debates about "still embryonic, feminist aesthetics" (1984, p. 105). Modleski points to the endlessness of the soap opera format, its use of mechanisms to retard narrative progression and the gap, in much of the dialogue, between what is spoken and what is intended. She is interested in these traits because of they have their equivalents in avant-garde practices. Modleski's critique of the female viewer's pleasure in soaps is balanced by the possibility that this pleasure might, if properly understood, be incorporated into de-centered feminist artistic practices. More than Allen, who forsees the next step as relating his "constructed reader positions" to "the experiences of actual soap opera viewers" (p. 182), Modleski was working in the realm of theoretical and psychic possibilities. She could be sharply criticized in terms of her theoretical arguments, as Gripsrud demonstrated, but a more common approach was to ignore the work on feminist aesthetics, criticize her negative view of the female viewer, and treat her descriptions of the housewife viewer as if they were hypotheses for testing out on the ground. It would have been more interesting perhaps if the results of work with female audiences which called Modleski's account into question had themselves fed back into work on feminist aesthetics but by the late 1980s/early 1990s that moment had been lost. For good or ill, analysis of women's television had separated from the avant-garde and, although Modleski's work is described as influential, this aspect of it had relatively few followers.

A more common approach was to see in soap a female-orientated narrative in which women were central. Feminist film theory had wrestled with the position of the female spectator, but soap opera seemed to offer women stories that could be understood from their viewpoint. Brunsdon's proposal that soap operas, far from being mindless, actually required feminine competences was highly influential – as was the notion that soap stories paid attention to the complexities of the private sphere which tended to be ignored in other genres. Soaps were valued for the way that they made the work of emotional relationships visible in what could be seen as "a woman's space," a term which drew on the feminist demand that women engaged in political or social activity needed their own space in which terms could be discussed and redefined before being taken out into the public world. Soaps were indeed part of a highly gendered cultural system, but this rather lowly format did offer space for women to reflect on what it felt like to be female in the contemporary world.

This association of soap opera with female competences and understandings was a highly influential one and persists in more recent accounts. O'Donnell (1999), concluding his extensive study of soaps in Europe, comments on the different representations of men and women in the genre and remarks that "the general life cycle constructed by European soaps . . . is one in which women appear much more competent and dynamic than men" (p. 223) although this does not mean that they always have happier outcomes. Hayward (1997) disagreed with Modleski over the argument that US soap narratives were shaped by the rhythm of women's lives, but did accept that content and theme were marked by gender; she argued that "soaps remain unique both in positively portraying women and being in a form still produced primarily by women" (p. 143). In her study of viewers of Indian serials Purmina Mankekar (2002) comments that "an astonishing number continue to deal centrally with women's issues" and adds that even when gender is not the main theme it continues to be a "critical sub-text" (p. 303). Tsai (2000) compares Taiwanese serials from the historic genre which center on "male authority" with the rural soaps in which stories "often detail conflicts and power struggles between females" (p. 178) and which are marked also by "an attempt to promote a positive image of modern Taiwanese women" (p. 181).

The centrality of women, and in particular the predominance of stories about families, was an important element in work that sought to situate soap operas into the larger category of melodrama. Christine Gledhill, among others, demonstrated that melodrama was a term of considerable complexity, but it could be used to describe soap's emphasis on women's voices and domestic spaces, the use of heightened *mise-en-scène* and music to express what could not be spoken, the value placed on feeling and on moral judgments which clarified, if only temporarily, good and evil actions. The use of melodrama to describe soap opera as a genre had advantages. It allowed soap opera to be constructed alongside the women's film, the romance, and the costume drama as a distinctive form of popular culture. There was also a pull in the another direction, that is away from women's culture. Since much of genre television could be associated with the broad terms of melodrama it allowed for soaps, potentially at least, to be seen as a fundamental form of television rather than a separate women's space. And, since melodrama, as an element in popular culture, was a distinctive phenomenon outside anglophone cultures, the use of the term helped to acknowledge the crucial importance of the *telenovela* in the assessment of television drama.

One difference between content analysis and textual work on soap opera was the way in which theorists such as Brunsdon implicated the audience in a study of the text. Nevertheless, the shift to work with audiences and to notions of the female viewer constructed not from the programs but from the responses generated by qualitative research was recognized as significant. Although Ien Ang's work on *Dallas* viewers tried to explore their unconscious feelings and desires, much of this work, based as it was on interviews and questionnaires, reflected the conscious statements of the respondents. Although the ambivalences (and indeed guilt) of the female viewer were explored, what emerged from accounts such as those by M. E. Brown, Andrea Press, and Dorothy Hobson backed up Brunsdon's earlier textually based account of the ideal soap viewer as competent, and capable of making decisions about what stories and characters she engaged with. This positive view of the woman viewer has continued in more recent studies. Baym (2000) describes a US daytime soaps website as being "not only a place in which female language styles prevail but also a place in which there is considerable self-disclosure and support on the very types of female issues that provoke flame wars (if raised at all) in so many other groups" (p. 139). And Mankekar (2002) concluded her study of male and female viewers of soaps in a North Indian city by arguing for the importance of seeing "women viewers as active subjects in the light of the tendency to depict 'Third World' women as passive victims" (p. 317).

If soaps then were women's fiction, these studies revealed that it was not just because of the stories they told or the heroines (and villainesses) they offered but because of the way their viewers felt about these programs. It was a relationship comparable to that generated by particular forms of women's reading, providing women viewers with something that was specifically theirs. The notion of a woman's space re-occurred with accounts of women watching US soaps in a "distinctly female space . . . characterised by the absence of men" (Seiter, 1991, p. 244). For many, soap viewing was accompanied by female-dominated talk, a process which Brown (1994) found often linked mothers, daughters, and friends and could be described as "a woman's oral culture that bridges geographic distances" (p. 85). Even when men watched (and some studies included male viewers), it was claimed that women viewers defined the way in which the programs were understood and their role in everyday life beyond the viewing schedule. It should be noted, though, that women's possession of soaps was generally something that had to be worked for. It was vulnerable to changes in storylines and characterization in pursuit of other audiences; to self-criticism and guilt; and to critical pronouncements from male members of the family.

The concept of soaps as women's fiction is open to the criticism that the proposition depends on an essentialist account of gender. Myra MacDonald (1995) argues that "feminist romanticism about soap opera" (p. 72) helps to preserve gender distinctions in relation to the myth of femininity that should instead be challenged. But the proposal that there is a specific relationship between soap opera and women viewers has been deemed problematic in itself and the question posed in the title of this section has indeed been taken from one such account. Gauntlett and Hill (1999), on the basis of a five-year study of 450 British viewers, argue that the programs themselves have changed with less emphasis on women's stories; that the network of talk and discussion which surrounds the program is not exclusive to women and that men generally did not find it difficult to admit to liking soaps; and more broadly that their respondents did not want to distinguish between "'women's' and 'men's interests'" (p. 219) in television viewing, except in relation to sport. To some extent, this kind of critique has been backed up by other audience studies – Buckingham's work on children's viewing, Lull's on families watching television, Gillespie on teenagers, Tulloch on the elderly – though there has been relatively little work specifically on men and soap opera. (For an interesting account of the relationship between male viewers and (Greek and US) soaps, see Frangou, 2002.) More trenchantly, Gauntlett and Hill concluded that "academics (and others) should stop talking about soap opera as a 'women's genre'" (p. 246) though their own evidence ("whilst women were three times more likely than men to rate soap operas as 'very interesting', men do nevertheless watch soaps" (p. 227)) could perhaps be interpreted differently. Certainly, evidence from other surveys still tends to suggest that women are the most engaged viewers of soaps. A British survey published in 2002 by the Broadcasting Standards Commission found that the most strongly committed viewers of prime-time soap operas were predominantly younger, working-class women, many of whom were at home all day looking after small children. Much of the international work discussed here indicates a different kind of engagement by women, even when the programs are viewed as part of the family.

In part, some of this criticism comes from the tendency to read "feminist work" as a block, neglecting the reservations and differences in position which have now been traced out, for instance, by Brunsdon (2000) in her work on the relationship between feminist writers and their object of study. The body of work that associated women and soap opera has to be read in the context of feminist politics in which notions of, for instance, "women's space" had particular strategic connotations. It is not necessary to deny that soaps have been, and in certain situations still are women's fiction, in order to tell other stories based on different research into soaps.

Soap Opera and the Public Sphere

We have seen how the notion of the private sphere was important for the discussion of soap opera as women's fiction. In this section, however, I want to look more closely at the contribution soaps make in the public sphere. The binary opposition sometimes made between the public and private sphere is, I think, a misunderstanding of the function of the concept of the private sphere. As I have indicated, the term did not generally indicate a final retreat from the public sphere nor that women should disengage completely from the political and social activity. The intention was rather that what was deemed to be activity in the public sphere should at the very least be informed by experiences and feelings that were traditionally understood as private and personal. In this formulation soaps are better understood not as belonging to the private sphere but as operating on the boundary of the public and the private, negotiating over how terms might be used. Soaps tend to frame problems and solutions so that they offer a particular explanation that might be applied not just to the fictional world, but to the world in which the viewer might take action. This didactic quality led Buckingham (1987) to describe *EastEnders* as a "teacherly text," suggesting, for example, in a discussion of how the program handled ethnicity, that "the crucial question is not whether *EastEnders*' black characters are 'realistic', but how the serial invites its viewers to make sense of questions of ethnicity" (p. 102). It is this overt sense-making activity that leads us into the public sphere. Because this places an emphasis on the social context in which soaps are making an intervention, this didactic aspect also raises questions about how soaps play a role in the processes of globalization and modernization.

The capacity of soaps to create public debate about particular issues is well recognized. Brazilian *telenovelas* have "dealt with bureaucratic corruption, single motherhood and the environment; class differences are foregrounded in Mexican novelas and Cuba's novelas are bitingly topical as well as ideologically correct" (Aufderheide, 2000, p. 263). Controversial stories bring, for instance, sexuality into public debate and the sensationalist handling of stories dealing with sex and violence has made many programs vulnerable to the criticism made of US daytime soaps that "soap opera, while playing lip service to the feminist stance, actively popularizes the rape myths of patriarchal culture" (Dutta, 1999, p. 35). In addition, soaps often have the function of representing groups or figures who tend to be under-represented in other dramas – characters whose political attitudes, ethnicity, sexuality, or age makes them different from the standard hero. Again, this tends to raise complex debates about whether and how this is done. While soap producers claim to be pushing boundaries forward, the groups represented and academic critics tend to brand such attempts as tokenistic representation; as Judith Franco (2000) remarked when discussing a Flemish soap, *Thuis* "represses differences of sexual preference and ethnicity" (p. 460), despite the token presence of a working-class Moroccan character and a bisexual Dutchman. Nevertheless, soap operas have provided a way of widening television's representational field and some critics have been more sympathetic to the attempt. Hayward has indeed argued that US daytime soaps have a positive social role in exploring shifting and marginal identities and that they privilege "difference over homogeneity, understanding over rejection" (p. 191).

But we can discern broader patterns here than controversies about representation. Audience work indicates that soap opera's use of social issues and minority figures in their storylines is incorporated into the broader processes of making complex social identities. The British *Queer as Folk* is an example of a soap which deliberately set out to represent gays in an assertive and challenging way and, as a myriad websites show, found international audiences which identified strongly with its characters and stories. Similarly, Chris Barker (1999), in a study of the "production of multiple and gendered hybrid identities among British Asian and Afro-Caribbean girls," found that talk about soaps was appropriate for this kind of research because of their emphasis on "interpersonal relationships intertwined with social issues" (p. 119). But such overt storylining is not always necessary, as Marie Gillespie found in her study of how young British Asians used the rather anodyne *Neighbours* for such activity. Soaps, with their intertwining of the public and private, may be a particular appropriate resource for work on identity which frequently involves the presentation of the self in public and the Foucauldian notion of self-production.

Gillespie's example shows how this "teacherly" mode of soaps may go well beyond the intention of the producers, but another body of work makes a stronger association between soap opera's capacity to engage with the public sphere and certain modes of production. Although, as we have seen, some US critics have claimed a progressive function for their daytime soaps, other commentators have associated the social inclusivity of soaps with the traditions of public service broadcasting. Such work thus tends to distinguish between US soaps and other programs, as Liebes and Livingstone (1998) do when they set out a model for identifying three types of soaps: community soaps, dynastic soaps, and what they term dyadic soaps. The first two groups offered ways of articulating social relations based, in the case of community soaps, on class and, in the case of dynastic soaps, on generation and family. By contrast, in the dyadic model, in which stories centered almost exclusively on the establishment and breakup of romantic couples, there was less sense of stability and cultural relations. Programs of this kind, predominantly US daytime soaps, were less engaged with the social and "less expressive of any particular cultural environment" so they concluded that, if this were the form more generally adopted, it would be "more difficult for nationally produced soap operas to reflect the cultural concerns of their country" (pp. 174–5).

James Curran (2002) makes a similar connection between British public service broadcasting and the community orientation of some of its soaps, suggesting that a system which does not depend on the market is more likely to support "a sense of social cohesion and belonging" and to extend the traditional social realist acknowledgment of the working class to groups such as the elderly, single parents, the unemployed, and "some ethnic minorities." For Curran, such inclusivity is specifically contrasted with the "glamourised, 'upscale' settings that dominate much of American domestic drama." His consequent claim that "public service broadcasting promotes sympathetic understanding of the other" is a bold one, but certainly speaks of the possibilities of social intervention at a very direct level (p. 207). One specific example of this might be the BBC's use of *EastEnders*, alongside other materials in documentaries and websites, to draw attention to issues of domestic violence, an integrated approach underpinned by the BBC's public service remit of education as well as entertainment. O'Donnell (1999) extends this by arguing that European soaps have a strong relationship with certain strands of political culture; he suggests that many of Europe's soap operas and *telenovelas* explicitly promote the social democratic "values of solidarity, caring for and about others, defending other people's rights, compromises and co-operation" which are being abandoned by their governments in the management of social welfare. Indeed, he goes so far as to suggest that soap audiences are, by their engagement with such programs, being helped to keep such values alive and that the luminaries of the public sphere – politicians, activists, teachers – might benefit from paying attention to the lessons of these programs (pp. 222–3).

As these examples suggest, soap opera has been widely used in debates about US domination of global television. In the 1990s the resilience of indigenous soaps tempered some fears and the *telenovelas* of Latin America offer an interesting example of how US programs could be successfully displaced by more popular and local forms. The companies producing such programs could use this success as the basis for their own export strategies that included the sale of Mexican *telenovelas* to the US market. It should be noted, however, that the need to root such stories in the local and national experience means that not all *telenovelas* can be exported and, for many critics, the process of making programs for the more undifferentiated audiences of the export market has led to a "tendency to dissolve cultural difference into cheap and profitable exoticism" (Martin-Barbero, 1995, p. 284).

Martin-Barbero (1995) attributed the success of the *telenovela* to "its capacity to make an archaic narrative the repository for propositions to modernize some dimensions of life" (p. 280). This association with a modernizing agenda is seen as a key element in the success of a soap. O'Donnell (1999) suggests, for instance, that younger women "provide much of the zest of European soaps and . . . represent a modernizing force" in their stories (p. 222). This chimes with the didactic project assigned to soaps in other countries in which production (often state-controlled), text and reception come together in different ways to present a version of the modern state. Purnima Mankekar (2002) and Lila Abu-Lughod (2002), for instance, offer rich ethnographic accounts of how soap opera serials were used in north India and Egypt respectively to offer a particular account of national identity and modernity. Mankekar describes men and women viewing prime-time serials, "a cross between American soap operas and popular Hindi films", which carry "explicit 'social messages'" (p. 303). She observes how the programs are discussed by those watching who simultaneously identify with the emotional storylines and criticize acting or shot set-up. The programs have a specific agenda in terms of encouraging what are seen as modern attitudes within a nation, but Manekekar observes how the viewers are conscious that "the serials they watched had been selected, censored and shaped by the state" (p. 314). The process of viewing and discussion meant that modernizing messages might be dismissed or more personal interpretations made central to viewing pleasure.

A similar pattern of didactic modernity and complex reception is traced by Abu-Lughod (2002) in her study of the highly successful Egyptian program *Hilmiyya Nights*, first screened in 1988. Unusually, the program, which followed a group of characters from the late 1940s to the present day of the early 1990s, was shown on a yearly basis during Ramadan. It was, she suggests, produced by "a concerned group of culture-industry professionals" who constructed themselves "as guides to modernity and assume the responsibility of producing through their television programmes, the virtuous modern citizen" (p. 377). The serial employed spectacle, melodrama, and a realist attention to class and regional detail to embed the stories of individual characters in a historical narrative that "provided an explicit social and political commentary on contemporary Egyptian life" (p. 381). Abu-Lughod traces the response of the educated classes, including censors and intellectuals, who sought to protect the public from elements of this controversial history, but also points out how the:

Christine Geraghty

"uneducated public" with whom the programme was immensely popular nevertheless refused simply to absorb the secular vision of a modern Egypt with its emphasis on education and patriotism across classes. Such a portrayal is instead set in the context of a different kind of lived modernity marked by "poverty, consumer desires, underemployment, ill health and religious nationalism." (p. 391)

Both these examples vividly illustrate how the "teacherly" dimension of these local versions of serial drama is used for a modernizing agenda in public debate, but is experienced differently and unevenly by those for whom the modern state is a more ambivalent project. Both also illustrate the range of work undertaken in the study of soap opera/*telenovelas*/serial drama and how deeply it is embedded in the lived experience of television production and viewing.

References

Abu-Lughod, L. (2002) "The Objects of Soap Opera: Egyptian Television and the Cultural Politics of Modernity," in K. Askew and R. Wilk (eds.), *The Anthropology of the Media: A Reader*, Oxford: Blackwell, pp. 377–93.

Allen, R. (1985) *Speaking of Soap Operas*, Chapel Hill, NC: University of North Carolina Press.

Allen, R. (ed.) (1995) *To Be Continued . . . Soap Operas Around the World*, London: Routledge.

Askew, K. and Wilk, R. (eds.) (2002) *The Anthropology of the Media: A Reader*, Oxford: Blackwell.

Aufderheide, P. (2000) *The Daily Planet*, Minneapolis: University of Minnesota Press.

Barker, C. (1999) *Television, Globalization and Cultural Identities*, Buckingham: Open University Press.

Baym, N. K. (2000) *Tune In, Log On: Soaps, Fandom and the Online Community*, London: Sage.

Bourne, S. (1989) "Introduction," in T. Daniels and J. Gerson (eds.), *The Colour Black*, London: British Film Institute, pp. 119–29.

Brown, M. E. (1994) *Soap Opera and Women's Talk*, London: Sage.

Brunsdon, C. (1997) "Crossroads: Notes on Soap Opera," reprinted in *Screen Taste: Soap Opera to Satellite Dishes*, London: Routledge, pp. 13–18.

Brunsdon, C. (2000) *The Feminist, the Housewife, and the Soap Opera*, Oxford: Oxford University Press.

Buckingham, D. (1987) *Public Secrets*, London: British Film Institute.

Cantor, M. G. (1986) "Review," *Contemporary Sociology*, *15*(3), 386–7.

Creeber, G. (2001) *The Television Genre Book*, London: British Film Institute.

Curran, J. (2002) *Media and Power*, London: Routledge

Dutta, M. B. (1999) "Taming the Victim: Rape in Soap Opera," *Journal of Popular Film and Television*, *27*(1), 34–9.

Flitterman-Lewis, S. (1992) "All's Well that Doesn't End: Soap Opera and the Marriage Motif," in L. Spiegel and D. Lewis (eds.), *Private Screenings*, Minneapolis: University of Minnesota Press, pp. 217–26.

Franco, J. (2000) "Cultural Identity in the Community Soap," *European Journal of Cultural Studies*, *4*(4), 449–72.

Frangou, G. (2002) "Soap Opera Reception in Greece: Resistance, Negotiation and Viewing Positions," PhD thesis, University of London.

Gauntlett, D. and Hill, A. (1999) *TV Living: Television, Culture and Everyday Life*, London: Routledge.

Geraghty, C. (1981) "The Continuous Serial: A Definition," in R. Dyer et al. (eds.), *Coronation Street*, London: British Film Institute, pp. 9–26.

Geraghty, C. (1991) *Women and Soap Opera*, Oxford: Polity Press.

Gripsrud, J. (1995) *The Dynasty Years*, London: Routledge.

Hayward, J. (1997) *Consuming Pleasure: Active Audiences and Serial Fictions from Dickens to Soap Opera*, Lexington, KY: University Press of Kentucky.

Liebes, T. and Livingstone, S. (1998) "European Soap Opera: The Diversification of a Genre," *European Journal of Communications*, *13*(2), 147–80.

Liebes, T. (2003) "Herzog's 'On Borrowed Experience': Its Place in the Debate over the Active Audience," in E. Katz et al. (eds.), *Canonic Texts in Media Research*, Oxford: Polity Press, pp. 39–53.

Lu, S. H. (2000) "Soap Opera in China: The Transnational Politics of Visuality, Sexuality and Masculinity," *Cinema Journal*, *40*(1) (Fall), 25–47.

Macdonald, M. (1995) *Representing Women*, London: Arnold.

Mankekar, P. (2002) "National Texts and Gendered Lives: An Ethnography of Television Viewers in a North Indian City," in K. Askew and R. Wilk (eds.), *The Anthropology of the Media: A Reader*, Oxford: Blackwell, pp. 299–322.

Martin-Barbero, J. (1995) "Memory and Form in the Latin American Soap Opera," in R. Allen (ed.), *To Be Continued . . . Soap Operas Around the World*, London: Routledge, pp. 276–84.

Modleski, T. (1984) *Loving with a Vengeance*, New York: Methuen.

Mumford, L. S. (1995) *Love and Ideology in the Afternoon*, Bloomington: Indiana University Press.

Nelson, R. (1992) *TV Drama in Transition: Forms, Values and Cultural Change*, London: Macmillan.

O'Donnell, H. (1999) *Good Times, Bad Times*, London: Cassell/Leicester University Press.

Porter, M. J., Larson, D. L., Harthcock, A., and Nellis, K. B. (2002) "Redefining Narrative Events: Examining Television Narrative Structure," *Journal of Popular Film and Television*, 23–30.

Seiter, E., et al. (1991) "'Don't Treat Us like We're So Stupid and Naïve': Towards an Ethnography of Soap Opera Viewers," in E. Seiter et al. (eds.), *Remote Control*, London: Routledge, pp. 223–47.

Stempel, M. L. (1995) *Love and Ideology in the Afternoon: Soap Opera, Women and Television Genre*, Bloomington, IN: Indiana University Press.

Tsai, Y. (2000) "Cultural Identity in the Era of Globalization: The Structure and Content of Taiwanese Soap Operas," in G. Wang et al. (eds.), *The New Communications Landscape*, London: Routledge, pp. 174–87.

Williams, R. (1989) "Most Doctors Recommend," in A. O'Connor (ed.), *Raymond Williams on Television*, London: Routledge, pp. 81–3.

CHAPTER SEVENTEEN

The Shifting Terrain of American Talk Shows

Jane M. Shattuc

By 1995 there were an average of 15 hour-long talk shows on the air during the daytime in the major television markets in the United States. This new genre had ended the near 50-year reign of soap operas as the most popular daytime "dramatic" form, and talk had become the most watched television genre for women. In May 1993 *The Oprah Winfrey Show* attracted a larger number of women viewers than network news programs, nighttime talk shows, morning network programs or any single daytime soap opera. Every day more than 15 million people were tuning in to watch Oprah Winfrey and her female audience debate personal issues with as much passion as an old-time revival meeting or the recent balanced budget deliberations in Congress (see Oprah, 1994, pp. 1–5).

Yet within talk shows, there was a historical shift signifying major changes in the genre and American culture in the mid-1990s. After half a decade of the dominance of four programs with a general political commitment – *Phil Donahue* (1967), *Oprah Winfrey* (1984), *Sally Jessy Raphaël* (1985) and *Geraldo* (1987) – talk shows seemingly lost their tie to the public sphere. Scores of new talk shows aired: *The Jerry Springer Show* (1991), *Maury Povich* (1991), *Montel Williams* (1991), *Jenny Jones* (1991), *Ricki Lake* (1993), *Gordon Elliott* (1994), *Carnie* (1995), *Tempestt* (1995). Topics moved from personal issues connected to a social injustice to interpersonal conflicts that emphasized the visceral nature of confrontation, emotion, and sexual titillation. The expert disappeared as the number of guests proliferated – each program staging a whirlwind succession of five-minute sound bites of conflict, crisis, and resolution.

Everything also became younger – the guests, the studio audience, the host, and even the demographics. Increasingly, the hosts – Danny Bonaduce (*The Partridge Family*), Gabrielle Carteris (*Beverly Hills 90210*), Tempestt Beldsoe (*The Cosby Show*), Carnie Wilson (*Wilson Phillips*) – came from the entertainment industry instead of news from which the first generation of talk show hosts came. Suddenly, they were promoted as "experts" based on claims of their "average" status as products of middlebrow commercial culture. The studio audience moved from its earlier role of citizens making commonsense judgments to spectators hungering for confrontation. Talk shows came increasingly to resemble a televised coliseum where the screaming battles of an underclass were carried out as a voyeuristic spectacle, in stark contrast to their earlier role as a venue for social change and personal development. "Go Ricki!!" had become the rallying cry for not only the death of the public sphere, but also the private sphere – nothing seemed taboo.

Faced with this shift toward the increased tabloidization in talk shows, both the political left and the political right in America reacted vigorously. Liberal to left magazines (such as *Ms.*, *The New Yorker*, and *The Nation*) decried the lack of social consciousness exhibited by the programs. William Bennett, the neo-conservative former Secretary of Education, launched a campaign against the new talk shows in October 1995. He labeled them a form of "perversion," while oddly praising the old liberal programs, such as *The Oprah Winfrey Show* and *Phil Donahue* (once considered purveyors of abnormality), for upholding family values. How could talk shows have reached this contradictory moment?

Such a significant difference in program content and reception allows us to examine not only the historical change within the genre, but also the role that identity politics played as it was popularized within a commercial medium. Traditional theories of generic evolution argue that a genre begins in a naïve state, before evolving toward greater awareness of its own myths and conventions (Feuer, 1992, p. 156). Here, a self-consciousness of the medium has produced programs in the 1990s that communicate at a number of levels, commenting on its own history and methods. Due to the phenomenal popularity of *The Oprah Winfrey Show*, the programs produced in the 1990s emerged out of the networks' historical need to repeat success. In an attempt to create a different market, the new programs reached out to a younger audience television and constructed a new form based on the sheer pleasure of breaking social taboos – especially those maintained by an older generation of "serious" talk shows.

More so, TV talk shows are dynamic cultural objects; they have signaled major changes in American culture, the television industry, and other TV programs. In the 1990s, they abandoned the unwritten guidelines that identity politics once provided talk shows in defining social injustice in terms of race, gender, class, and sexual preference. On one hand, these new programs stem from a general shift in the American political temperament as gay rights, affirmative action, and abortion rights have come under fire with the neo-conservative attack in the 1990s. But, on the other hand, the new talk shows spring from the voracious appetite of commercial television, where innovation has led to a self-conscious gutting of the feminist notions of empowerment based on confession, testimony, and social conversion. Rather *Ricki* et al. attest to the power of the broader pleasures of youthful rule-breaking and renegade individualism in the face of the social regulation imposed not only by the identity politics of the 1990s, but also by the established talk shows of the 1980s.

Defining the Genre

The talk show is much older and more complex than might be implied by its straightforward identification with "simple" pop culture. Daytime TV talk shows are a subgenre of a larger form known as "talk shows" that is as old as broadcasting in America and borrows its basic characteristics from nineteenth-century popular culture, such as tabloids, women's advice columns, and melodrama. Today the term encompasses programs as diverse as *Larry King Live*, *The Oprah Winfrey Show*, *The 700 Club*, *The Tonight Show*, *Rush Limbaugh*, *The Television Show*, *Ricki Lake*, and *Good Morning America*, as well as talk radio and a host of local TV talk shows which are united by their emphasis on informal or non-scripted conversation as opposed to the scripted delivery of the news.

Nevertheless, until 1994, when the form started to change, the issue-oriented daytime talk show of the first generation was what a majority of Americans understood by the term talk show. As a historical subgenre, it was divided from the other types of talk shows by the following five distinct characteristics. First, it was "issue-oriented." The content of the program emanated out of social problems or personal matters that had a social currency – such as rape, drug use, or sex change. Second, it was distinguished by the centrality of active audience participation. Third, the subgenre was structured around the moral authority and educated knowledge of a host and an expert who mediate between guests and audience. Fourth, they were constructed for a female audience in that women were the overwhelming majority of viewers. Finally, this kind of talk show was an hour-long syndicated program produced by non-network production companies for sale to network-affiliated television stations. The first generation of talk shows, including four top Nielsen-rated programs in the 1980s, *Geraldo*, *The Oprah Winfrey Show*, *The Phil Donahue Show*, and *Sally Jessy Raphaël*, fit these generic traits. Such unanimity allows them to be characterized as a distinctive cultural group.

The issue-oriented content of the first generation differentiated daytime talk shows from other interactive TV forms such as game shows and other talk shows. Talk shows were not the news. But even at their most personal and emotional, the topics of these talk shows emanated from a current social problem or issue. They could be considered the fleshing out of the personal ramifications of a news story: the human interest story. There needed to be a cultural conflict around which the drama of the show was staged. These subjects were culled from current newspaper and magazine articles, viewer mail, and viewer call-ins. The producer considered whether or not the issue had opposing sides and also if it was socially broad enough to be of interest to a large audience. In fact, for their license renewals in the 1980s local stations categorized these talk shows as "informational" programs. On one end, programs staged classic social policy or public sphere debates – for example, "Mystery Disease of the Persian Gulf War" featuring army personnel (*Phil Donahue Show*, March 23, 1994), "Press Actions on Whitewater" featuring reporters (*Phil Donahue Show*, March 16, 1994), "Strip Searching in Schools," with school administrators (*Sally Jessy Raphaël*, March 14, 1994), or even "Do Talk Shows and Self-Help Movements Provide Excuses?" including lawyers and cultural critics (*Oprah Winfrey Show*, February 22, 1994).

More typically, the social issue was placed in a domestic and/or personal context – for example, "Arranged Marriages" (*Oprah Winfrey Show*, March 10, 1994), "When Mothers Sell Babies for Drugs" (*Geraldo*, March 17, 1994), "Custody Battles with Your In-Laws" (*Sally Jessy Raphaël*, April 22, 1994) and "Domestic Violence" (*Phil Donahue Show*, February 1, 1994). Such domestic social issues were often further broadened to deal with perennial behavior problems – for example, "You are not the Man I Married" (*Geraldo*, February 14, 1994), "Broken Engagements," (*Oprah Winfrey Show*, January 31, 1994), "Ministers Who Seduce Ladies" (*Sally Jessy Raphaël*, April 19, 1994) or "Jealousy" (*Phil Donahue Show*, March 3, 1994). These topics were still social; they involved the breaking of a cultural taboo (such as infidelity, murder, seduction, or non-procreative sex).

Formally, the convention of audience participation also divided daytime talk shows from other programs that bore the appellation "talk show." Spectacles were traditionally defined by a separation between an active presentation on a stage and a passive viewing audience, as in Aristotelian theater, classical Hollywood film, and network drama.[1] The fiction is maintained through the fourth wall convention (the imaginary separation over which we peep as a seemingly "real" drama unfolds). The role of the viewing public is effaced. Within fictional television, sitcoms are the only genre that offers a role for the audience. It is configured as the laugh track ("canned laughter") or with the declaration "taped before a live audience." Both function as an attempt to signal and encourage the correct viewer reaction to the fiction.

In fact, Robert Allen (1992, pp. 122–4) argues that the audience of daytime talk shows share a stronger similarity with game shows than with any other genres. (*Commonweal* magazine calls it a "sibling" relationship, 1992, p. 18.) Both genres highlight the studio audience directly by lighting the stage and audience with equal emphasis. The studio audience becomes part of the performance just "on the other side of the screen" (Allen, 1992, p. 123). Here, the live studio audience is represented as an "ideal audience" that listens respectfully and asks the questions (or guesses the answer) for the viewer at home. Despite appearing to have greater spontaneity than celebrity talk shows, these responses are still highly regulated through the host's selection, prior coaching, and the general production process of camerawork, miking, and segmentation.

More specifically, both talk shows and game shows blur the line between audience and performance. They allow the audience member to shift from a characterized viewer to a performer. Allen argues that many game shows depend on this change for their entertainment. He cites the renowned example of Johnny Olsen's invitation to "come on down!" on *The Price is Right* (1956–74) as the audience member switches from audience member to participant. Here, the audience member becomes an active agent and a central actor (no matter how much pre-rehearsal occurs).

Yet the audience participation of daytime talk shows was held in check by the third characteristic that is similar to radio: the moral authority of the host and experts who regulate the discussion. Whether it was Oprah, Phil or Sally or Dr. Joyce Brothers, Senator Donald Riegle, or Rabbi Richard Simon, these people represented the educated middle class (or bourgeois public sphere). The hosts used the authority of their experience from many programs on the topic to establish her/his authority. Oprah Winfrey often began: "What I have gotten from doing a number of shows on the subject . . ." The expert's power was based on higher education and/or a specialized occupation often within the healthcare industries. Consider the continual tagging of "PhD" after the expert's name in an attempt to deflect questions of credibility away from experts of whose "knowledge" we knew little or nothing. Nevertheless, through their advice and mediation of the conflict, host and expert communicated the "established" position or "culturally acceptable" course of action for handling these issues. Even hairstylists and dressing consultants in makeover shows communicated the socially acceptable or middle-class fashion look to a lower class as Armani clothing and $300 haircuts were paraded as socially normative. The expert and the host provided the ballast that kept the talk show's conflict in check. Their authority distinguished this first generation from the 1990s generation of talk shows, such as *The Ricki Lake Show* (1993–), *The Richard Bey Show* (1993–), *Gordon Elliot* and *Jenny Jones* (1991–) that depend more on a "free-for-all" style in the absence of expertise.

Fourth, talk shows were constructed primarily for female audience in the 1980s and early 1990s. According to the A. C. Nielsen Company, women were the majority viewers (80 percent) of these daytime talk shows, as compared to talk radio where they represented less than half (42 percent).[2] However, much like the soap operas (their fictional cousins), talk shows not only appealed to women at home, but the majority of the studio audiences, as well as the production staffs, were women. This is one distinction that divided TV talk shows from issue-oriented radio talk shows whose audiences were disproportionately male.

Fifth, these talk shows were hour-long first-run programs made by independent producers, but financed and distributed by syndicators. They were sold to local network affiliates and independent stations to fill their fringe schedule that is not dominated by network feeds. This institutional distinction separates them from the more prestigious network productions, such as *Today*, *Good Morning America*, *The Tonight Show*, and *Late Night With David Letterman*, as well as soap operas and news. Daytime talk shows stood as an ancillary form of programming. Their independence from the networks, combined with their high profits, low production costs and daytime placement, allowed these daytime programs the latitude to produce content that would normally be censored on network television. As a "degraded" form, the programs addressed topics that were not only sensational and "impolite," but were also politically and socially controversial and thus rarely aired on network television (for example, homosexuality, abortion issues, and incest).

Other characteristics of the first generation of talk shows were less automatic features. The talk shows tended to be scheduled during the daytime – midmorning and late afternoon – as a transition from news programming in the morning to the soap operas in the afternoon and a transition back to the evening news. However, many shows were rerun late at night when the newer talk shows also air. They also tended to run five days a week in dependable daily "strips," in industry parlay. Other countries, such as Great Britain, aired them only once or twice a week.

Even though a number of similar talk shows began in the 1990s, the highly politicized period of the 1960s and 1970s, when America experienced the civil rights, anti-war, women's and gay rights movements, left its stamp on the content and structure of these talk shows. Many of these shows began before 1990, including *The Phil Donahue Show* (1967–), *Sally Jessy Raphaël* (1985–), *The Oprah Winfrey Show* (1984–), and *Geraldo* (1987–). The form originated in November 1967 when Phil Donahue's show first aired in Dayton, Ohio, with a program devoted to atheism featuring activist Madalyn Murray O'Hairn. The other three programs began in the mid-1980s, with a mix of sensationalism and a liberal political agenda that champions the rights of the disenfranchised. The coherence of this group of programs has became even clearer with the rise of the "youth" talk shows of the 1990s, which shifted the genre away from identity politics toward a more apolitical and ironic treatment of social issues.

The Shift: The Second Generation

These new talk shows did not suddenly appear *deux ex machina*, out of the sky spreading their destructive power over an innocent America, as some of the press would lead one to believe. Rather they continue an ongoing trend involving the slow commercial airing of American private lives, which began with nineteenth-century tabloids and the rise of the penny press. As Republican administrations began deregulating American television in the 1980s, *Geraldo* was able in 1987 to break open the talk show, moving it beyond "restrained" discussions of the relation of the personal and the political. It now emphasized the confidential world and its conflicts – a sensibility implicit in the concept of the personal as political. By 1994 *Oprah Winfrey* generated $180 million in revenues, proving the tremendous financial reward possible from this new inexpensive combination of the sensational with the political. However, *The Ricki Lake Show*, launched in 1993, is most often cited for innovating the "no-holds-barred" youth format of talk shows in the 1990s.

The history of *Ricki* undercuts the industry discourse that the new tabloid talk shows grew out of audience demand or popular tastes. Rather, *Ricki Lake* is a program whose creation and success were self-consciously engineered by its creators to capitalize specifically on a youth market and the expansion in cable channels. When the program premiered in 1993, there were twenty other talk shows on the air which played to the generalized demographic market of women 18 to 49 years old. The producers of *Ricki* chose to split up the viewing public by targeting a younger generation and creating a talk show whose core viewers were to be 18 to 34, thus counter programming against *The Oprah Winfrey Show*'s older audience.

Such a programming strategy stems from the general narrowcasting that evolved in the late 1980s and early 1990s as the opening of the broadcast spectrum and cable disrupted the general "network" audience. Garth Ancier, creator of *Ricki Lake*, was one of the founding fathers of the Fox network, where he was head of programming. Along with his programming team, he innovated: *Married . . . with Children*, *The Simpsons*, *In Living Color*, *21 Jump St.*, and *The Tracy Ullman Show* – all programs which marked Fox's reputation for youthful irreverence. *Ricki Lake*, with its penchant for asocial and humorous personal confrontations, found a welcoming home on the Fox affiliates across the country.

The producers used the established talk shows as models, but appealed to a more exuberant and rebellious sense of youth. The program format was altered by adding more guests to create a faster pace and more people of color were added in order to broaden the appeal of the show. The other *Ricki Lake* executive producer, Gail Steinberg (a producer on *The Phil Donahue Show* for six-and-a-half years), stated in 1994:

> We saw no reason why [a youthful approach] couldn't be successful in daytime and a lot of people didn't agree. They said these people aren't home to watch daytime TV. They said it would never work. . . . You can't be a better "Oprah." She is the best. So we decided we wouldn't even try. We would carve out a new niche for ourselves. ("Star Talker," 1994, p. 57)

They hunted for a young host and picked 26-year-old Ricki Lake, who argued that "this show is geared toward a totally different audience, which could not relate to talk shows before we came along. So in that sense I guess we are a voice for younger people" ("Star Talker," 1994, p. 57). Yet Ricki Lake's model was the established shows of the 1980s; she prepared to be a host by watching *Oprah Winfrey* and *Phil Donahue* shows with a trainer.

By creating more general topics and less in-depth interviews through increasing the number of guests from three to six, *Ricki Lake* staged broad conflicts at breakneck pace. For example, for a program on unwed mothers and fathers-to-be (September 9, 1995), *Ricki Lake* set up the antagonism through a series of prerecorded statements by the family of a woman (Danielle) even before she speaks. Each family member, in a pre-rehearsed style, states "I hate Max . . ." and gives their individual reasons. The conflict is clear and inflamed within the first five minutes. Described as the "*Melrose Place* of talk shows" (Grant, 1994,

p. B5), the program creates each segment around a new pair of guests in conflict. As host, Ricki moves quickly to the stated controversy implicit in the chosen topic. She prods the guests to fill in the details of their tension in the first half of the interview. Then, she asks about the more inflammatory issue of "how does it feel?" with the hope that things will explode in the second half of this eight-minute interview (e.g. "Danielle, how does that make you feel to hear them say what they have to say?"). Even former talk show host, Morton Downey Jr., sees a parallel to his highly explosive program: "Ricki Lake is the female Morton Downey Jr. except there was more meanness in [his] show than in 'Ricki.' Her show is designed to show the niceness as well as the outrageousness. I think that she is doing it right" (Grant, 1994, p. B5). The first-season success of the *Ricki Lake* formula of social shock with therapeutic overtones was phenomenal. Beginning in the fall of 1993, the program was seen on 212 stations, breaking Oprah's record of 179 stations in little over one season.

The triumph of *Ricki Lake* changed the talk show industry dramatically, with old shows becoming more confrontational and new programs popping up "trying to out-Ricki Ricki," using the faster pace, more crowded rosters of guests, glitzier graphics and other elements that "Ricki" made famous" (Grant, 1994, p. B13). There were two historical stages in the talk show market of the 1990s. In 1991, a number of talk shows premiered which were similar to the 1980s programs, but less overtly political and placing more emphasis on entertainment and sensation: *Maury Povich, Montel Williams, Richard Bey, Jerry Springer*, and *Jenny Jones*. As a whole, they were hosted by personalities who were over 40 and not clearly tied to the news industry.

But it is the 1995 youth-oriented talk shows, or the second wave of talk shows, which caused a national debate. In part, the controversy emanated from the overwhelming number of these mass duplicated programs. They were youthful, yet stylistically uniform as they attempted to replicate the glitzy, sensationalism and twenty-some-year-old host of *Ricki Lake* (Jacobs, 1995; Graham, 1995). According to *Broadcasting and Cable*, approximately 13 nationally released talk shows were scheduled to appear in the fall of 1995 (Tobenkin, 1994). The ones which made it onto national television were *Carnie, Tempestt, Charles Perez, Gabrielle, Danny!*, and *Mark*. By 1995, many television markets in the United States had as many as 15 "issue-oriented" talk shows running on any given day. These programs represented not only a noticeable generic shift, but also an industrial difference from the talk shows of the 1980s.

Television history was marked by a period of consolidation in the 1980s as large numbers of television companies were taken over by huge corporations, such as General Electric's acquisition of NBC. Sony, as the parent company for Columbia TriStar Television Distribution, is typical of this corporate change. As a Japanese-based international media conglomerate, Sony still centers its holdings on the production of electronic media equipment. It has diversified into the recording industry with Sony Music Entertainment Corporation (Columbia and Epic labels), and Sony Pictures Entertainment (Columbia TriStar Pictures),

Sony Television Entertainment (Columbia TriStar Television, Columbia TriStar Television Distribution, Columbia TriStar International), the Culver Studios, and the Sony Theater chain.

Sony Television Entertainment, through Columbia TriStar International Television, has substantially invested in HBO Ole in Latin America, VIVA Television in Germany, Galaxy/Australis in Australia, HBO Asia in the Far East, and CityTV in Canada. It has launched the Game Show Network, as well as producing the top game shows *Wheel of Fortune* and *Jeopardy!*, and the soap operas *Days of Our Lives* and *The Young and the Restless*. Columbia TriStar Television Distribution also distributes *Seinfeld* and *Mad About You*. In its 1995 annual report, Sony described *Ricki Lake* as having charted "new ground" as "the fastest growing talk show in the history of television" (Sony, 1995, p. 27). Interestingly, the report fails to mention *Tempestt*, which it obviously views as a less than important product.

Given the sheer size of these conglomerates, television companies have more money to invest on speculative productions with the potential for a very high yield if the program is successful. For instance, many of the 1995 talk shows were launched with investment of less than $6 million. In other words, the companies were willing to take a chance on a $6 million loss for the possibility of a $40 million gain after one year of success (based on the histories of *The Oprah Winfrey Show* and *The Ricki Lake Show*).

Another determining factor in this growth was the expansion of the broadcasting environment with the launch of new national networks involving Fox and UPN affiliates and independent stations. These new stations have become relatively easy launching pads for these cheaper and more "tawdry" talk shows, whereas talk shows had been the exclusive domain of network affiliates in the 1980s. Because of the larger number of talk shows competing to be chosen by these cable stations, the new programs had to deliver respectable ratings within two or three weeks of their debut or the program was quickly dropped. *Variety* stated that "there is a sizeable contingent of competitors waiting to pounce on the weak links in the cable ecosystem, each with its own notion of what works" (Benson, 1993, pp. 27, 36).

It is too simple to paint the change in production between the early talk shows and later ones as the loss of social scruples under the corrupting influence of profit and competition. Oprah Winfrey's astute announcement in September 1994 trades on this assumption, when she declared she would no longer do "trash TV." In a cover story in *TV Guide*, she "confessed":

> I understand the push for ratings caused programmers to air what is popular, and
> that is not going to change. I am embarrassed by how far over the line the topics
> have gone, but I also recognize my contribution to this phenomenon. ("Truth . . . ,"
> 1994, pp. 15–16)

On one level, she failed also to confess that her pledge to take the high road serves as a shrewd business move to differentiate her talk show in the face of indistinguishable tabloid shows. She will reign over her self-created moral uplift market for talk shows while the other shows will take a smaller and smaller share of the increasingly crowded tabloid market.

On a more important level, Winfrey's statement perpetuates the myth that television merely follows popular tastes toward tabloidization instead of actually creating tastes. However, she promised to create a new market with her oath to create "shows with images of what we would like to be" (Winfrey, 1994, p. 18). As a result, *The Oprah Winfrey Show* turned to bourgeois knowledge as the source of programming; the program is now based on upbeat interviews with "exemplar" celebrities and self-help discussions with experts and a quiescent audience nodding in approval.[3]

Meanwhile, the youth-oriented talk shows continue the tradition of soliciting the active audience by constructing their vision of a younger, more fun-loving consumer. This process has been vaguely called "sensationalizing" the medium – a rather inexact term given that the genre has always been sensational. A more exacting analysis of the production and publicity process in the 1990s reveals the construction of what I would call an implied "self-conscious" or "playful" viewer/consumer. On one level, they could be read as straight talk shows offering "individual" solutions to the underclasses' everyday personal dilemmas. For instance, some *Ricki Lake* titles for one week in 1995 were:

> "Listen, Family, I'm Gay . . . It's Not a Phase . . . Get Over It!" (11/20)
> "Girl, You're Easy Because You're Fat . . . Respect Yourself ASAP" (11/21)
> "My Family Hates My Friend Because She is Black" (11/22)
> "Someone Slap Me! Today I Meet My All-Time Favorite Star" (11/23)

Ricki Lake (and Jerry Springer) end their programs with the host moving off to one side to comment on the conflict and offer advice. After the angry program on unwed parents-to-be (September 9, 1995), Ricki Lake offered the following: "It is difficult when the one you love is hated by everyone else in your family. It is important not to close the doors on your family because they do love you and want the best for you. But they just may not be able to express their feelings." Platitudes and biblical style commands ("honor your mother and father") have replaced the language of socially conscious language of Freudian therapists as talk shows have sought to broaden the serious appeal of their audience.

Although the salacious content of older talk shows (*Geraldo*, for instance) looked indistinguishable from the new programs, there were still important differences in the shift in the 1990s. The newer programs maintained the six-segment program structure, but instead of three guests discussing an issue in some depth, the programs timed each segment around a new guest pair. The performance was choreographed around the introduction of the complaining or aggrieved individual while the problem person(s) waited "unknowingly" off-stage with a camera trained on him/her. Under the host's prodding the complainant revealed a series of intimate details about the other's deviant sexual activity. Together, the host and the guest came to the mutual conclusion that this person has a problem and, thus, the program has its necessary "conflict."[4]

However, these newer shows invite a second reading, which deconstructs the serious or liberal "do-good" intentions of 1980s talk shows, by inviting an ironic reading. Here, the methods of the earlier programs are so excessively overplayed that they highlight the contradictions of any talk show – particularly the first generation of Oprah Winfrey and Phil Donahue (even Geraldo Rivera) – that unctuously attempts to help disadvantaged people while they simultaneously profit from the act. According to Alan Perris, senior vice president of programming at Columbia TriStar Television Distribution, "Geraldo does things that are pretty much on the edge a lot, but he call himself a journalist. We have an actress and it's fun" (Grant, 1994, p. B6). Although these new programs flaunted their asocial values to the point of a callous lack of care, they cannot be subsumed within right-wing ideology. Their style and content derived from the sexual liberation movement of identity politics (gay, lesbian and bisexual culture, and feminism).

Conclusion

In many ways, today *The Jerry Springer Show* remains the last successful vestige of the second generation of talk shows. Along with the 1990s talk shows, it can be read as a celebration of outlaw culture – gay, black, and/or female. None of the newer talk shows were as simplistic or mindless as the established press claimed. They may have been amoral, but the shifts in tone, political points of view, and styles are vertiginous and contradictory. At their worst, they exploited and made people inured to social injustice. But at their best, they finally offered an active, even aggressive, "in your face" identity to people who have been represented either as victims or perverts by a dominant culture.

On one level, this argument supports a classical reading in cultural studies: the carnivalesque. Through their emphasis on physical excesses, scandal, and offensiveness to the status quo, talk shows reproduce much of the same liberatory pleasures that carnival provided in Rabelais's world. Bakhtin (1968, p. 10) writes: "Carnival celebrated the temporary liberation from the prevailing truth of the established order: it marked the suspension of all hierarchical rank, norms, and prohibitions." However, Umberto Eco (1985) cautions that such rule-breaking could also be a form of "authorized transgression." He argues that although comedy is built around breaking rules, the rules are always in the background. Eco writes: "Our pleasure is a mixed one because we enjoy not only the breaking of the rule, but also the disgrace of an animal-like individual [the transgressor]" (p. 6). Talk shows repeat this form of law enforcement because we are able to play out our transgressive desires through identification with this "low" culture of talk show guests without ever having leave our orderly complacent world.

Nevertheless, due to the growing social controversy about talk shows, *Richard Bey* was cancelled in Boston in December 1995. In January 1996, *Gabrielle* and *Richard Perez* were also cancelled (Biddle, 1996). During the same month, Phil Donahue quit and Geraldo Rivera announced that he would like to be a network anchorman and changed the title of *Geraldo* to *The Geraldo Rivera Show* featuring more hard news stories. Only *The Jerry Springer Show* carries on the tabloid tradition with any success. With programs entitled "Advice to Oprah Letter Writers" and "Tipping and Gift Anxiety," as well as her Book Club, Oprah Winfrey has gone on a "spiritual quest of moral uplift" – returning to the values associated with the bourgeois tradition of feminine advice. Daytime talk shows are returning to the *Mike Douglas* style of the 1960s featuring the light entertainment of celebrity talk with the popularity of Rosie O'Donnell and Ellen DeGeneres as comic hosts. It appears that we have gone full circle; talk is returning to the banter of an apolitical America in the mid-1960s (Biddle, 1996).

While there is now less bad taste and incivility on American television, who or what has William Bennett protected? The disenfranchised from exploitive convoluted images of itself? Or the American middle class from dealing with the impolite and impolitic behavior of its underclass? What is sadly lost is an important venue for average people to debate social issues that affect their everyday lives. Those discussions may not have taken place in total freedom as in the idealized public sphere imagined by Jürgen Habermas. Nor did they follow the language and habits of civil debate. Nevertheless, the shows "empowered" an alienated class of women to speak. In the end, we need to consider how our notions of "good taste" mask power and stop debate.

Notes

1 There are some other exceptions to the Aristotelian division between active stage and passive audience such as WWF wrestling, the carnival, and Brechtian theater. But these forms integrate limited audience participation. Most often the audience's actions are highly prescribed and used to comment on the spectacle's construction.

2 According to a recent poll in *TALKERS* magazine 30 percent of the audience members for talk radio define themselves as "conservative to ultra conservative" whereas only 18 percent of the audience members define themselves as "liberal to ultraliberal." See "The Talk Radio Audience" (1995).

3 As evidence, here is a week's worth of program titles from October–November 1995: "Oprah's Child Alert: Children and Guns, Part I" (10/30), "Are You Still Haunted by Your First Love" (10/31), "Caroline Kennedy on Privacy," (11/1), "Loni Anderson" (11/2), and "What Your House Says about You" (11/3).

4 The competitive pressure for new and more revelatory conflict shows led to a series of ongoing confessions by ex-producers of talk shows. For example, Joni Cohen-Zlotowitz (former *Charles Perez* producer) states: "We were told by the corporate people we had to put conflict into every segment . . . And the conflict had to be wild. Get a fight going the first segment or you're going to lose [the audience]. Those are the guidelines." She also recalls producing a show called "I

have a Surprise I want to Tell You" in which a young woman told her father that she worked as a stripper. "She said that she refused when she was told to have the woman perform in front of her dad. The supervising producer was screaming at me: 'That's what makes this show.'" Chuck Sennett (lawyer for the *Charles Perez* show) acknowledged conflict is a key element in the program, noting that it "is a part of every talk show from *Charles Perez* to *The McLaughlin Group*" (Berkman, 1995).

References

Allen, R. C. (1992) "Audience-Oriented Criticism and Television," in R. C. Allen (ed.), *Channels of Discourse Reassembled: Television and Contemporary Criticism*, Chapel Hill, NC: North Carolina Press, pp. 101–37.

Bakhtin, M. (1968) *Rabelais and his World*, Cambridge, MA: MIT Press.

Benson, J. (1993) "Talk Shows Reaching Saturation Point," *Variety*, October 11, pp. 27, 36.

Berkman, M. (1995) "Liars Send in Clowns for Sicko Circuses," *New York Post*, December 4.

Biddle, F. (1996) "TV's New Shocker: The Talk Turns Tame," *Boston Globe*, January 5, p. 1.

Commonweal (1992) "TV Talk Shows," February 14, p. 18.

Eco, U. (1985) "The Frame of Comic 'Freedom,'" in T. A. Sebeok (ed.), *Carnival*, Berlin: Mouton, pp. 1–9.

Feuer, J. (1992) "Genre and Television Studies," in R. C. Allen (ed.), *Channels of Discourse Reassembled: Television and Contemporary Criticism*, Chapel Hill, NC: North Carolina Press, pp. 138–60.

Grant, T. (1994) "Ricki Lake's TV Talk-show Formula," *Globe*, December 18, p. B5.

Graham, J. (1995) "Early Word on New Talk Shows," *USA Today*, October 3, p. 3D.

Jacobs, A. J. (1995) "Ricki Wannabes," *Entertainment Weekly*, March 25, p. 52.

The Oprah Winfrey Show Press Packet (1994) "The Oprah Winfrey Show's Ability to Attract Female Viewers . . . ," September, pp. 1–5.

Sony Corp. (1995) *Sony Annual Report*, p. 27.

"Star Talker: The Next Generation," *Broadcasting and Cable* (1994) December 12, p. 57.

"The Talk Radio Audience," *Talkers* (1995) October, p. 7.

Tobenkin, D. (1994) "Bumper Crop of Talk Shows Hopes to Tap 'Ricki's' Success," *Broadcasting and Cable*, December 12, p. 47.

"Truth, Trash, and TV," *TV Guide* (1994) November 11–17, pp. 15–16.

CHAPTER EIGHTEEN

Television and Sports

Michael R. Real

The televising of sports events may be the best thing that ever happened to both sports and television, at least in terms of commercial growth and profitability. The benefits for sports from television contracts have been huge financial rewards for sports teams and athletes at the same time as television has greatly increased public access to sports and sporting events. The benefits to television from televising sports have included a relatively inexpensive source of program content. Sports feature physical action that is perfect for television's visual quality and are unscripted dramas ideally suited to attract and retain viewers – especially those with the demographic and financial characteristics desired by television advertisers. The combination seems a marriage made in heaven.

Analyzing Telesport: Background and Players

Yet, while analysts have spelled out these positive elements, many have also detected strong negatives in the combination. Is the relationship really symbiotic, in which each gains from the other, or is it parasitic, one in which television sucks the best out of sports? This latter negative position is the one adopted by Benjamin Rader (1984) in his book-length history of the relationship, *In Its Own Image: How Television Has Transformed Sports*. Steven Barnett (1990), in *Games and Sets: The Changing Face of Sport on Television*, views the relationship with less nostalgia for an idealized past but with increasing anxiety about a cable- and satellite-driven future for televised sport. Without question, televised sports have contributed greatly to sport-related dysfunctions: bloated player salaries, unstable sports franchises ever ready to pull up roots, over-hyped extreme competitions, unbalanced demands for victory at any cost, passivity and escapism, boozy and sexist advertising and viewing environments, gambling addictions, the commodification of fan paraphernalia, obsessions with trivia, and a feeling of distance from the integrity of traditional, balanced, coherent sporting activity, the classic ideal of "mens sans in corpore sano." What are the known truths behind the glamorous and controversial linkage of television and sport?

The central concepts and areas of research on sports and television cover the process from start to finish. Hundreds of studies have examined:

- the production of televised sport in economic, technological, and political context
- the texts and audiences of televised sports, especially in reference to competition, sex, race, class, commercialization, and geographical identities and relations
- the hyper-commercial corporatization and globalization of televised sports

Historically, sports events have been firmly established as valued program content for television for more than a half century. From the first shadowy telecasts of baseball from Columbia University in May of 1937 and tennis from Wimbledon a month later, sports programming has become increasingly prominent, until today they are almost ubiquitous. Today's many dedicated satellite and cable sports channels combine with regular sports telecasting on broadcast networks and stations to make sporting events, stars, and controversies a constant feature of daily life and a frequent topic across all communications media. The union of television as a distribution medium and sports as program content is reminiscent of lucrative and popular alliances from Alexander's unifying the Mediterranean world, Charlemagne's melding into one the Roman and Christian empires, and President Monroe's proclaiming that all the Americas are one sphere of influence. The combination is, in short, powerful – even, if you will, imperial.

Critics dispute the benefits and damages of televised sports almost as vehemently as combatants on the field fight over ball, territory, and victory. Writers of stature – Nick Hornby, David Halberstam, James Michener, Michael Novak, and many others – praise the contribution of sport to human emotion, discipline, ambition, and achievement. Baron Pierre de Coubertin, founder of the modern Olympics, watched the first proto-television sports broadcast on kiosk monitors in Hitler's Berlin in 1936 and maintained, as he always had, that sport promoted virtue, character, idealism, and international understanding. Coming from another angle, sociologists and economists document the system functions that televised sport fulfills as a major social force, creating regional identities, moving consumer goods, providing mythologies of upward mobility and meritocracy.

But televised sport's more dubious achievements include creation of "the sports geek" and "the sports nut," self-labels employed by those whose waking hours are dominated by a quest for the latest and greatest sports conquests and information. Are they the beer-guzzling, chip-chomping, passive sponge pitied by the critics of television sports, or are they rather the self-realized, brilliantly informed worshipers of high achievement championed by defenders of "mediasport"? Are they caught up in a web of misogyny, racism, violence, and destructive escapism, or are they celebrating human excellence expressed through the body athletic and the competition most noble? Do television and sports serve each other symbiotically?

Commentaries on television and sport now flourish throughout the world in many languages and offer a variety of critical positions – the negative, the positive, the functionalist, and more. Historically, the first decades of major televised sports in the middle of the twentieth century generated very few scholarly studies. Then, as the impact of television on sports and of sports on television became apparent, excellent analyses emerged throughout the English-speaking world. In England, Ed Buscombe (1975) published his pioneering book about soccer on television in the same year that my study of the American Super Bowl appeared. Major British studies of television and sport followed from Alan Tomlinson, Garry Whannel, and Steve Barnett. Australian research also emerged, led by Jim McKay, David Rowe, Geoffrey Lawrence, John Goldlust, and Toby Miller. Key Canadian texts were produced by Richard Gruneau, David Whitson, Hart Cantelon, Jean Harvey, and Margaret MacNeill. US efforts featured Lawrence Wenner, Pamela Creedon, Jennings Bryant, Dolf Zillman, Leah Vende Berg, and Nick Trujillo. American work developed empirical findings about specific elements in the television text and audience responses and also introduced the aspects of history and culture. Many British, Australian, and Canadian studies also examined distributions of representational power, finance, politics, and culture.

Central to much of the best work on television and sports has been the argument, developed in a 1983 book by Gruneau (1999, p. 114), "for a critical approach to the study of sport that combines social theory with history, interpretive cultural analysis, and political economy." This avoids "the one-dimensional perspectives that reduce the analysis of sport to purely material (e.g. technological, economic) or idealist (e.g. cultural/linguistic) determinants" (p. 115). In 1984 Sut Jhally drew attention to the central role that television plays in the "sports/media complex." In the 1950s and 1960s, males with peak incomes were drawn to televised sports, advertisers paid significant amounts to reach that audience, broadcasters offered large television rights fees, television marketing began to influence decisions about sports schedules, times, locations, and more, the attraction of cross-ownership of sports and media businesses increased, and there was a massive all-round infusion of capital. From its beginnings as a mere program category on television, the coverage of sports on television, supplemented by newspaper, radio, and other media, grew in the following decades into a vast and powerful sports/media complex.

Recent decades have seen an increasing abundance of research studies and scholarly examinations. Consequently, it is now possible to draw from the extensive scholarship to outline a developed understanding of the dynamics within television and sports, the vast scale, technologies, and financing associated with televised sports, the varied and far-flung audiences, and the cultural role of "telesport" as an articulator of personal and global forces, warts and all.

Michael R. Real

Inside the Appeal of TV Sports: Arledge, Murdoch, and "Auteurs"

Televised sports have developed under the leadership of many broadcast and sport groups and individuals. Innovations and expansions are generally relatively anonymous, following the potential of the medium and subject-matter into new techniques, graphics, speed, representational strategies, and promotional marketing. However, Morris and Nydahl (1985) suggest that individual "auteurs" have developed techniques of image, language, and technology to maximize the potential of the televised sports spectacle as drama. In the same way that an Alfred Hitchcock, Federico Fellini, or Ingmar Bergman gave his personal stamp to feature films, according to French auteur film theory, so a television producer contributes a personal stamp to televised sports. While film is considered a director's medium, television is very much a producer's medium. If this is true, in the United States, it can plausibly be argued that television sports are indebted to one "auteur," namely Roone Arledge, the late president of Sports (and, after 1977, Sports and News) at ABC Television. As a Columbia University graduate with a degree in English and with many career options, in the early 1960s, he was often asked why he chose a career in sports and television. He explains:

> Sports were life condensed, all its drama, struggle, heartbreak, and triumph embodied in artificial contests. To play a game well, endless practice was required, just as it was in mastering life. Sports always contained the unexpected – a catch that should have been made and wasn't, a bar that shouldn't have been leaped and was. So did life – chaos intruding on the orderly patterns of civilization. Sports could bring tears or laughter, in wonderment over its sometime absurdity.

> Television could capture it all, and in the 1960s, there was a chance to do it creatively. I wanted to make the game more intimate, and a lot more human. (2003, p. 28)

Arledge was dedicated to moving beyond the few fixed cameras that gave comprehensive, but unexciting views of sporting events. In a long memo for ABC at the time, he pitched how sports could be done better, specifically college football:

> We will utilize every production technique that has been learned in producing variety shows, in covering political conventions, in shooting travel and adventure series to heighten the viewer's feeling of actually sitting in the stands and participating personally in the excitement and color of walking through a college campus to the stadium to watch the big game.

> . . . In addition to our fixed cameras we will have cameras mounted in Jeeps, on mike booms, in risers or helicopters, or anything necessary to get the complete

story of the game . . . all the excitement, wonder, jubilation, and despair that make this America's number one sports spectacle, and a human drama to match bull-fights and heavyweight championships in intensity.

In short – WE ARE GOING TO ADD SHOW BUSINESS TO SPORTS! (pp. 30–1; emphasis in original)

But Arledge not only championed major American sports on television; he also pioneered the addition of world sports, even fairly obscure ones, to the American television schedule. On a flight back from London during its first season in 1961, Arledge came up with the famous credo that opened each weekly episode of *Wide World of Sports*:

Spanning the globe to bring you the constant variety of sport . . .
The thrill of victory and the agony of defeat . . .
The human drama of athletic competition . . .
This is ABC's Wide World of Sports.

Wide World traveled to 53 countries in the next four decades, televising 4,967 events in more than 100 different sports. The show's success was such that it spawned imitators around the globe. In Australia, Kerry Packer's Nine Network dominated national television ratings for years led by strong sports programming, including a four-hour Saturday afternoon clone actually called *Wide World of Sports*. A top British magazine-format sports show of the time was called *World of Sport*. In his long tenure with ABC, Arledge also pioneered such televised sports staples as instant replay, Monday Night Football, announcer-stars like Jim McKay and Howard Cosell, and the up-close-and-personal style of Olympic television coverage. His autobiography, *Roone* (2003), offers an engaging look behind the scenes of television and sports.

Subsequent empirical research vindicates Arledge's attention to detail. Announcers are important to fan interest, as Arledge emphasized. Jennings Bryant, Dolf Zillman, and their co-authors (1977, 1982) have measured how fans find more excitement in a match in which announcer commentary highlights conflict and drama, for example, by stressing the off-the-field conflict between competing players. They also find that up-close viewing of televised sports violence increases stimulation, involvement, and enjoyment for viewers (1998). They test the enjoyment of spectator sports (1989) and show how close matches enhance enjoyment (1994), validating Arledge's manipulation of game schedules. Research also indicates that television does bring fans into the sports ritual and promotes a psychological identification with its participants (Breen and Corcoran 1982, 1986; Real, 1989, 1996; Moragas et al., 1995), consistent with Arledge's emphasis on bringing television fans right into the excitement of the game through personal emotions.

The most auteur-like presence in televised sports in recent years has been Rupert Murdoch. Because of his worldwide media empire and his television programming emphasis on sports, Murdoch has reshaped televised sports much as Arledge had done in an earlier era. Murdoch's contribution, however, has been less in techniques of presentation, Arledge's forte, and more in buying, combining, packaging, promoting, and selling televised sports on national and global dimensions. His controversial role is considered in more detail in the final section of this chapter.

The Telesport Text and Audience: Narrative Spectacle Intensified

Central to the power of sports as presented on television are the same elements that are identified as "news values" in the press in general and as "entertainment values" in commercial media. The sporting event is programmed for television because it is "a story." It has a narrative sequence in which protagonists and antagonists, heroes and villains (even if only arbitrarily identified as such by viewers or announcers), engage in direct conflict issuing in victory and defeat. The news values of conflict, recency, human interest, prominence, and localness direct the selection of news for newspapers and broadcasting. These also dominate the selection and presentation of televised sports.

Telesport narratives: dramatic stories and open-endings

Like news, movies, and most of entertainment media, sports on television are presented through the conventions of narrative drama. Because sporting events have a beginning, middle, and end, they are narrative events, but television is able to intensify the feeling of dramatic story build-up in a variety of ways. Whannel (1984, p. 102) explains that

> the insistence that television does not simply cover events, but transforms them into stories – is to raise questions about the polarity between actuality and fiction. Television sport can clearly be seen in terms of dramatic presentation and analyzed as a form of narrative construction.

Gruneau (1989) examined how this works in practice in the way that a World Cup ski race is covered. He found that the CBC director of the Whistler race consciously built his coverage around narrative conventions aimed at entertainment value. These were: "spectacle, individual performance, human interest, competitive drama, uncertainty, and risk" (p. 148). Camera placement, skier profiles, announcer commentary, sequencing of shots and information – all elements aim to intensify the narrative impact. The several hours between the "live to tape" videotaping and the actual airing on *Sportsweekend* were used to heighten the drama. The race was shot without announcers so their "live" commentary during the telecast could maximize the drama.

> Everything about their commentary was geared to create the illusion that they were seeing the action for the first time. The use of active verb tenses, emotion, and even prediction ("He should do well here!") was all staged – it was pure show business. (p. 149)

Such storytelling is omnipresent in telesports; Kinkema and Harris (1998) cite 16 studies on how narratives are self-consciously employed by the media to dramatize sports. The well-established power of narrative to attract and retain human interest makes such storytelling a fundamental human activity in all spheres of life. Televised sports provide this in intensified, vivid forms.

Television vs live attendance: sports up-close and personal

The power of the narrative drama in televised sports is further intensified by the way that television "personalizes" the experience. Television personalizes both the way the competing athlete is represented and the way the viewer feels connected to the game. The most obvious version of the former is in what ABC dubbed the "up close and personal" profiles of Olympic competitors. In this technique, the program will break away from the live competitions to present a one- or two-minute biography of an athlete, often focusing on a valiant effort to overcome handicaps or hardships in order to reach this world-class level. Such profiles convey a sense of the person's home, family, and native environment, as well as uniquely dramatic elements in the athlete's life. When the program comes back to the live competition, the viewer has an invigorated sense of understanding and caring about the featured athlete. The viewer's emotional investment is intensified, suspense is increased, and the outcome has more impact.

Television personalizes the sport experience better even than can live attendance at a sporting event, because it can intensify and personalize the way the viewer feels connected to the competition. The television viewer sees close-ups of players, hears anecdotes about them, reads statistics on screen, accesses instant replays, watches parallel action from simultaneous games elsewhere, and enjoys interviews with victors and/or losers at the end, none of which is part of "traditional" attendance at games. Stadiums have also been forced to adjust to this "superior" reality of the televised version by adding huge in-stadium television screens – measuring 100 feet by 40 feet or more in many cases, and featuring high-resolution and vivid color. These screens mimic the television experience for stadium fans and remove the disincentive that otherwise dictates against attending games. Because live crowd interaction and expensive ticket sales are still essential to the emotion and economy of major sports, telesport games still attempt to draw maximum large live attendances and avoid any appearance of becoming fake television studio events. Their validity and power require the full stadium experience, even though the vast majority of viewing and income result from the televised version.

Televised sports also tailor products to the demographics of specific audiences. They "personalize" the experience for targeted consumers by creating symmetry among game, players, announcers, advertised products, associated entertainment, and similar stylistic ingredients. These elements will come out very differently in telecasts of a Nascar auto race and a figure-skating competition. Telecasts of snookers, darts, billiards, poker, and similar competitions featuring individual precision without physically violent encounters seek a different tone for their viewers than do football, soccer, boxing, and faux-sports like wrestling. The viewer can further individualize the relationship to the television competition by accessing the websites of teams or players, by playing related videogames like the Madden NFL annual software package, and by consuming sports pages, specialized magazines, and even attending pubs, bars, or other public locales associated with their favorites. For example, when the Chicago Bears played in Super Bowl XX, hundreds of Bears fans in the Windy City Sports Club located thousands of miles away in San Diego rented a large ballroom at the Hilton Hotel, set up large-screen televisions, featured cheerleaders and other fan favorites, consumed the requisite food and liquor, and together relished the Bears' 46 to 10 triumph over the New England Patriots (Real, 1989). Major sports telecasts now feature simultaneous interactive websites where additional statistics are available, viewer play selection can be tested against the actual game, and other features also enhance the live television experience.

The telesport celebrity

The huge and odd phenomenon of "celebrity" in contemporary culture plays very effectively into this intensification and personalization. Because the fan already knows a great deal about Beckham, Annika, Shaq, or Tiger, they relate to their performance in a particular competition with something of the personalized sensations that a player who is a family member or friend evokes for a fan. The player is not merely a player, but is now someone the fan has come to "know" and care about, however strange it may be that this happens without the possibility of face-to-face interaction between player and fan ever occurring. Cunningham and Miller (1994, p. 77) note the connection between narrative and celebrity in televised sports: "TV sport is an individualizing genre, announcing, auditing and ending the careers of stars." Teams and television broadcasters heavily market this "star" quality as a generator of both fan devotion and the sales of jerseys and other sports paraphernalia. The huge salaries of major sports figures, topping more than $10 million per year, is a result of the marketing and promotion of individual sports celebrities. The price tag also confirms the success of this marketing. If contemporary culture is celebrity-obsessed, the world of televised sports plays a leading role.

Assemblage technique

The putting together of a television sports show involves more than narrative sequencing. Whannel (1992, p. 105) notes that "assemblage" techniques dominate in televised sports, following in the heritage of Arledge and ABC's *Wide World of Sports*. In the UK, Whannel finds assemblage at the core of the two programs that dominated television sport for two decades, *Grandstand* and *World of Sport*. This sports television magazine format "assembles" taped footage from games and events, live studio commentary, pre-produced features, and a variety of sporting competitions. Even live game coverage now includes assemblage techniques, as it cuts from live action to replay to pre-taped player introductions to advertisements to promotional spots to more live action to sideline shots and so on.

Spectacle

Television emphasizes the drama and grandeur of its sports programming with a result that places major telesports events in the class of "spectacle," a grandly overblown visual extravaganza that captures and mesmerizes audiences not with its content and quality as a human experience but with its overwhelming sensory stimulation and associations, sometimes reminiscent of Leni Riefenstahl's *Triumph of Will*. Television, for example, favors the one-day match form of cricket and has introduced numerous rule changes that, in the words of David Rowe (1999, p. 154), "are designed to ensure that the televised one-day cricket match is fast and furious, encouraging high scores and high drama." Harriss (1990, p. 118) notes, "The one-day spectacle is packaged in much the same way as a one-hour television melodrama." The slowness of five-day Test Match cricket is avoided and the entire experience accelerated. Rowe sees television sports capitalizing on rapid movement when it occurs and producing a sense of rapid movement when it does not. This makes "the spectacle of sports television louder and more frenetic" in order to attract, distract, and transfix viewers (Rowe, 1999, p. 154). Televised sports readily lend themselves to exaggeration and overblown rhetoric and production values, seeming to expand an event, but actually reducing it to a mere spectacle.

Yet, when responsibly directed without hyper-commercialism, television can take sports competitions and make them more intense, personal, and dramatic. Through specific techniques, the sports television text is thus able to convey a sense of the viewer becoming an actual participant in an entertainment event, of being an active presence in breaking news. Very few other media experiences, or even non-media cultural practices, no matter how hard they try, are able to achieve such vividness and involvement.

345

Michael R. Real

Sex, Race, and Class: Are Telesports Biased or Fair?

Many of the sports featured on television originated and/or grew to prominence in the nineteenth century. Baseball, basketball, football, volleyball, and many others took formal shape during the period when culture was becoming industrialized, urbanized, and technologically interconnected through telegraph, telephone, phonograph, film, and other precursors of television in the later decades of the 1800s. Patriarchy and racial discrimination were well-established at the time, and the sports emerged as largely male and white in their publicly celebrated forms. Even sports with older traditions – such as cricket, bowling, soccer, and billiards – were reconfigured into a modern Weberian "rationalized" form of organized competition with a stuffier, white, male aura. When television emerged as a powerful venue for sports in the mid-twentieth century, the world of sports had long been dominated by males and whites and, as it commercialized, was also owned by the wealthy entrepreneurial class.

In the first decade of major television sports in the 1950s, the producers and executives in charge of sports television decision making were white males. Golf was not a great ratings buster, but network executives loved the game so early on it established its continuing presence on television. Network decisions were not yet based on impersonal ratings, demographics, and profit margins. *Variety* television writer Les Brown (1971) recalls how a network executive during that period insisted that Sunday afternoon television programming was unimportant "because everyone is at the polo matches." Polo matches! The exotic cultural bias of such an attitude also indicates the upper-class status of leading television sport decision-makers of the time. The decades since have witnessed a long struggle for gender parity, racial justice, and class egalitarianism in televised sports. While numerous studies in England, Australia, and Canada (e.g. Gruneau, 1999) have specifically examined the role of class in relation to sports, the majority of literature about *television* and sport restricts the issue of equality to sex and race. This is especially true in the case of the United States where "class" is nearly as absent from scholarship as it is from political discourse.

Gender

In her pioneering anthology, *Women, Media and Sport*, Pamela Creedon (1994, p. 13) noted "Virtually nothing had been done to explore audience preferences for televised sports involving women until 1985." In that year, she and colleague Lee Becker initiated a series of experimental and survey research projects that, against their hopes, revealed the lack of interest in women competing on television. In retrospect, what their methods could not measure was the collective cultural valuation that only emerged later for televised women's sports. By 1992 women were the majority of the television audience for the ABC Winter Olympics from

346

Albertville, France. By 1997, the WNBA began its first season with more television coverage than had ever occurred in the history of women's sports (Wearden and Creedon, 1999). By July 1999 record television audiences followed the final game of the Women's World Cup soccer match between the United States and China, climaxed by Brandy Chastain edging in the winning penalty kick and notoriously stripping to her sports bra. By the early twenty-first century, women's NCAA basketball was competing with men's for television airtime and news reports.

But huge inequalities remain in the televised coverage of men's and women's sports. The media coverage of women athletes pales in comparison to the coverage of men; Tuggle et al. (2002) list ten studies that confirm this. Daddario (1998) finds "Women athletes in almost any sport receive a fraction of the coverage given to men athletes" (p. 16). The reason? "The ideology in sport is almost always masculine and also is used as a standard for inclusiveness" (p. 13). "Masculinity" is a powerful force in televised sports (Messner, 1992) creating a masculine hegemony (Bryson, 1984; Connell, 1995), a hegemony that is opposed but not eliminated by feminist criticisms (Hall, 1996; Griffin, 1998).

Both exclusion and improvement in the representation of women competitors in sports television can be seen in the Olympics, college sports, and major weekend television tournaments. The first modern Olympic Games in Athens, Greece, in 1896, allowed only males to compete and featured only white athletes. Neither women nor athletes from outside Europe and North America competed. In 1900 women were allowed to compete, but, by 1912, women still accounted for only 2 percent of the competitors. By the time the Games reached major television exposure in 1960, women provided a major portion of the competitors; that figure has remained at about 25 percent of the total athletes in the last four quadrennia of Olympic Summer and Winter Games.

At the collegiate level, since the creation of Title IX support for female athletes in 1971, television has slowly increased its interest in female competition. By the beginning of the twenty-first century, for example, the NCAA basketball "March Madness" and "Final Four" play-offs for *women* received consistent major network scheduling, exposure, promotion, and audiences. In weekend televised sports tournaments, sporadic individual high-visibility events – Billie Jean King defeating Bobbie Riggs, Annika Sorenstam or Michelle Wie playing in a PGA event – have drawn major television attention to women athletes. Women's tennis, golf, and figure-skating have become significant parts of television sports. Martina Navratilova, Serena and Venus Williams, Katarina Witt, and other women have become telesport celebrities, i.e. received major television exposure giving them access to its legion of fans and the attendant product endorsements and sponsorships.

However, Rowe (1999) notes that women have been under-represented in both organizational complexes, the media corporations and the sports organizations. Feminist groups and governments have pressed for an end to female exclusion from televised sports (Hargreaves, 1994; Hall, 1997), women in sports media have mobilized to improve their positions in media organizations (Cramer, 1994), and women's sports organizations have demanded more air time, sponsorship, and a larger share of broadcast rights revenue (Crosswhite, 1996). The stakes are large here. The under-representation of women athletes on television reduces the financial and psychological validation of women athletes. Stereotyping of women as secondary destroys the positive role modeling that athletes offer. Sports masculinity on television preserves the "male gaze" that objectifies and restricts female participation.

Controversies continue over the textual inclusiveness of televised sports. Duncan and Messner (1998) summarize the range of issues and problems. The enigma of women's soccer in the United States remains. Female youth soccer has thrived for a full generation, but no substantial television contract has resulted. The demise of the WUSA in 2003 revealed its failure yet to secure a stable media profile and audience. Women, and blacks, have been the subject of different treatment by television announcers. Women's physical attractiveness is mentioned far more than men's. The famous incident of Brandy Chastain whipping off her jersey after her winning kick in the 2002 women's soccer world championship epitomized the "sexiness" of the team that television and press reports did not resist commenting on. It is clear that gender will continue to be a major issue in televised sports.

Race

Growth in the inclusion of athletes of color in television sports has been especially dramatic. The first Negro major league player, Jackie Robinson, broke into then newly-televised baseball only in 1948. Through the 1960s many college football and basketball programs, particularly those dominant schools in the South that featured regularly on television, still fielded white-only teams. In 1968, when Olympic medal-winners Tommy Smith and John Carlos raised their fists in a black power salute in full view of the global television audience, they were quickly stripped of their medals.

Today, African Americans dominate many college and professional sports in the United States, but media representations remain problematic. Davis and Harris (1998, p. 156) summarize their review of research on this subject by writing, "In short, African-American athletes are receiving increased media coverage, although not at levels comparable to their European-American peers." In early studies of television play-by-play and commentary (Rainville and McCormick, 1977), African-American athletes were more often referred to as gifted with god-given talent, while Euro-American athletes were referred to as disciplined and self-actualized. More recent studies found improvement in such representations (Sabo et al., 1996). The dominance of the National Basketball Association by black athletes has occasioned charges that television and marketers are less enthusiastic about the NBA than they are about the NFL where, for example, four of the last six Most Valuable Players in the Super Bowl have been white, despite two-thirds of the NFL football rosters being black. The sports television staple of major league baseball has seen a significant decline in the number of African-Americans on major league rosters, as documented by *Sports Illustrated* (Verducci, 2003). In 1975, blacks filled 27 percent of roster spots, in 1995 it had dropped to 19 percent, and in 2002 only 10 percent. This is in contrast to European soccer which seems to feature an increasing mix of nationalities and ethnicities moving back and forth among top clubs, despite the racist traditions of many European clubs (Sandvoss, 2003). And, of course, Olympic television features a rainbow of nationalities and ethnicities.

Audiences

Few aspects of sports and television are as intriguing as the degree of fan identification with televised sports, athletes, teams, and outcomes. Geertz's famous description (1973) of "deep play" in the participation of Balinese in the cockfight certainly can be applied to the participation of the sports fan in the televised event. For Geertz, the cockfight opened up the understanding of Balinese society and culture. The kinship patterns, the structure of bets and gambling, the loyalties and identities of men and women, the distinction of membership and outside threat, everything was there. So it is with the modern televised sporting event.

The external "acting out" of fan identification tends to be more extreme in the arena or stadium experience, but the television viewer may experience emotions that are every bit as powerful and excruciating. The stadium fan paints his or her face, dons team colors, waves pennants, makes noise in the extreme, jumps and stands for crucial moments, and has a full kinetic experience of fandom. In contrast, the television viewer is forced to internalize the experience of delight or dismay at the performance of a beloved team or athlete with fewer outlets for kinetic release. For the "big game," the home viewing environment comes more to approximate the stadium experience in colors, noise, and activity. But the solitary viewer of a televised game has fewer outlets and is isolated, left to feel awkward in acting out, despite the same or even greater intensity of feeling.

Nick Hornby's experience as a perennial fan of the London-based football (soccer) club Arsenal is one of the most vivid and extended personal accounts of what it means to watch sports on television. His book *Fever Pitch* (1992) recounts attending games in person, but also describes the television experiences of a die-hard sports fan. His final description of an Arsenal season and the last game captures the ecstatic potential a television sports fan always knows, or at least hopes, is there. Arsenal had not won the English League Championship for 18 years. Finally, the team seemed to have arrived. It had a good year – but faded at the end. Needing an almost superhuman performance in the final game, it came down to the final seconds. Hornby's emotions were extreme as the referee's extended time was Arsenal's final and only hope. He notes: "Finally, with the clock in the corner of the TV screen showing that the ninety minutes had passed,

I got ready to muster a brave smile for a brave team." Then, 92 minutes in, Michael Thomas broke loose for what could be the winning score. Torn between well-learned skepticism and fanatic hope, Hornby watched the screen. And then "I was flat out on the floor, and everybody in the living room jumped on top of me. Eighteen years, all forgotten in a second" (p. 229). Recalling that moment, Hornby argues that no analogy is adequate to describe the power of such a sport fan experience – not sex because orgasm is familiar, repeatable, predictable, and admittedly nicer but not as uniquely intense as the "once-in-a-lifetime" last-minute Championship winner. Childbirth, career successes, other important moments in life – these cannot match the "unexpected delirium" of actually achieving what one has coveted for two decades.

The challenge to research is to explain such experiences without losing the uniqueness and power – and, yes, let's admit it, the danger – they hold for the television sports fan. The most systematic look at audience experiences with viewing televised sports is "Watching Sports on Television: Audience Experience, Gender, Fanship, and Marriage" by Lawrence Wenner and Walter Gantz (1998). Motives for telesport viewing include fanship, learning, release, companionship, and filler. Behavioral and affective correlates of telesport viewing include reading and talking about the game in advance; snacking and, to a lesser extent, drinking during the game; talking, yelling, and getting angry during the game, much more so when in a group setting; and, afterward, basking in reflected glory after a victory by the viewer's team. These behaviors are more common among men than women and, for both sexes, more common with fans than with non-fans. Men and women, who are both telesport fans or who are both non-fans, show considerable compatibility, but Wenner and Gantz acknowledge that there are significantly more "sports impassioned" men and therefore that finding an evenly-matched sports spectating partner may be difficult. On average, women are more motivated to watch sports "as a last resort," while men are more motivated by the opportunity to have a good time, drink, experience excitement, and learn about players.

In the words of Wenner and Gantz (1998, p. 244) their research "both reinforces aspects of the football widow and raises questions about it." They find that many sports viewers are active, discerning, and engaged, unlike the couch potato stereotype. Typical sports fans manage their emotions admirably, rather than letting them interfere with normal life before, during, and after games. Of course, they do find women are more likely to do household chores during viewing and men are more likely to drink. While sports viewing can be an issue in a troubled marriage, they conclude (p. 251): "the armchair quarterback and the football widow have neither been typecast accurately nor are they experiencing many sport-related marital problems."

In all, given the realities of telesport and culture, the text and audiences of televised sports are incredibly varied around the world. Local, small-scale sporting events feature school teams, town teams, off-beat sports, erratic production values, and minimal financing from Anchorage to Zurich. With the increased quality and reduced price of digital video, parents or friends of players can record games and insert them into local television, sometimes with enthusiastic local audiences. The attraction of sport is so great that, when anthropologist Eric Michaels (1986, 2003) provided video equipment to Australian aborigines to use for their own purposes, he was surprised at the results. He expected the aborigines to document sacred tribal customs and legends, tales from the past and myths for the future, maybe even political, economic, and social problems of aborigines. Instead, they chose first to videotape the annual tribal basketball competition. In fact, he found, great controversy arose when, in the process of re-loading tapes, some scoring baskets were missed. As a consequence, in viewing the tape, tribal viewers engaged in heated debate over whether the original scoring was correct. Telesport texts and audiences are endlessly varied and complex.

Production and the Political Economy of Televised Sports

Televised sports have become big business in a big way. The presentation of a sporting event comes from a sophisticated array of technology for image capture and distribution, broadcaster and league contracts, advertising sales, product merchandising, event planning and crowd control, promotions and cross-promotions, audience segmentation and loyalty, and, somewhere in the midst of all this, the game itself.

The production of televised sport is a particular form of what Stuart Hall (1980) describes more generally as the process of "encoding into meaning structures," a process that is based on frameworks of knowledge, relations of production, and technical infrastructures that facilitate and shape the television program and that parallel similar frameworks in the receiving audience. The British Broadcasting Corporation, the Australian Broadcasting Corporation, the Canadian Broadcasting Corporation, Dardashan Television, commercial networks, the satellite channels of Rupert Murdoch, and the many dedicated cable sports channels are among the units that produce and distribute the current vast array of sports on television. These entities enter into contracts with the sporting association – whether team, league, federation, or conference – to gain the legal right to transmit the games and competitions over television. The fees paid for such television rights are often in the billions of dollars in multi-year contracts. For example, in the early 1990s the three commercial American networks had in place commitments for more than $6 billion in contracts for professional football, baseball, basketball, and the Olympics. More recently, one league, the National Football League, collected $17.8 billion for one television contract. That NFL contract with ABC/ESPN, Fox, and CBS ran for eight years through 2005.

Commercial broadcasters then recoup those costs and the actual expenses of cameras, announcers, production trucks, relays, and transmission through the sale of advertising in and around the broadcast. The cost of advertising in major sports often involves multi-million dollar contracts. The US Super Bowl charges more than $2 million for 30 seconds of airtime for commercials. In turn, the advertiser recoups those costs through the sale of products – sales which will have been maintained or increased by motivating the viewers of televised sports to purchase the products advertised. In a typical year, major corporations will spend many hundreds of millions of dollars on advertising in televised sports events. Anheuser Busch, Coors, and other major beer companies, Ford, Chevrolet, Toyota, and other automobile manufacturers, McDonald's and the fast food franchises, credit card companies, cell phone services, the US military, and many others value the association in the consumer's mind between their products and televised sports. They especially value the opportunity to reach the adult male demographic with its extensive expendable income, a demographic that is especially attracted to televised sports but not to a number of other program genres.

Television income has become central to sports budgets. The Olympic Movement provides a striking example of this phenomenon. It is surprising to recall that in the early 1970s, the International Olympic Committee (IOC) was nearly broke. Then, it conceived the policy of taking one-third of the television rights fees for its own purposes, rather than leaving it all with the host city as it had done previously. Since that decision, the IOC has enjoyed huge budgets. It is also surprising to recall that in 1978, the only cities bidding to host the Olympics were Los Angeles and Tehran, Iran, a city that was a bad risk and would soon see the overthrow of the Shah. Up to that time, cities had been incurring huge burdens of public debt in hosting the games and no one wanted them. The IOC did not like the Los Angeles bid because it replaced the host city's commitment of public financial support with private corporate sponsorship and commercialism. When the Los Angeles games of 1984 resulted in a surplus of funds from its half-billion-dollar budget, the IOC became a convert to such sponsorship and commercialization, all made possible by the global visibility of the games through television.

Nothing illustrates the growth of televised sports more vividly than the progression of fees paid to the Olympic Movement for the rights to televise the Summer and Winter Olympics. In a half century, they increased more than a thousandfold, as indicated in Table 18.1.

What does this chart tell us? First, it reveals growth from nothing to financial centrality for television in the Olympics. Secondly, it reflects the expansion of access to Olympic events in the form of live or live-to-tape audiovisual representation. Viewers across America and the world could "watch" the Olympics as never before; in fact, the Olympic media event becomes virtually compulsory viewing (Dayan and Katz, 1992). Thirdly, it explains America's increasingly dominant role within the Olympic Movement. In the beginning, the American television rights fees were 80 to 90 percent of the world rights fees. With other national broadcasters paying increasing fees, that percentage was gradually

Table **18.1** US Television Olympic rights fees, 1960–2010 (in $ millions)

	Summer Games			Winter Games		
1960	Rome	CBS	0.4	Squaw Valley	CBS	0.05
1964	Tokyo	NBC	1.6	Innsbruck	ABC	0.59
1968	Mexico City	ABC	4.5	Grenoble	ABC	2.5
1972	Munich	ABC	12.5	Sapporo	NBC	6.4
1976	Montreal	ABC	25.0	Innsbruck	ABC	0.59
1980	Moscow	NBC	85.0	Lake Placid	ABC	15.5
1984	Los Angeles	ABC	226.0	Sarajevo	ABC	91.5
1988	Seoul	NBC	305.0	Calgary	ABC	309.0
1992	Barcelona	NBC	401.0	Albertville	CBS	240.0
1994				Lillehammer	NBC	307.5
1996	Atlanta	NBC	456.0			
1998				Nagano	NBC	392.0
2000	Sydney	NBC	705.0			
2002				Salt Lake City	NBC	545.0
2004	Athens	NBC	793.0			
2006				Turin	NBC	613.0
2008	Beijing	NBC	894.0			
2010				Vancouver	TBD	

Figure 18.1 The 1960 Winter Olympics in Squaw Valley, California, were telecast in the US by CBS. Source: CBS/Library of American Broadcasting, University of Maryland.

Michael R. Real

reduced. For the four summer games from 1996 to 2008, the US portion remains slightly greater than the total for the rest of the world (IOC, 2004a). The net result is that from 1984 until 2008, the IOC signed broadcast agreements that dominated Olympic budgeting, agreements worth more than $10 billion, with the bulk of it provided by US commercial networks.

The televised Olympics selectively take on different central features and exclusions in different countries (Moragas et al., 1995), while television money provides a major financial base for every dimension of the world Olympic Movement. Approximately two-thirds of the television revenue is allocated to the local host organizing committee of the games – for example, to the host city Athens in 2004 or Beijing in 2008; the actual amount ranges from 60 to 68 percent. Approximately 10 percent goes to the International Olympic Committee and another 10 percent to the international sports federations. National Olympic committees receive about 15 percent of the fees, with more than half that going to the United States National Olympic Committee, an unofficial acknowledgement of the US role in generating the money (IOC, 2004b).

The massive global exposure that television provides for the Olympics also creates an opportunity for the Olympics to sign large sponsorship contracts with major corporations. In the four-year Olympiad period that included the 1998 Nagano Winter Games and the 2000 Sydney Summer Games, 11 official sponsors paid $303 million in cash to the Olympics and another $276 million through in-kind contracted services. That $579 million total was allocated more broadly than the television rights fees themselves. Approximately 38 percent went to the host organizing committee for the Summer Games, another 23 percent went to the host committee for the Winter Games, and 16 percent went to the National Olympic Committees, 15 percent to the United States Olympic Committee, and 8 percent went to the IOC itself (IOC, 2004c). These global Olympic sponsorships are in addition to whatever millions are spent to be an "offical sponsor" of an Olympic team, such as the US hockey team, or an Olympic broadcast in a given country, as in the catch phrase "Brought to you by Coca-Cola . . . Nissan . . . Budweiser . . . proud sponsor of this year's Olympic telecast."

Countries do not share equally in the Olympic television phenomenon, just as they do not share equally in the largesse of televised sports in general. The developed countries of North America, Europe, and parts of the Asia-Pacific region carry the largest number of hours of Olympic televised sports and also serve the largest audiences. By contrast, the developing countries of South America, Africa, and major portions of Asia carry only a fraction of the hours of Olympic television and reach much smaller audiences (Moragas, 1995; Real, 1989).

The closest parallel to the massive scale of television and the Olympics is television and the World Cup football (soccer) finals. Each quadrennia sees a massive build-up of World Cup television contracts, scheduled hours, and audience promotion in which the qualifying tournaments and the month of final play-offs increasingly capture the world's attention, culminating in what is the

354

most watched single sports event on earth, the World Cup Final. The Olympic Opening and Closing Ceremonies rival it for global audience, but the Olympics are not a single sport and there is no single final, so the World Cup Final stands alone atop global television sporting events. When Brazil or Italy or Germany emerges as the final victor, it is as the ultimate televised sports champion like no other in the world.

Critical Issues and Trends: Corporatization and Globalization

Recent debates on television and sports have centered on the issues of globalization and hyper-commercialism (Rowe, 1999). Gruneau (1999, p. 117) has observed that, "Sport in particular has been swept up in a globalizing commodity logic at a pace far in excess of what I would have predicted in the early 1980s." Similarly, Rowe (1999, p. 75) acknowledges, "There is a marked globalizing trend in media sport which makes it increasingly hard to insulate any aspect of sport and media in any particular country from external, disruptive forces." David Andrews (2003) has examined in detail the role of sport in the strategies of transnational media corporations to enter and operate within different national contexts.

As sports became increasingly important to television programming, major media corporations bought into sports and built enterprises that could exploit the full commercial potential of sport for television purposes (see Table 18.2). For example, the Walt Disney Company owns a complete array of media and entertainment subsidiaries that produced revenues in excess of $25 billion in 2000. Its assets include the ABC Television Network, television stations, radio stations, cable channels, film production and distribution companies, theatrical productions, and theme parks and resorts. Three of its major divisions include sports operations. Similarly, the Time Warner Corporation has vast holdings in film, magazines, books, cable television, online, recording labels, and television networks. Perhaps the most interesting transnational media/sports complex is that put together by Rupert Murdoch through his News Corporation and News Limited. His US holdings include the Fox television network, 34 television stations, and Direct TV satellite service (CJR, 2004).

But Murdoch's international holdings are what make him the major player in international sports media. In addition to his ownership of major newspapers in London, New York, and his native Australia, Murdoch's British-based Foxtel (40 percent Murdoch-owned) gives him control of television rights to British and European soccer, rugby league, rugby union, West Indies and Pakistani cricket, and American football. His British-based BSkyB (40 percent Murdoch-owned) gives him control of television rights to premier league soccer, boxing, auto racing, major tennis events, and British and American basketball. His German-based Vox (49.9 percent Murdoch-owned) gives him control of television rights to American football in Germany. His Australian-based Channel Seven

Table 18.2 Examples of Media conglomerates' sports-related holdings

Walt Disney	Time Warner	News Corp.
Cable Television and International Broadcasting	*Television and Magazines*	*Sport Media*
	Sports Illustrated	18 Fox Sports regional networks
ESPN Inc. (80% – Hearst Corp. = 20%), includes ESPN, ESPN2, ESPN News, ESPN Now, ESPN Extreme	12 leading sport magazines, including *Field & Stream, Golf, Ski, Snowboard Life, Transworld Skateboarding*, etc.	Madison Square Garden Network
		Fox Sports Radio Network
Classic Sports Network	TBS Superstation	*Professional Sports Franchises*
ESPN Inc. International Ventures	Turner Network Television (TNT)	Los Angeles Dodgers (MLB)
Sportsvision of Australia (25%)	*Professional Sports Franchises*	New York Rangers (NHL) & New York Knicks (NBA) 20%, with Cablevision
ESPN Brazil (50%)	Atlanta Braves (MLB)	
ESPN Star (50%) – sports programming throughout Asia	Atlanta Hawks (NBA)	Los Angeles Kings (NHL, 40%)
	Atlanta Thrashers (MLS)	
New STAR (33%) owners of The Sports Network of Canada	Turner Sports	Los Angeles Lakers (9.8%)
	Good Will Games	Staples Center (40%)
TV Sport of France	Philips Arena	
Japan Sports Channel		
Multimedia Internet		
ESPN Internet Group		
ESPN.sportzone.com		
Soccernet.com (60%)		
NFL.com		
NBA.com		
NASCAR.com		
Professional Sports		
Anaheim Sports, Inc.		
Mighty Ducks of Anaheim (NHL)		
Walt Disney World Sports Complex		

Source: CJR 2004.

(15 percent Murdoch-owned) gives him control of television rights to golf, tennis, Australian rules football, and the 1996 and 2000 Olympics. His Hong Kong-based Star TV (64 percent Murdoch-owned) gives him control of television rights to Chinese soccer, badminton, Japanese baseball, World Cup soccer, motorcycling, and table tennis.

Rowe and McKay (1999, p. 192) conclude their description of his holdings by observing: "The extent of Murdoch's global sports reach is probably the major reason why in 1995 he became the first person to top the *Sporting News* list of the one hundred most powerful people in sport for two consecutive years." It is also why Andrews (2003) examined Murdoch in "Sport and the Transnationalizing Media Corporation," in which Andrews details how Murdoch used sport as a core aspect of his transnational market entry strategy to successfully penetrate national television markets in the United States, the United Kingdom, and Australia. Murdoch also used sport to enter and manage local media markets and to redirect his corporation's organizational structure toward transnational sports.

Current analyses of telesport struggle with how the contrived presentations and transnational structures become "naturalized" in the minds of viewers and policy makers. Gruneau (1989, p. 152), among others, is concerned with how television sports treat "the existing structures and competitive promotional culture of the modern sports/media complex as natural – an example of 'common sense.'" Sports on television contribute to capital accumulation and win consent for a definition of sport that is suited to a capitalist consumer culture. He concludes,

> It is a definition in which sport is widely understood as a naturally open, achievement-based activity, conducted to further individual sports careers and to generate investment. Equally important are the notions that specialization is the modern definition of excellence, that enjoyment is tied to skill acquisition, and that economic reward is an integral and necessary component of sporting entertainment. (p. 152)

Certainly, in this view, there is something mutually parisitic in the way sport and television have changed each other into hyper-commercial vehicles over the years. The seemingly mutually beneficial symbiosis that has led to the exponential growth of each has not been without its price.

References

Andrews, D. (2003) "Sport and the Transnationalizing Media Corporation," *Journal of Media Economics*, *16*(4), 235–52.

Arledge, R. (2003) *Roone*, New York: Harper Collins.

Barnett, S. (1990) *Games and Sets: The Changing Face of Sport on Television*, London: British Film Institute.

Breen, M. and Corcoran, F. (1982) "Myth in the Television Discourse," *Communication Monographs*, *49*, 128–40.

Breen, M. and Corcoran, F. (1986) "Myth, Drama, Fantasy Theme, and Ideology in Mass Media Studies," in B. Dervin and M. Voigt (eds.), *Progress in Communication Sciences*, vol. 7, Norwood, NJ: Ablex, pp. 195–224.

Brown, L. (1971) *Television: The Business Behind the Box*, New York: Harcourt Brace Jovanovich.

Bryant, J., Brown, D., Comisky, P., and Zillmann, D. (1982) "Sports and Spectators: Commentary and Appreciation," *Journal of Communication*, *32*(1), 109–19.

Bryant, J., Comisky, P., and Zillmann, D. (1977) "Drama in Sports Commentary," *Journal of Communication*, *27*(2), 140–9.

Bryant, J., Rockwell, S. C., and Owens, J. W. (1994) "'Buzzer Beaters' and 'Barn Burners:' The Effects on Enjoyment of Watching the Game Go 'Down to the Wire,'" *Journal of Sport and Social Issues*, *18*(4), 326–39.

Bryant, J., Zillmann, D., and Raney, A. A. (1998) "Violence and the Enjoyment of Media Sports," in L. Wenner (ed.), *MediaSport*, New York: Routledge, pp. 252–65.

Bryson, L. (1994) "Sport and the Maintenance of Masculine Hegemony," in S. Birrell and C. L. Cole (eds.), *Women, Sport, and Culture*, Champaign IL: Human Kinetics, pp. 47–64.

Buscombe, E. (ed.) (1975) *Football on Television*, London: British Film Institute.

CJR (Columbia Journalism Review) (2004) "Who Owns What?" www.cjr.org/tools/owners.

Connell, R. W. (1995) *Masculinities*, Sydney: Allen and Unwin.

Cramer, J. A. (1994) "Conversations with Women Sports Journalists," in P. J. Creedon (ed.), *Women, Media, and Sport: Challenging Gender Values*, Thousand Oaks, CA: Sage, pp. 159–80.

Creedon, P. J. (ed.) (1994) *Women, Media and Sport: Challenging Gender Values*, Thousand Oaks CA: Sage.

Crosswhite, J. (1996) "Pay TV and Its Impact on Women's Sport," in R. Lynch, I. McDonnell, S. W. Thompson and K. Toohey (eds.), *Sport and Pay TV: Strategies for Success*, Sydney: School of Leisure and Tourism Studies, University of Technology, Sydney.

Cunningham, S. and Miller, T. (with D. Rowe) (1994) *Contemporary Australian Television*, Sydney: University of New South Wales Press.

Daddario, G. (1998) *Women's Sport and Spectacle: Gendered Television Coverage and the Olympic Games*, Westport, CT: Praeger.

Davis, L. R. and Harris, O. (1998) "Race and Ethnicity in US Sports Media," in L. Wenner (ed.), *MediaSport*, Newbury Park, CA: Sage, pp. 154–69.

Dayan, D. and Katz, E. (1992) *Media Events: The Live Broadcasting of History*, Cambridge, MA: Harvard University Press.

Duncan, C. D. and Messner, M. A. (1998) "The Media Image of Sport and Gender," in L. Wenner (ed.), *MediaSport*, Newbury Park, CA: Sage, pp. 170–85.

Geertz, C. (1973) *The Interpretation of Cultures*, New York: Basic Books.

Griffin, P. G. (1998) *Strong Women, Deep Closets: Lesbians and Homophobia in Sport*, Champaign, IL: Human Kinetics.

Gruneau, R. (1999) *Class, Sports, and Social Development* (Foreword by R. W. Connell), Champaign, IL: Human Kinetics.

Gruneau, R. (1989) "Making Spectacle: A Case Study in Television Sports Production," in L. Wenner (ed.), *Media, Sports, and Society*, Newbury Park, CA: Sage, pp. 134–54.

Hall, M. A. (1996) *Feminism and Sporting Bodies: Essays on Theory and Practice*, Champaign, IL: Human Kinetics.

Hall, M. A. (1997) "Feminist Activism in Sport: A Comparative Study of Women's Sport Advocacy Organizations," in A. Tomlinson (ed.), *Gender, Sport and Leisure: Continuities and Challenges*, Aachen: Meyer and Meyer Verlag, pp. 217–50.

Hall, S. (1980) "Encoding/Decoding," in S. Hall (ed.), *Culture, Media, Language*, London: Hutchinson, pp. 128–39.

Hargreaves, J. (1994) *Sporting Females: Critical Issues in the History and Sociology of Women's Sports*, London: Routledge.

Harriss, I. (1990) "Packer, Cricket and Postmodernism," in D. Rowe and G. Lawrence (eds.), *Sport and Leisure: Trends in Australian Popular Culture*, Sydney: Harcourt Brace Jovanovich, pp. 109–21.

Hornby, N. (1992) *Fever Pitch*, London: Victor Gollancz.

IOC (International Olympic Committee) (2004a) "Global Broadcast Revenue," www.olympic.org/uk/organization/facts/revenue/broadcast.

IOC (International Olympic Committee) (2004b) "Statistics on the 1998 Olympic Winter Games in Nagano. Section 6: TV Revenues," www.olympic.org/uk/organization/facts/revenue/broadcast.

IOC (International Olympic Committee) (2004c) "Statistics on the 1998 Olympic Winter Games in Nagano. Section 7: TOP Marketing Program Revenues." www.olympic.org/uk/organization/facts/revenue/broadcast.

Jhally, S. (1984) "The Spectacle of Accumulation: Material and Cultural Factors in the Evolution of the Sports/Media Complex," *Insurgent Sociologist*, *12*(3), 41–57.

Kinkema, K. M. and Harris, J. C. (1998) "MediaSport Studies: Key Research and Emerging Issues," in L. Wenner (ed.), *MediaSport*, Newbury Park, CA: Sage, pp. 27–54.

Messner, M. (1992) *Power at Play: Sports and the Problem of Masculinity*, Boston, MA: Beacon Press.

Michaels, E. (1986) "The Aboriginal Invention of Television," Presentation to the Annenberg Scholars Seminar, University of Southern California, October.

Michaels, E. (2003) "A Model of Teleported Texts (with Reference to Aboriginal Television)," in T. Miller (ed.), *Television: Critical Concepts in Media and Cultural Studies*, vol. 2, London: Routledge, pp. 103–23.

Moragas, M. (1995) *Television in the Olympics*, Luton: University of Luton Press.

Moragas, M., Rivenburgh, N. K., and Larson, J. F. (1995) *Television in the Olympics*, London: John Libbey and Company.

Morris, B. and Nydahl, J. (1985) "Sport Spectacle as Drama: Image, Language and Technology," *Journal of Popular Culture*, *18*(4), 101–10.

Rader, B. (1984) *In Its Own Image: How Television Has Transformed Sports*, New York: The Free Press.

Rainville, R. E. and McCormick, E. (1977) "Extent of Covert Racial Prejudice in Pro Football Announcers' Speech," *Journalism Quarterly*, *54*, 20–6.

Real, M. R. (1975) "Super Bowl: Mythic Spectacle," *Journal of Communication*, *25*(1) (Winter), 31–43.

Real, M. R. (1989) *Super Media: A Cultural Studies Approach*, Newbury Park, CA: Sage.

Real, M. R. (1996) *Exploring Media Culture: A Guide*, Thousand Oaks, CA: Sage.

Rowe, D. (1999) *Sport, Culture and the Media*, Buckingham, UK: Open University Press.

Rowe, D. and McKay, J. (1999) "Field of Soaps: Rupert v. Kerry as Masculine Melodrama," in R. Martin and T. Miller (eds.), *SportCult*, Minneapolis: University of Minnesota Press, pp. 191–210.

Sabo, D., Jansen, S. C., Tate, D., Duncan, M. C., and Leggett, S. (1996) "Televising International Sport: Race, Ethnicity, and Nationalistic Bias," *Journal of Sport and Social Issues*, *20*(1), 7–21.

Sandvoss, C. (2003) "Collapsing Boundaries: Football Fandom and the Public Sphere," paper presented to the 53rd annual conference, International Communication Association, San Diego, CA, May.

Tuggle, C. A., Huggman, S., and Rosengard, D. S. (2002) "A Descriptive Analysis of NBC's Coverage of the 2000 Summer Olympics," *Mass Communication and Society*, *5*(3) (August), 361–75.

Verducci, T. (2003) "Blackout: The African-American Baseball Player is Vanishing. Does He Have a Future?" *Sports Illustrated*, July 3. Accessed online.

Michael R. Real

Wearden, S. and Creedon, P. J. (1999) "'We Got Next:' Images of Women in TV Commercials during the Inaugural WNBA Season," paper presented at the national convention of the Association for Education in Journalism and Mass Communication, New Orleans, LA, August.

Wenner, L. A. and Gantz, W. (1998) "Watching Sports on Televison: Audience Experience, Gender, Fanship, and Marriage," in L. Wenner (ed.), *MediaSport*, Newbury Park, CA: Sage, pp. 233–51.

Whannel, G. (1984) "Fields in Vision: Sport and Representation," *Screen*, *25*(3), 99–107.

Whannel, G. (1992) *Fields in Vision: Television Sport and Cultural Transformation*, London: Routledge.

Zillmann, D., Bryant, J., and Sapolsky, B. (1989) "Enjoyment from Sports Spectatorship," in J. H. Goldstein (ed.), *Sports, Games, and Play: Social and Psychological Viewpoints*, 2nd edn., Hillsdale, NJ: Lawrence Erlbaum Associates, pp. 241–78.

Suggested further reading

All recommended further reading is included in the references above, with one exception. I would especially recommend the following:

Arledge, R. (2003) *Roone*, New York: Harper Collins.

Barnett, S. (1990) *Games and Sets: The Changing Face of Sport on Television*, London: British Film Institute.

Creedon, P. J. (ed.) (1994) *Women, Media and Sport: Challenging Gender Values*, Thousand Oaks, CA: Sage.

McKay, J., Messner, M., and Sabo, D. (2000) *Masculinities, Gender Relations, and Sports*, Thousand Oaks, CA: Sage.

Moragas, M., Rivenburgh, N. K., and Larson, J. F. (1995) *Television in the Olympics*, London: John Libbey and Company.

Rader, B. (1984) *In Its Own Image: How Television Has Transformed Sports*, New York: The Free Press.

Rowe, D. (1999) *Sport, Culture and the Media*, Buckingham, UK: Open University Press.

Wenner, L. (ed.) (1998) *MediaSport*, Newbury Park, CA: Sage.

Wenner, L. (ed.) (1989) *Media, Sports, and Society*, Newbury Park, CA: Sage.

Whannel, G. (1992) *Fields in Vision: Television Sport and Cultural Transformation*, London: Routledge.

CHAPTER NINETEEN

"Where the Past Comes Alive": Television, History, and Collective Memory

Gary R. Edgerton

A Different Kind of History Altogether

Those who don't understand history are doomed to repeat it.

Tony Soprano, 1999

Television specializes in odd juxtapositions. Take Anthony "Tony" Soprano (James Gandolfini), the lead character in the critically-acclaimed HBO hit, *The Sopranos* (1999–present), quoting George Santayana. Of course, Tony didn't utter his off-the-cuff version of this famous saying after having just read it in the philosopher's own writings (Santayana, 1905, p. 284). (The actual quotation is: "Those who cannot remember the past are condemned to repeat it.") Rather, he heard it first as a recurring tag line on The History Channel, a network he often watches while unwinding in his upper-middle-class suburban New Jersey home. "It means we're in the mainstream," explains Artie Scheff, chief marketing officer of The History Channel, "We have become part of pop-culture. We're Tony Soprano's favorite channel. Letterman and Leno talk about us on a regular basis" ("Making History with History," 2001). He could have also added how much Ozzy Osbourne, the 53-year-old former Black Sabbath singer, loves The History Channel, as his bemused wife Sharon teases him by referring to her husband's network of choice as "The War Channel."

In the debut episode of MTV's *The Osbournes* (March 5, 2002), for instance, Ozzy becomes panic-stricken when he is unable to operate the remote control keypad and finds himself hopelessly stuck on The Weather Channel. He is rescued by 15-year-old Jack who quickly shows Dad how easy it is to use the device. They then bond together on the sofa – father and son – watching The

History Channel. Something altogether different is obviously going on at the Osbournes, as it is in households across the country. What do the fictional Tony in the east, Ozzy in Beverly Hills, and Dave and Jay coast-to-coast, all have in common? They fit the target profile of The History Channel's core audience – upscale men 25 to 54. This is a highly coveted demographic in the television industry since it is traditionally very hard to reach. This group used to only tune in to news and sports on a regular basis. Now historical programming is very much a part of this cohort's viewing agenda.

Made-for-TV history is currently a vast enterprise, spanning feature films and television series, commercial and public networks, corporate and independent producers. The last decade, in particular, has witnessed a dramatic rise in historical programming on television screens all around the world – particularly in the United States, the rest of North and South America, Europe, and parts of Asia – mostly in the form of biographies and quasi-biographical fictional narratives and documentaries, which coincides with a marked increase of interest in history among the general population. "I think we're living in a time when history has reemerged as one of the popular forms of entertainment, and that's great," observes producer, director, and writer, Ric Burns. "It sort of slept for a couple of decades, in the '60s and '70s, and now it's really back, as it was before TV when historical novels and historical movies and historical poetry and history itself were mainstays of popular culture" (Flanagan, 1994).

Ric's brother Ken is probably the most recognizable television producer specializing in historical programming, primarily because of the unprecedented success of *The Civil War* (PBS, 1990) as well as the consistently robust showings of his other TV specials. Ken Burns actually became one of public television's busiest and most celebrated producers during the 1980s, a decade when the historical documentary held little interest for most American TV viewers. Since 1990, however, 70 million Americans have now seen *The Civil War*; 50 million have watched *Baseball* (PBS, 1994); 30 million *Jazz* (PBS, 2001); and all of his other made-for-TV histories over the last decade have averaged around 15 million viewers during their debut telecasts. The cumulative popularity of Burns's television histories is striking by virtually any measure, and he – more than anyone – has emerged as the signature figure of this far larger programming trend (Edgerton, 2001).

Histories on TV encompass much more than just documentaries, however – irrespective of Ken Burns's extraordinary success and influence as a television producer and popular historian. Made-for-TV histories also employ a wide array of news, reality, and entertainment formats. Any constructive evaluation of historical programming also needs to begin with the understanding that it is an entirely new and different kind of history. Unlike written discourse, the language of television is highly stylized, elliptical (rather than linear) in structure, and associational or metaphoric in the ways in which it portrays images and ideas. Overall, then, this chapter explores the broader parameters of made-for-TV history, describing its stylistic preferences, and proposing ten general assumptions about the nature of this widespread phenomenon. This chapter concludes with some preliminary observations concerning the enduring (if often uneasy) relationship between the proponents of popular and professional history and also the opportunities and challenges that this linkage poses for television producers and scholars alike.

Expand Globally, Program Locally

History is the new rock 'n' roll.
> Henry Becton Jr., President of WGBH Educational Foundation, 2001

Today, historical programming on TV is more popular than ever before. On October 19–22, 2001, WGBH in Boston hosted the first World Congress of History Producers, attracting nearly 400 participants from more than 20 countries to an event it cosponsored with BBC History. The four-day affair, nicknamed HISTORY 2001, and funded largely by the Banff Television Foundation, provided producers, commissioning agents, creative talent, broadcasters, and a few scholars with a wide-ranging lineup of plenary sessions, assorted panels (with titles such as "The Ethical Quagmire: Facing the Tough Questions," "Where Does News End and History Begin?," and "Biography: Hagiography or Hatchet Job?"), master classes conducted by leading producers of history and biography programs, screening opportunities, and plenty of informal networking and sales meetings. Held in the wake of September 11, most attendees agreed that "historical programming is needed more than ever" (Ramsey, 2001). In addition, they were generally upbeat about the future prospects of history on TV, especially since funding for such programming had increased 300 percent since 1993 (Stearn, 2002, p. 26).

Most conference participants enthusiastically welcomed Henry Becton, Jr.'s booster allusion comparing history to rock 'n' roll. After all, Becton is an experienced professional, having served in higher administration at WGBH since 1978, as well as twice being named a director of the Public Broadcasting Service (PBS) – from 1987 to 1993 and again from 1995 to 2001. Attendees also wanted to hear his message because they could see that HISTORY 2001 was exceeding all expectations as a trade summit, laying the groundwork for what would eventually develop into an annual world congress devoted entirely to the promotion, cultivation, and assessment of television histories.

The international dimension of HISTORY 2001 (and its successors) is especially important to recognize since *history on TV is first and foremost a global phenomenon*. Another case in point is the worldwide expansion and positioning of the A&E Television Networks (AETN), a joint venture owned by The Hearst Corporation, ABC, and NBC. Launched in 1984, AETN is the parent corporation of A&E, The Biography Channel, The History Channel, and The History

Channel International. The combined reach of AETN is currently 235 million homes telecasting in 20 languages across 70 countries. On both the domestic and international levels, "The History Channel is the fastest growing cable network ever" ("Making History with History," 2001).

Second, most television histories strongly affirm the local needs, concerns, and self-perceptions of those who are watching. Even though made-for-TV history is global in reach and popularity, it is typically produced and programmed to appeal to national, regional, and localized tastes and sensibilities – former Speaker of the House Thomas P. ("Tip") O'Neill is usually credited with the well-worn adage that "all politics is local." The same can be said of history on television. When The History Channel International was launched in November 1998, its penultimate goal was stated as being "to adapt programs to local needs, using dubbing or perhaps adding a new host" for any new affiliated region that chose to accept its signal (The History Channel, 1997). As the coverage of The History Channel International grew dramatically over the next five years, network executives also made a concerted effort to enter into a series of "joint ventures . . . acquir[ing] locally produced programs" from participating nations to "fill out the rest" of its 24-hour 7-day-a-week schedule (Grele, 2000; The History Channel, 1997; "Making History with History," 2001). This careful attention to the expectations and desires of its rapidly expanding audience base facilitated the quick adoption of The History Channel International on continents as widely diverse in cultural orientation as Europe, South America, Asia, and Australia.

Third, television is the principal means by which most people learn about history today. TV must be understood (although it seldom is) as the primary way that children and adults form their understanding of the past. Just as television has profoundly affected and altered every aspect of contemporary life – from the family to education, government, business, and religion – the medium's fictional and non-fictional portrayals have similarly transformed the way tens of millions of viewers think about historical figures and events. Most people, for example, recall the first Persian Gulf War and the more recent War with Iraq through the lens of television, just as their frame of reference regarding slavery has been deeply influenced by TV mini-series, such as the fictional *Roots* (ABC, 1977) and the non-fictional *Africans in America* (PBS, 1998), along with cinematic portrayals, such as *Amistad* (1997), which characteristically has been seen by more people on TV than in movie theaters.

Fourth, history on television is now big business. There are over 250 broadcast and cable networks in America alone, and roughly 98 percent of these services resulted from the dramatic rise of cable and satellite TV over the last 25 years. Scores of cable networks have become closely identified with documentaries as a profitable staple of their weekly schedules. As one veteran producer, Kate Coe, explains, "[t]elevision today is awash in nonfiction programming. A&E, the Discovery Channel, the History Channel, Bravo, Oxygen, and dozens more present hours of documentary programs . . . [o]n any given day, as many as 2.5 million viewers tune into these [individual] shows" (Coe, 2003, p. C3). Moreover, ever since "Ken Burns's *The Civil War* proved that history on TV could be engaging – and attract millions of viewers," historical "documentaries [have been] all over the dial" (Gabler, 1997, p. 18). In 2002 *Variety* even reported that history "is now the preferred fare not just for pubcasters (i.e. public broadcasters) with a mission to educate, but also for commercial channels, and thematic cable and satellite outlets" (Johnson, 2002).

The proliferation of historical programming on TV is centered primarily on three fundamental business and economic reasons. To begin with, cable television's emphasis on "narrowcasting – specialized avenues for specialized tastes – is an area in which the documentary can thrive" (Natale, 1992). Next, non-fiction is relatively cost-effective to produce when compared to most fictional programming (according to the latest estimates, per-hour budgets for a dramatic TV episode now exceed $1 million, while documentaries average $500,000 for prestige productions, $300,000 for reality-based programs and dramatic recreations, and $150,000 for standard non-fictional fare at networks such as those owned by AETN). Even more significantly, though, many of these shows which have some historical dimension are just as popular with audiences as sitcoms, hour-long dramas, and movie reruns in syndication (Bellafante, 1997; Johnson, 2002; Katz, 1999; Mahler, 1997; Romano, 2003).

Fifteen biographical programs are currently thriving on US television, with a half-dozen more already in preparation (Poniewozik, 1999; Lafayette, 2003). Most of these existing series are also among the most watched shows on their respective networks. The forerunner and acknowledged prototype is A&E's *Biography* (1987–present) with an average nightly viewership of nearly 3 million, spawning videotapes, CDs, a magazine called *Biography* with a 2 million readership, and The Biography Channel. The *Biography* franchise celebrated its 15th anniversary with its 1,000th episode in 2002. The index of historical (and contemporary) individuals and couples featured on *Biography*, from Thomas Jefferson to Jackie Robinson to John Travolta and Condeleeza Rice, is sweeping and diverse. At the same time, this series typically relies on highly derivative stylistics which are a pastiche of techniques borrowed from TV news, prime-time dramatic storytelling, and PBS non-fiction *à la* Ken Burns. All told, A&E's *Biography* is a representative example of how history is often framed in highly conventional and melodramatic ways on TV, mainly to be marketed and sold directly to consumers around the world as a commodity.

Fifth, history on TV is also subject to the same kinds of generic influences that are affecting the rest of television at any given moment of time. For example, the first historical reality series, *The 1900 House*, was produced in Britain during the fall of 1999, during exactly the same television season that *Big Brother* debuted in Holland. This so-called "living history" program, a co-production of the UK's Channel 4 and PBS's Thirteen/WNET in New York, was based on the guilty pleasure of having viewers observe a contemporary family adapt to and interact in a setting that approximates the accommodations and furnishings of a turn-of-the-century home. The voyeuristic appeal of watching *The 1900 House* shared

Gary R. Edgerton

much in common with *Big Brother*, as did the soap opera nature of the action that ensued. *The 1900 House* was so successful on both sides of the Atlantic, in fact, that it spawned similar PBS telecasts, such as *Frontier House* (2002), *1940s House* (2002), *Manor House* (2003), and *Colonial House* (2004). Not surprisingly, objections to these "you are there" made-for-TV histories have appeared in both popular and scholarly journals. As one British scholar concluded, "'[r]eality history' may be entertainment, but it is neither reality nor history" (Stearn, 2002, p. 27).

Cable television also jumped on the "living history" bandwagon. *Variety* started referring to a "'Survivor' after-effect," adding that "popular network reality shows have brought a new audience to 'really real' nonfiction programming on cable" (McDonald, 2001). The most recent example of this growing tendency is *Extreme History with Roger Daltrey*, which debuted on The History Channel in the fall of 2003. This half-hour series capitalizes on the strategy of marketing history alongside rock 'n' roll by casting of a well-known pop star as the show's featured host. A network press release even describes "Roger Daltrey, [the] lead singer of the legendary rock band The Who . . . [as] an avid history buff, [who] goes on location to demonstrate the challenge of surviving history's epic adventures, explorations, and battles" (The History Channel, 2003).

Episodes include Daltrey scaling the Montana Rockies like Lewis and Clark in 1805; driving steers through the Chisholm Trail of Texas and Oklahoma; and shooting the Colorado rapids in a wooden rowboat much like John Wesley Powell did in 1869. Daltrey's exploits as a celebrity surrogate reenacting a pre-fabricated historical narrative epitomizes The History Channel's branding claim that it is the niche network – "Where the Past Comes Alive." Reality histories such as *Extreme History with Roger Daltrey* also illustrate the ongoing negotiation between popular generic trends, commercial imperatives, and historicity that is always a part of producing historical programming on TV.

Sixth, the technical and stylistic features of television as a medium strongly influence the kinds of historical representations that are produced. History on TV tends to stress the twin dictates of narrative and biography which ideally expresses television's penchant toward personalizing all social, cultural, and, for our purposes, historical matters within the highly controlled and viewer involving confines of a well-constructed plot structure. The scholarly literature on television has established intimacy and immediacy (among other aesthetics) as intrinsic properties of the medium (Adler, 1981; Allen, 1987; Bianculli, 1992; Fiske and Hartley, 1978; Newcomb, 1974; Newcomb, 2000). In the case of intimacy, for instance, the confines of the relatively smaller TV screen which is typically watched within the privacy of the home environment have long ago resulted in an evident preference for intimate shot types (i.e. primarily close-ups and medium shots), fashioning most fictional and non-fictional historical portrayals in the style of personal dramas or melodramas played out between a manageable number of protagonists and antagonists. When successful, audiences closely identify with the historical "actors" and stories being presented, and, likewise, respond in intimate ways in the privacy of their own homes.

Television's immediacy usually works in tandem with this tendency toward intimacy. Both TV and film are incapable of rendering temporal dimensions with much precision. They have no grammatical analogs for the past and future tenses of written language and, thus, amplify the present sense of immediacy out of proportion. The illusion created in television watching is often suggested by the cliché, "being there," which is exactly what David Grubin, celebrated producer of such presidential documentaries as *LBJ* (1992), *FDR* (1994), *TR, The Story of Theodore Roosevelt* (1996), *Truman* (1997), and *Abraham and Mary Lincoln: A House Divided* (2001), is talking about when he says, "you are not learning about history when you are watching . . . you feel like you're experiencing it" (Grubin, 1999). Made-for-television histories, in this regard, are best understood as personifying Marshall McLuhan's eminently useful – though often misunderstood – metaphor, "the medium is the message."

Seventh, the improbable rise and immense popularity of history on TV is also the result of its affinity and ability to embody current concerns and priorities within the stories it telecasts about the past. Television's unwavering allegiance to the present tense is not only one of the medium's grammatical imperatives, it is also an implicit challenge to one of the traditional touchstones of academic history. Professional historians have customarily employed the rigors of their craft to avoid presentism as much as possible, which is the assumption that the past is being judged largely by the standards of the present. The revisionist work of postmodernist historians like Hayden White (1975) has challenged this principle in academic circles. White and others have argued that historiography is much more about telling stories inspired by contemporary perspectives, than it is concerned with recapturing and conveying any kind of objective truth about the past (Ermarth, 1992; Hutcheon, 1988; White, 1985). This alternative scholarly outlook has gained increased momentum in some quarters over the last generation, even calling into question whether or not there is an authentic, knowable history at all beyond the subjectivity of the present. Most popular historians for their part, such as television producers and filmmakers, take this postmodernist viewpoint one step further. They tacitly embrace presentism through the back door by concentrating only on those people, events, and issues that are most relevant to themselves and their target audiences.

The mid-1990s revising of the "prime-time Indian," ranging from fictions (e.g. CBS's *Dr. Quinn, Medicine Woman*, 1993–98) to docudramas (e.g. TNT's *Crazy Horse*, 1996) to documentaries (e.g. Kevin Costner's *500 Nations*, 1995; Ric Burns's *The Way West*, 1995; Ken Burns and Stephen Ives's *The West*, 1996), along with literally dozens of other programming examples, is a telling case in point. Televised (and filmic) representations from a decade ago largely employed Native American characters as emblems for a wide assortment of mainstream multicultural, environmental, and New Age spiritual concerns, rather than reconstructing the old small-screen stereotypes primarily on the basis of the existing historical record (Bird, 1996; Rollins and O'Connor, 1998). Television histories, in general, are less committed to rendering a factually accurate depiction as their highest priority, than animating the past for millions by accentuating those matters that are most relevant and engaging to audiences in the present. This preference, on the most elementary level, is commercially motivated, often resulting in an increasing number of viewers. In a deeper vein, though, the goal of most popular historians is also to utilize aspects of the historical account as their way of making better sense out of current social and cultural conditions.

Horace Newcomb recognized this tendency 30 years ago in his seminal article, "Toward a Television Aesthetic," when he identified a special sense of history as one of the representative characteristics of TV programming. Newcomb wrote that the "television formula requires that we use our contemporary historical concerns as subject matter . . . we [then] tak[e these] concern[s] and place [them], for very specific reasons, in an earlier time [when] values and issues are more clearly defined [and] certain modes of behavior . . . more permissible" (Newcomb, 1974, pp. 258–9). In contrast, professional historians regularly take issue with TV's application of presentism as a guiding principle. What is lost, they argue, is the fuller historical picture, or that part of the past that is most unlike the present, but is nonetheless a vital component of the way things actually were.

Eighth, TV producers and audiences are similarly preoccupied with creating a "useable past," a long-standing tenet of popular history, where stories involving historical figures and events are used to clarify the present and discover the future. There is a method behind the societal self-absorption implied by presentism. Ken Burns's *The Civil War*, for example, attracted nearly 40 million viewers during its initial telecast in September 1990. Much of this documentary's success must be equated with the way in which Burns's version of this nineteenth-century conflict, stressing the personal ramifications of the hostilities, makes the war comprehensible to a large contemporary audience.

Overall, this series addresses a number of current controversies which reflect the shifting faultlines in the country's underlying sense of itself as a national culture, including the questions of slavery, race relations, and continuing discrimination; the rapidly changing roles of women and men in society; the place of federal versus local government in civic affairs; and the individual struggle for meaning and conviction in modern life. In this way, *The Civil War* as useable past is an artistic attempt to better understand these enduring public issues and form a new consensus around them, serving also as a validation for the members of its principal audience (which skewed older, white, male, and upscale in the ratings) of the importance of their past in an era of unprecedented multicultural redefinition (Statistical Research Incorporated, 1990).

Ninth, collective memory is the site of mediation where professional history must ultimately share space with popular history. Interdisciplinary work in memory studies now boasts adherents in American studies, anthropology, communication, cultural studies, English, history, psychology, and sociology (Fussell, 1989;

Kammen, 1993; Le Goff, 1996; Lewis, 1975; Lipsitz, 1990; Schudson, 1992; Zelizer, 1992, 1998). The contemporary preoccupation with memory dates back to Freud, although recent scholarship focuses more on the shared, collective nature of remembering, rather than the individual act of recalling the past which is the customary realm of psychological inquiry into this topic area. Researchers today, most importantly, emphasize how collective memory "exists in the world rather than in a person's head, and so is embodied in different cultural forms" (Zelizer, 1995, p. 232). It "is, above all, archival," explains Pierre Nora, "[i]t relies on the materiality of the trace, the immediacy of the recording, the visibility of the image" (1989, p. 13).

For their part, professional historians "have traditionally been concerned above all else with the accuracy of a memory, with how correctly it describes what actually occurred at some point in the past" (Thelen, 1989, p. 119). "Less traditional historians have [recently] allowed for a more complex relationship, arguing that history and collective memory can be complimentary, identical, oppositional, or antithetical at different times" (Zelizer, 1995, p. 216). According to this way of thinking, more popular uses of memory have less to do with accuracy per se, than using the past as a kind of communal, mythic response to current controversies, issues, and challenges. The proponents of memory studies, therefore, are most concerned with how and why a remembered version is being constructed at a particular time, such as the aforementioned *The Civil War* in 1990, than whether a specific rendition of the past is historically correct and reliable above all else.

Rather than think of professional and popular history as diametrically opposed traditions (i.e. one more reliable and true; the other unsophisticated and false), it is perhaps more helpful to consider them as two ends of the same continuum. In his 1984 book, *Culture as History*, the late Warren Susman first championed this more sympathetic appreciation of the popular historical tradition. Susman noted that myth and history are intimately linked to each other. One supplies the drama; the other the understanding. The popular heritage holds the potential to connect people passionately to their pasts; the scholarly camp maps out the processes for comprehending what actually happened with richness and depth. Susman's fundamental premise was that popular history and professional history need not always clash at cross-purposes. Together they enrich the historical enterprise of a culture, and the strengths of one can serve to check the excesses of the other.

Many subsequent scholars from a wide variety of disciplines have concurred with Susman's basic thesis and continued to deepen his arguments in the intervening years. In his widely acclaimed book, *The Noble Dream* (1988), Peter Novick has skillfully examined the controversies that have fundamentally affected history as a field of study over the last generation. Current debates continue in the literature and at conferences concerning the relative merits of narrative versus analytic history, synthetic versus fragmentary history, and consensus versus multicultural history (Novick, 1988). Within this context, popular history

Gary R. Edgerton

and professional history are seen less as discrete traditions, and more as overlapping parts of the same whole, despite the many tensions which still persist. For instance, popular histories can nowadays be recognized for their analytical insights, while professional histories can similarly be valued for their expressive possibilities. Susman succinctly summed up this more inclusive vision with his often quoted affirmation: "History, I am convinced, is not just something to be left to historians" (1984, p. 5). He, of course, wrote this belief while also taking for granted that scholars were already essential to historical activity and would continue to be so in the future.

Finally, the flip side of presentism is pastism (a term coined by historian, Joseph Ellis) which refers to the "scholarly tendency to declare the past off limits to nonscholars" (Ellis, 1997, p. 22). Robert Sklar perfectly captured this long-standing bias in the context of "film and history" with his metaphor, "historian-cop," which alludes to the tone of policing that usually emerges whenever professional historians apply the standards they reserve for scholarly books and articles to motion pictures. In this specific instance, Sklar calls for a greater awareness of both the production and reception processes of filmmaking as a way of better appreciating how these more encompassing frameworks influence what audiences actually see and understand as history on the screen (Sklar, 1997).

Made-for-TV history is an even more tempting and incendiary target than film and history for the proponents of pastism, especially since its impact and popularity with the general public far outstrips anything that can ever be achieved in theaters. As a result, television histories are sometimes rejected out of hand for either being too biographical or quasi-biographical in approach, or too stylized and unrealistic in their plot structures and imagery. Occasionally, these criticisms are well-founded; historical programming certainly furnishes its share of honest "failures" or downright irresponsible and trashy depictions of the past. Other times, though, history on TV delivers ably on its potential as popular history, having even gained a degree of support in academe and increasing interest in the scholarly literature since the 1980s, no doubt reflecting the growing desire among many professional and popular historians to finally reconcile each other's traditions in a mutually respectful, if still cautious working relationship (Edgerton and Rollins, 2001; Ludvigsson, 2003; McArthur, 1978; O'Connor, 1983, 1988, 1990; Sobchack, 1996; Toplin, 1996).

Memory Makes for Strange Bedfellows

History is stuck with television as the primary mediator of memory.
Andrew Hoskins (2001, p. 345)

The mutual skepticism that sometimes surfaces between professional and popular historians is both understandable and unfortunate. Each usually works with different media (although some scholars do produce historical TV programs, videos, and films); each tends to place a dissimilar stress on the respective roles of analysis versus storytelling in relaying history; and each tailors a version of the past which is designed for disparate – though overlapping – kinds of audiences. These distinctions are real enough. Still the scholar and the artist, the expert and the amateur, can complement each other more than is sometimes evident in the expressions of suspicion, defensiveness, and even, on occasion, contempt, that too often arise on both sides.

The popular history tradition is actually as old as the historical impulse itself. The first historians, dating back to the ancient Hebrews and Greeks, were poets and storytellers; and their original approach to the past was to marshal whatever evidence and first-person stories they could into an all-inclusive historical epic. This master narrative was typically populated by heroes and villains who allegorically personified certain virtues and vices in the national character which most members of the general population recognized and responded to immediately. Television as popular history still adopts facets of this strategy at its most rudimentary level, although our small-screen morality tales about the past are far more seamless and sophisticated in their construction, thus rendering these formulaic elements invisible to most contemporary viewers.

Popular history is essentially artistic and ceremonial in nature. In the case of television histories, the act of producing, telecasting, and viewing historical programming becomes a large-scale cultural ritual in and of itself. This process, in turn, completes a number of important functions: it organizes together various viewing constituencies into a web of understandable relations which are defined mostly by their differing identities and positions of power; it loosely affirms majoritarian standards, values, and beliefs; and it facilitates a society's ongoing negotiation with its past by portraying those parts of the collective memory that are most relevant at any given time to the producers of these programs as well as the millions of individuals who tune them in.

Professional history, in contrast, is resolutely scientific and empirical in orientation. It developed gradually over the second half of the nineteenth century, mainly in reaction to the 2,500-year legacy of popular history. This new scholarly tradition recast the study of history inside the increasingly respectable and rigorous mold of science with its principal attachments to systematic inquiry, objectivity, and the pursuit of new knowledge. In effect, professional history rejected the obvious mythmaking of popular history and adopted a more modern and disciplined method of gathering historical facts and then testing and cross-checking them for validity and reliability.

By the turn of the twentieth century, history had become institutionalized as a full-fledged occupation in colleges and universities. Professional historians pioneered a wide array of specialty areas which they examined as impartially as they could, aspiring for a detached and truthful rendering of their subjects, independent of all personal tastes and biases. The ideal of objectivity has been modified considerably since the 1960s to take into account the inevitability that both scholars and their facts always come with very definite points of view. From the vantage point of a new century, moreover, the subjective excesses of popular history appear less like a difference in kind than a matter of degree, when compared against the ideological exuberance of contemporary scholarship.

Most surprisingly, America's pre-eminent examples of popular history currently originate on prime-time television, encompassing the full spectrum of actual and fictionalized presentations. Live "media events," such as TV coverage of the Kennedy assassination, Vietnam, Watergate, the Challenger disaster, the fall of the Berlin Wall, Desert Storm, the O. J. Simpson trial, the funeral of Princess Diana, the Clinton scandal, and the recent War in Iraq, "are in competition with the writing of history in defining the contents of collective memory" (Dayan and Katz, 1992, p. 213). "Early in the [twentieth] century, we thought history was something that happened temporally 'before' and was represented temporally 'after' us and our personal and immediate experience," recounts Vivian Sobchack. "Today, history seems to happen right now – is transmitted, reflected upon, shown play-by-play, taken up as the stuff of multiple stories and significance, given all sorts of 'coverage' in the temporal dimension of the present as we live it" (1996, p. 5).

The collective memory of 9/11, to cite yet another obvious example, is indistinguishably linked with the way in which this event and its aftermath was telecast continuously over four straight days to worldwide audiences numbered in the hundreds of millions. Viewer attention was effectively channeled into familiar narrative patterns featuring heroic public servants and villainous foreign terrorists. These slowly unfolding storylines were further enhanced by the shocking repetitive power of seeing the two World Trade Center towers burning and finally collapsing time and again. TV, therefore, transformed 9/11 into "instant history" by taking what was essentially a localized New York City catastrophe and turning it into a global media event with the whole world bearing witness (Dayan and Katz, 1998).

"In this sense, television act[ed] as an agent of history and memory, recording and preserving representations to be referenced in the future" (White and Schwoch, 1997, p. 771). In subsequent months and years, real-life footage from 9/11 has been regularly incorporated into numerous network documentaries produced both here and abroad, while fictionalized scenes of domestic terrorism have appeared on such widely diverse entertainment programs as *Law & Order* (NBC, 1990–present), *JAG* (CBS, 1995–present), *The West Wing* (NBC, 1999–present), *Third Watch* (NBC, 1999–present), and *Star Trek: Enterprise* (UPN, 2001–present), among many other series. "As historians who focus on popular memory have [long] insisted, we experience the present through the lens of the past – *and* we shape our understanding of the past through the lens of the present" (Rosenberg, 2003, p. B13).

Just as TV sometimes pre-empts the authority of professional historians in determining what exactly should be considered historic, scholars are likewise crossing over into the public sphere of popular history more than ever before.

Academics were well established as expert commentators on The History Channel starting from the network's first season in 1995. Today scholars are even finding themselves cast as featured players. PBS, for instance, premiered *History Detectives* in the summer of 2003 with four leads – two independent appraisers and two Ivy League professors – who all work independently on his or her own 20-minute case study to uncover the hidden history behind a found or purchased cultural artifact (e.g. Did Mark Twain once own the gold pocket watch that now belongs to Jack Mills of Portland, Oregon?); or a legend passed down about a famous historical figure (e.g. Did Ulysses S. Grant really sign the firehouse guestbook in Morristown, New Jersey, on the country's centennial, July 4, 1876?); or a city landmark named after a one-time founding father who is now all but forgotten (e.g. Why was the baseball stadium in Atlantic City named after Pop Lloyd?). *History Detectives* is an amalgam of dramatic and investigative techniques gleaned from shows as popular and different as *Antique Roadshow* (PBS, 1997–present), *CSI: Crime Scene Investigation* (CBS, 2000–present), and *60 Minutes* (CBS, 1968–present). The intent of this series is to go well beyond The History Channel's usual preoccupation with war, adventure, and politics by emphasizing the value and relevance of a bottom-up view of the past in the daily lives of the program's viewing audience.

As cooperation between professional and popular historians has increased over the last decade, a few select scholars have even emerged as modest television stars who are able to influence programming decisions in their own right. A case in point is Simon Schama, an art history and history professor at Columbia University, who wrote and presented the 16-part mini-series, *A History of Britain* (2000–2), based on his bestselling trilogy of books of the same name (Schama, 2000, 2001, 2002a). *A History of Britain*, co-produced by BBC Television and The History Channel, exhibits both Schama's exceptional dramatic talents at being able to design extended historical narratives for TV as well as his lively on-screen presence as a learned guide setting the appropriate context and sharing colorful anecdotes and asides with the viewer. This particular mini-series was such a success in the international media marketplace that Schama signed a "$4.6 million book and TV deal" in the summer of 2002 for his next three works: The first "'Rough Crossings' will examine Anglo-American relations; the second, 'Brushes with Death,' will deal with art when the artist is in crisis; and the third will focus on Hawaii and the notion of a tropical paradise" (Johnson, 2002).

Schama's approach to made-for-TV history combines the storytelling accessibility of the popular historian with the detailed rigor of his scholarly training and background as a professional historian. His most recent television history, entitled *Murder at Harvard* (2003), and produced under the auspices of PBS's flagship series *American Experience*, once again illustrates his unique ability to bridge both historical traditions without shortchanging either one. Based on the second half of his 1991 book, *Dead Certainties: Unwarranted Speculations*, *Murder at Harvard* retells the notorious tale of one of the most sensational murder trials in American history. In 1850 a Harvard chemistry professor, John White Webster, was tried and eventually hanged for supposedly killing Dr. George Parkman, a Boston Brahman from whom he had borrowed money but to whom he was unable to repay the loan. Schama fleshes out these two historical figures, along with several others, by re-evaluating all the known facts, the trial transcripts, the newspaper reports, and the mixed motives of each principal character to see if he can finally solve the still controversial homicide some 150 years later.

In raising the matter (and the dead) in *Murder at Harvard*, Schama and his collaborators, producers Melissa Banta and Eric Stange, recreate a number of key scenes complete with invented dialogue and presumed interactions between the major players. "I knew I was crossing a line historians don't usually cross," confides Schama who speaks often and directly to the audience throughout the docudrama, "the line that separates history from fiction. I felt free to let my imagination work to get me closer to the truth." Besides the murder mystery, moreover, this film also becomes an exploration into how historical methods are utilized and the past reconstructed. As Schama reveals at a later point in *Murder at Harvard*: "Maybe I thought what I was after was not a literal documentary truth, but a poetic truth – an imaginative truth – and for that I was going to have to become my own Resurrection man, I was going to have to make these characters live again."

No matter how enriched and tempered by scholarly knowledge and expertise, taking such poetic license is clearly more the province of popular rather than professional history. Schama, Banta, and Stange openly enlist the on-screen opinions of five distinguished historians – Pauline Maier, Ronald Story, Karen Halttunen, James Goodman, and Natalie Zemon Davis – to examine the historiographic implications of their approach in *Murder at Harvard*. Toward the beginning of the docudrama, for instance, Pauline Maier argues that "[h]e [Schama] is not writing a whodunit. He's trying to deal with a more philosophical issue, and that is how do we know about the past." As part of the film's conclusion, Natalie Zemon Davis adds:

> The historians' fictionalizing can help him or her ask new questions about his evidence, questions that might never have come up before. When you're trying to put yourself fully in the mind of your actors and see them moving through the streets of Boston, for instance, or moving through a trial, you suddenly think about things that never occurred to you before. You might even then be able to go back to the evidence and find the answers.

Such self-reflexivity in a made-for-television history suggests the increasing depth and potential of this programming genre and also the growing sophistication of the audiences it attracts. Even as historical films and TV series are becoming more accepted by the historical establishment, residual resistance still remains in the more traditional wing of the discipline. One published critique of *Murder at Harvard*, for instance, reflects this ongoing bias: "[F]or the film to have succeeded as a meditation on historical truth, it would have needed a third plot line: a discussion by the filmmakers of how to present the double stories of murder and history. They might have, for example, interspersed footage of applying for grants, writing a script, auditioning actors, or deciding what material to cut" (Masur, 2003). In response to such impractical criticisms, Simon Schama asserts that "[u]nderlying many of these complaints against the possibility of serious television history, given that the subject is to be left to bungling (as it is implied) 'amateurs'. 'Real' history is, apparently, the monopoly of the academy" (Schama, 2002b). Robert Toplin concurs that "[a] great deal of ink and airtime are wasted on angry indictments of cinematic history for engaging in practices of the genre or for inventing and manipulating evidence. These criticisms would not seem irrelevant if they were framed with an understanding of the way Hollywood drama works" (Toplin, 2002, pp. 201–2).

Despite these lingering tensions, the highly dynamic relationship between scholars and television producers these days feature three principal patterns of interaction: First of all, television histories are built upon the foundation of academic scholarship. They are essentially synthetic in nature and should not be judged on whether or not they generate new knowledge, as much as on how creatively and responsibly they shed additional light on the existing historical record. According to David Grubin in describing his own Emmy Award-winning work: "Historical documentary is a kind of poetry resting on a foundation of fact" (Grubin, 1997, p. 14). Second, professional historians are more involved than ever in the production processes of many television histories. They characteristically influence, but rarely control the end products of such programming. Third and lastly, television histories frequently provide professional historians with opportunities to introduce their scholarly ideas and insights to much larger audiences. Too often, made-for-TV histories are hastily misperceived as the last word on any given topic, simply because of the unprecedented power and influence of television as a medium. Rather than being definitive, television histories are probably best understood as dramatic alternatives to the many published histories that exist within a general subject area.

Overall, television producers as historians typically reverse the usual academic hierarchy, trusting first the lessons found in art (i.e. storytelling, video aesthetics, film clips, photography, period music, etc.), before turning to the scholarly record to fill in the details of their more public visions of history. This is admittedly a speculative approach; but then again, popular and professional historians alike are all amateurs when it comes to detecting the human traces of lives once lived among the emotional resonances of the past. In the final analysis, made-for-television histories enable unprecedentedly large audiences to become increasingly aware of and intrigued by the stories and figures of the past, spurring some viewers to pursue their newfound historical interests beyond the screen and into other forms of popular and professional history.

Gary R. Edgerton

References

Adler, R. P. (ed.) (1981) *Understanding Television: Essays on Television as a Social and Cultural Force*, New York: Praeger.

Allen, R. C. (ed.) (1987) *Channels of Discourse: Television and Contemporary Criticism*, Chapel Hill, NC: University of North Carolina Press.

Bellafante, G. (1997) "These are Their Lives," *Time*, March 17, p. 67.

Bianculli, D. (1992) *Teleliteracy: Taking Television Seriously*, New York: Continuum.

Bird, S. E. (1996) *Dressing in Feathers: The Construction of the Indian in American Popular Culture*, Boulder, CO: Westview.

Coe, K. (2003) "On the Prowl for Telegenic Experts," *The Chronicle of Higher Education*, August 1, C3–C4.

Dayan, D. and Katz, E. (1992) *Media Events: The Live Broadcasting of History*, Cambridge, MA: Harvard University Press.

Dayan, D. and Katz, E. (1998) "Political Ceremony and Instant History," in A. Smith and R. Paterson (eds.), *Television: An International History*, 2nd edn., New York: Oxford University Press, pp. 97–106.

Edgerton, G. (2001) *Ken Burns's America*, New York: Palgrave for St. Martin's Press.

Edgerton, G. and Rollins, P. C. (eds.) (2001) *Television Histories: Shaping Collective Memory in the Media Age*, Lexington, KY: University of Kentucky Press.

Ellis, J. J. (1997) *American Sphinx: The Character of Thomas Jefferson*, New York: Knopf.

Ermarth, E. D. (1992) *Sequel to History: Postmodernism and the Crisis of Representational Time*, Princeton, NJ: Princeton University Press.

Fiske, J. and Hartley, J. (1978) *Reading Television*, New York: Methuen.

Flanagan, J. (1994) "The Other Burns Comes Back East: With *The Way West*," *Current*, June, retrieved at http://www.current.org/hi411.html.

Fussell, P. (1989) *The Great War and Modern Memory*, New York: Oxford.

Gabler, N. (1997) "History's Prime Time," *TV Guide*, August 23, p. 18.

Grele, R. J. (2000) "An Interview with Charles Maday, Jr. of The History Channel," *Organization of American Historians Newsletter*, November, retrieved at http://www.oah.org/pubs/nl/2000nov/mayday.html.

Grubin, D. (1997) "From Story to Screen: Biography on Television," *Humanities*, May/June, 11–15.

Grubin, D. (1999) "Documentaries and Presidents," (ID#123826), C-SPAN 1 (telecast on July 4, 1999; appearance recorded on April 6, 1999), 81 minutes.

The History Channel (2000) "*Extreme History with Roger Daltrey*: Surviving History's Epic Challenges . . . One Day at a Time," *Press Release*, July 14, 2 pages.

"The History Channel: Making the Past Come Alive," *Video Age International* (1997) *17*(4, March–April), 22–3.

Hoskins, A. (2001) "New Memory: Mediating History," *Historical Journal of Film, Radio and Television*, *21*(4) (October), 333–46.

Hutcheon, L. (1988) *A Poetic of Postmodernism: History, Theory, Fiction*, New York: Routledge.

Johnson, D. (2002) "Docus Make History," *Variety*, August 12, p. 1.

Kammen, M. (1993) *Mystic Chords of Memory: The Transformation of Tradition in American Culture*, New York: Vintage.

Katz, R. (1999) "Bio Format Spreads Across Cable Webs," *Variety*, August 2–8, pp. 23, 27.

Lafayette, J. (2003) "DeBitetto Brings New Script to Drama at A&E Network Has Plans for Shoring Up 'Biography' Franchise," *CableWorld*, January 13, retrieved at http://www.kagan.com/archive/cableworld/2003/01/13/cwd03011308.shtml.

Le Goff, J. (1996) *History and Memory: European Perspectives*, New York: Columbia University Press.

Lewis, B. (1975) *History: Remembered, Recovered, Invented*, Princeton, NJ: Princeton University Press.

Lipsitz, G. (1990) *Time Passages: Collective Memory and American Popular Culture*, Minneapolis: University of Minnesota Press.

Ludvigsson, D. (2003) *The Historian-Filmmaker's Dilemma: Historical Documentaries in Sweden in the Era of Häger and Villius*, Uppsala, Sweden: Uppsala University Press.

McArthur, C. (1978) *Television and History*, London: British Film Institute.

McDonald, K. A. (2001) "Truth is as Strong as Fiction on Tube," *Variety*, July 25, p. 1.

Mahler, R. (1997) "Reality Sites," *The Hollywood Reporter*, Nonfiction Special Issue, April 8, pp. N8–N9.

"Making History with History," *Reveries* (2001) March, retrieved at http://www.reveries.com/reverb/media/scheff.

Masur, L. P. (2003) "Television: History or Fiction?," *The Chronicle of Higher Education*, July 11, p. B15.

Natale, R. (1992) "Cable Carries the Day (and Night) for Docus," *Variety*, October 21, p. 14.

Newcomb, H. (ed.) (2000) *Television: The Critical View*, 6th edn., New York: Oxford University Press.

Newcomb, H. (1974) *TV: The Most Popular Art*, New York: Anchor.

Nora. P. (1989) "Between Memory and History: Les Lieux de Memoire," *Representations*, 26 (Spring), 7–25.

Novick, P. (1988) *That Noble Dream: The "Objectivity Question" and the American Historical Profession*, Cambridge: Cambridge University Press.

O'Connor, J. E. (1988) *Teaching History With Film and Television*, revised edn., New York: American Historical Association.

O'Connor, J. E. (1990) *Image as Artifact: The Historical Analysis of Film and Television*, Malabar, FL: Krieger.

O'Connor, J. E. (ed.) (1983) *American History/American Television: Interpreting the Video Past*, New York: Ungar.

O'Neill, T. P. with Hymel, G. (1994) *All Politics is Local and Other Rules of the Game*, New York: Times Books.

Poniewozik, J. (1999) "Bio Sphere," *Time*, August 23, pp. 62–6.

Ramsey, A. (2001) "History Producers Hear the Call in Boston," *ICOM: Film & Video Production & Post-production Magazine*, December, retrieved at http://www.icommag.com/december2001/history_prod.html.

Rollins, P. C. and O'Connor, J. E. (eds.) (1998) *Hollywood's Indian: The Portrayal of the Native American in Film*, Lexington, KY: University of Kentucky Press.

Romano, A. (2003) "History Channel Reads Some Good Nielsen Books," *Broadcasting & Cable*, June 9, p. 30.

Rosenberg, E. S. (2003) "September 11, Through the Prism of Pearl Harbor," *The Chronicle of Higher Education*, December 5, pp. B13–B14.

Santayana, G. (1905) *Life of Reason, Reason in Common Sense*, New York: Scribner's.

Schama, S. (1991) *Dead Certainties: Unwarranted Speculations*, New York: Random House.

Schama, S. (2000) *A History of Britain: At the Edge of the World? 3500 BC–1603 AD*, New York: Miramax.

Schama, S. (2001) *A History of Britain, Volume II: The Wars of the British 1603–1776*, New York: Miramax.

Schama, S. (2002a) *A History of Britain, Volume 3: The Fate of Empire 1776–2000*, New York: Miramax.

Schama, S. (2002b) "Television and the Trouble with History," *BBC History Lecture* (delivered on May 29), final draft at http://www.bbc.co.uk/bbcfour/features/schama-lecture.shtml.

Schudson, M. (1992) *Watergate in American Memory: How We Remember, Forget and Reconstruct the Past*, New York: Basic Books.

Sklar, R. (1997) "Historical Films: Scofflaws and the Historian-Cop," *Reviews in American History*, 25(3) (September), 346–50.

Sobchack, V. (ed.) (1996) *The Persistence of History: Cinema, Television, and the Modern Event*, New York: Routledge.

Statistical Research Incorporated (1990) "1990 Public Television National Image Survey," Commissioned by the PBS Station Independence Program, September 28, pp. 2.1–2.8.

Stearn, T. (2002) "What's Wrong with Television History?," *History Today*, 52(12) (December), 26–7.

Susman, W. (1984) *Culture as History: The Transformation of American Society in the Twentieth Century*, New York: Pantheon.

Thelen, D. (1989) "Memory and American History," *The Journal of American History*, 75(4) (December), 1119.

Toplin, R. B. (ed.) (1996) *Ken Burns's The Civil War: Historians Respond*, New York: Oxford.

Toplin, R. B. (2002) *Reel History: In Defense of Hollywood*, Lawrence, KS: University Press of Kansas.

White, H. (1975) *Metahistory: The Historical Imagination in Nineteenth-Century Europe*, Baltimore, MD: Johns Hopkins University Press.

White, H. (1985) *Tropics of Discourse: Essays in Cultural Criticism*, Baltimore, MD: Johns Hopkins University Press.

White, M. and Schwoch, J. (1997) "History and Television," in H. Newcomb (ed.), *Museum of Broadcast Communications Encyclopedia of Television*, vol. 2, Chicago: Fitzroy Dearborn Publishers, pp. 770–3.

Zelizer, B. (1992) *Covering the Body: The Kennedy Assassination, the Media, and the Shaping of Collective Memory*, Chicago: University of Chicago Press.

Zelizer, B. (1995) "Reading the Past Against the Grain: The Shape of Memory Studies," *Critical Studies in Mass Communication*, 12(2) (June), 214–39.

Zelizer, B. (1998) *Remembering to Forget: Holocaust Memory Through the Camera's Eye*, Chicago: University of Chicago Press.

CHAPTER TWENTY

"How Will You Make it on Your Own?": Television and Feminism Since 1970

Bonnie J. Dow

"How will you make it on your own?
The world out there is awfully big, and girl this time you're all alone."

These words from the first-season theme song for *The Mary Tyler Moore Show* neatly encapsulate the feminist resonance of the program: the lead character, Mary Richards, is a woman "all alone" (meaning: without a man), thus positioning her as a "new" type of female character on television in 1970. The (feminist) difference of *Mary Tyler Moore* was established both through its sociohistorical context, because 1970 was also the year when the second wave of feminism became nationally visible in the United States, and also through its televisual context, as it challenged a tradition of female roles on US television that had largely relegated lead female characters in situation comedies (sitcoms) to stereotypical roles defined by familial relationships (wife, mother, daughter), feminine occupations (teacher, nurse, secretary), and/or feminine concerns (husband-hunting). Mary Richards was an unmarried woman living independent of family, working in a traditionally male-dominated setting (a television newsroom) and, while romantically active, she was not focused on finding a husband. The success of *Mary Tyler Moore*, which ran for seven seasons on CBS during which it received high ratings and numerous awards and generated three spin-offs, solidified its status in television history as the sitcom that set the standard for representations of independent womanhood. More than 25 years after it left prime time, it is still used as the basis for evaluation of single-woman shows and their feminist implications, having been compared over the years to programs as diverse as *Murphy Brown*, *Ally McBeal*, and *Sex and the City*.

This state of affairs is most usefully explained as the result of *Mary Tyler Moore*'s establishment of a new television character type that has been reworked continuously over the past 30-odd years in an array of television programming

379

and that has come to be the most visible indicator of the influence and representation of feminism on television: the independent working woman. More specifically, characters in this category are generally young, educated, white, middle-class professionals who are heterosexual and unmarried, who live in an urban setting, and who often work in a traditionally male occupation. These characteristics signify, in important ways, the visibility of certain kinds of liberal feminist arguments associated with the second wave: the need for women's access to higher education (especially professional schools), employment (especially traditionally male-dominated occupations), and equal pay, as well as the critique of traditional marriage and motherhood (and their incompatibility with careerism and self-fulfillment for women) and the undermining of the sexual double standard (under which men, but not women, are permitted to be sexually active outside of marriage). The character of Mary Richards and the feminist persona that would come to be associated with her, was a historically specific creation that grew from, was fed by, and fed into feminist discourses that were circulating in American culture in the late 1960s and early 1970s.

The creation of *Mary Tyler Moore* was part of a general turn toward more socially relevant programming that CBS was pursuing in the early 1970s and that led to the development of other topical shows, such as *All in the Family* and *M*A*S*H*. This strategy was designed not only to update the network's offerings, but also to attract a new and different kind of viewer: younger, urban, with more disposable income. Women, as primary consumers for their families, had always been a central target for television advertising, and the creation of new female characters contributed to the pursuit of new female viewer/consumers, especially upscale working women (Rabinovitz, 1989). The creation of programming that responded to the changing conditions of women's lives was a commercial strategy in important ways, and the success of *Mary Tyler Moore* led to a host of other situation comedies that continued to explore previously uncharted territory. In 1974, Rhoda, Mary Richards's best friend, got a sitcom of her own as she moved to New York (as a single woman) (*Rhoda*, 1974–78). Another friend of Mary's, Phyllis, began to explore life as a young widow in San Francisco in her own show (*Phyllis*, 1975–7), and the same year also saw the launch of the first successful sitcom about a divorced woman (*One Day at a Time*, 1975–84). Exploring life "on your own" as a single woman, whether never-married, divorced, or widowed, was also the premise of *Fay* (1975–6), *Alice* (1976–85), and *The Sandy Duncan Show* (1972).

The general parameters of these programs – the focus on a young woman (usually signaled by the title), who has moved to a new city to start a new life – show the outlines of what I term television's "lifestyle feminism," in which a character's connection to the changes wrought by the women's liberation movement is indicated by how she lives rather than by how she thinks or by the presence of any kind of explicit feminist content in the narrative of the sitcom. For example, despite Mary Richards's independence as a single working woman, her role within the narratives of *Mary Tyler Moore* was alternately maternal and daughterly: she nurtured other characters in her work family, solving their problems, and she received guidance and protection from the sitcom's father figure, Lou Grant, her boss in the newsroom. Thus, *Mary Tyler Moore* took its central character out of a traditional family setting only to recreate a conventional family dynamic in the workplace, altering a female character's circumstances to fit changing times while retaining the traditional functions of women familiar from previous domestic sitcoms. Whenever Mary Richards attempted to challenge this situation, she would be put back in her "place" by the end of the episode, such as when she tried to strike a blow for women by hiring a female sportscaster over Lou's objections. Yet, when the sportscaster refused to report on contact sports (presumably because of her feminist pacifism), Mary was forced to fire her. Lou provided the lesson learned from the incident, telling Mary that she had proven "that a woman has the chance to be just as lousy in a job as a man" (Dow, 1996, pp. 40–4).

One Day at a Time, as the first sitcom to focus on a divorced woman, continued to stretch the boundaries of women's roles on television, and it also engaged somewhat more directly with issues raised by feminism. Its central character, Ann Romano, was in her 30s and had two teenage daughters; after her divorce, she moved to Indianapolis from the suburbs to start a new life. She had married young into a very traditional marriage, and the narrative of the series focused on her development into an independent woman, learning how to parent on her own, to negotiate the job market, and to construct egalitarian relationships with men. Indeed, as the sitcom premiered in 1975, after several years of visible feminist activity and rhetoric, its indebtedness to women's liberation was more explicit than that of *Mary Tyler Moore*. For instance, Ann took back her maiden name after her divorce, preferring to be called "*Ms.* Romano," and the sitcom highlighted issues that had been raised by feminists, such as the unequal economic consequences of divorce for men and women, the difficulty of collecting child support, and women's search for self-fulfillment outside the roles of wife and mother. Moreover, in important ways, the sitcom reflected the influence of feminist consciousness raising on American culture, as Ann's discourse and her realizations about herself and her situations were often couched in the language of self-awareness.

By 1984, its last year on prime time, *One Day at a Time* was the only survivor of those "single woman on her own" sitcoms that had introduced lifestyle feminism to American viewers in the 1970s. The exception to the dominance of lifestyle feminism in 1970s sitcoms was a shorter-lived, but much discussed, spin-off from the popular *All in the Family*. *Maude* (1972–8), focusing on a cousin of Edith Bunker, the wife of Archie Bunker, *All in the Family*'s patriarch, was a sitcom about a large, outspoken, 40-something, suburban housewife who had a loudly voiced opinion on everything, especially the status of women. Maude was smart, sarcastic, and in control of her life – the counterpart to the uncertain women exploring their independence on other sitcoms. Her bravada was entertaining, and funny, and her analyses of women's issues such as sexual harassment, menopause, and abortion were often right on target for feminist viewers, especially in the controversial two-part episode from the show's second season, when she decided to have an abortion after finding herself pregnant at the age of 47. Yet, as Susan Douglas has pointed out, "Maude was also fearsome, an often unyielding battle-ax" who "reinforced the stereotype of the feminist as a strident, loud, unfeminine bruiser who could afford to be a feminist because she was older, less needy of male approval, and financially comfortable" (Douglas, 1994, p. 203).

As the above examples demonstrate, feminism had its earliest articulations on television in the form of situation comedy, a state of affairs that makes sense given comedy's ability to defuse the anxiety and controversy attached to social change. As Taylor (1989, p. 27) has observed: "comedy is a more flexible form than drama because it can create multiple, conflicting and oppositional realities within the safe confines of the joke." Situation comedy brings social issues into a family ("real" or metaphorical) and personalizes those issues, making them the problems of individual characters rather than tying them to social and political circumstances. The solutions to those problems, even when they are tied to larger issues, are also individualistic, resulting from personal choices made within a context of familial understanding and support. Thus, beginning in the 1970s, situation comedy domesticated feminism by attaching it to figurations of the "new woman," demonstrating how "political conflict is rerouted into individual 'lifestyle choices'" (Rabinovitz, 1999, p. 147).

Although situation comedy built around some version of lifestyle feminism was the "preferred fictional site for a 'feminist' subject position" in the 1970s (Rabinovitz, 1989, p. 3), dramas reflecting feminist influence also emerged, albeit later in the decade. The crucial difference between the feminist representations of sitcom and of drama is that while the former are usually noted as somewhat progressive, the latter are almost always regarded as primarily exploitative. Yet 1970s dramas featuring lead female characters display many of the same characteristics as the "new woman" sitcoms that preceded them: they featured women in traditionally male occupations, they signaled their focus on a female lead in their titles, and they centered on single, young, heterosexual, white (with one exception) working women. All were crime-fighting dramas of one sort or another: *Charlie's Angels* (1976–81), *Police Woman* (1974–8), *Get Christie Love* (1974–5), *Wonder Woman* (1976–9), and *The Bionic Woman* (1976–8). The insertion of lead female characters into a television genre overwhelmingly coded as masculine in previous iterations indicates television's preference for featuring the discourses of liberal feminism, which highlight issues of women's access to and opportunities in the public sphere. Indeed, the mid-1970s, when all of these dramas appeared, was a time when liberal feminist ideals were a powerful presence in American culture, as the campaign for the ratification of the Equal Rights Amendment (ERA) was in full swing. In addition, *Get Christie Love* broke the color barrier as well, as its title character was African American, although it was the least successful of all of the female-centered dramas and lasted barely a season.

The most successful of these programs was *Charlie's Angels*, which was dismissed by feminists at the time as simply an excuse to feature beautiful, sexy, scantily clad women in dangerous situations. Indeed, the lead characters in all these dramas displayed their bodies regularly, whether while working undercover as prostitutes (*Police Woman, Get Christie Love*), wearing a sexy superhero costume (*Wonder Woman*), or posing as prison inmates or beauty contestants to expose some criminal plot (*Charlie's Angels, Bionic Woman*). Moreover, while all were positioned to some degree as action heroines who were bright, active, used weapons, and, in the case of *Wonder Woman* (a superhero with supernatural talents) and *The Bionic Woman* (who had bionic legs and a bionic arm and ear of extraordinary strength), quite powerful and capable of physically besting male villains, their agency and their independence from male authority varied. Pepper Anderson, the focus of *Police Woman*, often had to be rescued by her male colleagues, *Wonder Woman*'s Diana Prince went to great lengths to hide her powers, and was always following the orders of her male boss (whom she secretly loved), and Jaime Sommers, the *Bionic Woman*, also kept her talents hidden except when working as a secret agent, and she, too, took her orders from a man (Douglas, 1994).

Generally, as Susan Douglas has argued, the most powerful women in these dramas were a superhero and a science experiment, who appeared "only in comic book settings that could never be mistaken for reality" (1994, p. 218). Yet she and other feminist critics have argued for the overlooked feminist resonance of *Charlie's Angels* (see also Inness, 1999; Womack, 2003). Indeed, while it certainly earned its label as "jiggle television," the show was also a representation of female competence (while they took orders from Charlie, whom they and viewers only experienced as a disembodied voice, they generally solved the cases using their own initiative), female solidarity (the women were good friends who supported and protected each other), and even female independence (the show was about their work, not their love lives, and they were seemingly unconcerned with finding husbands). Finally, the show overtly referenced the presence of sexism, both in its premise, that the Angels became private investigators because of the sexism they had faced as police officers, and in its narratives, in which the Angels often refer to male villains as "male chauvinist pigs." Ultimately, "feminism and antifeminism stood in perfect suspension [in *Charlie's Angels*]. In seeking to have it both ways – to espouse female liberation and to promote the objectification of women's bodies – *Charlie's Angels* offered a compromise with empowering and thwarting effects" (Douglas, 1994, pp. 218, 216).

By the 1980s, when the public visibility of feminist activism had waned, the ERA had failed, and national politics had taken a conservative turn, television's earliest form for representing feminism, the single woman sitcom, was also in decline. *One Day at a Time* was on the air until 1984, but, with the exception of *Kate and Allie* (1984–9), single woman shows did not fare well in the early 1980s. *Kate and Allie* focused on two divorced women with children who blended their families and shared a New York apartment. The sitcom was a twist on the traditional family, as Kate worked to support the family while Allie stayed at home. It had many of the themes of *One Day at a Time*, emphasizing the "conflicts of female independence and the search for female self-actualization" (Rabinovitz, 1989, p. 12). The sitcom reflected feminist influence in other ways as well; for instance, it provided a vision of sisterhood, and it offered an alternative to the patriarchal nuclear family. Until 1988, when *Murphy Brown* premiered, *Kate and Allie* was the only truly successful 1980s heir to the single-woman television tradition. Another situation comedy important for its representation of feminism, *Designing Women* (1986–93), did not focus on a single woman, but, rather, on a group of women who worked together in an interior design business. All were single at the outset of the show (although two had children who were occasionally seen), all were in their thirties or forties, all were white and middle-class. Thus, the characters bore some similarity to earlier "feminist" characters, but the single-woman sitcom theme of exploring independence after a life change was absent.

Yet *Designing Women* was probably the most explicitly feminist sitcom since *Maude*, and its vision of feminism emerged largely through actual feminist discourse rather than through the lifestyle of its characters. Although the show was set in the workplace, that workplace was also the home of the senior partner, Julia Sugarbaker. Thus, the setting was implicitly a domestic one; and the characters were only occasionally seen doing actual work; rather, they spent a great deal of time sitting on the couch and the upholstered chairs, discussing their lives and experiences. Often, their conversation would turn to discussion of a feminist issue, such as sexual harassment, sexual assault, wife battering, pornography, or body image, precipitated by an event or experience of one of the characters. In some ways, the structure and content of the characters' conversations mimicked the dynamics of feminist consciousness raising, in which personal experiences are shared and analyzed to reach a conclusion about women's political status. Indeed, as I have argued elsewhere, one of the noteworthy characteristics of *Designing Women* is its frequent resistance to the television tendency to reduce social issues to individual problems. Situation comedy "rarely acknowledge[s] the role that cultural biases play in the problems faced by their characters. *Designing Women*, in its acknowledgement of cultural responsibility for women's problems, challenges this generic characteristic" (Dow, 1996, p. 117).

The 1988 debuts of both *Roseanne* and *Murphy Brown* would mark a watershed year in US television's representation of women and feminism, especially if judged by the amount of popular and critical attention these sitcoms would receive. Yet, there were other trends in 1980s television that clearly indicate the influence of feminism in American culture. For example, the 1980s saw the rise of the "superwoman" character type, in popular family sitcoms such as *The Cosby Show* (1984–92), *Family Ties* (1982–9), and *Growing Pains* (1985–92). These were married women with children who also had a professional career and yet seemed to effortlessly combine the two (Lotz, 2001). Clair Huxtable of *Cosby* was a lawyer, Elyse Keaton of *Family Ties* was an architect, and Maggie Seaver of *Growing Pains* was a journalist. All had a number of children, and all had husbands who also had professional careers, yet the narratives of these shows seldom, if ever, featured work/family conflicts. The appearance of these characters signaled the advent of "postfeminism," a term coined by popular media to refer to the aftermath of the second wave of feminism, in which it was presumed that feminism was over, having achieved its goals of integrating women into public life, a process that, according to these shows, was relatively painless. Thus, although these programs did not necessarily include storylines linked to feminism as *Designing Women* had, they clearly featured the gains of previous feminist activism in their portrayals of female characters.

Yet, there was also a dark side to postfeminism. Another theme raised in popular discussions of women's lives after the second wave focused on the price of progress, indicated by a large amount of media attention being given to such issues as infertility afflicting women who had put off childbearing in favor of careerism, the dangers of daycare, the difficulty of marrying after thirty, and the conflicts and competition between women who "chose" to stay at home and women who "chose" to work (see Dow, 1996; Press, 1991). This other side of postfeminism was often featured in the popular professional serial dramas of the 1980s, such as *Hill Street Blues* (1981–7), *L.A. Law* (1986–94) and *St. Elsewhere* (1982–8) (which focused on a police station, a law firm, and a hospital, respectively), as well as in the drama *thirtysomething* (1987–91), which focused on a group of friends and their relationship and career struggles (see Heide, 1995; Probyn, 1997). These shows all featured women in their thirties in powerful professional roles who were often coded as "feminist" through their career focus, single status, and their traditionally masculine occupations. Yet, at the same time, they were often portrayed as characters with evident anxieties related to being taken seriously at their jobs, to their lack of satisfying personal lives, and to their desire for children, reinforcing the postfeminist "conflict between careerism and personal health and happiness" (Dow, 1996, p. 98).

The exception to generalizations about women's roles in dramas in the 1980s is *Cagney and Lacey* (1982–8) which, while it possessed some of the same characteristics as earlier feminist-influenced dramas (the lead characters – Christine Cagney and Mary Beth Lacey – were signaled by the title and were white, female police officers in their thirties in New York City), has been hailed as a drama that represented feminism visibly and positively. Julie D'Acci's (1994) thorough book-length study of *Cagney and Lacey* details the considerable press attention to the drama that focused on its implications for the status of women, and she provides a history and analysis of both the production context and the content of the series. D'Acci concludes that while the show began with a clear feminist sensibility that was meant to appeal to the coveted working-women audience, its lead characters were softened and its politics watered down over time, as more attention was given to their personal relationships and problems (including Cagney's alcoholism and dating life and Lacey's family life). Linking the process to the backlash against feminism in the 1980s, as well as to commercial pressures, she argues that *Cagney and Lacey*'s representation of feminism "changed . . . from a criticism of institutional inequities (sexism, racism, and, to a lesser degree, classism) to an examination of women's issues (or what the industry imagined as such issues) that had the potential for dramatic intensity and exploitability" (D'Acci, 1994, p. 103). The "women's issue" feminism D'Acci identifies in *Cagney and Lacey* emerged in episodes centering on such issues as rape, sexual harassment, abortion, incest, child molestation, and breast cancer, topics that were proven to attract female viewers, and she notes that "even when a script was not explicitly feminist, general audience recognition that the treatment of the subject resulted from a women's movement interpretation and intervention made the feminism at least fairly obvious to many viewers" (1994, p. 160).

Importantly, part of the reason that programs such as *Cagney and Lacey* and *Designing Women* received media attention during the 1980s for their progressive representations of women is that their creators were female, a rarity at the time, and press treatments of these shows assumed a link between the gender of their creators and the viewpoint of the programming. This factor was also important in reactions to the sitcom *Murphy Brown* (1988–98), a show which debuted the year that *Cagney and Lacey* left prime time and signaled the resurrection of the single-woman sitcom. *Murphy Brown* was often compared to *Mary Tyler Moore*, since it also focused on a woman working in a television newsroom, although rather than the struggling local news station in *Mary Tyler Moore*, *Murphy Brown* focused on a highly successful network news magazine, *FYI*, and its staff. The title character was easily recognizable as a feminist representation – she was single, white, professional, and working in a man's world – although, as I have argued previously, she is more accurately a postfeminist character, in that she was depicted as a woman who had paid a price for her success in terms of the barrenness of her personal life and her lack of traditionally feminine attributes. A recovering alcoholic, depicted as competitive with other women rather than supportive of them, Murphy had no friends outside the workplace and was entirely defined by her career. Although played by the beautiful Candice Bergen, Murphy Brown was a masculinized character who wore severely tailored clothing, had a brash and aggressive physical and interpersonal style, and was generally depicted as a failure at the kinds of womanly functions that had defined Mary Richards's role in the workplace. Indeed, particularly in the early years of the sitcom, a sharp contrast was drawn between Murphy's feminism (remininiscent, in several ways, of Maude) and the stereotypical femininity of her co-worker, Corky Sherwood, an ex-beauty queen who was also a reporter for *FYI*. Several episodes focused on clashes between the two women, in which Murphy became a comic buffoon, punished for her lack of traditionally feminine qualities, which Corky had in abundance, although she was much less professionally competent. As I have argued elsewhere,

Murphy is a vision of liberal feminist success, and the scapegoating of her character is a recurring reminder of the problems that success creates. Murphy is too abrasive, confident, outspoken and powerful (for a woman) to be left unchecked. In *Murphy Brown*'s postfeminist vision, patriarchy is no longer the problem, feminism (and the problems it creates for women) is. (Dow, 1996, p. 149)

In 1992, *Murphy Brown* became the focus of national controversy, because of a storyline in which Murphy became accidentally pregnant by her ex-husband (to whom she had been married for five days 20 years earlier) and decided to bear and raise the child herself.

Then-Vice-President Dan Quayle used this storyline as an example in a speech, commenting that positively portraying single motherhood as "just another life-style choice" was evidence of the lack of family values promoted by Hollywood. Various commentators weighed in on the issue (including the sitcom itself, which featured Quayle's remarks in an episode in which Murphy responded to them), yet the most regrettable outcome of this controversy was that it elided the race and class politics that were raised by Quayle's original comments, which came in a speech about poverty in the inner city, implicitly raising the specter of the Black welfare mother. Instead, the issue became just another conflict over women and their "choices," with little recognition of the ways that racial and class privilege figure in how those choices are made and how they are evaluated (see Dow, 1996; Rabinovitz, 1999). Ultimately, Murphy's child made little difference to the narrative patterns of the sitcom. Although episodes in the season after she gave birth focused on the trials and tribulations of single motherhood, these storylines functioned much as earlier ones had, in that they showcased Murphy's lack of traditionally maternal skills and exploited her unsuitability for motherhood for comic effect. Within a couple of seasons, the sitcom reverted to its focus on the workplace, and Murphy's child was almost never seen.

1988, the year of *Murphy Brown*'s debut, also saw the debut of *Roseanne* (1988–97), a sitcom also noted for its feminist sensibility, but which had a very different premise. Focused on a working-class family, *Roseanne* did not have the traditional markers of feminist television: its lead was married with three children, she worked in a factory, she did not have a college degree. *Roseanne*'s feminist resonance stemmed primarily from the persona of its star: Roseanne Barr, who was the co-creator, writer, and lead actor for the sitcom and was a well-known stand-up comedian, whose routines contained strong critiques of class, gender, and ethnic ideologies. These themes were carried over into *Roseanne*, which was praised by feminist critics for its refusal to idealize family life. Moreover, Roseanne Barr was a large woman, unashamed of her size, who did not fit conventional expectations for female beauty on television but refused to be de-sexualized as a result. She was what Kathleen Rowe (1997) has termed an "unruly woman," who flouted conventions of feminine appearance, behavior, and speech, both in public life and within the narratives of *Roseanne*. For example, Rowe analyzes an episode of the sitcom in which Roseanne dreams that she kills her children because they nag her for attention, interrupting a long-awaited bath following a difficult workweek. When accused of mistreating her husband, she replies "The only way to keep a man happy is to treat him like dirt once in a while." Although, like traditional family sitcom, *Roseanne* generally concluded its episodes with scenes of reconciliation and restoration of equanimity, its narratives offered interrogations of conventional wisdom about class, about family life, and about gender roles and relationships. As Rowe concludes, the sitcom is distinguished by its pattern of "representing the unrepresentable. A fat woman who is also sexual; a sloppy housewife who's a good mother; a 'loose' woman who is also tidy, who hates matrimony but loves her husband, who hates the ideology of 'true womanhood' but considers herself a domestic goddess" (p. 82).

Murphy Brown and *Roseanne* remained on prime time well into the 1990s, and they were followed by a number of imitators that did not fare as well. *Grace Under Fire* (1993–8) was an attempt to capitalize on the success of the working-class woman sitcom. It starred another former stand-up comedian, Brett Butler, as a white, divorced mother (and formerly battered wife) who worked in an oil refinery. Like Roseanne, Grace was outspoken about issues of gender and class. *Murphy Brown*'s resurrection of the single-woman sitcom led to programs such as *Ellen* (1994–8), which focused on a single, white, 30-ish bookstore owner in Los Angeles and her group of friends and co-workers. *Ellen*, a decidedly postfeminist and apolitical comedy, struggled in the ratings for much of its run, receiving its greatest notoriety when its lead character came out as a lesbian, a plotline preceded by the coming-out of the sitcom's star, stand-up comedian Ellen DeGeneres (see Dow, 2001; Walters, 2001). *Cybill* (1995–8), a sitcom about a white, divorced, 40-something working actress in Los Angeles with two daughters, was another reiteration on the single-woman premise, and was given some feminist resonance because of the politics of its star, Cybill Shepherd, and because of its attention to the plight of aging women, particularly in youth-obsessed Hollywood. Finally, *Living Single* (1993–8), a sitcom focused on four unmarried African-American women living together in New York City, arguably owed its premise more to the success of *Designing Women* than to that of *Murphy Brown*. *Living Single* did not address feminist issues in the visible fashion that *Designing Women* had, and its narratives were generally driven by relational and career struggles, with only occasional attention to gender or race politics; however, it is noteworthy for its disruption of the generally all-white focus of single-woman sitcoms. Another comedy focusing on the personal and professional travails of four single women in New York City appeared on the cable network HBO in 1998. Somewhat non-traditional in its eschewal of conventional sitcom structure, *Sex and the City* (1998–2004) was also distinguished by its ability, because it appeared on pay-cable, to address sexual issues in a frank fashion. Indeed, its depiction of women's sexual agency and its emphasis on the strong friendship between its main characters, particularly when contrasted with their generally less successful relationships with men, were *Sex and the City*'s most progressive characteristics (Kim, 2001).

In the late 1990s, media attention on feminist representation galvanized around a new entrant to the single-woman tradition: *Ally McBeal*. A show focusing on a young woman lawyer (the title character) working in a law firm in Boston, *Ally McBeal* is hard to categorize: an hour-long show with serial (continuing) plotlines, it hardly fits the generic outlines of situation comedy. On the other hand, it was decidedly comic in tone, and blended elements of fantasy into its narrative (e.g. when Ally sees a sexually attractive man, an oversized, cartoonish tongue pops outs of her mouth, and she hallucinates a dancing baby that reminds her of her biological clock). It was often referred to as "dramedy" by reviewers, although it was categorized as comedy for the purposes of Emmy nominations. Regardless, *Ally McBeal* possessed the familiar markers of feminist representation with its focus on a white, unmarried, professional working woman living in an urban area; indeed, the popular press often compared it directly to *Mary Tyler Moore*.

There are several reasons why *Ally McBeal* was seen as engaging with feminism: it had storylines that dealt with issues like sexual harassment or gender discrimination, and its characters were consistently embroiled in different skirmishes in an ongoing battle of the sexes. But it was the character of Ally herself that received the most attention from those who wished to analyze the show's relationship to the current state of feminism. In important ways, *Ally McBeal* was a vision of what feminism has accomplished for privileged white women who came of age after the second wave: like Ally, such women have not had to fight for an elite education, have benefited from the breaking down of professional barriers to women, and with these battles won, have been free to obsess about their relational lives, which was the primary focus of the show's narratives. Although often spurred by cases that her firm would handle, the issues that drove the narrative of the show centered on the neurotic and self-absorbed Ally's search for a mate, her desire for children, and/or her relationship with her married ex-boyfriend (who also worked at the firm, as did his wife). Indeed, Ally's obsession with finding a relationship was a sign that feminism had not really managed to dislodge the conviction that a woman is not complete without a man (or that women cannot get along, as the program also showcased female sexual competition and cattiness between the females at the firm). What *Ally McBeal* offered was a postfeminist vision in which feminism is pretty much over because it did what it needed to do, and in which the women who benefited most from it – young, white, heterosexual women like Ally – feel discouraged because their pursuit of professional success has somehow diverted them from finding the perfect mate (see Dow, 2002; Dubrovsky, 2002; Ouellette, 2002).

The 1990s also saw the continuing representation of competent professional women in ensemble dramas such as *ER* (1994–present), *The X-Files* (1993–2002), and *The Practice* (1997–present), in which the portrayal of working women was arguably much more complex and balanced. The same can be said of the female-dominated ensemble dramas created by the Lifetime network in the last few years, including *Any Day Now* (1998–present), *Strong Medicine* (2000–present), and *The Division* (2001–present), all of which feature highly competent

female professional characters and storylines indicating a gendered consciousness (and which are also praiseworthy for their incorporation of women of color
as major characters). Yet, as has always been the case, media attention to the
feminist implications of popular television tends to focus on programs with a
central female protagonist, rather than on shows that feature women as part of an
ensemble cast. Generally, media associate feminism with the "exceptional woman"
premise rather than with the depiction of women as part of a group (although the
latter premise, because it often demonstrates female solidarity, has some powerful feminist implications). For example, the 1990s drama *Dr. Quinn, Medicine
Woman* (1993–8), which focused on a woman doctor practicing on the frontier in
the post-Civil War period, received a fair amount of attention for its politics,
which blended feminism (as Dr. Michaela Quinn fought sexism from townspeople skeptical of a woman doctor) and anti-racism (as she took up the cause
of Blacks and Native Americans who faced discrimination). Indeed, because it
was set in the past, *Dr. Quinn* was able to incorporate liberal feminist discourse
in some rather overt ways (see Dow, 1996; White, 2001). Moreover, a drama
focused on a female lead was something of a watershed, as it premiered at a
moment when the only other female-headed drama remaining on prime time was
the long-running *Murder, She Wrote* (1984–96).

The drought in woman-centered dramas was broken by some unlikely candidates, both of which received enormous attention for their portraits of a "new"
kind of womanhood on television. *Xena: Warrior Princess* (1995–2001) and *Buffy
the Vampire Slayer* (1997–2003) departed from the tradition of representing
women's progress through placing female characters in formerly masculine professional contexts (law, law enforcement, and medicine primarily), instead offering fantasy female figures with supernatural powers who used those powers in
the fight against evil. A far cry from *Wonder Woman* or *The Bionic Woman*, *Xena*
and *Buffy* exemplified what some reviewers labeled "tough girl feminism," in
which women's physical strength and "take-no-prisoners" attitude were taken as
indications of their enactment of feminist ideology. Moreover, reviewers often
compared the shows to *Ally McBeal* to make the point that *Xena* and *Buffy*
offered far more satisfying depictions of female strength and solidarity.

Buffy the Vampire Slayer centered on Buffy Summers, a teenager who moved
to the amusingly named Sunnydale, California, with her divorced mother, only
to discover that Sunnydale was once also known as "the Hellmouth," and it is an
area prone to mystical occurrences involving a variety of demons, including
vampires, werewolves, zombies, etc. who wreak havoc on the community. Buffy,
it turns out, was destined to come to Sunnydale, as she is the anointed Slayer,
the "Chosen One," whose responsibility it is to keep the dark forces in check.
Importantly, only females are chosen to be Slayers. The show began in 1997, and
over the first few seasons, Buffy comes to understand the dimensions of her
responsibility as a Slayer. In various episodes, aided by her group of close
friends, she fights off the demons with her extraordinary strength and slayer
skills, while still trying to live the life of a normal teenager. The conflict between

these two tasks provides much of the humor and pathos of the show, as Buffy is helped by her loyal circle of friends, male and female (one of whom comes out as a lesbian). Buffy is an unlikely superhero: she is truly a sort of ordinary girl caught in extraordinary circumstances. It is her special slayer powers, extraordinary strength, and keen perceptual skills, coupled with her basic street smarts, that allow her to effectively perform her mission as a Slayer. And she does it within a community of friends who provide strong support and solidarity.

In many ways, *Buffy* offered a more progressive message than *Ally McBeal*, given its sense of social responsibility, its portrait of solidarity, and its self-assured title character. These characteristics are also evident in *Xena: Warrior Princess*, and the similarities do not end there.

Both were shows with huge cult followings, made up in large part by young women. Both were about strong female characters embedded in fantasy, both of these lead characters were beautiful women who wore skimpy outfits, had enormous physical strength, and regularly battled nasty villains. Finally, both had female sidekicks, and both shows encouraged acceptance of gay sexuality. Yet while *Buffy* was set in a recognizable contemporary world, *Xena* was a completely mythical character, existing in a different time and space.

Xena: Warrior Princess, a spin-off of the popular syndicated show *Hercules: The Legendary Journeys*, was set in an ambiguous pre-Christian era in which the Greek and Roman Gods struggle for domination along with a variety of other warring kings and emperors. The Amazons, a mythical all-female tribe, also play a large role in the narrative. Xena was often seen fighting mythical characters like Minotaurs and Cyclops, outwitting Ares, the Roman God of War (a former lover) or saving Aphrodite, the Goddess of Love, or winning a chariot race against Caligula, a Roman emperor. Xena was originally a villainous character who was set on killing Hercules. But when her army betrayed her, she underwent a conversion and dedicated herself to making amends for the violence of her past. Along with her young sidekick, Gabrielle (whose implicit romantic connection to Xena was the source of much speculation for fans and critics [see Helford, 2000]), Xena travels the world using her powers to defeat various oppressive characters.

Ultimately, the problems of portraying feminism through fantasy characters were confronted by a number of critics; as one put it, "Buffy's strength and confidence are not learned from the vast experience of generations of women; rather, they are her mystical birthright as a slayer" (Owen, 1999, p. 31). Moreover, the exercise of physical power to triumph over your enemies is not the same as social and institutional change. Yet, given their greater independence and agency, *Buffy* and *Xena* represented a kind of progress in comparison to the supernaturally skilled women of 1970s dramas, such as *The Bionic Woman* and *Wonder Woman*.

Indeed, the history of feminist representation on American television over the past 30-plus years encourages such attenuated estimations of progress, coupled with recognition that much stays the same. With rare exceptions, television offers visions of feminism that are white, middle- and upper-class, and hetero-sexual, and the medium prefers to represent feminist progress through lifestyle rather than politics, a pattern begun in the 1970s and continuing to the present day. Thus, women are depicted in male-dominated professions, married women with children work outside the home, and single women are often portrayed as persons in their own right, rather than only as aspiring wives. Yet, particularly after the rise of postfeminism in the 1980s, only rarely is explicit mention made of the feminist activism that spurred such changes, and equally rare is any admission that feminist activism is ongoing. Moreover, in contrast to most male characters on television, romantic (heterosexual) and familial relationships are still featured as the primary concern for most female characters, regardless of marital or maternal status. However, a focus on relationships between women, as friends (although not as romantic partners), provides some balance to this characteristic; programming such as *Designing Women*, *Cagney and Lacey*, *Any Day Now*, and *Sex and the City* emphasizes the benefits of female support, giving implicit (and sometimes explicit) credence to feminist arguments about the need for women's solidarity.

In many ways, the biases of entertainment television reflect those of media generally. Liberal feminism, and its focus on women's equality with men within existing social structures, has always been the easiest form of feminism for media to understand and incorporate. In news and non-fiction media, this takes the form of celebrating the accomplishments of individual exceptional women (especially women who succeed in male-dominated arenas). Entertainment television follows a similar pattern, promoting as progressive those representations that feature individual women succeeding in a "man's world" (lifestyle feminism) or that focus on women with traditionally masculine attributes ("tough girl" feminism). Television's visions of feminism thus become equated with the practice of individualism by women, eliding the importance of collectivity in feminist movements, ignoring the complexities of race, class, and sexuality in women's lives, and disregarding the structural problems that impede women's progress. Finally, television also plays a role in demonizing feminism by offering representations that foreground the problems that the movement has supposedly created for women, by, for example, encouraging them to focus on career at the expense of childbearing. Such representations fail to acknowledge that the difficulties of combining career and motherhood are hardly feminism's fault; rather, they are the result of the larger culture's failure to effectively respond to the needs of women in the workplace, an issue that feminists continue to pursue.

Yet, to conclude that feminism has had no positive effects on television's representation of women is too pessimistic; the variety and scope of representations of women's lives continues to expand, aided in recent years by the growth of women's networks such as Lifetime, WE, and Oxygen (see Meehan and Byars, 2000). Indeed, to call the programming discussed here "feminist" is not inaccurate: within the limits of US commercial television, these programs offer a version of feminist ideology. However, that ideology is largely suited to television's commercial needs rather than to the needs of a feminist politics committed to collective action and to the recognition of the intersections of race, class, and sexuality. Ultimately, for feminist critics, continuing to interrogate television's representations of feminism is a crucial task in the quest to improve them.

References

D'Acci, J. (1994) *Defining Women: Television and the Case of Cagney and Lacey*, Chapel Hill, NC: University of North Carolina Press.

Douglas, S. J. (1994) *Where the Girls Are: Growing Up Female With the Mass Media*, New York: Random House.

Dow, B. J. (1996) *Prime-Time Feminism: Television, Media Culture, and the Women's Movement Since 1970*, Philadelphia: University of Pennsylvania.

Dow, B. J. (2001) "*Ellen*, Television, and the Politics of Gay and Lesbian Visibility," *Critical Studies in Media Communication*, *18*(2) (June), 123–40.

Dow, B. J. (2002) "*Ally McBeal*, Lifestyle Feminism, and the Politics of Personal Happiness," *Communication Review*, 5(4), 259–64.

Dubrovsky, R. (2002) "*Ally McBeal* as Postfeminist Icon: The Aestheticizing and Fetishizing of the Independent Working Woman," *Communication Review*, 5(4), 264–84.

Heide, M. (1995) *Television Culture and Women's Lives: thirtysomething and the Contradictions of Gender*, Philadephia: University of Pennsylvania Press.

Helford, E. R. (2000) "Feminism, Queer Studies, and the Sexual Politics of Xena: Warrior Princess," in E. R. Helford (ed.), *Fantasy Girls: Gender in the New Universe of Science Fiction and Fantasy Television*, Lanham, MD: Rowman & Littlefield, pp. 135–62.

Inness, S. A. (1999) *Tough Girls: Women Warriors and Wonder Women in Popular Culture*, Philadelphia: University of Pennsylvania.

Kim, L. S. (2001) "*Sex and the Single Girl* in Postfeminism: The F Word in Television," *Television and New Media*, *2*(4) (November), 319–34.

Lotz, A. D. (2001) "Postfeminist Television Criticism: Rehabilitating Critical Terms and Identifying Postfeminist Attributes," *Feminist Media Studies*, *1*(1), 105–21.

Meehan, E. R. and Byars, J. (2000) "Telefeminism: How Lifetime Got its Groove, 1984–1997," *Television and New Media*, *1*(1) (February), 35–51.

Ouellette, L. (2002) "Victims No More: Postfeminism, Television, and *Ally McBeal*," *Communication Review*, 5(4), 315–35.

Owen, A. S. (1999) "*Buffy the Vampire Slayer*: Vampires, Postmodernity, and Postfeminism," *Journal of Popular Film and Television*, *27*(2) (Summer), 24–31.

Press, A. (1991) *Women Watching Television*, Philadelphia, PA: University of Pennsylvania Press.

Probyn, E. (1997) "New Traditionalism and Postfeminism: TV Does the Home," in C. Brunsdon, J. D'Acci and L. Spigel (eds.), *Feminist Television Criticism: A Reader*, New York: Oxford, pp. 126–37.

Rabinovitz, L. (1989) "Sitcoms and Single Moms: Representations of Feminism on American TV," *Cinema Journal*, *29*, 3–19.

Rabinovitz, L. (1999) "Ms.-representation: The Politics of Feminist Sitcoms," in M. B. Haralovich and L. Rabinovitz (eds.), *Television, History, and American Culture: Feminist Critical Essays*, Durham, NC: Duke University Press, pp. 144–67.

Rowe, K. K. (1997) "Roseanne: Unruly Woman as Domestic Goddess," in C. Brunsdon, J. D'Acci and L. Spigel (eds.), *Feminist Television Criticism: A Reader*, New York: Oxford, pp. 74–83.

Taylor, E. (1989) *Prime-Time Families: Television Culture in Postwar America*, Berkeley: University of California.

Walters, S. D. (2001) *All the Rage: The Story of Gay Visibility in America*, Chicago: University of Chicago.

White, M. (2001) "Masculinity and Femininity in Television's Historical Fictions: *Young Indiana Jones Chronicles* and *Dr. Quinn, Medicine Woman*," in G. R. Edgerton and P. C. Rollins (eds.), *Television Histories: Shaping Collective Memory in the Media Age*, Lexington: University Press of Kentucky, pp. 37–58.

Womack, W. (2003) *Disco Divas: Women and Popular Culture in the 1970s*, Philadelphia: University of Pennsylvania Press.

CHAPTER TWENTY-ONE

Television and Race

Sasha Torres

Introduction

This essay will sketch two trajectories along which to think about the intersection of race and television. First, I will suggest that an important shift has occurred in US network television's depictions of race since 1990. Briefly, I contend that network television of the early 1990s tended, as it had done for the previous 40 years, to produce race as a spectacle that bore particular social content understood to be proper to African American (and occasionally Chicano/a or Latino/a) communities, whether as a result of racism or racial pathology. In other words, these representations tended to collate "race" with black or brown persons either mired in or courageously transcending such social contexts as poverty, drug addiction, criminality, or racial oppression. Indeed, in the universe of this televisual discourse about race, black or brown persons, whether professional actors or those appearing "as themselves," rarely appeared without being linked, sooner or later, to one or another of these contexts. Here black or brown bodies enter representation in order to signify race, and race is understood as a problem. But by the end of the decade, this minoritizing logic of race was increasingly displaced by a competing understanding in American television: racialized bodies – black, brown, and yellow – seem to become decoupled from undesirable social contexts and instead become both themselves commodified and linked to other commodities. Race in this period signifies not so much a set of social problems as a new set of consumer choices. In the first part of this chapter, I will outline this shift, trace its connections to the political economy of the cable era, and chart its implications for representation.

In the second part of this chapter, I will take up three influential conceptual models for thinking about race and representation in the contemporary academy, and suggest their usefulness and limitations for thinking about this shift. Specifically, I will consider the focus on the stereotype, the understanding of race as a social construction, and the study of racial representation in relation to capitalist ideology.

Sasha Torres

The Spectacle of Race as Problem

How does it feel to be a problem?

W. E. B. Du Bois, *The Souls of Black Folk*

With the possible exception of some situation comedy, and the deployment of musicians in variety programming, television has most persistently portrayed race using strategies borrowed from ethnographic documentary practice, the Hollywood "social problem" film, or both. In both its fictional and non-fictional forms, then, American television has tended historically to anchor its depiction of raced bodies – and this has meant, for the most part, African-American bodies – to particular social conditions which it understands to be both undesirable and inextricably linked to racially marked communities. These conditions have included racist oppression and life in the "ghetto," with its attendant signifiers: poverty, unemployment, substandard housing, female-headed households, and drug trafficking and abuse, and violence. Whether African Americans are represented as falling victim to such conditions or bravely rising above them, such representation attaches them inexorably to the social problem.

A quick survey of some of the landmark examples of African-American television representation will bear this out. For example, some of television's earliest non-fictional representations of blacks arose in relation to the emergence of the civil rights movement. As I have argued elsewhere (Torres, 2003), news coverage of the movement tended both to generate and to record spectacular violence against black bodies engaged in nonviolent protest. Such coverage threw the problem of southern racism into stark relief, thus collating the black body in pain with that of a suffering *national* body still wounded by sectionalism and southern intransigence. During the same period, network documentaries like CBS's *The Harlem Temper*, NBC's *Sit-In*, and ABC's *Cast the First Stone* explored the effects of racism in an extended exercise of national soul-searching. In general, then, informational programming during the early 1960s tended persistently to link depictions of racial markedness to broad social ills.

The first dramatic programming to address race in a sustained way followed suit, often borrowing implicitly from information coverage of the period. Indeed, particular episodes of the dramatic series *The Defenders* (CBS, 1961–5), *East Side/West Side* (CBS, 1963–4), *Mr. Novak* (NBC, 1963–5), *The Lieutenant* (NBC, 1963–4) and others aired in the wake of the Birmingham movement might be read as an early version of the "ripped-from-the-headlines" aesthetic, taking up issues like red-lining, job discrimination and racist incidents in schools (Watson, 1994). For example, a celebrated 1964 episode of the short-lived CBS series *East Side/West Side*, "Who Do You Kill?," took its storyline from a short article in the *New York Times* about a child in Harlem who had been bitten by a rat in her crib. In the television version, the child's parents, played by James Earl Jones and Diana Sands, encounter the series regulars, social workers played by

George C. Scott and Cicely Tyson, as they wait at the hospital for word of their daughter's condition. The episode, made in a *cinema verité* style, dwells on the neighborhood, buildings, and inhabitants of its Harlem location in shots that in some cases seem drawn directly from those of *The Harlem Temper* or *Cast the First Stone*.

If "Who Do You Kill?" was ultimately rather pessimistic about the prospects for real racial progress, even shows which were much more optimistic about race relations tended to link individual black characters to the fortunes of their race. *The Bill Cosby Show* (1969–71), Cosby's follow-up to *I Spy*, told the story of Chet Kincaid, an athletic coach in an integrated urban high school. In an episode of *The Bill Cosby Show* called "The Lincoln Letter," Chet learns that he has inherited an elderly relative's prize possession, a letter written by Abraham Lincoln. Though the particular significance of Lincoln for blacks is never mentioned explicitly, Chet's family legacy is insistently bound to the histories of slavery and reconstruction. Consider also the happy integrationist comedy *Julia* (1968–71), which featured Diahann Carroll as a widowed nurse raising a young son entirely (and apparently unproblematically) surrounded by white people, and which studiously avoided the race problems that dominated national news throughout its run. Even *Julia*, though, couldn't entirely sever its main character from the racial meanings swirling around it, as evidenced by a key element of the show's back-story. Julia's husband had been a military pilot who was killed in Vietnam. Not only did this open the series to criticisms that it reproduced the stereotype of absent black fathers promulgated by the 1965 Moynihan Report (Bodroghkozy, 1992); it further alluded to the disproportionate losses suffered by blacks in Vietnam. An episode called "The Champ is No Chump," in which Julia wins a date with the heavyweight boxing champion on a TV dating show, inadvertently reinforces these meanings in its evocation both of Muhammad Ali and of his protests against the war.

In the 1970s, the various sitcoms produced by Norman Lear's Tandem Productions, including *All in the Family* (CBS, 1971–9), *Sanford and Son* (NBC, 1972–7), *Good Times* (CBS, 1974–9), and *The Jeffersons* (CBS, 1975–85), establish definitively the link between racial representation and the social problem. Indeed, this link is established despite, or more accurately because of these series' "liberal" intentions, as the project of representing racism (*All in the Family*); racial striving, deracination and assimilation (*The Jeffersons*); or raced poverty (*Sanford and Son*, *Good Times*) effectively renders the collation between racial subjects and race-as-problem total. This tendency was reinforced by a series of made-for-TV movies and mini-series on racial topics that punctuated the 1970s. These included *The Autobiography of Miss Jane Pitman* (1974), *Roots* (1977), *A Woman called Moses* (1978), *King* (1978) and *Roots: The Next Generations* (1979). In each of these, blackness was equated with its capacity to suffer or transcend the trials imposed by racism.

The rightward swing in the United States that brought Ronald Reagan to power in 1980 coincided with a broad renegotiation of the meanings of race.

With considerable success, conservatives sought to undo many of the gains made by racial minorities during the previous two decades by broadly undermining the legal bases on which programs like affirmative action had been based (Crenshaw et al., 1995). Scholars of television, including Herman Gray, Richard Campbell and Jimmie L. Reeves, have argued that television played a crucial role in this process through, on the one hand, news, documentary and "reality" coverage aligning African Americans in particular with welfare fraud, drug abuse, bad parenting and a "culture of poverty," and, on the other, fictional representations of middle class or affluent blacks like *The Cosby Show*, which suggested that the fight for racial equality had been won (Gray, 1989, 1995; Reeves and Campbell, 1994). Indeed, Gray has argued convincingly that these two very different types of programs, which conveyed dramatically different pictures of race in America, relied on each other as dialectical opposites which somehow produced the same social meaning: that, whether because blacks have "made it" or because we are hopelessly corrupt culturally or morally, aggressive civil rights measures were no longer necessary or appropriate (Gray, 1989). Thus while the terrain of the nation's racial imaginary shifted considerably during the Reagan–Bush period, television's racial representation was still very much bound to race-as-problem.

I would suggest that the culmination of the televisual tendency I have been describing came in 1992, with the Rodney King video, the coverage of the uprisings following the verdict in the Simi Valley trial, and the reinscriptions of the riots to be found in fictional television's recycling of these events through the 1992–3 television season (Caldwell, 1998; Torres, 1998). It was then that video technology and television institutions came face to face with the results of the policies of the Reagan–Bush administrations, and in particular with the urban poverty those policies had deepened. Race came to mean, depending on your point of view, either desperate urban poverty or senseless urban menace. In ways whose importance will become clearer in the next section, the obsessive coverage of the looting during the riots situated the residents of South Central Los Angeles *in opposition to* the commodity form, which, viewers were to understand, they could not obtain through legal means.

The Spectacle of Race as Commodity: Extreme Makeover: Home Edition

> The glamour of difference sells well.
>
> Paul Gilroy, *Against Race*

The deployment of the figure of race-as-problem in televisual representation relied on a particular political economy of the medium that has now largely been superseded. Race-as-problem, in other words, is a minoritizing gesture whose effect is to draw a boundary around the racial "other" for the amusement or education of the majority white viewer. As television historian Michael Curtin has noted, even the network television civil rights documentaries like those I've mentioned were directed at white audiences; quoting a 1957 CBS study of documentary viewers arguing that "social and political problems pertaining to a minority should always be presented in terms of their importance to the majority," Curtin notes that network documentaries "systematically marginalized the African American viewer" (Curtin, 1995). Clearly, the tendency Curtin documents persisted long past the late 1950s. And insofar as treating race as a problem presumes a "national" audience that is majoritarian in terms of class as well as race, this tendency corresponded well with the broadcasting strategy of the major networks when they were still the only game in town. But in the new, demographically fragmented universe in which networks must compete with the narrowcasting strategies of cable as well as with home movie viewing, however, US television must continually invent new strategies to garner newly constituted audiences, audiences which now might plausibly be conceived as composed of multiple minoritarian segments, each with merely a partial claim to the status of "national" audience. How else to understand the recent explosion of gay-themed programming (from *Ellen* and *Will and Grace* to *Queer Eye for the Straight Guy*, *Boy Meets Boy*, *The L-Word*, etc.) than to understand that minoritarian audiences, identities, and subjectivities have now become commodities to be traded among programmers and advertisers? How, as well, to understand a show like the episode of *Extreme Makeover: Home Edition* that aired on April 15, 2004, and that illustrates television's new approach to racial representation by attempting to decouple race from its social contexts and to resituate it as one commodity among others.

Extreme Makeover: Home Edition was spun off from the highly successful reality show *Extreme Makeover* (ABC, 2003–), which plucks its makeover candidates out of their mundane lives and sends them to Beverly Hills, where they are attended, over a period of weeks, by plastic surgeons, cosmetic dentists, Lasik experts, personal trainers, hairdressers, makeup artists and image consultants. "Re-done" from tip to toe, the candidates then return home, where they are "revealed" to their friends and family, who, as if coached to do so, invariably scream and/or weep at the sight of their newly glamorous loved ones. *Extreme Makeover: Home Edition* takes a similarly drastic approach to the problem of home repair. In every episode, a family in need of renovations is chosen to have its house demolished, rebuilt and refurnished in seven days by a crew led by series host Ty Pennington, formerly of The Learning Channel's sleeper hit *Trading Spaces* (2000–). Families are chosen based on the urgency of their need: a couple with two children under five expecting triplets, a family in which the eldest son has recently become wheelchair-bound, a woman with nine children whose husband has recently died of cancer. Despite the differences between their ostensible objects – bodies and houses – *Extreme Makeover* and *Extreme Makeover: Home Edition* share an abiding confidence in the power of commodified goods and services to transform lives. As one *Extreme Makeover: Home*

Edition viewer cannily observed in the show's forum on the website televisionwithoutpity.com, the show is a weekly hour-long commercial for its biggest sponsor, Sears.

The particular episode of *Extreme Makeover: Home Edition* that interests me is set in Watts, the neighborhood of South Central Los Angeles that saw some of the earliest and worst urban rioting of the 1960s. Indeed, the episode wastes no time in reinscribing the area's almost mythic status in the national racial imaginary: as the massive bus that – we are to believe – ferries the crew from renovation to renovation approaches the house in question, Ty announces "Alright, guys, so welcome to Watts. It's not exactly Pleasantville. And I think a lot of people are quite frightened of it, to be honest with you." Six short shots accompany Ty's speech: from a shot of sneakers tied together and flung over electrical wires, with Simon Rodia's Watts Towers in the background, we cut first to the interior of the bus on "Welcome," to a street sign reading "Watts." As Ty notes the area's dissimilarity to Pleasantville, we see a police car drive by before the camera pans to the right and tilts to take in the height of the Towers. Then, as Ty remarks that "a lot of people" are frightened of the neighborhood the camera offers us an image of a young black man riding a bicycle. On "to be quite honest with you," the camera passes over a pile of rubble – concrete and rebar – to the *Home Edition* bus going by in the background.

But the episode does not rely solely on Ty's introduction and the visual signs of the neighborhood – the street sign, the towers and, inevitably, the police car, which alludes not only to the neighborhood's current distress, but also to the incident of LAPD brutality that sparked the 1965 uprising. The episode goes further to set the scene, appealing to "official" civic history to make its point. From Ty's introduction, we cut to LA mayor James K. Hahn, who asserts, over images of street scenes and a menacing outdoor mural featuring a human skull, "Everyone remembers the Watts riots of 1965." He goes on to note "Obviously there's problems that happen in this community and continue to happen. It's what people call a high crime area. The good news is that it's starting to change." On the word change, we see the *Home Edition* bus coming up the street, and hear one of the show's signature upbeat musical themes. Hahn goes on: "Over the last year [there's been a] big reduction in crime in this area. A lot of that is due to people like Sweet Alice Harris." Thus *Home Edition* links its own arrival in Watts with the positive change associated with reduced crime. And in the process, it neatly links its project of familial and physical "renovation" with the object of this episode's beneficence, "Sweet Alice" Harris.

Sweet Alice Harris is a serious and committed African-American activist and resident of Watts, who began her work before the 1965 riots. Named the Woman of the Year in 2002 by California Lieutenant Governor Cruz M. Bustamante, Harris is founder and Executive Director of Parents of Watts, an organization dedicated to a range of anti-poverty work on issues related to education, health care, housing, parenting, and drug abuse. In announcing the award, Bustamante called her "one of the most influential community leaders of her generation."

The show, on the other hand, is more apt to describe her as "America's grand-mother," "a living saint" and "a giver" who "loves children, loves people." In other words, the show downplays Harris' political savvy and experience, prefer-ring to construct her simply as a loving and God-fearing grandma who likes to keep an eye on the kids in the neighborhood and happens to run the local community center. Bracketing the political implications of that decision, how-ever, what is more crucial here for my purposes is that *Extreme Makeover: Home Edition* has not been spurred to come to Watts to renovate Harris's house by her forty-plus years of anti-poverty work. Rather, they have come to renovate the house because a flood has recently damaged it.

The "freak storm" that flooded Harris' home serves to de-couple even Harris from race-as-problem, rendering her and Watts the victim of a "natural disaster" rather than a series of political ones – the decisions that have kept her and her community poor. As a flood victim, Harris becomes a kind of Everycalifornian. Her damaged house in Watts becomes as much a stand-in for the Malibu man-sions that have slid into the ocean during mudslides or the Oakland homes ravaged by fire as a sign of the urban unrest so often signified by "Watts" or "South Central." By depicting Harris as "America's grandmother" and by em-phasizing that, while she may *deserve* help for her work in the community, she *needs* help because of the flood, *Extreme Makeover: Home Edition* does its best to sever this organic intellectual from the effects of race that have so thoroughly organized her life and work.

Crucially, though, this sleight-of-hand cannot succeed completely. It is a testament to the depth of Harris's engagement with her community, and with African-American culture, that the episode, in representing her, must also repre-sent the social and political conditions in which she and her (mostly African-American) neighbors live. We do catch glimpses of the work carried out by Parents of Watts, and the show understands that to care for Sweet Alice is to care for her neighborhood. Thus, the show brings in a number of volunteers from Sears to distribute free mattresses and bedding to replace what had been damaged by the flood, repair damaged appliances, and install fences and plant trees and flowers in neighboring yards. Sears also provides a roomful of comput-ers for the community center, while the *Extreme Makeover: Home Edition* crew renovates its public spaces.

But the most effective counter to the show's tendency to remake and de-racialize Harris is Harris herself, who insistently punctuates her screen time with reminders of her location. "Do y'all know you're in *Watts*?" she asks the crew when they arrive, and we are to understand *Watts* to situate the show and its viewers with her in a location that is not merely geographic, but also historical, socio-economic and racial as well. During the "reveal," when she and her family are walking through the re-done house for the first time, Harris breaks down in tears in her new dining room. Pennington asks, "You OK, sweetie?" And she replies, "I just never thought nobody would do this, would give us this down in Watts. I just didn't believe. And now I got what everybody else in the magazines,

TV have. I didn't have to leave Watts. Lord, thank you. Couldn't leave. Couldn't leave." We cut to a talking head interview with her: "I didn't think nobody would do that for me comin' from Watts. And they *knew* I was from Watts. But they wasn't afraid to come to Watts." Clearly, for Harris, Watts is a space that is discontinuous with the rest of America, the America depicted in magazines and television, the America that is saturated with the commodity form. In insisting on her enduring connection to Watts – she is both "from Watts" and "couldn't leave" – Harris resists both the show's premise and its promise: that the commodity form is universally available, as close as the nearest Sears.

Thinking Race: Theoretical Models

Scholarly accounts of racial representation in American television have been dominated by the conceptual category of the "stereotype." A good example – though one might cite many others – of this tendency is Jannette L. Dates and William Barlow's collection, *Split Image: African Americans in the Mass Media*, which treats many aspects of mass-mediated culture and contains several sections on television (Dates and Barlow, 1990). In the general theoretical introduction to the volume, Dates and Barlow trace a number of African-American stereotypes back to their historical origins in antebellum popular culture, arguing that the versions persisting in contemporary commercial culture may be meaningfully linked to these origins in minstrelsy. For Dates and Barlow, widely circulated stereotypes such as the comic Negro, the Jim Crow figure, the pickaninny, the tragic mulatto and the Aunt Jemima are perpetuated by whites in an effort to secure and maintain cultural power. The history of African-American mass-mediated representation, then, is the history of a "split image," in which "the dominant trend in African American portraiture has been created and nurtured by succeeding generations of white image makers, beginning as far back as the colonial era," while "[i]ts opposite has been created and maintained by black image makers in response to the omissions and distortions of the former" (p. 3). The intellectual and political purchase of accounts like that of Dates and Barlow is considerable, for a number of reasons. The stability of the white oppressor/black victim binary is always tempting, and often accurate. In addition, such readings have historically been successful in organizing aggrieved collectivities of (usually middle-class) African-American spectators into counterpublics, as in the NAACP's campaign against the television version of *Amos 'n' Andy* (Cripps, 1983; Ely, 1991). Finally, these accounts are extremely efficient at replacing the pain of outrage and indignation with the pleasures of self-assured knowledge.

But, as I have suggested at greater length elsewhere (Torres, 2003), the analytic focus on the stereotype may produce as many conceptual problems than it solves. For one thing, the focus on "negative images" ignores the complex, and often resistant, spectatorship engendered by the sheer egregiousness of such stereotypes among minoritarian subjects. For another, as a mode of reading, the exclusive attention to stereotypes tends to flatten its textual objects to such an extent it almost always under-reads their complexities, and tends also to sever particular texts from the televisual flow which overdetermines and complicates even the most blatantly offensive TV moments. Finally, by taking what is only the most obvious form of televisual racism – the stereotype or "negative image" – as the medium's singular or even dominant form of racial ideology, stereotype-focused accounts risk drastically underdescribing other problematic representational modalities in which racial types figure marginally, if at all. Such forms are more subtle and may be just as insidious.

In a televisual world in which racial difference is being made to serve the medium's commodifying logic in new ways, this last point is crucial. *Extreme Makeover: Home Edition* is nothing if not respectful, even valorizing, of Sweet Alice Harris. Indeed, the problem with the program's racial politics has little to do with its depiction of Harris at all. Rather, what's problematic about the deployment of race here lies more precisely in the program's – dare I say it? – colonization of Harris and the landscape of Watts in the service of the belief that poor people need consumer choice rather than political choice, more commodities rather than more power. That *Extreme Makeover: Home Edition*'s worldview dovetails so neatly with the interests of its advertisers should remind us that Harris, and the "difference" for which she stands in, are deployed here not only in the service of capitalist ideology generally, with consumer choice offered as a compensation for political stagnation, but also specifically in the immediate economic interests of particular capitalist firms: ABC, Sears, and the program's other sponsors.

Thus, the focus on stereotyping is inadequate, theoretically and methodologically, to address the complexities of racial representation now. More useful is what we might call a "racial formation" approach. This approach, which has been elaborated in the humanities by scholars like Henry Louis Gates, Jr. and Kwame Anthony Appiah, and in the social sciences by such thinkers as Michael Omi and Howard Winant, understands race, not as a set of physical characteristics, but rather as the historically-variable *cultural meanings* assigned to such characteristics (Gates, 1986; Appiah, 1995; Omi and Winant, 1994). Such an approach stresses, first, that race, as a *biological* category, is meaningless. In Appiah's words, "there is a fairly widespread consensus in the sciences of biology and anthropology that the word 'race,' as least as it is used in most unscientific discussions, refers to nothing that scientists should recognize as real" (p. 277). Similarly, according to Omi and Winant, "there is no biological basis for distinguishing among human groups along the lines of race. Indeed, the categories employed to differentiate among human groups along racial lines reveal themselves, upon serious examination, to be at best imprecise, and at worst completely arbitrary" (p. 55).

But if race has no biological meaning, it is replete with cultural meaning. Indeed, it is precisely because race cannot be materially fixed in the body that it can be so richly imbued with meanings of other kinds. As Gates has written, "[r]ace is the ultimate trope of difference because it is so very arbitrary in its application" (p. 49). It is as a trope, or figure of speech, that common uses of the term race are able to trade on the biological connotations of the term in order to assign characteristics differentially to different groups of persons, and to make those assignments seem "natural" and inevitable. Thus Omi and Winant define race as "a concept which signifies and symbolizes social conflicts and interests by referring to different types of human bodies" (p. 55). They use the term "racial formation" to describe the sociohistorical process by which such racial meanings are created, altered and destroyed.

It must be emphasized here that to say that race is not a biological fact, but a cultural construction is not to suggest that our cultural understandings do not bear important or even deadly consequences for those whose "types of human bodies" are made to bear negative meanings by the cultures in which they live. Race may not be "real," but racism certainly is. "In this respect," as Appiah points out, "races are like witches. However unreal witches are, *belief* in witches, like belief in races, has had – and in many communities continues to have – profound consequences for human social life" (p. 277). In emphasizing the "consequences for human social life," Appiah here points to the ways in which racial understandings, however specious, have tended to *make themselves "true"* by generating concrete social practices. In this way, race links the realms of culture and society. Or, in Omi and Winant's formulation, "race is a matter of both social structure and cultural representation" (p. 56).

Thus, as an interpretive tool for thinking about television, this approach to race directs our attention to the social uses to which race has been put both on TV and in relation to it. It allows us to attend, in other words, not only to television's representations of race, but also to the ways in which social practices like television viewership, or social institutions like telejournalism, have been understood as raced within the television industry and by scholars and the public. In addition, such an approach has the crucial advantage of making visible the ways in which constructions of race change over time. The notion of racial formation, in particular, emphasizes the fact that racial definitions are continually being contested and struggled over, a simple fact that focus on the stereotype tends to elide. Perhaps most crucially, a racial formation approach insists on the centrality of race to our cultures' ways of producing and conveying meaning, across representations, institutions and practices. If race is one of the crucial ways in which we order the social, then it behooves us to notice and analyze its work systematically and thoroughly across the cultural field, and not just where its appearance happens to offend us.

A number of scholars have deployed approaches very much like the ones I have been discussing here in their analysis of television's racial projects. Herman Gray's invaluable *Watching Race* (1995), for example, considers the televisual meanings of "blackness" during the 1980s, a period in which racial formations in the United States were being broadly contested on both the right and left. Aniko

Bodroghkozy's work on *Julia* and *Good Times* situates these series, and the meanings of race they generated, within broader cultural discussions of race during the periods in which they aired (Bodroghkozy, 1992, 2003). And Phillip Brian Harper's essay, "Extra-Special Effects: Televisual Representation and the Claims of the 'Black Experience,'" elaborates the ways in which television's representations of African Americans have historically been subject to two contradictory desires: the wish for "realistic" and "authentic" mimetic depictions of working-class black life, and the craving for simulacral depictions that would uplift the race. All of this work is particularly valuable in that it attends to the meanings of race in play in communities of color as well as in dominant culture. In addition, this work insists on elaborating the connections between the representational and the social in careful and illuminating ways.

Thinking the Commodity

I have suggested, by way of a reading of *Extreme Makeover: Home Edition*, that television's uses of race have shifted recently, with the medium now deploying race, racial identities, and racial styles not to signify social problems, and not to display minoritized others for a majoritarian audience, but rather to garner as many fragments of the post-cable audience as possible in order to sell them to advertisers. Within these representations, racial difference becomes a commodity on offer, much like those on offer in the commercials, for today's hip, progressive consumer, for whom race is presumably no big deal. Raced bodies signify not an enduring legacy of oppression with which all Americans must struggle, but a newly reassuring form of consumer choice. Crucially, such representations attract not only audiences of color, but majoritarian ones as well.

Consider a recent national ad campaign in the United States for the telecommunications giant Verizon, designed to look like mini-sitcoms and featuring three families, the Elliots, the Davises, and the Sandovals (as of this writing, the ads could be found on the Verizon website at http://www.verizonmarketing.com/families/tv_ads.asp?movie=5). What's interesting about these commercials is not their cannibalization of the sitcom, but their alignment of racial difference with consumer choice. The Sandovals are Latino. The Davises are black. And the Elliots are a mixed-race family, in which the mother is Latina and the father is white. Seth Stevenson, who reviewed the ads for *Slate*, notes that, despite their shared theme music and format, they were produced by three different ad agencies, "each . . . a specialist in its chosen demographic." This assertion might appear to beg the question of what demographic, exactly, the Elliots are presumed to represent, and here is where things get interesting: as Stevenson puts it, "while the Davises and Sandovals are narrowly-targeted to specific demographics, the Elliots are Verizon's 'mass market family' . . . They're meant to appeal to everyone, nationwide, as a flagship symbol of Verizon's brand." Thus the mixed-race family has become, in television's new racial order, the figure for consumption generally. Stevenson, who sees the ads as "PC utopias," seems to have fallen prey to their logic; I would argue that they seem utopian to Stevenson not because they depict people of color without stereotyping them, but rather because they suggest that people of color have the same access to commodities as whites. But would such a state of affairs, even if it in fact existed, really constitute a utopia? In this respect, these ads share with *Extreme Makeover: Home Edition* a misplaced fantasy of and faith in the transformative power of the commodity.

We need more work on television that takes on the medium's changing collation of race and the commodity form directly and explicitly. A useful foundation for such work can be found in studies of television's production of capitalist ideology, the ideas that shore up existing capitalist social relations. The notion of ideology has a long and complex history within Marxist studies; the most pertinent strain of that tradition for the effort to think television's commodification of race would be that which uses Marx's notion of "commodity fetishism," elaborated in the first volume of *Capital* (1976). The process Marx describes, in which social relations are obscured by the relations among commodities in the marketplace, has served for many later thinkers as a model of how ideology both emerges out of and conceals the material relations of production (Lukács, 1971; Žižek, 1989).

Mimi White's "Ideological Analysis and Television" (1992) provides a useful overview of work on television and ideology. An early account focusing on television's ideological operations in the US context is Todd Gitlin's "Prime Time Ideology: The Hegemonic Process in Television Entertainment" (1979). In addition, much work in the British cultural studies tradition has engaged the notion of ideology; in particular, Stuart Hall has returned a number of times to the complexities of these issues. See, for example, his classic and widely reprinted "Encoding/Decoding" (1980); also relevant are "Culture, the Media, and the 'Ideological Effect'" (1977), "The Rediscovery of 'Ideology': Return of the Repressed in Media Studies" (1982); and "The Problem of Ideology: Marxism without Guarantees" (1983). Hall takes up the issue of ideology specifically in relation to race in "Gramsci's Relevance for the Study of Race and Ethnicity" (1986), and "The Whites of Their Eyes: Racist Ideologies and the Media" (1990).

Conclusion

Contemporary scholars and viewers of the medium find ourselves in a changing televisual landscape with respect to race. Though televisual racism is certainly still part of the terrain, it is less likely than ever before to take the familiar stereotyped forms. Instead, the dominant mode of television's racism now lies in the ways its representations tend to wrench persons of color out of the still-pervasive political context of white dominance and out of the still-relevant social context of communities of color. In the process, real political, social, and economic inequalities disappear. In their place, television offers up racialized figures as consumers first and last, producing a powerful if misleading alibi both for existing racial formations and for capitalism itself.

References

Appiah, K. A. (1995) "Race," in F. Lentricchia and T. McLaughlin (eds.), *Critical Terms for Literary Study*, Chicago: University of Chicago Press, pp. 274–87.

Bodroghkozy, A. (1992) "'Is This What You Mean by Color TV?': Race, Gender and Contested Meanings in NBC's *Julia*," in L. Spigel and D. Mann (eds.), *Private Screenings: Television and the Female Consumer*, Minneapolis: University of Minnesota Press, pp. 143–68.

Bodroghkozy, A. (2003) "Good Times in Race Relations? CBS's *Good Times* and the Legacy of Civil Rights in 1970s Prime-time Television," *Screen*, 44(4) (Winter), 404–28.

Caldwell, J. (1998) "Televisual Politics: Negotiating Race in the L.A. Rebellion," in S. Torres (ed.), *Living Color: Race and Television in the United States*, Durham: Duke University Press, pp. 161–94.

Crenshaw, K., Gotanda, N., Peller, G., and Thomas, K. (eds.) (1995) *Critical Race Theory: The Key Writings that Formed the Movement*, New York: The New Press.

Cripps, T. (1983) "*Amos 'n' Andy* and the Debate over American Racial Integration," in J. E. O'Connor (ed.), *American History/American Television: Interpreting the Video Past*, New York: Frederick Ungar, pp. 33–54.

Curtin, M. (1995) *Redeeming the Wasteland: Television Documentary and Cold War Politics*, New Brunswick, NJ: Rutgers University Press.

Dates, J. L. and Barlow, W. (eds.) (1990) *Split Image: African Americans in the Mass Media*, Washington, DC: Howard University Press.

Du Bois, W. E. B. (1969) *The Souls of Black Folk*, New York: New American Library.

Ely, M. P. (1991) *The Adventures of Amos 'n' Andy: A Social History of an American Phenomenon*, New York: Free Press.

Gates, H. L. (ed.) (1986) *"Race," Writing, and Difference*, Chicago: University of Chicago Press.

Gilroy, P. (2000) *Against Race: Imagining Political Culture Beyond the Color Line*, Cambridge, MA: Belknap Press of Harvard University Press.

Gitlin, T. (1979) "Prime Time Ideology: The Hegemonic Process in Television Entertainment," *Social Problems* (February), 251–66.

Gray, H. (1989) "Television, Black Americans, and the American Dream," *Critical Studies in Mass Communication*, 6, 376–86.

Gray, H. (1995) *Watching Race: Television and the Struggle for "Blackness,"* Minneapolis: University of Minnesota Press.

Hall, S. (1977) "Culture, Media, and the 'Ideological Effect,'" in J. Curran, M. Gurevitch and J. Woollacott (eds.), *Mass Communication and Society*, London: Edward Arnold, pp. 315–48.

Hall, S. (1980) "Encoding/Decoding," in S. Hall, D. Hobson, A. Lowe and P. Willis (eds.), *Culture, Media, Language*, London: Hutchinson, pp. 128–38.

Hall, S. (1982) "The Rediscovery of 'Ideology': Return of the Repressed in Media Studies," in M. Gurevitch, T. Bennett, J. Curran and J. Woollacott (eds.), *Culture, Society, and the Media*, London; New York: Methuen, pp. 56–90.

Hall, S. (1983) "The Problem of Ideology: Marxism Without Guarantees," in B. Matthews (ed.), *Marx: 100 Years On*, London: Lawrence & Wishart, pp. 57–84.

Hall, S. (1990) "The Whites of Their Eyes: Racist Ideologies and the Media," in M. Alvarado and J. O. Thompson (eds.), *The Media Reader*, London: British Film Institute, pp. 7–23.

Sasha Torres

Harper, P. B. (1998) "Extra-Special Effects: Televisual Representation and the Claims of the 'Black Experience,'" in S. Torres (ed.), *Living Color: Race and Television in the United States*, Durham: Duke University Press, pp. 62–81.

Lukács, G. (1971) *History and Class Consciousness: Studies in Marxist Dialectics*, Cambridge, MA: MIT Press.

Marx, K. (1976) *Capital*, 3 vols, New York: Penguin.

Omi, M. and Winant, H. (1994) *Racial Formation in the United States: From the 1960s to the 1990s*, New York: Routledge.

Reeves, J. L. and Campbell, R. (1994) *Cracked Coverage: Television News, the Anti-Cocaine Crusade, and the Reagan Legacy*, Durham: Duke University Press.

Stevenson, S. (2004) "Ad Report Card: Customers Like Me," *Slate* (April 26), http://slate.msn.com/id/2099476/.

Torres, S. (1998) "King TV," in S. Torres (ed.), *Living Color: Race and Television in the United States*, Durham: Duke University Press, pp. 140–60.

Torres, S. (2003) *Black, White, and In Color: Television and Black Civil Rights*, Princeton, NJ: Princeton University Press.

Watson, M. A. (1994) *The Expanding Vista: American Television in the Kennedy Years*, Durham, NC: Duke University Press.

White, M. (1992) "Ideological Analysis and Television," in R. C. Allen (ed.), *Channels of Discourse, Reassembled: Television and Contemporary Criticism*, Chapel Hill, NC: University of North Carolina Press, pp. 161–202.

Žižek, S. (1989) *The Sublime Object of Ideology*, New York: Verso.

PART SEVEN

Television/The Public and Audiences

CHAPTER TWENTY-TWO

Television, Public Spheres, and Civic Cultures

Peter Dahlgren

Television is constantly in transition, in terms of its industrial structures, legal frameworks, technological parameters, and programming output. In terms of its sociocultural significance, it remains analytically a moving target for researchers. While television's programming has been an object of popular and academic criticism since its beginnings, there has been intensified debate, discussion, and research in recent years as observers try to come to terms with such global phenomena as reality TV and docusoaps (cf. Hill, 2004; Lochard and Soulez, 2003) and, more generally, the "trash" and "dumbing-down" character of the medium (cf. Dovey, 2000; Glynn, 2000). In Europe, the ongoing political integration of the region is juxtaposed to the still prevalent national character of television broadcasting and issues of national/cultural identity (Richardson and Meinhof, 1999; Weiten, Murdock, and Dahlgren, 2000; Ward, 2002; see also Schlesinger, 2004). The evolution of broadcast journalism is also being addressed, not least in terms of the "rolling" 24-hour news format of the satellite TV news channels (cf. Robinson, 2002). The implications of post-9/11 and the "war on terrorism" also constitute important current themes (Kellner, 2003; Thussu and Freedman, 2003; Zelizer and Allan, 2002).

In these diverse contexts, and also in many others, the role of television is analytically framed – explicitly or implicitly – by the thematic of the public sphere. Studies in these areas address, in a variety of ways, television's role in maintaining and fostering this communicative space that is so vital for the health of democracy. The public sphere as an analytic concept, deriving from the work of Habermas in the early 1960s (and translated fully into English in Habermas, 1989) emerged as a useful analytic perspective for making both the empirical and the normative sense of the media from a standpoint of democratic ideals. With its emphasis on the historical evolution of the media under capitalism, its insistence on the norm of unconstrained access to information, on wide ranges of views, and on the role of discussion among citizens as the foundation for political opinion formation, this concept has inspired countless studies of the media. It has been explored not least specifically in relation to television (Dahlgren, 1995; Price, 1995), and, especially in the United Kingdom, was put to use as a theoretical foundation for the defense of public service broadcasting (Garnham, 1983). Habermas himself has subsequently modified some of his earlier views (Habermas, 1996), toning down some of his earlier pessimism, as well as updating them in keeping with the ever-sprawling character of media culture of late modernity. In regard to the latter point, we can note that to speak of *the* public sphere has become all the more inaccurate; we have instead a landscape of many – often overlapping – public spheres. The plural form is often linguistically awkward, so we tend to use the singular; nevertheless, we should keep in mind its multifarious character.

Like the notion of "democracy" itself, the concept of the public sphere has a strong normative dimension. It is invoked to remind us of ideals and visions, to suggest a trajectory to follow. The concept has entered into wide circulation; while it originally was understood as a product of Frankfurt School critical theory, it is now often used as a generic shorthand for the democratic goals and responsibilities of journalism and of civic life.

In this regard, it may well be that the concept has become rather jaded. While the notion has become common, I think a point was reached when many observers felt that the media, in particular television, seemed – with some notable exceptions – to be galloping in the opposite direction of the ideals it embodies. There emerged a new wave of literature – from both academic and journalistic corners – cataloguing the problems and deficiencies of journalism, not least on television (cf. Fallows, 1997; Franklin, 1997; Bourdieu, 1998; McChesney, 1999; Downie and Kaiser, 2002; Gans, 2002). This literature often accentuates how the political economic logic of market forces impact on media output (see also McManus, 1994; Baker, 2002). While the public sphere theme has in recent years been applied – with various degrees of optimism – to the Internet (cf. Thornton, online; Anderson and Cornfield, 2003; Jenkins and Thornburn, 2003; Dahlgren, 2001, 2004), it seemed that in regard to the traditional mass media, and television in particular, there was little new to say, and certainly not much new that we might find encouraging.

In this chapter, I want to return to the topic of television and the public sphere, updating my earlier perspectives (Dahlgren, 1995). The concept of the public sphere remains a central construct in our ongoing efforts to understand the relationship between democracy and the media. And television remains the dominant medium for most people in Western democracies, despite the profound media revolution associated with the Internet. Yet the conceptual premises of the public sphere become all the more problematic in the evolving historical situation, as the media, with television in the vanguard, seemingly drift ever further in the direction of entertainment and consumption. How do we orient ourselves in this reality via the concept of the public sphere? Can the concept help us generate anything more than a lament over the present situation?

Concurrent with these developments, we have been witnessing a rising international chorus pointing to the dilemmas and crises of Western democracy more generally: the formal political arenas are stagnating, with declines in civic engagement. Also, at the margins of the formal political system we observe new modes of involvement, loose networks of civic activism, and new forms of politics, e.g. life politics, identity politics and single-issue movements. In this emergent context, public sphere theory does not have much to say about the origins or character of civic engagement. Public sphere theory has seemingly operated with an implicit view that most citizens, if given the chance, are pleased to participate in formal democracy. Such, of course, is not necessarily the case. Likewise, it has little to say about the newer forms of engagement. There is a decided sociocultural blind-spot here in the strongly normative character of public sphere theory. I try to address this by taking up what can be seen as the cultural prerequisites of the public sphere, the theme of why and how citizens become motivated to participate in the public sphere – and, by extension, politics – in the first place. I label these prerequisites civic cultures (i.e. again we are dealing with plurals, as I shall discuss).

Building on a good deal of literature from the past decade, I begin the discussion by briefly reviewing some of the largely familiar ways in which the late modern media milieu is seemingly at odds with traditional conceptions of the public sphere. Not least, I underscore how the boundaries between a set of polarities that we can subsume under the headings of "public" and "private" have become destabilized, with their boundaries becoming all the more porous, evoking the issue of how we define politics of the political in relation to the non-political, especially the popular. The challenge becomes how to make this forcefield analytically useful. With a starting point in John Ellis' (2000) notion of television as a "working through," I try to chart a path that avoids the rigid dichotomies yet retains the analytic utility.

Also, I underscore that the vision of the public sphere points to more than just media output; it also involves interaction and engagement on the part of citizens. Engagement in the public sphere can be seen to have cultural origins, which I refer to as civic culture. Civic cultures can thus be understood as a set of preconditions for populating the public sphere. Such cultures are important in facilitating engagement in the broad domain of what we might term the politically relevant, in creating a climate that is conducive to citizen participation in the shaping of society's political life, and in fostering fluid communicative borders between politics and non-politics.

Many factors impact on civic cultures, including all the forces of social structure, yet the media, and not least television, also play a very important role. Thus, from this analytic angle of vision, television has a dual role in terms of the public sphere. On the one hand, it serves as an important (and problematic) institution – a site of the sprawling public sphere. On the other hand, in contributing more broadly to the character of civic cultures, television also serves, seen in social constructionist terms, to pre-structure civic dispositions to the public sphere that it offers. In my discussion, I will often use the ubiquitous genres of reality television as an empirical referent. I don't think we necessarily end up with a cheerier picture of the present situation, but hopefully we will have a more nuanced understanding of what is actually involved – and at stake – when we speak about television and the public sphere.

Television, Public Spheres, and the Late Modern Media Milieu

In today's world it has become quite difficult to generalize empirically about television, and equally challenging to grasp it as a totality via some theoretic discourse. Television has simply become too heterogeneous in too many ways in terms of its technology, structures, programming, and uses. Ellis (2000) suggests that having passed through the ages of scarcity and availability, television in the new century is entering the "age of plenty," with an enormously abundant output. From a variety of perspectives, it is a powerful medium, yet when we speak of the *power of* television, we must not lose sight of the fact that we also have to take into account the *power over* the medium that various elite groups in society can exercise. Television is an institutional actor, but is also acted upon – a distinction that is no doubt empirically blurry in many cases, given television's embeddedness in the political and economic power structures of society. And whatever arguments we care to make about the responsibility that television has to society, it cannot ignore – even in its public service mode – the impact of audience preferences. Thus, television's output, the topics and modes of representation, are shaped to no small extent by the impact of circumstances and relations beyond its own organizational settings; it operates in a configuration of sociocultural forces.

The medium is also becoming increasingly connected to telecommunication and computer technologies, with digital transmission and interactivity; in particular, we are seeing various forms of television-Internet technical convergence. Just as significant, I would argue, is the ever-increasing integration of television and the larger *media culture*. By this I mean that via its forms, contents, and increasingly self-referential character, the late modern media milieu is gradually moving toward enhanced interconnectedness. Thus to talk about "television" means that we often get inexorably involved in talking about other elements of this media milieu as well, even if television is its most prominent element.

For example, *Big Brother* and other docusoaps on television generate programs *about* these programs on television. They are covered in various newspapers, weekly and fan magazines; on the Internet one can follow plot developments and take part in discussions about different series. A number of the participants become mini-celebrities and we can intertextually follow their (albeit usually short-lived) careers in stardom in various media. Shrill tabloid headlines announce "scandals," naming various docusoap participants, with the understanding that for the general public there is no mystery as to who these people are. Such developments do not mean that television has ceased to exist as a distinct cultural institution, but it does suggest that we need to keep in mind its increasing fusion – and central position – within the larger media landscape.

Other salient attributes of the medium we should mention here include abundance, speed, the prevalence of the image, and its specific modes of cognition. These qualities are by now quite familiar to us, and hardly controversial. Yet, when juxtaposed to the framework of the public sphere, they retain a challenging relevance as the debates over television's role in democracy have indicated. The public sphere perspective, grounded in the legacy of the Enlightenment, has favored such notions as the transmission of information, the development of knowledge, and the centrality of dialog, reflection, and reason. Television, and the contemporary media milieu in general, is in part supportive of this legacy, yet in part presents an obstacle to its realization.

The abundance of television offers the viewer enormous amounts of information (using the term in the broadest sense) from which to choose. Yet this abundance easily becomes disorienting, competing intently for our attention, hovering in its massiveness beyond our capacities for making sense. There is a veritable deluge of output, with the sensory stimulation of some programming approaching the threshold of normal human perception and comprehensive capacities. Gitlin (2001) speaks of "media torrents" and "super-saturation," and reminds us that under such circumstances, we (must) devise strategies for navigating the flood, for sorting and selecting from an output infinitely larger than we can meaningfully deal with.

Also, while the speed – manifested not least in television's sound bites and visual cuts – can help us keep abreast of the very latest developments, the velocity of mediated communication also leaves us little time for critical thought. Thus, according to theorists like Virilio (cf. Virilio, 2000; see also Armitage, 2000), the speed built into modern media culture threatens in the long run to subvert modernity and democracy. From our public sphere angle, such high-speed communication is deemed to be out of sync with the tempo of human reflection and discussion, generating stress and unreason, not least in situations of decision making. This speed also tends to contribute to a spatial-temporal "media time" at odds with the experiences of the non-mediated world. The media world thereby takes on the attributes of a separate symbolic reality. Moreover, that television – and much other media output – is so visual in character has from its beginning been a problem in terms of Enlightenment rationality. While images can be very informative, they traditionally are also associated with emotional affect. That television has largely become a "pleasure machine" in our civilization further exacerbates these tensions with the Enlightenment-based understanding of the public sphere.

More concretely, in the last decade or so, we have witnessed dramatic developments in the formats of television, not least within factual television, or journalism (very) broadly understood. The docusoaps, reality programs, talk shows, various forms of infotainment magazine shows, and even the traditional news programs themselves, manifest rather specific and obvious trajectories of development toward the sensationalist. These developments mesh with the deregulation and the increased power of market forces, and have evoked considerable critical response, as I mentioned above. One might argue that such tendencies toward tabloidization are nothing new per se – their antecedents appeared concurrently with the rise of the mass media in the nineteenth century – but the rapidity and extent of the changes in the past decade or so have been impressive. We see a decisive move away from the traditions that strive for "objective" rendering of the world toward approaches that underscore the personal, the sensational, the subjective, the confessional, the intimate.

The horizons of Enlightenment rationality thus seemingly sit uncomfortably with many of the features that define contemporary television. One might, however, make a conceptual distinction between the general characteristics of the medium – its technology and fundamental media logic – and the manner in which it is used. Striving to retrieve television from its postmodern vortex, Graber (2001) draws on empirical studies with viewers to conclude that television *can* be an excellent medium of traditional news and information – if it is properly used. The point is that under the present circumstances of political economy, it is being misused. While it may be argued that such a position is consequently not of practical relevance, I would say that it at least keeps the door open for critical reflection on – and professional experimentation with – the possibilities of the medium.

Television's Leaky Hegemony

If we pull together such critical analysis and commentary on television, it may actually seem odd, as Dovey (2000) points out, to mobilize the Habermasian tradition of the public sphere at all in regard to television, if the reality seems so far removed from the ideal. Yet Dovey reminds us that there are in fact other ways of looking at television's contribution to the public sphere, and he cites Scannell (1996) as an excellent example of this. Scannell argues that television, at least in its national, public service context, has done a great deal to resocialize private life toward a shared public culture. Over the years the trajectory of public service broadcasting has been to continually expand the topics and perspectives that can be aired and uttered in public, rendering more and more terrain as familiar and accessible to larger audiences, to be talked about and interpreted by them. In the process, television has been increasingly inclusive in incorporating audiences-as-citizens into the communicative space of the public sphere. The golden age of public service has, of course, faded into history, so we can only go so far with that type of argument. Yet Dovey indicates that there is in this line of reasoning a potentially useful analytic kernel we should hang on to – namely the idea of extending the public sphere, and making it accessible; I will come back to these themes.

There are, of course, also voices questioning the very premises of the Habermasian scheme. In the course of the debates over the years, much of the drift was toward alternatives to the traditional laments his and similar views have engendered. It was argued – without necessarily going over the postmodernist deep-end – that the strict rational premises of the Habermasian public sphere were not only unrealistic but indeed even undesirable. They wall off awareness and appreciation of other, alternative ways of knowing and relating to the world, which were also of importance for the life of a democratic society, such as the imaginary, the symbolic, the poetic, the ironic (cf. my discussion in Dahlgren, 1995; Corner, 1999, also has useful overviews of many of the key debates). In trying to steer some reasonable middle course between the seemingly stern rigor demanding discursive rationality, and popular postmodernist relativism that celebrates whatever programming people choose to watch, I find that Ellis's (2000) view of contemporary television provides a useful frame. It realistically incorporates the basic attributes of the medium, takes seriously its public functions, is sociologically well anchored, and avoids the pitfalls of the extreme conceptual positions. It also allies itself with the idea of television's potential for extending the public sphere, as illustrated by Scannell, and yet is not blind to the tradition of compelling critiques.

Ellis talks of television as a "working through," of television not providing any ultimate or definitive point of view, but rather offering its viewers vast amounts of transitory glimpses, preliminary meanings, multiple frameworks, explanations, and narrative structures for working through private and public concerns. It is largely oriented to the present moment, to the experiential rather than the analytic, to the personal rather than the structural. As he puts it, "Television attempts definitions, tries out explanations, creates narratives, talks over, makes intelligible, harnesses speculation, tries to make fit, and, very occasionally, anthematizes" (Ellis, 2000, p. 79). In the meanings that television offers viewers, uncertainty prevails over certainty; there is a perennial tentativeness in the voice of television. This working through operates via all genres and their hybrids, but is predicated on viewer familiarity with generic formulas as a key vehicle for the communication of meaning.

Such a view does not preclude mechanisms of mainstreaming; the working through of television largely takes place within broadly hegemonic boundaries. Television's working can also take massive and intensely politicized turns, as it has done in the post-9/11 era in the United States. And in a more long-term perspective, programs such as *The Apprentice* are sitting ducks for critique of ideology in the neo-liberal era. In this reality soap, led by the New York casino and building magnate Donald Trump, the participants compete to demonstrate their entrepreneurial skills by trying to eliminate their competitors, with the winner being rewarded with a high position in the Trump empire. Yet

Peter Dahlgren

the point of the working through perspective is that much of the meanings offered up are anchored in a transitory present; new issues and angles will emerge, there will always be a bit more cacophony than coherence – at least on the surface level that is most immediately relevant for opinions, attitudes, and practices. The more deeply rooted ideological dimensions are less transitory in character, and can be mobilized in new contexts. However, even political rhetoric based on deep ideological dimensions can diminish in their impact: for example, the grip that the Bush administration's patriotic post-9/11 rhetoric had on the US political climate is considerably less at this writing (spring 2004) than when it was first launched.

We can briefly contrast Ellis's view with that of Scheuer (2001), who makes the argument that there is a conservative, right-wing bias in television's modes of representation. In this analysis the tendency toward personalizing, dichotomizing, putting style over substance, the emotional and physical over the intellectual and the moral, and the general aversion towards complexity, all serve to support social and political conservatism. While it is true, for instance, that the medium tends not to promote analytic engagement, and that personification and psychologism may work against perspectives of collective well-being and responsibility, it strikes me that this type of reading founders precisely in its adherence to a traditional textual rationality. One can certainly appreciate Scheuer's sense of exasperation with television, and it cannot be denied that the modes of representation he criticizes often indeed carry a conservative import. However, I see no deterministic imperative that in principle precludes using such modes for progressive messages. Rather, I would here refer to the external social forces that impact on the contemporary messages of television to explain its ideological slant. And certainly his call for better "media literacy" – to be taught in the schools and elsewhere – to enhance "critical viewing" reflects an inadequate view of television that misses something essential about its cultural power and the nature of the modern media milieu.

Three more elements can be factored in to expand Ellis's low-key orientation. First, any critical approach to the ideological dimensions of televisual discourses in the late modern context must take into account an array of different trajectories that deal with, for example, not only class, but also gender, ethnicity, technology, environment – and that these articulate with each other in complex and at times contradictory ways in local, national, regional and global settings. We cannot posit one unified, singular emancipatory path; moreover, today the political dichotomies of left and right do not always provide clear guidelines. This does not undercut the possibility of critical analysis, but does suggest that it should retain a certain degree of modesty: it will always inevitably be partial, situated, and contingent. Secondly, we should keep in mind the lessons of reception research since the early 1980s, that different groups can make different ideological sense of the same programming (cf. Tulloch, 2000). Thus, "working through" does not deny the hegemonic character of television, but recognizes that such hegemony is loose, leaky and always at risk. And thirdly, though Ellis

418

does not make a big point of it, his approach avoids the unproductive polarization of the rational and the affective. In our everyday lives we make sense of our experiences, ourselves, and the world around us largely via what I call an arational mode, a combination of using our head and our heart. There is no reason why the public sphere should – or even could – be any different.

The Public, the Private, and the Political

Yet, even if we accumulate these analytic gains that can help us to approach the public sphere in a conceptually less stressed manner, questions remain. Certainly one of the central quandaries of public sphere theory is that social and cultural evolution continues to scramble the distinction between public and private. This is a development that is abundantly visible not least on television (e.g. the most private intimacies between docusoap participants become public; the private lives of celebrities and even politicians become public). The traditionalist stance is to strive to focus narrowly on "political communication" in the formal political arena, in official mainstream media, and in ritualized, disciplined modes of communication. In the process it thereby shuts its eyes, so to speak, to a lot of reality, including – as I touched upon above – the dominant attributes of the television medium and the major trends in programming genres, as well as to the major sociocultural trends of late modern democracy.

The concepts of public and private – without here going into formalized definitions – encompass an ensemble of notions that readily align themselves into sets of polarities, in a sense organizing some of the topics I have presented above. The idea of "public" in traditions like the Habermasian is, as I noted, implacably associated with reason, rationality, objectivity, argument, work, text, information, and knowledge. One might also add: the discursively dominant, the authoritative, the masculine, the Caucasian. The private resonates with the personal, with emotion, intimacy, subjectivity, aesthetics, style, image, and pleasure. (There is a large literature on these themes as they pertain to the media; for some recent treatments, see the collection by Corner and Pels, 2003). The "public," of course, is also associated with the conventional perception of politics, while the private is often linked to the popular, as in popular culture. In the late modern media milieu, the private is also profoundly related to consumption.

At a fundamental level, what is at stake in the public sphere perspective is the question of where the political resides, and how it is positioned against that which is deemed non-political. There has been a flood of discussion and debate around this issue. Depending on circumstances, the seemingly private can often harbour the political, a point that has often been forcefully made, not least by feminist political theorists (Voet, 1998; Dean, 1997; Lister, 1997; Meehan, 1995). And certainly cultural studies, since its inception, solidly affirms the always potentially political character of popular culture, a view that has also entered into some corners of political science (cf., Street, 1997). The possibilities for topics to

419

become politicized are key elements of the open, democratic society. In the final instance, it can be said that politics has to do with decision-making, but the realm of "political relevance" is larger, always shifting – and can never be fully specified. Yet we might distinguish the steps in a schematic sequence, keeping in mind that the differences between the steps depend upon context rather than any innate essence: the non-political, the para-political (i.e. that which is in some way in the process of becoming contested), the fully political (which points to issues that are overtly being contested in society), and finally, established politics. The latter indicates the structured arena of formalized political conflict, a domain dominated by party politics. Such a sequence helps to loosen the imagery of a water-tight boundary around politics and the political, and emphasizes the democratic communicative processes whereby issues can potentially emerge in everyday talk among citizens, gain momentum, enter the public sphere, become contested, and – possibly but not necessarily – enter into the formal arena of politics. The women's movement is certainly a paradigmatic example of this sequence.

In today's world, of course, the public communicative space for citizen talk in the public sphere consists of a vast, sprawling social field of almost immense variety, criss-crossed by mass media as well as interactive media. It encompasses many different communicative contexts and styles, cultural frameworks, and power relationships. Yet it is easy to insert into this overall picture the understanding of television as a working through, a daily environment where topics are raised in various ways, and where themes become contested issues, taking on the character of the political. If decision making is the final step in the communications view of politics, there is always a good deal of contestation in the preceding steps that constitute the politically relevant. This is the public sphere seen from the perspective of the life-world, where through the processes of talk people can thrash out just what are important questions within the realm of the political and what positions they should take on them.

Certainly, some public spheres are closer to centers of power and decision making, e.g. the spheres constituted by writers, opinion-leaders in the elite press that comment on national and international politics. Other public spheres are more marginal, or even marginalized. Those on television are, of course, more broadly based, more popular. Television plays an important, if difficult to calibrate, role here, not least in the case of identity politics. Seen in the long term of the working through perspective, many of the issues raised and positions taken by various groups find their way into television and can achieve visibility and enhanced legitimacy, not least via entertainment programs. Here we must most decidedly move beyond the boundaries of reasoned deliberation and consider the impact of the arational dimensions. For example, the 1970s sitcom *All in the Family*, with its main character the working-class bigot Archie Bunker, through its humor and irony challenged discriminatory views on gay people and problematized the traditional view on women's social position. (I realize that critics pointed to a certain polysemy here – that viewers who agreed with Archie could get their prejudices validated – but I'd say in the long run the program undermined those positions.)

At the time, the controversies the program touched upon could reverberate with considerable electricity. Three decades on, the visibility and legitimacy of gay men and lesbian women on television has been enormously enhanced, not least in sitcoms, and gender issues are continuously framed in many programs. This is not to say that social discrimination or oppression is a thing of the past, but in terms of television's representations in these areas, the public sphere has been transformed, and center of gravity has shifted in society at large. The underlying assumptions that might frame specific issues relating to homosexuality today (for example, in regard to the right to adopt children) now depart from a different set of dispositions in the media – and in the larger society.

I should emphasize that the argument here has nothing to do with the issues of "taste" or high and low culture. Nor does my view veer toward a celebration of everything on television: it is not the case that since television is "popular," it thus simply "okay." Such is a populist argument. Instead, I suggest that we have to take seriously television's popularity, and consider how in its very unHabermasian modes, it can promote – as well as constrict – the public sphere. Simons expresses this view cogently: ". . . there is a structural and necessary relation between the popularisation of culture and the democratisation of politics . . ." (Simons, 2003, pp. 186–7). Yet, how do we analytically go further with this? If the popularity of television's modes of representation is relevant to our understanding of its role in the public sphere, what else might be of relevance? How can we probe further into its role, particularly when it comes to questions of fostering engagement among citizens? This is where I hope that the notion of civic cultures can be useful.

Citizenship, Social Agency, and Civic Cultures

For all its compelling qualities, the perspective of the public sphere still leaves a number of important themes unattended. Perhaps most significantly, it does not have much to say about *why* people actually participate in the public sphere. Yet it is precisely this theme of participation that has moved to the center of discussions in Western democracies. As the tensions and dilemmas of democracy have become increasingly debated in the past decade, among the key questions being asked is why is political participation on the decline, at least within the contexts of formalized politics? And, contrapositively, how are we to understand the rise of newer forms of engagement, as manifested, for example, in transnational movements? How is that many citizens have "bailed out" of formal politics and begun generating their own "counter public spheres" (cf. Fraser, 1992; Fenton and Downey, 2003; Warner, 2002), contesting not only specific issues, but also the very arrangements of communicative space? More generally, we could pose the question as follows: what factors contribute to people *acting* as citizens, i.e.

engaging themselves in the public sphere and political life (however we may care to define the parameters of the political)? From the standpoint of television, we would ask how its working through character, the attributes of its various representational modes, together with its reception and uses, may foster or hinder engagement. My sense is that a culturalist turn in the public sphere tradition might pen some analytic doors. With television in my sights, I here offer an overview of the civic cultures perspective that I have been developing (Dahlgren, 2000, 2002, 2003) as a way to conceptualize the factors that can enhance or impede political participation. I can here only sketch the contours, but I hope it may offer some suggestive new ways to think about the public sphere, not least in relation to television.

My starting point is the notion of political agency, to see citizenship as a mode of individual and collective action, and to begin to probe the cultural conditions of such agency. As a concept, citizenship traditionally builds upon a set of rights and obligations, historically evolved in society, and underscores universalism and equality. In the modern world it has usually been linked to the nation–state. More recently, citizenship has become an object of contemporary social theory (cf. Isin and Turner, 2003). For my purposes, a major strand highlights the subjective side of citizenship, as a dimension of our identities. (cf. Preston, 1997; Isin and Wood, 1999; Clarke, 1996; Mouffe, 1993, 1999). This perspective argues that for civic agency to be actualized, it is necessary that one can see oneself as a citizen, that this social category be a part of one's sense of self – even if the actual word "citizen" may not be a part of one's vocabulary. People's identities as citizens, that is, their sense of belonging to – and their perceived possibilities for participating in – societal development becomes a crucial element in the life of democracy. While citizenship still generally evokes the notion of a subjectivity that is positioned publicly, with the growing intertwining of public and private, citizenship as an identity can become increasingly interlaced with other dimensions of the composite, multidimensional self.

Agency is always situated and circumstantial, so to highlight citizenship in terms of civic agency requires that we refract it through a larger analytic prism. We have to ask how the preconditions for civic agency, for political engagement, might be understood. And if we further emphasize identity as a key element in civic agency, we need to ask what are the cultural factors that can impinge on this identity, and promote (or hinder) among people such perceptions of their (multifarious) civic selves – as well as promote their civic engagement. Cultures consist of patterns of communication, practices, and meaning; they provide taken for granted orientations as well as other resources for collective life. They are internalized, intersubjective; they exist "in our heads," as it were, guiding and informing action, speech, and understanding. Against that backdrop, we can conceptually launch the notion of civic culture, as referring schematically to those cultural patterns in which identities of citizenship, and the foundations for civic action, are embedded. Civic cultures, to the extent that they are compelling, operate at the level of taken for granted, everyday reality. It is more accurate to speak of cultures – in the plural – since the argument rests on the assumption that in the late modern world there are many ways in which civic culture can be accomplished – and citizenship enacted.

Civic cultures have in part a normative status; conceptually, robust civic cultures are necessary prerequisites for viable, civic-based public spheres – and, thus, the functioning democracy. They are also empirical and can be observed and analysed (see below). The perspective of civic cultures is interested in the processes of how people develop into citizens, how they come to see themselves as members and potential participants in societal development (as opposed, for example, to merely taking as given that people become legal citizens at a certain age), and such senses of self are maintained. The key here is to insist on the processual and contextual dimension: the political and politics are not simply given, but are constructed via word and deed, as noted above. I should emphasize that, at bottom, the idea of robust civic cultures resonates with notions of "strong" or "radical" democracy; it emphasizes participation and "deliberation," as well as a view of citizenship that is associated with neo-republicanism (cf. Barber, 1984; Benhabib, 1996; Eschele, 2001; van Gunsteren, 1998; Mouffe, 1992).

Civic cultures are potentially both strong and vulnerable. They can shape citizens; they can serve to empower or disempower. Oppressive power often results in an oppressive (or atrophied) civic culture, but we should avoid determinism: power is never fully guaranteed, and resistance can emerge via the mediation of revitalized democratic civic cultures (as we saw during the collapse of communism). Yet civic cultures sit precariously in the face of political and economic power. They are shaped by an array of factors: the nature of the legal system, factors of social structure, economics, education, organizational possibilities, infrastructure, spatiality, can all have their impact. And not least, the media, both the traditional mass media such as television and the newer interactive media, impact directly and routinely on the character of civic cultures via their form, content, specific technical logics, and modes of use.

We can think of civic cultures as resources, as storehouses of assets which individuals and groups draw upon and make use of in their activities as citizens. This calls to mind Putnam's (2000) notion of "social capital"; in his framework, such resources reside in the social connections within networks of reciprocal relations (Putnam, 2000, pp. 21–4). With his "bowling alone" metaphor, Putnam captures not least the decreasing communicative interaction among citizens, as well as declines in shared values, trust, and reciprocity. Increased fragmentation and atomization leads to a decline in social trust, which further inhibits participation. Putnam and others look to the sociocultural landscape for explanations – and find for example, the dumbing-down effects and monopolization of time that is associated in particular with television (a view that has elicited much debate; cf. Edwards, Foley and Diani, 2001). This view of social capital underscores its character as a resource, but the sociological framework as such does not lend itself readily to analysis of the processes of meaning making and subjectivity that

I have in mind with the notion of civic cultures. We need to incorporate some of the political sociology involved here, but go beyond it and make a "cultural turn." The following section attempts do this by presenting civic cultures as a circuit with five dimensions that lend themselves to empirical investigation. I will illustrate each with references to television.

Dynamic Dimensions: As Seen on TV

The notion of civic cultures that I am developing explicitly blends elements from a number of different intellectual traditions, as will become obvious. Such synthesizing can sometimes be a risky business, but my sense is that the modest melange I come up with serves its purpose. The first obvious element is, of course, the public sphere perspective; it takes considerable conceptual initiative, yet, as I have suggested above, carries within it a number of problems, such that it seems at times oddly removed from the everyday sociological realities. Secondly, I draw upon some currents from political science, or, more specifically, political communication. While it can be criticized for often being too bound to the prevailing political/institutional arrangements, and too wedded to constrictive methodologies, it does address the important realm of formal, democratic politics and offers a number of useful analytic concepts. A third tradition builds on various kinds of cultural theory; I'll simply call it "culturalist" (to avoid yet another discussion of "cultural studies"). The culturalist orientation, for its part, can turn our attention to such topics as the production of meaning and subjective realities. It generally does not deal with the structural, institutional dynamics of democracy, nor, until recently, has it really been concerned with identities pertaining to citizenship. Each tradition has its own coherence and its own limitations, yet together, and used judiciously, they seem to offer a large toolkit.

Conceptually, civic cultures can be modeled as integrated circuits with five mutually reciprocal dimensions. For the purposes of our discussion, identities as citizens is the key dimension, but it will quickly become apparent how this dimension interplays with the other four. While integrated, each dimension can be individually thematized and studied in empirical cases, in relation to both the mass media and the newer interactive media. These dimensions can be seen as starting points for empirical inquiry about the media's significance for civic cultures, and for our specific concerns here, on the role of television. The five dimensions could be seen as starting points for analysing the ways in which television's "working through" takes place under a given period or regarding a particular issue. The examples and comments I offer should be taken as suggestive, rather than definitive. We do know a good deal already about television that we can mobilize for the purposes of illuminating its contributions to civic cultures, but there is still a lot we don't know, and about which we tend to speculate.

424

Knowledge and competence

Knowledge, in the form of reliable, referential cognisance of the social world, is indispensable for the vitality of democracy. A subset of knowledge is competencies, and, in particular, the skills to deal communicatively in the sociopolitical world are pivotal. Some degree of literacy is essential; people must be able to make sense of that which circulates in the public sphere, and to understand the world they live in.

While it seems rather obvious that people must have access to reliable reports, analyses, debates and so forth about current affairs, it is also becoming more challenging to specify and assert any unified set of legitimate parameters of public knowledge. That is, what kinds of knowledge and competencies are required by whom for the vitality of civic cultures cannot be established once and for all. As sociocultural heterogeneity increases, and as the media matrix of late modern society evolves, issues of linguistic capacity and cultural proximity become highly relevant. This is reiterated in the growth of multiple (and counter) public spheres; different groups know different things, to some extent. However, as I mentioned earlier, the increasing fragmentation of the public sphere must also be seen as a problem for democracy. For its part, television can be seen as a forcefield between centripetal and centrifugal forces. Despite its own growing multiplicity in the age of television abundance, it can still serve as a significant mainstream for public knowledge.

The popularity of the medium is bound up with the pleasures that it offers, even in knowledge acquisition. Given the attributes of television mentioned earlier, it has obvious limits as to what it can reasonably provide in terms of traditional notions of knowledge, even in its most pedagogic modes. The laments about reality and trash television cannot be blithely dismissed, but such trends can, of course, be seen from different angles. Glynn (2000) makes the point that these developments represent different regimes of knowing, that speak, in different ways, to different groups – in particular, those with less social power. But such programs also articulate with late modern emphasis on the individual and their horizons as a platform for political reflection, and notions of power having to do with the micro-contexts of everyday life. This is not an argument for the simple redemption of such programs, but rather a plea for the importance of seeing their potential relevance – in both positive and negative terms – for civic cultures.

Basically, I would suggest that the importance of the knowledge and competence dimension, as it pertains to television, should be recognized, but not exaggerated. The public sphere approach, with its heavy emphasis on rational knowledge, tends to blast away at television – while missing some of the other important features the medium can contribute to civic cultures. In other words, these other dimensions can conceptually serve partly to "balance" the limitations of television's knowledge profile. Moreover, we should keep in mind Schudson's (1999) point, namely, that it has become a modern pitfall to judge civic qualities

425

and democracy more generally exclusively in terms of information and knowledge, ignoring the other important dimensions involved that have manifested themselves historically. Citizens are more than talking heads.

Values

In terms of a more political science view of civic cultures, the values of relevance are those that obviously support democracy. Further, it should be underscored that values must have their anchoring in everyday life; a political system will never achieve a democratic character if the world of the everyday reflects anti-democratic normative dispositions. We can distinguish between substantive values such as equality, liberty, justice, solidarity, and procedural ones, like openness, reciprocity, responsibility, accountability, and tolerance. In short, civic cultures need to underscore commitment to both the rules of the game and the larger visions of a democratic society.

In this regard, television offers a mixed bag. The personnel of television, as well as the representatives of power who appear in its programming, rhetorically reiterate many of the basic values and virtues of democracy, such as individual rights and respect for the law. Without such a backdrop, political scandal would prove difficult – it is predicated on the existence of normative boundaries that are perceived to be broken (cf. Thompson, 2000; Lull and Hinerman, 1998). Yet, in late modernity, it may well be that it is becoming more difficult to generate scandal, partly because of the perceived routine character of deviation from espoused norms on the part of the powerful. Skepticism and cynicism take root. Such deviation is in some cases amplified by the dramaturgy of television, and in other cases it is ignored. One can say that, given the patterns of journalistic coverage, scandal in the media mostly concerns the political elite, with less attention paid to the economic elite (though there have been some particularly interesting exceptions in the past couple of years). In the realm of values we see also the blurring of public and private. Depending on the country and its public culture, the values reflected in the personal lives of political elites may at times take on more importance than their public actions.

Moving to a more culturalist angle of vision, it is clear that much of television's output concerns values that pertain to private life, individual choice-making, and social relations. It is not simply the case that it is easier to render such contexts as popular. In the evolution of late-modern societies we observe the emergence of more personal modes of civic engagement, life politics, identity politics, and other sets of issues that hover outside the formal political system. The processes whereby television has worked through the vicissitudes of values over the years would suggest that values as such can become the site for contestation over civic culture and democracy itself.

Just which values are more in keeping with democracy, and which are more conducive to civic relations, can be – and is – disputed. We should also mention here that a tension exists between values that promote private solutions –

consumption and market relations – and those that foster collective cooperation, social involvement, and civic relations. While common sense may suggest that people's well-being will always involve some kind of balance in these matters, it must be said that the voices of contemporary television unsurprisingly tilt these value polarities strongly in the neo-liberal direction.

Affinity and trust

I have in mind here something less ambitious than "community" – rather, a minimal sense of commonality among citizens in heterogeneous late modern societies, a sense that they belong to the same social and political entities as citizens – despite all other differences. They have to deal with each other to make their common entities work, whether at the level of neighborhood, nation-state or the global arena. If there exists a nominal degree of affinity, for example, conflicts can then become enacted between "adversaries" rather than "enemies," as Mouffe (1999) puts it, since an awareness of a shared civic commonality is operative.

Civic affinity blurs into civic trust. Here too we should aim for a modest level. Certainly, a degree of trust in government and other major institutions is important, but in the civic context we must also add trust between citizens. Putnam (2000, p. 136) distinguishes between "thick" trust based on established personal relationships, and "thin" trust, the generalized expectations of honesty and reciprocity that we accord people we don't know personally, but with whom we feel we can have satisfactory exchanges. I have in mind the latter, as a kind of pre-condition in, for example, entering into civic networks. Such trust may often be group-specific rather than universal, but at least such cohesion can be seen as starting point for civic engagement.

Television tends to show very few examples – in fiction or non-fiction – of the "thin" trust that typifies civic social bonds, and even less of cooperation or activism that makes a political impact. The social bonds displayed lean toward romance, male bonding such as sports or action/military, social friendships, etc. (A large-scale content analysis of this theme might yield interesting results . . .) Television is also known to promote fear and suspicion in some audience groups because of its emphasis on crime, linking danger with public space and thereby contributing to the uncivic retreat into the private domain. On the other hand, through its continual processes of making visible a wider and wider array of social sectors, lifestyles, and generally unconventional personalities, one might hypothesize that the medium does an important job of rendering familiar elements of society that many people would otherwise never encounter.

Practices

Democracy must be embodied in concrete, recurrent practices – individual, group, and collective – relevant for diverse situations. Such practices help

427

generate personal and social meaning to the ideals of democracy. They must have some element of the routine, of the "taken-for-granted" about them, if they are to be a part of civic cultures, yet the potential for spontaneous interventions, one-off, novel forms of practice, also needs to be kept viable. Elections can be seen as a routine form of practice in this regard, but civic cultures require many other practices, for example, running campaigns, holding meetings; even the practices involved in maintaining different organizational forms. Across time, practices become traditions, and experience becomes collective memory. Today's democracy needs to be able to refer to a past, without being locked in it. New practices and traditions can and must evolve if we are to ensure that democracy does not stagnate.

No doubt one of the key practices of civic culture is discussion, a cornerstone of the public sphere and neo-republicanism, as I discussed earlier. We can empirically investigate civic talk by examining, for instance, its various discursive modes, its spatial and contextual sites and settings, and its social circumstances. We might consider what tacit rules are operative in these contexts, and how mechanisms of social etiquette about talk can either promote or hinder the practices of public discussion. Eliasoph (1998), for example, reveals the troubling cultural patterns that inhibit discursive civic practices in American civic culture, contributing to what she calls the "evaporation" of the public sphere.

Different kinds of programming situate the viewing subject differently; in some cases, viewers are very much invited to respond actively, take positions, even argue, as we see in some talk shows and reality TV programs. Other programming leans more toward positioning the viewer as a non-dialogic receiver of information from on high (e.g. traditional news programs). These observations relate to the televisual text; what real viewers actually do is, of course, an empirical question. Some viewers will be left speechless by some talk shows, while others will be provoked into argumentative response by a serious documentary. Taking a broad historical sweep, some may claim that we talked more with each other about political matters before we had television. Others will argue quite the contrary, that the medium is a catalyst for talk, that it actually stimulates us to discursively do the working through that Ellis (2000) writes about.

The patterns of watching television – our overall working relationship with the medium – can also be treated as a kind of practice having civic relevance. In their major study of television audiences in Britain, Gauntlett and Hill (1999) found that guilt was an important feature of many viewers' experiences. Often using a work ethic frame, they felt that television was a waste of time, and that they would be better off doing something else. The sense of television as trash showed itself to correlate with social class, with more middle-class viewers troubled by this sense of losing their control to something they deem unworthy. In the American context, Lembo (2000), based on a large study of qualitative interviews, mapped out what he sees to be the basic components of a viewing culture. These include the actual decision to watch television, the interaction with specific programming imagery and information, and the incorporation of this experience back into daily life. He also identifies a typology of use: "discrete" use is selective and focused; viewing has a cohesion and coherence in relation to people's social location and to their sense of accomplishment. "Undirected" and "continuous" use are diffuse, less the product of conscious decision. Moreover, Lembo articulates discrete use with a solid, modernist sense of self-formation, a sociality that is action-oriented, and productivity. It is linked productivity and power. Undirected and continuous use relates less to consistency and coherence; the viewing is characterized by more context-free involvement with imagery where normative frameworks are weak. This type of engagement with television's image world generates disjunctures, where fleeting meaning does not cumulatively add up to much. The flows, eruptions, and interruptions of the viewing experience situate viewers squarely in the domain of corporate commodity culture, according to Lembo.

Even if this analysis in part rediscovers the consequences of differences in class-based cultural capital, it suggestively invites attention to be paid to the issue of watching as a practice that relates to individual autonomy, personal development, and empowerment. Set in relation to civic culture, we might say that the issue is not between watching television or not, but rather where the locus of control and definitional power lie – i.e. who's in charge, so to speak. We would conclude from Lembo's study that the way we relate to television in our everyday lives correlates with the civic horizons we may develop and the sense of empowerment we may experience.

Identities

This brings us back to identities. For democracy to work, people need to see themselves at least in some way as citizens – though, again, few people may find that the word "citizen" actually gets their adrenaline flowing. What is at stake is not a label, but the subjectivity of membership and efficacy. It should be clear that this sense of self is predicated on the interplay with the other dimensions of civic cultures. Thus, values may reinforce affinity and trust, which in turn can encourage certain practices, and these practices may have positive impact on knowledge, and so forth. We can envision both positive and negative spirals in this regard.

Generally, it can be said that the mass media have become less and less effective in fostering identities of citizenship among its audiences. Where do we find the sense of civic "we-ness" in contemporary television? Dovey (2000) suggests that it lies in the construction of universal collectivity still associated with public service television (and even there it is dwindling), but more generally with the journalism of high modernity. But beyond that – and in particular as we move to the realm of subjective experience in late modern societies – it becomes quite thin. Increasingly, television and the rest of the media milieu position us as consumers. This is the prime identity promoted especially by television; it is in the domain of consumption where we are to be empowered, where we make choices, where we create ourselves. To be sure, being a citizen and a consumer are not always antithetical: citizens need to consume, and consumption at times can be politically framed. Yet there is a fundamental distinction between consumption, which is predicated on the fluctuations of the market, and the principles of universality embedded in the notion of the citizen. Normatively, democracy cannot be reduced to markets – even if today's political discourse often makes use of rhetoric derived from market relations.

The cultural prerequisites television offers for the public sphere are equivocal. Looking at the medium from the perspective of civic culture we find a series of tensions and ambiguities. It may be that television offers largely shrivelled, voting-oriented versions of civic identities. On the other hand, it provides a continuous flood of topics that touch people in various ways. Some of these topics can, especially if processed by discussion, resonate with core values, encourage affinity, suggest practices, mobilize identities, and generate engagement in the public sphere. They can evoke contestation, and further develop the terrain of the political, thereby pumping blood into the body of democracy. Or not: we can no doubt just as easily find evidence for negative spirals. The point is not that we should try to arrive at some ultimate, once-and-for-all evaluation, since that obviously would be so grandly sweeping as to be meaningless. Rather, the idea is that we can look beyond the formal and normative visions of the public sphere to see how civic cultures may operate in specific program contexts, at particular points in time, within a given society. The five dimensions discussed here hopefully can open some doors for empirical probes, helping us to elucidate the continually shifting boundaries between the public and the private, and how the political is engendered and defined.

References

Alasuutari, P. (ed.) (1999) *Rethinking the Media Audience*, London: Sage.
Anderson, D. A. and Cornfield, A. (eds.) (2003) *The Civic We: Online Politics and Democratic Values*, Lanham, MD: Rowman & Littlefield.
Armitage, J. (ed.) (2000) *Paul Virilio: From Modernism to Hypermodernism*, London: Sage.
Asen, R. and Brouwer, D. (eds.) (2001) *Counterpublics and the State*, Albany: SUNY Press.
Baker, C. E. (2002) *Media, Markets, and Democracy*, New York: Cambridge University Press.
Barber, B. (1984) *Strong Democracy: Participatory Politics for a New Age*, Berkeley: University of California Press.
Benhabib, S. (ed.) (1996) *Democracy and Difference*, Princeton: Princeton University Press.
Bourdieu, P. (1998) *On Television*, New York: New Press.
Clarke, P. B. (1996) *Deep Citizenship*, London: Pluto.
Corner, J. (1999) *Critical Ideas in Television Studies*, Oxford: Oxford University Press.
Corner, J. and Pels, D. (eds.) (2003) *Media and the Restyling of Politics*, London: Sage.
Dahlgren, P. (1995) *Television and the Public Sphere*, London: Sage.
Dahlgren, P. (2001) "The Public Sphere and the Net: Structure, Space and Communication," in L. Bennett and R. Entman (eds.), *Mediated Politics in the Future of Democracy*, Cambridge: Cambridge University Press, pp. 33–55.

Dahlgren, P. (2002) "La formation d'identité et les citoyenne d'UE: les dilemmes démocratiques et les conditions mediales," in S. de Proost and J. Ferry (eds.) *Éducation, médias et citoyenneté post-nationale*, Brussels: Éditions del'Université Libre de Bruxelles, pp. 99–123.

Dahlgren, P. (2003) "Reconfiguring Civic Culture in the New Media Milieu," in J. Corner and D. Pels (eds.), *Media and Political Style: Essays on Representation and Civic Culture*, London: Sage, pp. 151–70.

Dahlgren, P. (2004) "Internet, Public Spheres and Political Communication: Dispersion and Deliberation," in *Political Communication* (forthcoming).

Dean, J. (ed.) (1997) *Feminism and the New Democracy*, London: Sage.

Dovey, J. (2000) *Freakshow: First Person Media and Factual Television*, London: Pluto Press.

Downie Jr., L. and Kaiser, R. G. (2002) *The News About the News*, New York: Random House.

Edwards, B., Foley, M. W., and Diani, M. (eds.) (2001) *Beyond Tocqueville: Civil Society and the Social Capital Debate in Comparative Perspective*, Hanover, NH and London: University Press of New England.

Eliasoph, N. (1998) *Avoiding Politics*, New York: Cambridge University Press.

Ellis, J. (2000) *Seeing Things: Television in the Age of Uncertainty*, London and New York: I.B. Tauris Publishers.

Eschele, C. (2001) *Global Democracy, Social Movements, and Feminism*, Boulder, CO: Westview Press.

Fallows, J. (1997) *Breaking the News*, New York: Vintage Books.

Fenton, N. and Downey, J. (2003) "Counter Public Spheres and Global Modernity," *Javnost/The Public*, *10*(1), 15–32.

Franklin, B. (1997) *Newzak and News Media*, London: Edward Arnold.

Fraser, N. (1992) "Rethinking the Public Sphere: A Contribution to the Critique of Actually Existing Democracy," in C. Calhoun (ed.), *Habermas and the Public Sphere*, Boston: MIT Press, pp. 109–42.

Garnham, N. (1983) "Public Service versus the Market," *Screen*, *5*(1), 13–14.

Gans, H. (2002) *Democracy and the News*, New York: Oxford University Press.

Gauntlett, D. and Hill, A. (1999) *TV Living: Television, Culture and Everyday Life*, London: Routledge.

Gitlin, T. (2001) *Media Unlimited: How the Torrent of Images and Sounds Overwhelms Our Lives*, New York: Metropolitan Books/Henry Holt and Company.

Glynn, K. (2000) *Tabloid Culture: Trash Taste, Popular Power, and the Transformation of American Television*, Durham and London: Duke University Press.

Graber, D. (2001) *Processing Politics: Learning From Television in the Internet Age*, Chicago: University of Chicago Press.

Habermas, J. (1989) *The Structural Transformation of the Public Sphere*, Cambridge: Polity.

Habermas, J. (1996) *Between Facts and Norms*, Cambridge, MA/London: MIT Press.

Hill, A. (2004) *Real TV: Audiences and Popular Factual Television*, London: Routledge.

Isin, E. F. and Turner, B. S. (eds.) (2003) *Handbook of Citizenship Studies*, London: Sage.

Isin, E. F. and Wood, P. K. (1999) *Citizenship and Identity*, London: Sage.

Jenkins, H. and Thornburn, D. (eds.) (2003) *Democracy and New Media*, Cambridge, MA: MIT Press.

Kellner, D. (2003) *From 9/11 to Terror War*, Lanham, MD: Rowman & Littlefield.

Lembo, R. (2000) *Thinking Through Television*, Cambridge: Cambridge University Press.

Lister, R. (1997) *Citizenship: Feminist Perspectives*, London: Macmillan.

Lochard, G. and Soulez, G. (eds.) (2003) Theme issue: "La télé-réalité, un débat mondial," *MédiaMorphoses*, 9, Paris: INA/PUF.

Lull, J. and Hinerman, S. (eds.) (1998) *Media Scandals: Morality and Desire in the Popular Culture Marketplace*, New York: Columbia University Press.

Lunt, P. and Stenner, P. (2005) "The Jerry Springer Show as an Emotional Public Sphere," *Media, Culture and Society*, *27*(1), 59–81.

McChesney, R. (1999) *Rich Media, Poor Democracy: Communication Politics in Dubious Times*, Champaign, IL: University of Illinois Press.

McManus, J. H. (1994) *Market-Driven Journalism*, London: Sage.

Meehan, J. (ed.) (1995) *Feminists Read Habermas*, London: Routledge.

Morley, D. (1992) *Television and Cultural Studies*, London: Routledge.

Mouffe, C. (ed.) (1992) *Dimensions of Radical Democracy*, London: Verso.

Mouffe, C. (1993) *The Return of the Political*, London: Verso.

Mouffe, C. (1999) "Deliberative Democracy or Agonistic Pluralism?," *Social Research*, 66(3), 745–58.

Preston, P. W. (1997) *Political/Cultural Identity*, London: Sage.

Price, M. E. (1995) *Television, the Public Sphere, and National Identity*, Oxford: Oxford University Press.

Putnam, R. (2000) *Bowling Alone: The Collapse and Revival of American Community*, New York: Simon & Schuster.

Richardson, K. and Meinhof, U. (1999) *Worlds in Common? Television Discourse in a Changing Europe*, London: Routledge.

Robinson, P. (2002) *The CNN Effect: The Myth of News, Foreign Policy, and Intervention*, London: Taylor and Francis.

Scannell, P. (1996) *Radio, Television and Modern Life*, London: Blackwell.

Scheuer, J. (2001) *The Sound Bite Society*, London and New York: Routledge.

Schlesinger, P. (2004) "The Babel of Europe? An Essay on Networks and Communicative Spaces," Paper presented at symposium on One EU – Many Publics? University of Stirling, February 5–6.

Schudson, M. (1999) *The Good Citizen: A History of American Civic Life*, Cambridge, MA: Harvard University Press.

Simons, J. (2003) "Popular Culture and Mediated Politics: Intellectuals, Elites and Democracy," in J. Corner and D. Pels (eds.), *Media and the Restyling of Politics*, London: Sage, pp. 171–89.

Stevenson, N. (2002) *Understanding Media Cultures*, London: Sage.

Street, J. (1997) *Politics and Popular Culture*, Cambridge: Polity Press.

Thompson, J. B. (2000) *Political Scandal: Power and Visibility in the Media Age*, Cambridge: Polity Press.

Thornton, A. (2004) "Does Internet Create Democracy?," accessed January 14 at http://www.zip.com.au/~athornto//...

Thussu, D. K. and Freedman, D. (eds.) (2003) *War and the Media*, London: Sage.

Tulloch, J. (2000) *Watching Television Audiences*, London: Edward Arnold.

Turner, B. S. (1999) "McCitizen: Risk, Coolness and Irony in Contemporary Politics," in B. Smart (ed.), *Resisting McDonaldization*, London: Sage, pp. 83–100.

Van Gunsteren, H. R. (1998) *A Theory of Citizenship*, Boulder: Westview Press.

Virilio, P. (2000) *The Information Bomb*, London: Verso.

Voet, R. (1998) *Feminism and Citizenship*, London: Sage.

Ward, D. (2002) *The European Union, Democratic Deficit, and the Public Sphere: An Evaluation of EU Media Policy*, Amsterdam: IOS Press.

Warner, M. (2002) *Publics and Counterpublics*, New York: Zone Books.

Weiten, J., Murdock, G., and Dahlgren, P. (eds.) (2000) *Television Across Europe*, London: Sage.

Zelizer, B. and Allan, S. (eds.) (2002) *Journalism After September 11*, London: Routledge.

CHAPTER TWENTY-THREE

Television and Public Opinion

Justin Lewis

Uncertainty is the inspiration of academic research. While certainty can be left well alone, uncertainty needs to be explored, investigated, mapped out, tested, and confirmed. Uncertainty is so pivotal to academic endeavors that it is difficult to erase. Research that has the audacity to diminish uncertainty will itself be scrutinized and interrogated – until, perhaps, uncertainty is triumphantly restored, and attempts by researchers to make claims about the world are left to wither on the vine. The academic world thus has a bias in favour of uncertainty – which is, after all, so much easier to establish than certainty.

The growth of television, which became the dominant information and entertainment technology in industrial societies so rapidly that we scarcely had time to consider its significance, was a phenomenon with profoundly uncertain outcomes. If some of those uncertainties had been anticipated by the growth of radio, cinema, and the mass circulation press, the sheer presence of television – its ubiquity, its dominance of most people's leisure time, and its appropriation of national and political rituals – raised weighty questions about its power and its influence.

Indeed, the uncertainty created by television was such that it provoked a whole string of anxieties. Would television provoke violence, delinquency, or moral decadence? Would it be a tool for mass propaganda and persuasion? Would it create cultural homogeneity and banality? If some of these anxieties were, in retrospect, exaggerated or misplaced, they set in train a now considerable body of research that aimed to gather and measure the nature of television's social significance. But for all these efforts, the history of research into the social impact of television on its audience has been plagued by a persistent uncertainty.

This is partly because establishing certainty in the investigation of television's social impact is difficult. Television can, after all, assume any number of forms. It can be commercial or educational. It can be driven by ideology or by ratings. It can be bland and formulaic or shocking and provocative. There is, in other words, no essential "television" and hence no definitive, uncontested form of influence. So, for example, debates about whether television is "good" or "bad" for children will always be irresolvable partly because, to put it bluntly, it depends *what* they're watching.

But even if we can define television's characteristics at a particular moment, unraveling and isolating these from the various other ideological influences in social life is a methodological challenge that exceeds the capabilities of most research budgets. So, for example, television's ubiquity means that there is no possibility of comparing a group of people who watch TV with a similar group who do not. And academics, as is their wont, tend to be more inclined to point out the theoretical or methodological limits of a research project than to emphasize its strengths. Hence uncertainty itself, in one sense, has become the dominant paradigm for assessing television's influence. Most summaries of audience studies thus tend to leave a great deal of room for speculation.

As a consequence, one of the best-rehearsed debates in media studies is between those who assert that the media *do* play a significant role in shaping public perceptions, and those who stress the audience's ability to create their own meanings from media. This debate is played out in a variety of contexts. So, for example, while some see television as part of a hegemonic system, in which the majority give consent to a set of dominant ideas, others point to the way which even large, commercial media corporations can produce television programs that challenge dominant ideas. And while some researchers point to moments when television does appear to influence consciousness (e.g. Gerbner, Gross, Morgan and Signorielli, 1980; Heide, 1995; Iyengar, 1991; Jhally and Lewis, 1992; Lewis, 2001; McKinley, 1997; Ruddock, 2001), others focus on moments of audience power and play (e.g. Hills, 2002; Hobson, 1982; Hodge and Tripp, 1986; Jenkins, 1992). And on a global scale, some see the global expansion of transnational media corporations as a force for ideological uniformity and/or a form of cultural imperialism (e.g. Herman and McChesney, 1997), while others stress the power of local contexts, enabling people to interpret media content in specific and useful ways (e.g. Ang, 1996).

In this context, it would be easy to imagine that we have learnt little about television's social impact over the last fifty years. Indeed, it might be argued the 1950s debate – between those who focused on what the media do to people ("media effects") and those who looked at what people do with media ("uses and gratifications") – is still being played out half a century later.

And yet what these debates conceal is not only the volume of accumulated evidence on the social impact of television, but the extent to which that evidence does tell us a great deal about the nature and extent of television's influence. In what follows, I shall try to summarize how far we have come by focusing on a specific aspect of television's impact – namely, the role television plays in influencing public opinion.

The Construction of Public Opinion

Governments have always been aware of television's power, and have thus made efforts to place it in the hands of those it trusted, whether they be cultural elites (as in Britain), business elites (as in the United States) or political elites (as in the former Soviet bloc), so that television's democratic potential might be harnessed and controlled. And with every passing decade, political parties have become increasingly attuned to the importance of sophisticated media campaigns. Television, in particular, is seen by politicians and opinion formers as the central arena for the battle for hearts and minds.

Nonetheless, any evaluation of the relationship between television and public opinion must begin by acknowledging the degree to which "public opinion" is, itself, a construction (Herbst, 1993; Lewis, 2001; Salmon and Glasser, 1995). Our most well-known manifestation of public opinion – the public opinion poll – often purports to be a way of *measuring* public opinion. The problem with this idea is that it implies that the way people think about politics and public affairs corresponds entirely to the terms of the pollster's questionnaire, which then simply records what is already there. And yet the conversation between the pollster and the respondent bears little relation to the kinds of conversations people actually have about the world. Indeed, members of the public play only a small part in that conversation, having no say about what topics will be chosen or the terms in which they will be discussed. They are very much respondents rather than discussants.

Opinion polls are, in this sense, not so much a measurement of public opinion as a representation of it. While this representation occurs with public consent, we should see polls as a matter of technical convenience rather than an attempt to replicate everyday political conversations. The poll allows us to use the wonders of sampling methods to *transform* commonplace thoughts and conversations into the precise language of columns and percentages.

But this does not mean that there is some "real" or "authentic" public opinion that might be revealed to us by some other method (such as the "focus group" interview, which is, perhaps, closer to a "normal" conversation). Any method representing public opinion will necessarily impose its own definitions and assumptions. And while the public opinion survey may be more prescribed and contrived than other methods of information gathering, it has the advantage of dealing with meaningful numbers that allow us to generalize across populations. Indeed, polls allow far more scope for indicating the complexity and subtlety of public attitudes than elections (where samples are self-selecting and which generally only offer people the opportunity to express themselves through one, fairly ambiguous question).

Moreover, the opinion survey is so well established as the dominant signifier of public opinion that, in practical political terms, television's ability to influence *how people respond to opinion surveys* is, in and of itself, of key importance. Since opinion polls are often produced in response to prominent media discourses, opinion polls can be seen, in part, as a way of gauging the extent to which people accept or reject those discourses.

Television, Knowledge, and Opinion

One of the problems with early attempts to explore television influence – often referred to as the "effects" tradition – was the assumption that this would take the forms of people simply *replicating* attitudes or behaviors they had heard or seen on television. This approach was often developed in spite of, rather than in response to, the nature of television content. But television content is often descriptive rather than proscriptive. A news program, for example, offers a set of descriptions or impressions about the world, while journalistic notions of object-ivity and balance mean that news is *not* an overtly opinionated discourse. A simple "monkey see, monkey do" approach may be applicable to unambiguously propagandist forms of television, but it is unlikely to shed much light on a television content predicated around notions of impartiality and fairness. And yet this conception – that influence is a matter of imitation – still tends to dominate public discussions of media influence.

Such an approach also fails to capture the complexity of opinion formation. Both journalism and media research have often assumed a simple demarcation between "facts" and "opinions," or between information and values. At the heart of this demarcation is the laudable democratic ideal that the function of news is to provide people with neutral information, allowing viewers the opportunity to make independent judgments about the world. If only it were so simple!

While it may be possible to construct a continuum with "facts" at one end and "opinions" at the other, it is hard to conceive of information that is "value free," or opinions that are not predicated upon certain "facts" or assumptions about the world. So, for example, an apparently descriptive, "value-free" statement such as "US defense spending rose last year, in response to 9/11" would pass most journalistic tests as being "merely" a statement of fact. And yet this statement is replete with ideological assumptions. The term "defense spending," for exam-ple, implies that such spending is predicated on the need to "defend" the United States from attacks. This idea may be popular (most support a country's right to self-defense) but it is – given the global, interventionist ambitions of US foreign policy – highly contestable.

Similarly, the phrase "9/11" – the common abbreviation for the terrorist attacks on the Pentagon in Washington and the Twin Towers in New York in 2001 – contains within it an assumption that the events of that date were of such monumental significance that all that is necessary to signify them is the date. This shorthand is denied to other egregious events in countries lower down the hierarchy of news value.

Moreover, the linking of the two suggests firstly that the obvious response to terrorism is military expansion – a notion that could be seen as akin to suggesting that a larger sledgehammer is needed to swat a mosquito. It also implies that US's military was, hitherto, *insufficiently* funded – an idea which, given the exceptional size of the US military budget, is, at best, a matter of opinion.

In short, innocent statements of "fact" invariably contain value-laden assumptions about the world. Thus the realms of "fact" and "opinion" are, in practice, mutually dependent and interconnected. This makes investigation into television's influence on public opinion a far more complex endeavor than simply isolating the effect of television's "opinionated" moments. It requires us to consider how information presented on television might be ideologically inflected, and, in turn, how certain patterns of information might make some opinions seem plausible and others seem irrelevant.

Ironically, one of the research traditions to undermine the dichotomy between fact and opinion was one predicated on precisely this distinction. The agenda-setting approach, developed in the 1960s and 1970s, was partly a response to the failure of earlier "effects" research to show anything other than a "minimal" media influence on public behavior or opinion. Agenda-setting thus declared a more modest ambition, to focus not on what people thought, but on what people thought *about* (Cohen, 1963). Agenda-setting research was, in turn, highly successful in showing that television (and the news media generally) does indeed "set the agenda," and that surveys asking people which issues they were concerned with were often directly related to which issues dominate the media agenda – regardless of the "real world" salience of those issues (Iyengar and Kinder, 1987). So, for example, public concern about crime is prompted by the volume of media coverage of crime, rather than by people's personal experience of crime or by data on crime levels.

But as agenda-setting research became more sophisticated, it became clear that the "agenda-setting" phenomenon was not ideologically innocent. So, for example, an agenda dominated by crime, immigration and taxes is easier for right-wing politicians to exploit, while an agenda dominated by healthcare, education, and the environment will tend to favor those on the left. More profoundly, it became apparent that the kinds of information made available by television – and the frameworks used to articulate that information – were the building blocks for opinion formation (McCombs, Danielian, and Wanta, 1995).

Meanwhile, another body of research, based on the "cultural indicators project" at the University of Pennsylvania, looked beyond news and current affairs to explore whether television in the United States might be cultivating certain views of the world. Unlike much of the media effects research that preceded it, the project began by analysing the *nature* of the world represented or described by television (Signorielli and Morgan, 1990). Their premise was that television's influence was likely to be found less in discrete or specific messages than in its repetitions and generalities. If the television world could be identified as featuring some stories while neglecting others, it seemed plausible to suppose that heavy viewers of television would be more likely than less frequent viewers to draw upon these stories to understand the world.

This form of "cultivation analysis" has, like agenda-setting, produced data that suggests that many people do rely upon television to describe the world, and that they draw upon that information to form opinions. So, for example, across most genres, studies found that US television features more men than women (by a ratio of around 3 to 1), as well as repeatedly reproducing a number of gender stereotypes. It is perhaps not surprising, then, that heavy TV viewers were likely to draw upon this deep well of symbolic representations and to replicate the assumptions therein. Thus, cultivation analysis found that, regardless of class, gender, education or other key variables, heavy viewers are more likely than light viewers to express traditional and stereotypical views about gender roles (Morgan, 1989).

What both cultivation analysis and agenda-setting research tells us is that television *does* influence public opinion – not by proselytizing, but by representing the world in certain ways, and by favoring some stories over others. Television's influence is, in this respect, neither absolute nor guaranteed. So while US television appears to cultivate the view that the world is a violent and dangerous place, British television does not. This is partly because British and US television, overall, tend to tell different stories about violence. Television's influence is, in this sense, culturally specific. It also depends upon certain conditions.

So, while television may coalesce around certain ideas, themes, or stories, it inevitably allows for degrees of ambiguity and heterogeneity. A range of qualitative studies have shown how audiences have appropriated television texts and, depending on factors such as class, race or locale, made them meaningful in different ways. Other studies have focused on the textual richness of certain forms of "cult" television, many of which can sustain both idiosyncratic and subversive readings (Hills, 2002). At the same time, qualitative research has also fleshed out some of the forms of influence suggested by cultivation and agenda-setting research (McKinley, 1997).

It is also increasingly difficult to consider television in isolation from other media. In an era of media conglomeration, where huge global corporations like Time Warner, Disney and Rupert Murdoch's News Corporation have a stake in a wide array of cultural industries (which include, among others, television, film, the press, the music industry, books, magazines, and the Internet), television's stories may interconnect across a range of other media. This intertextuality makes it increasingly difficult to see television as operating outside the wider synergy of the media world in general. Television's importance, in this respect, is its prominence and its reach. Put simply, more people spend more time with it than they do with any other medium. But the frameworks or stories that dominate television are likely to spill over into other media, meaning that any attempt to explore television's influence must, at some point, consider these interconnections.

Specifying the Nature of Television Influence

Television is often discussed as if its content is inevitable or irrelevant – so, for example, television is seen as "good" or "bad" in itself, hence its social impact is seen, in absolute terms, as benign or negative. Although there may be instances in which the "medium is the message," this view is generally unhelpful in specifying media influence. In short, whether television is seen as a cultural, public service activity or as an economic, commercial activity will have consequences on the form and content of the stories it tells. Some television programs are unpredictable and full of possibilities, others are not. There are some areas where television provides viewers with a diversity of stories and images, and other areas where the information provided is limited and monolithic. Some television systems support a wide array of perspectives, while others tend to be narrower, formulaic, and homogenous. Television can deal with subjects where our knowledge and experience allows us to make critical judgements, or with topics where television is our only or our primary source. Television is, in this sense, a site of political and cultural struggles. Its content matters, because the nature of that content has a social impact.

Because we have research indicating that different audiences can behave in different ways, or that television can be interpreted in ways the producers might not have imagined or intended, this does not "disprove" research showing that television promotes or encourages certain views of the world. Both are not only possible, they reveal the complex and situated nature of television's presence. Moreover, the fact that the social impact of television is often complex and always conditional does not make it any less significant or profound.

Indeed, it is because different audiences can do different things at different times in response to different kinds of television that many have chosen to read the now voluminous body of research on the subject as inconclusive. This is, I would argue, both a lazy and misleading conclusion. On the contrary, the research suggests that we can be more precise than ever about the social impact of television. In sum, we can say that television *does* influence public opinion in specific ways and with certain limits. In what follows, I will sketch out the conditions that make this influence more or less apparent.

The Informational Context

The ability of television to shape public opinion will depend partly upon the quality and variety of information available to us. So, for example, television's authority to speak about a subject will partly depend upon what else we know about that subject (Condit, 1989). What we think about the place we live in, for example, will often be informed by a range of personal experiences, as well as a variety of information sources. We will therefore view television's references to

439

that place – be it a county, city, or country – in the context of a comparative wealth of information. What we think about a country farther from home, on the other hand, may be entirely dependent on the news media, making it more difficult for us to understand that media coverage critically or in any meaningful external context. So, for example, research suggests that people will often depend on extremely limited media information to make – often misleading – judgements about other countries (Lewis, 2001).

The quality of information available to us is central to the whole notion of media literacy. Media literacy refers to the ability to be able to view television in a critical context. This ability, in turn, relies upon the availability of a rich informational climate in which we are able to contextualize what we see or hear: information thereby gives us the power to situate television messages within a broader framework (Lewis and Jhally, 1998).

Research into the relationship between media and public opinion has repeatedly suggested that media influence in modern democracies is less a matter of overt forms of persuasion than a question of providing *a context or framework in which opinions are formed* (Philo 1990; Iyengar, 1991; Gamson, 1992). Thus we tend to rely on media sources for information, and, in turn, "one's store of information shapes one's opinions" (McCombs, Danielian, and Wanta, 1995, p. 295).

Television is, of course, capable of providing us with a rich informational climate – regardless of other information sources. For those subjects we know little about – especially if they involve politics, social or public affairs – television is often our primary source of information. The *nature* of that informational climate will be critical in shaping the range of responses available to people. So, for example, during the 1991 war with Iraq, those Americans who were aware that Saddam Hussein had been, a year earlier, a US ally in the region, were less likely to support going to war than those who assumed US opposition to Saddam Hussein had been a consistent policy (Morgan, Lewis, and Jhally, 1992).

Similarly, before the US-led war with Iraq in 2003, we might have known that

- Saddam Hussein was a brutal dictator, who terrorized his own people;
- Government and intelligence reports seemed to suggest he possessed weapons of mass destruction;
- Saddam Hussein appeared unwilling to cooperate fully with efforts to disarm Iraq;
- Awareness of this information may incline people to support a war to liberate the Iraqi people and to disarm a dangerous regime.

On the other hand, we may have been aware that:

- When Saddam Hussein used chemical weapons against his own people and against Iran, he had been armed and supported by both the United States and the United Kingdom;

- Leading members of the US administration – notably Colin Powell and Condoleeza Rice – had previously argued that the policy of containment of Iraq had been successful, and that Saddam Hussein was no longer a threat;
- The United States and the United Kingdom continue to arm and support many undemocratic and brutal regimes around the world.

This information, in turn, may lead people to conclude that the case for war was rather less compelling. While the relationship between information and opinion here is not inevitable, it is clear that *how* we are informed will influence the opinions we are able or likely to construct.

It is no surprise, in this context, that public attitudes toward the 2003 Iraq war broadly reflected the weight of information available in a country's broadcast media. So, for example, research by Media Tenor suggested that the media in different countries provided different informational climates. German media, for example, were more likely to present information discrediting the case for war: US TV networks, by contrast, tended to present information commensurate with a pro-war view, with the tone of British broadcast media networks coming somewhere between the two (Schreiner, 2003). These informational climates were reflected in polls, which showed most Germans taking anti-war positions, and most Americans taking more pro-war positions, with the British more evenly divided. This is not say that there are no other explanations for the international divergence of public opinion, simply that the media coverage in different countries made certain opinions more plausible – and easier to hold – than others.

The case of the United Kingdom is particularly interesting in this respect. Research suggests that broadcast media in the UK took care to make a range of information available before and after the war, but that this plurality narrowed during the war itself, when the information presented tended to favor pro-war positions. So, for example, during the war broadcasters tended to assume that Iraq possessed weapons of mass destruction and that the Iraqi people welcomed the invasion (Lewis and Brookes, forthcoming). Public opinion, in turn, went from tilting toward anti-war positions before the war (and, again, by the end of the summer), to suggesting a pro-war majority during and immediately after the war.

The Prominence of Information

This fact that this shift in UK attitudes toward the war occurred *after* a fairly "open," "information-rich" period of coverage also suggests that, when opinions are not well established or deeply grounded, the *prominence* of information will be instrumental in informing how people respond to surveys. Both agenda-setting and cultivation research have repeatedly shown that the *sheer quantitative presence* of images, message, or ideas is significant in informing how people respond to questions about the world.

Thus it is, for example, that environmental problems – such as global warming, pollution and the thinning of the ozone layer – became newsworthy in the early 1990s. When these problems received a high level of media coverage, polls, accordingly, suggested that the environment was a matter of significant public concern. More than ten years on, most of these problems – especially global warming – have become more critical, and little has been done to address them, and yet, in countries like the United Kingdom and the United States, surveys show that concern about the environment has significantly diminished.

This quantitative point – that the volume of coverage makes a difference – is often missed by news broadcasters, who often assume that viewers are attentive and respond to the information available, rather than to the quantity of that information. So, for example, references to the background or history of events tend not to be repeated, on the basis that this information has already been made available and is thus already in the public domain (Lewis, 1991). And yet the failure to repeat this information makes it more likely that it will disappear from public consciousness.

The power of the media to set the agenda – and thus to shape concerns and priorities – is partly based on deeply held assumptions about television's transparency. While many people are aware, in their more critical moments, that television is not simply a reflection of the world, this awareness is easily forgotten – especially given the centrality of television to contemporary society, culture, and politics. So while most of us will say, if asked, that we are concerned about issues like global warming, it is hard to resist the idea that the apparent lack of prominence or urgency in media coverage of the issue means that there is no need to concern ourselves *too* much about it.

Conversely, while, for most people, the day-to-day risk to life posed by international terrorism is extremely low – especially when compared to the scale of risk posed by lack of access to clean water, by preventable diseases, poor diets or, indeed, by climate change – most of us assume that the prominence of this threat in day-to-day news coverage makes the risk of terrorism both more imminent and more palpable.

A high degree of media literacy is required to avoid susceptibility to this form of influence. This may come from a well-grounded set of beliefs, strong enough to resist television's shifting priorities. But it also requires us to understand the many commercial, ideological, and cultural forces that shape television coverage. So, for example, terrorism is more newsworthy than climate change *not* because it is relevant to more people, but because of certain news conventions. Terrorism has news value for various reasons: first, government elites have chosen to make it a priority; secondly, those elites link it to wars in Afghanistan and Iraq – also highly newsworthy events; and thirdly, because it tends to involve dramatic, distinctive events rather than gradual trends.

More generally, since television's political agenda takes its broad cue from the interests of and debates between political elites, it is hard for us to imagine a politics that lies outside the dominant media framework of concerns. For the increasing numbers of people who feel that representative politics is irrelevant to them, it is much easier to see themselves as uninterested in politics, rather than to imagine a politics that lies outside dominant definitions and agendas.

Reinforcing Popular Narratives

One of the points made by cultivation analysis is that we only tend to acknowledge media influence when it involves a *change* of attitude or behavior. This assumes that consent for the status quo is a natural state of affairs that requires no explanation. One of the most important arguments made by cultivation research is to counter this assumption, to see that television's sphere of influence often involves *reinforcing* or legitimating certain prevalent attitudes. Consent for any ideological position, in other words, needs to be won and won again if it is to be sustained, and television is often active in this process.

So, for example, cultivation analysis has discovered that television in the United States has a "mainstreaming" effect, encouraging heavy viewers to coalesce around a set of narrow, prevalent opinions, while lighter viewers are able to embrace a broader range of views (Gerbner, Gross, Morgan, and Signorielli, 1980). Similarly, it appears to be easier for heavy viewers to endorse traditional, patriarchal views about gender than less traditional feminist views. In both cases, US television appears to be instrumental in reinforcing certain attitudes. In so doing, its social impact may have more to do with resisting or limiting social change, rather than encouraging it. Thus, while television has undoubtedly responded to changes in society's attitudes to men and women, cultivation research indicates that its representations in the United States err on the side of reinforcing rather than undermining traditional attitudes about sex roles.

There is a broader point here about television's influence: when television replicates well-known stories about the world, those stories are easy for viewers to make meaningful. So, for example, when a dictator – such as Saddam Hussein – is compared to Adolf Hitler, this comparison resonates with a well known and oft-repeated narrative about World War II. If, on the other hand, a dictator is compared to President Suharto or Pol Pot, these claims have less popular resonance, and are thus more difficult to interpret or make meaningful. As a result, television is more likely to have an impact on its audience when it connects with popular narratives.

Of course, television can also play a part of the process of popularizing these narratives. This is certainly the case with World War II, whose popular meaning has been solidified by countless television dramas, films, documentaries, and news bulletins. But television can also create these narratives. So, for example, Jenny Kitzinger (2000) has described how major news events can function like latter-day popular folk tales, such that references to those events carry certain meanings, and that these meanings can then frame subsequent events.

So, for example, the "meaning" of the BSE/CJD story in the United Kingdom – involving the spread of "mad cow" disease from cattle to humans – became one in which the government and scientific establishment were seen to mislead the public by insisting on the safety of British beef. Any reference to the BSE story thereby triggers this framework of understanding, and can influence how later events are understood. The subsequent coverage of competing claims about the safety of the MMR vaccine in Britain, for example, made many references to the BSE story, thereby reinforcing public doubts about government assurances about the MMR vaccine (Lewis and Speers, 2003).

Associations

While television news may – or may not – resonate with popular narratives, the use of narrative on television news is often truncated and difficult to comprehend (Lewis, 1991). Television news often tends to assume that viewers have some knowledge of the background to an event, while ignoring some of the central components of storytelling. Viewers are generally given the key points of the story, followed by further details, rather than a narrative-driven account – one which might introduce viewers to the key characters and draw them into a storyline.

Partly for this reason, many viewers miss much of the detail of television news. What remains are a residual series of impressions, and it is these – rather than the manifest content of the news – that tend to influence opinion formation. This means that those opinions registered in surveys are often based upon a fairly small number of informational building blocks. Television's influence thus takes place in the context of this rather threadbare informational environment.

Some have argued that this is not a problem for a functional democracy. Sniderman, Brody, and Tetlock (1991), Popkin (1991) and Zaller (1992) have all argued that, even in a threadbare informational climate, citizens can use "information short-cuts" or "low-information rationality" to construct reasonable and rational opinions. As Sniderman, Brody, and Tetlock (1991, p. 19) put it: "people can be knowledgeable in their reasoning about political choices without necessarily possessing a large body of knowledge about politics."

What this sanguine view fails to take into account, however, is the way the power to construct opinions is limited by lack of information. Delli Carpini and Keeter's comprehensive study of political knowledge in the United States (1996) indicates that the distribution of political knowledge is not merely a matter of personal effort or predilection, but the consequence of structural conditions like social class, race or gender. Like Pierre Bourdieu's analysis of "cultural capital" (Bourdieu, 1984), their data suggest that the power and legitimacy that come with knowledge of contemporary politics is linked to social position. The authors argue that the ability to participate effectively in democratic decision making is dependent upon one's access to accurate information, and that this information is distributed unequally to disadvantaged members of the least powerful social groups.

Delli Carpini and Keeter's research also suggests that many people are not so much *uninformed* as *misinformed*. So, for example, significant proportions of the electorate appear to have voted for candidates while assuming those candidates stood for policies that were contrary to their actual positions or records (Lewis, 2001). So, while it may be possible to make rational democratic decisions without knowing details of institutional politics, absorbing a limited amount of information may contribute to misleading conclusions.

This often involves the use of ideologically powerful *associations*, which are then used to make misleading assumptions about the world. This associative process is part of the process of political persuasion, and political elites will repeat certain kinds of associations in order to try to influence public consciousness. They take their lead here from advertisers, who have long understood that you make claims not by argument but by juxtaposition. You don't say that a product will make you attractive or popular – such a claim is untenable – you simply associate the product with attractiveness and popularity. In the same way, those making the case for war with Iraq would juxtapose it – vaguely, intangibly, but repeatedly – with the war on terrorism. For a public with a limited knowledge of geopolitics, these associations become the building blocks for making sense of the world.

In the case of the 2003 Iraq war, the nature and influence of this rhetorical strategy was laid bare in Kull's compelling study of public beliefs and support for the war in Iraq (Kull, 2003). The study established that support for the war against Iraq was clearly connected to the discourse of anti-terrorism, regardless of the lack of evidence to support a link between Iraq and international terrorism. Hence the belief that Saddam Hussein had connections to the 9/11 attacks or to Al Qaida was not only widely held, *but directly linked to support for the war in Iraq*. As the study soberly points out, there is no evidence to support a link between Saddam Hussein and Al Qaida, or the attack on the Twin Towers (on the contrary, the Iraqi regime was notably antagonistic toward this kind of fundamentalist politics).

Belief in this link came from an associational logic that was very clearly present in news coverage. Thus, political leaders did not need to state a *connection* between the Iraqi regime and terrorism, they merely had to *juxtapose* the two. In this way, the Bush administration and its allies exploited the elliptical nature of the news media's influence on public opinion.

We can see the same process at work in the coverage of the environment. In the case of climate change, this involves turning associations (between environmental problems like the thinning ozone layer and global warming) into *causal* links. In the last few years, media reports have often mentioned global warming and the ozone layer in the same breath, as in:

> A new satellite which will provide scientists with a kind of health check for the earth was launched today. The satellite, costing £1 and half billion, will orbit earth for the next 5 years studying things like the hole in the ozone layer and global warming. (*ITV News*, March 1, 2002)

While these reports do not assert a causal connection, *in the absence of any other explanation offered*, most people tend to assume one. When asked about the greenhouse effect, for example, most people suggest that greenhouse gases affect the climate by "thinning the ozone layer" (Hargreaves, Lewis, and Speers, 2003). Thus, once again, an association between two things provides the building blocks for the construction of a misleading view of the world.

Conclusion

Fifty years ago, many scholars facing the spasmodic and contradictory body of evidence about media influence came to the conclusion that the media's effect on public opinion was "minimal." Since then, more sophisticated approaches to the issue have indicated that, in various ways, the media *do* have a significant effect on the way that many people understand the world. Surveying this accumulation of evidence, John Zaller (1992) has argued that the era of "minimal effects" has long gone, and that we now have a body of research that clearly identifies the importance of television in the social construction of reality.

David Morley (1992, p. 36) suggests that "it remains necessary to analyze and understand the pleasures that popular culture offers its consumers if we are to understand how hegemony operates through the processes of commercial popular culture." While there is no disputing the importance of pleasure in this process, we might also substitute it with the word "information." Since television plays a crucial role in constructing the informational climate that allows us to make sense of the world, it is central to the formation of public opinion.

This does not mean that people are passive recipients of media influence. On the contrary, they actively construct meanings based on media information. But the form and nature of those constructions will often be limited by the nature of the information television – or other media – does or does not provide. And television itself is caught up in a complex set of economic, social and cultural relations that will shape its output. Indeed, television's relationship with dominant power structures will be critical in informing the nature of its influence.

References

Ang, I. (1996) *Living Room Wars: Rethinking Media Audiences for a Postmodern World*, New York: Routledge.

Bourdieu, P. (1979) "Public Opinion Does Not Exist," in A. Mattelart and S. Siegelaub (eds.), *Communication and Class Struggle*, New York: International General, pp. 124–30.

Bourdieu, P. (1984) *Distinction: A Social Critique of the Judgment of Taste*, Cambridge, MA: Harvard University Press.

Cohen, B. (1963) *The Press and Foreign Policy*, Princeton: Princeton University Press.

Condit, C. (1989) "The Rhetorical Limits of Polysemy," *Critical Studies in Mass Communications*, 6(2), 103–22.

Delli Carpini, M. and Keeter, S. (1996) *What Americans Know about Politics and Why It Matters*, New Haven, CT: Yale University Press.

Gamson, W. (1992) *Talking Politics*, Cambridge: Cambridge University Press.

Gerbner, G. and Gross, L. (1976) "Living with Television: The Violence Profile," *Journal of Communication*, 28(3), 173–99.

Gerbner, G., Gross, L., Morgan, M., and Signorielli, N. (1980) "The Mainstreaming of America: Violence Profile No. 11," *Journal of Communication*, 30(3), 10–29.

Gerbner, G., Gross, L., Morgan, M., and Signorielli, N. (1986) "The Dynamics of the Cultivation Process," in J. Bryant and D. Zillman (eds.), *Perspectives in Media Effects*, Hillsdale, NJ: Lawrence Erlbaum, pp. 17–40.

Hargreaves, I., Lewis, J., and Speers, T. (2003) *Towards a Better Map: Science, the Public and the Media*, London: Economic and Social Research Council.

Heide, M. (1995) *Television, Culture and Women's Lives: thirtysomething and the Contradictions of Gender*, Philadelphia, PA: University of Pennsylvania Press.

Herbst, S. (1993) *Numbered Voices*, Chicago: University of Chicago Press.

Herman, E. and McChesney, R. (1997) *The Global Media*, London and Washington: Cassell.

Hills, M. (2002) *Fan Cultures*, London: Routledge.

Hobson, D. (1982) *Crossroads: The Drama of a Soap Opera*, London: Methuen.

Hodge, B. and Tripp, D. (1986) *Children and Television*, Cambridge: Polity Press.

Iyengar, S. (1991) *Is Anyone Responsible?*, Chicago: University of Chicago Press.

Iyengar, S. and Kinder, D. (1987) *News that Matters*, Chicago: University of Chicago Press.

Jenkins, H. (1992) *Textual Poachers: Television Fans and Participatory Culture*, New York: Routledge.

Jenkins, H. (1995) "Out of the Closet and into the Universe," in J. Tulloch and H. Jenkins (eds.), *Science Fiction Audiences*, London: Routledge, pp. 239–65.

Jhally, S. and Lewis, J. (1992) *Enlightened Racism: The Cosby Show, Audiences, and the Myth of the American Dream*, Boulder, CO: Westview.

Kitzinger, J. (2000) "Media Templates: Patterns of Association and the (Re)construction of Meaning over Time," *Media, Culture and Society*, 22(1), 61–84.

Kull, S. (2003) *Misperceptions, the Media and the Iraq War*, October, Program on International Policy Attitudes, University of Maryland.

Lewis, J. (1991) *The Ideological Octopus: Explorations into the Television Audience*, New York: Routledge.

Lewis, J. (2001) *Constructing Public Opinion: How Elites Do What They Like and Why We Seem To Go Along With It*, New York: Columbia University Press.

Lewis, J. and Brookes, R. (2004) "How British Television News Represented the Case for the War in Iraq," in S. Allan and B. Zelizer (eds.), *Reporting War: Journalism in Wartime*, London and New York: Routledge.

Lewis, J. and Jhally, S. (1998) "The Struggle Over Media Literacy," *Journal of Communication*, 48(1), 109–20.

Lewis, J. and Speers, T. (2003) "Misleading Media Reporting? The MMR Story," *Nature Reviews Immunology*, 3(11), 913–18.

McCombs, M., Danielian, L., and Wanta, W. (1995) "Issues in the News and the Public Agenda," in C. Salmon and T. Glasser (eds.), *Public Opinion and the Communication of Consent*, New York: Guilford Press, pp. 281–300.

McKinley, E. G. (1997) *Beverly Hills, 90210: Television, Gender, and Identity*, Philadelphia: University of Pennsylvania Press.

Morgan, M. (1989) "Television and Democracy," in I. Angus and S. Jhally (eds.), *Cultural Politics in Contemporary America*, New York: Routledge, pp. 240–53.

Morgan, M., Lewis, J., and Jhally, S. (1992) "More Viewing, Less Knowledge," in H. Mowlana, G. Gerbner, and H. Schiller (eds.), *Triumph of the Image: The Media's War in the Persian Gulf: A Global Perspective*, Boulder, CO: Westview Press, pp. 216–33.

Morley, D. (1992) *Television, Audiences and Cultural Studies*, London: Routledge.

Philo, G. (1990) *Seeing and Believing: The Influence of Television*, London: Routledge.

Popkin, S. (1991) *The Reasoning Voter: Communication and Persuasion in Presidential Campaigns*, Chicago: University of Chicago Press.

Ruddock, A. (2001) *Understanding Audiences*, London: Sage.

Salmon, C. and Glasser, T. (eds.) (1995) *Public Opinion and the Communication of Consent*, New York: Guilford Press.

Schiller, H. (1989) *Culture, Inc.: The Corporate Takeover of Public Expression*, New York: Oxford University Press.

Schreiner, W. (2003) "The War on TV: 'Better' is not Good Enough," *Media Tenor 2*. On line at http://www.mediatenor.com/index1.html.

Signorielli, N. and Morgan, M. (1990) *Cultivation Analysis*, Beverly Hills, CA: Sage.

Sniderman, P., Brody, R., and Tetlock, P. (1991) *Reasoning and Choice: Explorations in Political Psychology*, New York: Cambridge University Press.

Stromer-Galley, J. and Schiappa, E. (1998) "The Argumentative Burdens of Audience Conjectures: Audience Research in Popular Culture Criticism," *Communication Theory*, *81*, 27–62.

Zaller, J. R. (1992) *The Nature and Origins of Mass Opinion*, Cambridge: Cambridge University Press.

CHAPTER TWENTY-FOUR

Reality TV: Performance, Authenticity, and Television Audiences

Annette Hill

Reality TV is a catch-all category that includes a wide range of entertainment programs about real people. Sometimes called popular factual television, reality TV is located in border territories, between information and entertainment, documentary and drama. Originally used as a category for law and order popular factual programs containing "on scene" footage of cops on the job, reality TV has become the success story of television in the 1990s and 2000s. There are reality TV programs about everything and anything, from healthcare to hair-dressing, from people to pets. There are reality TV formats sold all over the world, from the United States to Uruguay.

The debate about what is real and what is not is the million-dollar question for popular factual television. In this chapter, I explore the twin issues of perform-ance and authenticity, since the performance of non-professional actors often frames discussion about the authenticity of visual evidence in popular factual television.[1] The way real people and their stories are represented on television is closely connected to how we judge the truthfulness of visual evidence. To invoke the work of Brian Winston (1995), "claiming the real" is a common practice of reality programming, but there is little interrogation of these truth claims in the programs themselves. Television audiences are certainly aware of the ways tele-vision "puts reality together" (Schlesinger, 1978), and talk about how various formats, or editing techniques, can create different degrees of "reality" in popu-lar factual television. Viewers of reality programming are most likely to talk about the truth of what they are seeing in relation to the way real people act in front of television cameras. The more ordinary people are perceived to perform for the cameras, the less real the program appears to be to viewers. Thus, performance becomes a powerful framing device for judging reality TV's claims to the real. And, television audiences are highly skeptical of the truth claims of much reality programming precisely because they expect people to act up in order to make entertaining factual television.

Reality TV

The reality genre has mass appeal. Popular series such as *American Idol* in the USA or *I'm a Celebrity* . . . in the United Kingdom have attracted up to and over 50 percent of the market share, which means more than half the population of television viewers tuned into these programs. To achieve such ratings these reality series have to be all-round entertainers. In 2004, the reality cable channel, Reality Central, signed up more than 30 reality stars to appear on and promote the channel. According to Larry Namer, the co-founder of E! Entertainment and Reality Central, there is a large base of reality TV fans: "to them reality TV is television. It's not a fad" (Cozens, 2003).

In 2000, the reality game show *Survivor* rated number one in American network primetime (27 million viewers) and the final three episodes earned CBS an estimated $50 million in advertising revenue. In 2002, the finale of the reality talent show *American Idol* (Fox, United States) attracted 23 million viewers, and a market share of 30 percent, with almost half the country's teenage female viewers tuning in to watch the show.[2] In January 2003, *American Idol* drew nearly 25 million viewers two nights running, making it "the most watched non-sports show in the network's history."[3] By February 2003, Fox had another winner, this time with the finale of reality dating show *Joe Millionaire*, which drew 40 million viewers, making it almost as popular as the broadcast of the Academy Awards, and "the highest series telecast on any network since CBS' premiere of *Survivor II* in January 2001."[4] In comparison, only 15 million viewers watched the number one crime drama series *CSI: Crime Scene Investigation* (CBS), or sitcom *Friends* (NBC), during the same period. Reality programs regularly win the highest ratings for the majority of half-hour time slots during prime-time American television.[5]

Reality TV is just as popular in the United Kingdom. In 2000, over 70 percent of the population (aged 4–65+) watched reality programs on a regular or occasional basis (Hill/ITC, 2000). The types of programs watched most often by the public in 2000 were: police/crime programs (e.g. *Police, Camera, Action!* [ITV]) watched either regularly or occasionally by 72 percent of adults and 71 percent of children; "places" programs (e.g. *Airport* [BBC]) watched by 71 percent of adults and 75 percent of children; and home/garden shows (e.g. *Changing Rooms* [BBC]) watched by 67 percent of adults and 84 percent of children. Amongst the under-16s (in particular, the under-13s), pet programs (e.g. *Animal Hospital* [BBC]) were as popular as the categories cited above – watched by 83 percent of children and 63 percent of adults (Hill/ITC, 2000). All of these reality programs have performed strongly in prime-time schedules, and have attracted up to and over a 50 percent market share.

The picture is the same in many other countries around the world. In Holland, the first *Big Brother* "became one of the country's top-rated shows within a month, and drew 15 million viewers for its climax on New Year's Eve 1999."[6] In

Spain, more people tuned in to watch *Big Brother* in 2000 than the Champions League semi-final match between Real Madrid and Bayern Munich (Hill, 2002). Similarly, in 1997 the finale of *Expedition Robinson* (the Swedish version of *Survivor*) was watched by half the Swedish population in 1997.[7] In Norway, *Pop Idol* (2003) received 3.3 million SMS votes, in a country with a population of 4.3 million.[8] The French version of *Big Brother* (*Loft Story*, 2002) was a ratings hit, with over 7 million viewers, despite regular demonstrations by "Activists Against Trash TV" calling for the series to be banned, and carrying placards which read "With trash TV the people turn into idiots."[9] The pan-African version of *Big Brother*, produced in Malawi, involved ten contestants from ten different countries, and despite calls by church groups in several African countries for it to be banned, the show remained popular with viewers, who praised it for bridging cultural gaps.[10] The Russian reality game show *The House* (*Dom*) enthralled Russian television viewers in 2003, as they watched contestants build a £150,000 five-bedroom house (the average wage in Russia is less than £150 a month).[11] When a woman won *Big Brother 3* in Australia, Channel Ten attracted twice as many viewers as its main rival, Channel Nine, the number one rated channel (2003).[12] More than three million people, about half the population of television viewers in Australia, tuned into the hit reality property series *The Block* on Channel Nine. The series featured the renovation of apartments in Sydney by four couples, who were given a budget and 11 weeks to renovate their properties. After 12 weeks, the apartments were auctioned, and the couple with the highest bid won. The conclusion to *The Block* was "Australia's most watched TV show since the 2000 Sydney Olympics. Only the funeral of Princess Diana drew a bigger audience for a non-sport related program."[13] The format has been sold to the US Fox network, ITV1 in the UK, and TV2 in Denmark, as well as being picked up by broadcasters in Belgium, France, the Netherlands, and South Africa. The *Herald-Sun* called *The Block* "a runaway smash that shows no sign of losing steam."[14]

Although examples of reality TV can be found throughout the history of television, reality programs arrived on mass in prime-time television schedules during the 1990s. Docusoaps, also called "fly-on-the-wall" documentaries, "soap-docs," or "reality-soaps," became the "motor of primetime" during the mid- to late 1990s in Britain (Phillips, 1999, p. 23). There were as many as 65 docusoaps broadcast on the main channels between 1995 and 1999, attracting audiences of up to 12 million. Docusoaps were so popular that the term even made it into the *Oxford Dictionary* (Phillips, 1999, p. 22). The docusoap is a combination of observational documentary and character-driven drama. One TV producer explained: "We'd seen that flashing bluelight documentaries could work, but many of the latest ones are factual soaps, very character-led . . . nothing seems to be too mundane. It's the technique of a soap opera brought into documentaries" (Biddiscomb, 1998, p. 16). Although there had been predecessors to the docusoap, namely Paul Watson's *The Family*, it was its "prioritisation of entertainment over social commentary" that made the docusoap so different from observational documentary, and perforce popular with general viewers (Bruzzi, 2001, p. 132).

Since its arrival in 2000 the reality game show has become an international bestseller. The birth of the reality game show format can be traced to British producer Charlie Parsons, who developed the idea for *Survivor* in the early 1990s, and sold an option on the rights to Endemol, before a Swedish company bought the format and renamed it *Expedition Robinson*. In the meantime, Endemol had been working on a similar idea, *Big Brother*, the brainchild of Dutch TV producer John de Mol, who described the format as

> the voluntary locking up of nine people during a hundred days in a house, watched continuously by 24 television cameras, to which the viewers, at the intercession of the inmates, once in two weeks vote against one of the inmates who has to leave the house, until the last person to stay in can be called a winner. (Costera, Meijer, and Reesink, 2000, p. 10)[15]

Surprisingly, *Big Brother* was a hit. More than three million people watched the finale in Holland (the program first appeared on RTL in 1999) and voted by telephone for the winner. The fact that the format worked well with converging media, such as websites and telephones, only added to its strong economic performance in the television marketplace. After the "smash hit" of *Survivor*, noted earlier, the networks scrambled to glut the market with a winning formula of game show, observational documentary, and high drama.

Debating Reality TV

Since the early days of reality programming, critics have consistently attacked the genre for being voyeuristic, cheap, sensational television. Articles such as "Danger: Reality TV can Rot Your Brain," "Ragbag of Cheap Thrills," or "TV's Theatre of Cruelty" are typical of the type of commentary that dominates discussion of reality programming.[16] In a UK report for the Campaign for Quality Television in 2003, reality TV was singled out by Michael Tracey of the University of Colorado as the "stuff of the vulgate," encouraging "moral and intellectual impoverishment in contemporary life."[17] Robert Thompson of Syracuse University suggests that reality TV is popular "because it's stupid and moronic."[18] In his book *The Shadow of a Nation*, the broadcaster Nick Clarke states that the popularity of reality TV has led to a dangerous blurring of boundaries between fact and fiction, and as a result reality TV has had a negative effect on modern society. As one critic commented: "In essence, this may as well be network crack: reality TV is fast, cheap and totally addictive . . . the shows [are] weapons of mass distraction . . . causing us to become dumber, fatter, and more disengaged from ourselves and society" (Conlin, 2003). The mixed metaphors of drug addiction and war indicate how the reality genre is often framed in relation to media effects and cultural, social, and moral values.

Reality TV may be popular, but audiences are able to make distinctions between what they perceive to be good and bad reality programming. After public protest of a proposed real-life version of *The Beverly Hillbillies*, CBS President Les Moonves admitted there are limits to public taste in reality programming.[19] When audiences watch reality TV they are not only watching programs for entertainment; they are also engaged in critical viewing of the attitudes and behavior of ordinary people in the programs, and the ideas and practices of the producers of the programs. As John Ellis points out, audiences of reality programming are involved in exactly the type of debates about cultural and social values that critics note are missing from the programs themselves: "on the radio, in the press, in everyday conversation, people argue the toss over 'are these people typical?' and 'are these really our values?'" (Ellis, 2003).

Early academic studies into the then emerging phenomenon of reality TV focused primarily on the definition of the genre, and its relationship with other types of television genres. Work by Bill Nichols (1994), John Corner (1995, 1996), and Richard Kilborn (1994, 1998) on the status of reality programming within factual television is particularly useful in highlighting early debates about the factual and fictional elements of the reality genre. In many ways such early debates about the "reality" of reality TV raised important questions about actuality, and the epistemology of factual television, that have still not been answered today. Much of the work of Nichols, Corner, and Kilborn was related to positioning an emergent and hybrid genre within the arena of documentary television, and within existing academic debates about documentary studies. The issues Corner and Kilborn raised about the characteristics of reality programming and the impact of popular factual television on the future of documentary television are issues they have continued to address in their contemporary work. Both scholars have written extensively about the changing nature of audiovisual documentation, and the role reality TV has to play in opening up debate about the truth claims of factual television (Corner, 2002a, 2002b; Kilborn, 2003). Although they are critical of aspects of reality programming, they recognise its popularity over the past decade cannot be ignored by scholars in documentary studies.

Recent work by scholars in documentary studies and cultural studies suggests that the reality genre is a rich site for analysis and debate. In his book *Lies, Damn Lies and Documentary* (2000), Brian Winston addresses the legal and ethical framework to documentary television, and argues for greater responsibility in the making and regulating of factual programs. Jon Dovey, in his book *Freakshow* (1999), considers genres such as true confessions and docusoaps as examples of first-person media, a type of media that often foregrounds private issues at the expense of wider public debate about social and political issues. John Ellis (2000, 2002) in his book *Seeing Things* argues that genres such as chat shows or documentaries invite us to witness the modern world, and through this process understand the world around us. John Hartley in his book *The Uses of Television* (1999) suggests popular factual programs can teach us how to become "do-it-yourself" citizens, how to live together in contemporary society. By contrast, Gareth Palmer (2003) considers the surveillance context of many popular factual programs, and argues that television's use of CCTV raises important issues about our civil liberties. Jane Roscoe and Craig Hight (2001) examine mock-documentary as an example of popular factual forms that play with boundaries of fact and fiction, and question the status of audiovisual documentation. Finally, Holmes and Jermyn (2003) examine the economic, aesthetic, and cultural contexts to the genre.

These selected examples of research in the emerging genre of reality TV illustrate how debate about the genre need not be dominated by arguments about "dumbing down," or voyeur TV. Whilst these debates can be found in media discussion of reality TV, many academic scholars have moved the debate to fresh terrain. My own research contributes to the body of existing work on the production, content, and reception of reality TV. My previous research in crime and emergency services reality programming (Hill, 2000a, 2000b), along with an edited collection on *Big Brother* (Hill and Palmer, 2002; Hill, 2002), represents a move to situate the audience in debate about reality TV. Along with a variety of other scholars in media studies, such as Arild Fetveit (2002), Nick Couldry (2002), Frances Bonner (2002), Gay Hawkins (2001), Ib Bondebjerg (2002), Friedman (2002), and Mathjis, Jones, Hessels, and Verriest (2004), it is now clear that the discourse around reality TV is now rich and varied.

A Note on Methods

The research presented in this chapter is drawn from a multi-method research project conducted during 2000–1. The research aim was to provide information and analysis regarding viewing preferences and strategies across all age ranges for a variety of reality programming, available on terrestrial, satellite, cable, and digital television in the UK. The research was funded by the public organization the Economic and Social Research Council, the regulatory body the Independent Television Commission (now Ofcom), and the television company Channel 4. The research also received support from the Broadcasting Standards Commission (now Ofcom), the BBC, and Channel Five. I used quantitative and qualitative audience research methods, in conjunction with analysis of the scheduling, content, and form of reality programs. The data from the quantitative survey, conducted using the national representative sample (over 9,000 respondents aged 4–65+) of the Broadcaster's Audience Research Board (BARB), enabled me to gather a large amount of information on audience preferences for form and content within reality programming, and audience attitudes to issues such as privacy, accuracy, information, and entertainment. On the basis of what I learnt about audience attitudes and preferences to reality programming in the survey, I used qualitative focus groups to explore key issues such as authenticity and performance, information and entertainment, and the social context to watching reality programming. I used quota sampling to recruit (self-defined) regular viewers of a range of reality programming. There were 12 groups, consisting of male/female viewers, aged 12–44, in the social category C1C2DE, living in the South East of England. I also conducted family in–depth interviews over a six-month period, observing family viewing practices, and the relationship between scheduling, family routine, and the content of reality programs. There were four visits to ten families living in the South East of England.

Performance

At the heart of the debate about the reality of reality TV is a paradox: the more entertaining a factual program is, the less real it appears to viewers. Viewers of reality programming are attracted to various formats because they feature real people's stories in an entertaining manner. However, they are also distrustful of the authenticity of various reality formats *precisely because* these real people's stories are presented in an entertaining manner.

In his article "Reality TV in the Digital Era: a Paradox in Visual Culture" Arild Fetveit argues:

> The advent of digital manipulation and image generation techniques has seriously challenged the credibility of photographic discourses. At the same time, however, we are experiencing a growing use of surveillance cameras, and a form of factual television that seems to depend more heavily on the evidential force of the photographic image than any previous form: *reality* TV. (2002, p. 119)

Fetveit's argument draws on the development of photographic practices to understand the growth of reality programming in the 1990s. The history of photography suggests the way we look at photographic images has changed over time, from viewing images as illustrative of real objects or people, to viewing images as evidence of real objects or people. The introduction of digital manipulation as a photographic technique during the 1990s has ensured we are more likely to look at digitally enhanced photographic images as illustrations of real objects or people. Take, for example, the playful photographs of famous people that regularly feature on the front pages of tabloid newspapers; readers are likely to view a photograph of President George Bush and Prime Minister Tony Blair kissing as illustrative of their close relationship as political leaders of the USA and UK, rather than as actual evidence of a romantic relationship. Fetveit argues that it is precisely at this moment of change in our "belief in the evidential powers of photographic images" (2002, p. 123) that reality TV has flourished. For Fetveit, it is no coincidence that reality programming directs viewers to its television images of reality, showing caught on camera footage of car crashes, or rescue operations again and again in order to draw attention to the evidential powers of on-scene reality footage. With one eye on photography, and another on reality

TV, Fetveit suggests our loss of faith in the evidential nature of digitally enhanced photography has been replaced by our faith in the evidential nature of reality television.

Fetveit's argument is useful in understanding why viewers may trust the type of on-scene footage, or surveillance footage, which is so familiar from reality programming such as *Cops*, or *Neighbours from Hell*. Audiences place a great deal of trust in the ability of television cameras to capture real events, as they happen. However, audience trust in the authenticity of reality television is complex, and dependent on the ways in which each reality format is set up to capture the stories of everyday people. In fact, audiences are likely to distrust visual evidence in reality programming – "I'm not quite sure I trust that what we're seeing is not being staged . . ." (31-year-old housewife). Just as the development of photographic techniques is connected with the changing ways we look at photographic images, so too is the development of production techniques within reality programming connected with the changing ways we look at television images.

Viewers expect particular types of factual television to offer them visual evidence of real life. News and documentary are the two most common genres within factual television where viewers place a great deal of trust in the truth claims of audiovisual documentation. If we look at research carried out in the United Kingdom in 2002 by the television regulatory bodies the Independent Television Commission (ITC) and Broadcasting Standards Commission (BSC), we can see that over 90 percent of the UK population were interested in watching news, and nearly 90 percent believed that television news provided accurate information (ITC/BSC, 2003, p. 60). In relation to documentary, almost 80 percent of the public were interested in watching documentary television, and nearly 60 percent believed documentaries provided accurate information. With regard to popular factual television, audience trust in the honesty of the situations portrayed was lower than news or documentary, and varied according to different types of reality programming. Less than half of the UK population (42 percent) believed docusoaps were accurate, and only 20 percent believed reality game shows were accurate.

Audiences are distrustful of reality formats that may appear to encourage nonprofessional actors to "act up," such as docusoaps, or reality game shows. The ITC/BSC research indicates reality game shows like *Big Brother* scored low (20 percent) in audience assessment of the honesty of situations portrayed in these programs. One of the reasons this is the case is that the format is designed to promote performance. Contestants are engaged in a popularity contest, where they are on display in the performance space of the *Big Brother* house (Corner, 2002b, p. 257). As one viewer commented on a contestant in *Big Brother*: "I don't think she ever forgot that the cameras were there . . . she was plucking her . . . pubic hairs with her tweezers! In the garden with everyone else watching!" (31-year-old housewife). In this section, I consider how the "criteria of truthfulness" (Ellis, 2002) applied by viewers to reality television is often associated with their belief that the more people act up in front of cameras, the less real a program appears to be.

There are certain types of factual genres, such as docudrama, in which real people perform for the cameras, and their performance is taken as evidence of the truthfulness of the program itself (Paget, 1998). However, in terms of reality TV, many programs are judged by viewers as unreal precisely because of the performance of non-professional actors. Audiences have a high degree of cynicism regarding the portrayal of real people in popular factual television. In 2000, 73 percent of the public thought stories in reality programming were sometimes made up or exaggerated for the purposes of entertainment, and only 12 percent thought stories about real people actually happened as portrayed in the programs (Hill/ITC, 2000).

The general public's lack of trust in the actuality of popular factual television is partly explained by documentary fakery scandals, and partly by the use of formats associated with fictional genres in popular factual series. During a fakery scandal in the United Kingdom, several docusoaps were accused of faking certain scenes for dramatic effect. According to Bruzzi (2000), *Driving School* (which attracted over 12 million viewers) reconstructed certain scenes, and manipulated others in order to maximize the drama of Maureen's story, as we saw her struggle to pass her driving test. The producers intervened in the outcome of the story: "they were concerned that Maureen, the series 'star' subject, would not pass her manual driving test, an event they felt would be the series natural and desired conclusion, and so suggested that she learn instead in an automatic" (Bruzzi, 2000, p. 88). Despite, or perhaps because of, press discussion about the truthfulness of *Driving School*, Maureen became a celebrity in her own right. This example of the blurring of the boundaries between documentary and soap opera, and between non-professional actors and television celebrities, is only one example of the type of public discussion surrounding reality programs. This public discussion fuels audience skepticism about the authenticity of reality TV, and leads to a high degree of anticipation that "real" people perform in popular factual television. In 2000, 70 percent of adult viewers thought that members of the public usually overacted in front of cameras in reality programs (Hill/ITC, 2000).

The way viewers talk about ordinary people in reality programs illustrates their inherent distrust in particular types of reality formats. This 41-year-old male carpenter makes a clear distinction between hidden camera programs, and other types of reality formats:

I just think that they're two entirely different programs – the ones with the hidden camera and the ones where, like *Big Brother*, where they're actually acting to the camera – to me, they're entirely different categories, I don't even think they're in the same thing. I mean . . . it's the same with *Changing Rooms*, these people on there they're just playing to . . . I'd rather watch something where the camera's hidden and you actually see people . . . I mean just through the day, talking to people . . . that is something natural, you ain't acting, you know. That is actually what happens, that is a true to life thing.

This viewer's perception of hidden camera formats as more "true to life" than formats "where they're actually acting to camera" was common to all the respondents in my research. Even though hidden camera programs involve a high degree of construction, where people are set up and filmed without their prior knowledge or consent, the very fact that they do not know they are being filmed is a clear indication for audiences that the programs are authentic. Here are a group of adult viewers discussing a particular hidden camera series, *House of Horrors*, which attempts to shame dishonest builders by secretly filming them on the job:

Esther: This is a fairly real one 'cos nobody knows the camera's there, so they're just being natural and they're being caught out and if they knew the camera was there, they'd behave completely differently, I believe.

Eric: I think we'd all behave completely differently, wouldn't we, if there are cameras there.

Esther: But if you're being a crook, which is what these people are, then they're really being caught out then, it would be a completely different story if they knew the cameras were there.

Pantelis: They'd be completely honest if they knew the camera was there.

Their discussion of the naturalness of ordinary people on television encapsulates the way audiences make judgments about the "honesty" of people, and programs, according to prior knowledge of filming. The program might be "set up," as is the case of *House of Horrors*, where the builders work on a site pre-fitted with hidden cameras. But people's reactions are natural. In *House of Horrors* dishonesty is actually a sign of honesty because the program is about crooked builders who pretend to be trustworthy.

A reality format like *Big Brother* can be understood in terms of the tensions and contradictions between the performance of non-professional actors, and their authentic behavior in the *Big Brother* house. In an article on *Big Brother* titled "Performing the Real," John Corner (2002b, pp. 263–4) comments on the "degree of self-consciousness" and "display" by the various personalities in the "predefined stage" of the *Big Brother* house. As Corner notes, the performance of contestants gives television audiences the opportunity for "thick judgemental and speculative discourse around participants' motives, actions and likely future behaviour" (2002a, p. 264). I want to focus on the way audiences speculate and judge moments when the performance of non-professional actors breaks down, and they are "true to themselves." Corner (2002b, pp. 263–4) sums up this viewing process as follows:

One might use the term "selving" to describe the central process whereby "true selves" are seen to emerge (and develop) from underneath and, indeed, through, the "performed selves" projected for us, as a consequence of the applied pressures of objective circumstance and group dynamics. A certain amount of the humdrum and the routine may be a necessary element in giving this selving process, this unwitting disclosure of personal core, a measure of plausibility, aligning it with the mundane rhythms and naturalistic portrayals of docusoap, soap opera itself, and at times, the registers of game-show participation.

Other researchers have also discussed this notion of "performed selves" and "true selves" co-existing in hybrid formats within the reality genre. Roscoe and Hight (2002, p. 38) discuss the "performed" nature of docusoaps, and how this type of construction of documentary footage can open up space for debate about the documentary genre. Roscoe comments on how *Big Brother* is "constructed around performance" (2001, p. 482), with participants involved in different levels of performance, based on the roles of "housemate," "game show contestant," and "television personality," and how audiences are invited to join in with these performances "across the formats of the different shows." Mikos et al. (2000), in their research on *Big Brother* in Germany, also suggest audiences are engaged in an assessment of performance and authenticity. In my earlier research on *Big Brother*, I noted that the tension between performance and authenticity in the documentary game show format invites viewers to look for "moments of truth" in a constructed television environment (Hill, 2002).

Audiences frequently discuss the difference between performed selves and true selves in reality programming, speculating and judging the behavior of ordinary people, comparing the motives and actions of people who choose to take part in a reality program. And they discuss the behavior of ordinary people in a reality program on an everyday basis. Here is a typical example of the way viewers talk about acting in *Big Brother*:

> Sometimes, I think, can you really act like your true self when there's a camera there? You know. Maybe in *Big Brother* a little bit more you can act yourself because you're going to forget after a while aren't you? But I'm a bit dubious about people acting themselves . . . The way they were all acting, the way of their body movements and all that, it just looked too fake . . . to me. (21-year-old male dairy worker)

This viewer's tentative question about being able to "act like your true self" in front of a television camera opens the door to speculation about levels of acting in the *Big Brother* house, and to judgment of individual contestants' true or "fake" behavior.

I want to highlight several examples of audience discussion about the improvised performances of contestants in *Big Brother* in order to explore how viewers engage with the inherent contradictions between fact and fiction in this type of hybrid genre.

There is a common mode of engagement when watching *Big Brother* and this is characterized by discussion that goes backwards and forwards between trust and suspicion of the behavior of ordinary people in the house. In the following debate, a group of male and female adult viewers discuss the various "selves" on display in the *Big Brother* house:

Rick: With *Big Brother* you don't know if they're playing up, yeah, it's just, it's a weird scenario for them to be in, you must just think . . . well, you don't know what's going on inside their head.

Paul: Maybe you put yourself in that situation and, see, it's like I watch it and if, if I was on *Big Brother*, I'd want everyone to like me or . . . I think of myself as an alright person but then if I was on there I'd, I'd be acting different, thinking "I've got to do this 'cos people are going to like me," so maybe that's, that's why, maybe, I think they're acting up.

Peter: They must have thought about everything they've done and said before they actually said or done it. Not like real life, just someone coming out with a comment, but, this could get me out this week – I better not say that, I better just say "does anyone want a cup of tea?" Not 'cos I want to make it but I better ask them to look good.

Pauline: 'Cos at the end of the day, it's a competition, isn't it? There was seventy grand on the line, wasn't there? I'd act up for it! [laughs]

Their discussion is characterized by a cautious assessment of the abilities of *Big Brother* contestants to "act up." A point to remember is that the *Big Brother* contestants are strangers to each other and to viewers. Unlike celebrity reality game shows, such as *Celebrity Big Brother*, or *I'm a Celebrity . . .* , where we know the "personality" of the contestants beforehand, in the case of ordinary people reality formats the participants are strangers to us. When audiences attempt to judge the difference between the contestants' performing selves and their true selves in *Big Brother*, they cannot refer to past performances but must rely on their own judgment of the contestants' behavior and "what's going on inside their head." Inevitably, viewers turn to their own experience, and speculate about how they might behave in a similar situation. The discussion therefore becomes one based on hypothetical situations – "if I was on *Big Brother*" – interspersed with knowledge of the format, and the effect of the game on contestant's behavior – "they must have thought about everything they've done and said before they actually said or done it."

In his book *The Presentation of the Self in Everyday Life*, the sociologist Erving Goffman claims we are all performing all of the time on various different stages, such as work or home, to various different audiences, such as our boss, or our family. For Goffman, our houses, cars, clothing, and other such everyday items are "props" and "scenery" required for the "work of successfully staging a character" (1969, p. 203). In any social encounter, a performer will be aware of their audience and vice versa. The process of communication between the performer and audience is an "information game," where performers will reveal and conceal their behavior to others (1969, p. 20). On the *Big Brother* stage there are two types of audience, one that is inside and another that is outside the house. The inside audience has first-hand knowledge of the performance of individuals within the group, but this knowledge is only partial, as the contestants cannot witness all the actions, or performances, of the other members of the social group. The outside audience has second-hand knowledge, but is witness to, in

Goffman's terms, the "front" and "backstage" behavior of the housemates via the 24-hour surveillance cameras. By front and backstage, Goffman (1969, p. 34) refers to moments in social interaction when an individual ceases to play a part convincingly, when we see beyond a "personal front" to the real person inside the performer. Viewers of *Big Brother* have a privileged position in the "information game," and are able to anticipate future incidents or behavior based on prior knowledge of the front and backstage behavior of housemates. However, although viewers do feel they have a bird's-eye view of events in the *Big Brother* house, there is a general questioning of how viewers can really get to know these performers at all. According to Goffman, when social interaction occurs there is a "natural movement back and forth between cynicism and sincerity" (1969, p. 31) on behalf of performers and audiences. In audience discussion of *Big Brother* there is a "natural movement back and forth" in their talk of how viewers judge the sincerity of ordinary people in reality game shows. I would argue it is in the act of trying to judge the scene change from performing self to true self that audiences draw on their own understanding of social behaviour in their everyday lives. As Goffman (1969, pp. 241–2) indicates, when we do not have full information of a factual situation we "rely on appearances . . . and, paradoxically, the more the individual is concerned with the reality that is not available to perception, the more he must concentrate his attention on appearances." Although, when we watch a reality game show such as *Big Brother* we rely solely on representations of real people, we also rely on our knowledge of social interaction. In the final part of this chapter, I consider how we judge authentic performances in popular factual television.

Authenticity

According to Van Leeuwen (2001), authenticity can mean different things to different people. Authenticity can mean that something is not an imitation or copy, but the genuine article, as in an *authentic* Picasso painting. It can also mean that something is reconstructed or represented just like the original, as in a translation of Homer's *The Iliad*. Authenticity can mean that something is authorized and has a seal of approval, as in ephemera sold as part of the Elvis Presley estate. And finally, authenticity can mean that something is true. It is the final definition of authenticity that most concerns us here, as an ordinary person in a reality program is often perceived as authentic if they are "thought to be true to the essence of something, to a revealed truth, a deeply held sentiment" (Van Leeuwen, 2001, p. 393).

Here is an example of the way audiences typically talk about authentic "performances" in reality programming:

Peter: It's real life, innit, I mean . . .
Rick: I don't think it is though . . . The ones on holiday are more real life than these people, I don't, I don't believe anything now, I think it's all an act

but on holiday they might be acting a little bit more but because they're drunk as well it's real life, innit.

Nancy: It's not real life really, is it? 'Cos real life doesn't happen like that?

Rick: If it was real life you, you'd have to not know that the cameras were there and that's never the case in any of those programmes.

Paul: If it was real life I'd be watching someone sitting down watching telly all day.

This group of adult viewers were discussing travel reality formats, such as *Ibiza Uncovered*, that often feature tourists behaving badly. Variations on the word "reality" are echoed in each turn in the conversation ("real," "real life"), and this points to a critical examination of the truth claims of these programs. As with other examples of audience discussion, the authenticity of reality programming is examined in relation to the performances of the people featured in the programs themselves. These viewers question how the talk and behavior of ordinary people being filmed on holiday in Ibiza can be judged as authentic, given that they are under the influence of alcohol. For one viewer, the fact that British tourists are drunk is a good indication of the reality of their behavior in the program – the more drunk, the less control these tourists will have of their behavior. But, for other viewers in the group the fact that these tourists know they are being filmed for a reality program is a good indication of the falseness of their behavior. One viewer refers to the commonsense belief that in order to create entertaining television you need people to be entertaining – "if it was real life I'd be watching someone sitting down watching telly all day."

Montgomery (2001, pp. 403–4) argues there are three types of authentic talk in broadcasting:

First there is talk that is deemed authentic because it does not sound contrived, simulated or performed but rather sounds natural, "fresh", spontaneous. Second, there is talk that is deemed authentic because it seems truly to capture or present the experience of the speaker. Third, there is authentic talk that seems truly to project the core self of the speaker – talk that is true to the self of the speaker in an existential fashion.

Although for Montgomery, the second type of authentic talk is most common to television, in particular reality programming, audience talk about reality programming illustrates all three aspects of authenticity, not just in the way ordinary people talk, but also, perhaps more importantly, in how they behave on television. I'd like to return to the reality series *Ibiza Uncovered* in order to illustrate how audiences assess authentic performances of ordinary people according to what appears natural, what appears true to the situation portrayed, and what appears true to the self of the people portrayed.

The following discussion is based on a story in *Ibiza Uncovered* about two married men on holiday with their wives and children. The two men are "Jack the lads," who are out on the town, looking for some action. We follow them as they drink in bars, flirt with single girls, some of whom flash their breasts or bottoms at the men (and the cameras), and stagger home at the end of the night, somewhat the worse for wear. From the point of view of the program itself, the authenticity of the talk and behavior of these two men is presented very much as true to their experience – "this is what we are normally like on holiday in Ibiza." From the point of view of the audience, the program's truth claims are treated with suspicion, but not rejected outright. These male viewers (aged 18–44) draw on their own experience to assess the authenticity of the behaviour of the two men on holiday:

Max: You go to Southend, it's like filming Southend on a Friday or Saturday night, you see exactly the same thing.

Shaun: I think that's rubbish what they've put on there, if the camera's there, everybody's going to act up.

Max: Yeah, that's right, especially on holiday . . .

Shaun: They were in the bar, had a drink, turned around and that was it, straight away. It doesn't work, not so quick as that, but because the cameras are there, the girl sees the camera, thinks "Oh, I want to be on TV" . . .

Max: And the thing is, it starts them off sober and we're going out clubbing, you can see them as they get . . . as they're getting a little bit tipsy but they're getting a little bit, they're getting tipsy a little too quick for my liking . . . and then it shows them being childish . . .

Brian: I think they'd be worse if the cameras weren't there!

Shaun: They were in a different skin . . .

Max: In fact, I think it could have got naughtier . . . they were being a little bit the boys . . . people go out there and doing what they were doing to those girls, they wouldn't still be on that dance floor, I tell you that now. Not a chance, not a chance . . .

Terry: I've been to Spain and all that, with the boys and everything, and I've never seen anything like that . . .

Max: Let's face it, if you had two other guys who weren't two guys who were coming across Jack the lad. I mean, all us guys have been Jack the lad at some stage, most probably some of us still are, but, if they picked another two guys that were more, er, nervy, then how would it have gone. The entertainment might not have been there.

Brian: But they might have been actors, mightn't they?

Max: But you won't get . . . I don't think they were actors 'cos any guy that they says "Right, there's a camera, we're making this, do you mind us filming you?" and they would have looked at these guys and said "Well, like, they're a bit Jack the lad, they're game and we're in there." Boom, that's what they got.

Barry: You've got to find someone whose wife, who'd let them go and film them anyway. I mean my wife wouldn't let me do that. I mean I'd love to go and do it. I mean it'd be great crack . . . [laughs]

Terry: She'd know exactly what you were going to get up to!

There are several overlapping points being made by these male viewers about the authenticity of this scene from *Ibiza Uncovered*. Most of the viewers referred to their own experience of being out for a night on the town, being "a bit Jack the lad," to make sense of the scene. They all agreed that the scene was not authentic, albeit for different reasons. For Shaun, the two men attracted an unnatural (i.e. instantaneous) interest from girls precisely because there were cameras present. This meant the situation was unnatural, and the men weren't themselves – "they were in a different skin." For Max, the scenario seemed false because the men didn't act the way he imagined they would act – they were drunk too quickly, they didn't flirt enough, "it could have got naughtier." Thus, the scene wasn't true to this viewer's experience of similar situations (Southend on a Friday night), and the men weren't true to themselves, in the sense of being red-blooded males. There was certainly agreement the men performed well, and provided entertainment – one viewer even suggested the men were actors. But, the fact that the men gave such good performances drew attention to how the program was constructed. The final reality check comes from one viewer who judges the scene untrue in relation to his own experience of being married – were those men really given permission by their wives to behave badly?

Van Leeuwen (2001, p. 397) argues that authenticity is in crisis because it can mean different things to different people. We have come to question the concept of authenticity, "just as the idea of the reality of the photograph came into crisis earlier." At the start of this chapter I discussed the relationship between photography and reality programming in the work of Fetveit (2002). Is the authenticity of visual evidence in reality programming in crisis, just like the authenticity of the digitally enhanced photograph? Audiences are likely to question the authenticity of ordinary people and their behavior in highly constructed reality programming, such as reality game shows, or docusoaps, where the format is designed to encourage self-display. We can see from the way viewers talk about the characters in this type of reality programming that they are certainly sceptical of the authentic behavior of ordinary people in televised situations. However, just as Van Leeuwen suggests that although authenticity may be in crisis, it has not lost its validity, I would argue television audiences may question the authenticity of people's performances in reality programming, but reality programming has not lost its validity altogether. Speculation about the performance of ordinary people can lead to critical viewing practices, in particular regarding the authenticity of certain types of reality programming.

Conclusion

The twin issues of performance and authenticity are significant to our understanding of popular factual television. Much contemporary reality programming, especially documentary game shows, or docusoaps, is concerned with self-display. The sites we associate with reality formats such as *Big Brother* are stages where ordinary people display their personalities to fellow performers and to audiences. The fact that reality game shows are set up to encourage a variety of performances (as contestants, as TV personalities) ensures such programs are viewed as "performative" popular factual television. The manner in which ordinary people perform in different types of reality programs is subject to intense scrutiny by audiences. Discussion tends to focus on general home truths about "acting up" in front of television cameras, and the unreality of television about real people. As one viewer put it: "if it was real life I'd be watching someone sitting down watching telly all day." Most viewers expect ordinary people to act for the cameras in the majority of reality programming. These expectations do not, however, stop audiences from assessing how true or false the behavior of ordinary people can be in reality programming. Audiences gossip, speculate, and judge how ordinary people perform themselves and stay true to themselves in the spectacle/performance environment of popular factual television. Audience discussion is characterized by a natural movement backwards and forwards between trust and suspicion of the truthfulness of ordinary people and their behavior on TV. Inevitably, audiences draw on their own personal experience of social interaction to judge the authenticity of the way ordinary people talk, behave, and respond to situations and other people in reality programs. Whether people are authentic or not in the way they handle themselves in the *Big Brother* house, or on holiday in Ibiza, is a matter for audiences to debate and critically examine on an everyday basis. When audiences debate the authenticity of performances in reality programming they are also debating the truth claims of such programs, and this can only be healthy for the development of the genre as a whole.

Notes

1   This chapter is adapted from a book-length study, *Real TV: Audiences and Popular Factual Television* (London: Routledge, 2004).
2   Online at http://media.guardian.co.uk/story/0,7493,787312,00.html (accessed June 27, 2003).
3   Online at http://www.cbsnews.com/stories/2003/01/25/entertainment/main537964.shtml (accessed June 27, 2003).
4   Online at http://www.cbsnews.com/stories/2003/02018/entertainment/main541100.shtml (accessed June 27, 2003).
5   In the last week of January 2003 reality programs won 15 out of 18 half-hour time periods on Monday–Wednesday evening. Online at http://www.cbsnews.com/stories/2003/01/25/entertainment/main537964.shtml (accessed June 27, 2003).
6   Online at http://news.bbc.co.uk/1/hi/entertainment/tv_and_radio/1346936.stm (accessed August 26, 2003).
7   Online at http://news.bbc.co.uk/1/hi/entertainment/tv_and_radio/1346936.stm (accessed August 26, 2003).
8   Online at http://www.cbsnews.com/stories/2003/02/18/entertainment/main541100.shtml (accessed June 27, 2003).
9   Online at http://news.bbc.co.uk/1/hi/entertainment/tv_and_radio/1341239.stm (accessed August 26, 2003).
10   Online at http://news.bbc.co.uk/1/hi/world/africa/3110681.stm (accessed August 26, 2003).

11 Online at http://news.bbc.co.uk/1/hi/entertainment/tv_and_radio/3141021.stm (accessed August 26, 2003).
12 Online at http://newsstore.f2.com.au/apps/news (accessed August 26, 2003).
13 Online at http://newsstore.f2.com.au (accessed August 26, 2003).
14 *Broadcast*, October 31, 2003, p. 11.
15 This quotation was cited in translation in de Leeuw (2001).
16 See *The Times*, December 20, 2002, pp. 4–5; *Financial Times*, November 11, 1999, p. 22; *The Observer*, August 20, 2000, p. 15.
17 See *Broadcast*, June 20, 2003, p. 2.
18 Cited in "Reality TV Takes Off." Online at http://www.cbsnews.com/stories/2003/01/16/entertainment/main536804.shtml (accessed June 27, 2003).
19 Online at http://www.cbsnews.com/stories/2003/01/16/entertainment/main536804.shtml (accessed June 27, 2003).

References

Biddiscomb, R. (1998) "Real Life: Real Ratings," *Broadcasting and Cable Television International* (January), pp. 14, 16.
Bonner, F. (2003) *Ordinary Television*, London: Sage.
Bruzzi, S. (2000) *New Documentary: A Critical Introduction*, London: Routledge.
Bruzzi, S. (2001) "Observational ('Fly-on-the-wall') Documentary," in G. Creeber (ed.), *The Television Genre Book*, London: British Film Institute, pp. 129–32.
Bondebjerg, I. (2002) "The Mediation of Everyday Life: Genre, Discourse and Spectacle in Reality TV," in A. Jerslev (ed.), *Realism and "Reality" in Film and Media*, Copenhagen: Museum Tusculanum Press, pp. 159–92.
Conlin, M. (2003) "America's Reality TV Addiction," Online at http://aol.businessweek.com/bwdaily/dnflash/jan2003/nf20030130_8408.htm (accessed January 31, 2003).
Corner, J. (1995) *Television Form and Public Address*, London: Edward Arnold.
Corner, J. (1996) *The Art of Record: A Critical Introduction to Documentary*, Manchester: Manchester University Press.
Corner, J. (2001a) "Documentary Realism," in G. Creeber (ed.), *The Television Genre Book*, London: British Film Institute, pp. 126–9.
Corner, J. (2001b) "Form and Content in Documentary Study," in G. Creeber (ed.), *The Television Genre Book*, London: British Film Institute, pp. 125–6.
Corner, J. (2002a) "Documentary Values," in A. Jerslev (ed.), *Realism and "Reality" in Film and Media*, Copenhagen: Museum Tusculanum Press, pp. 139–58.
Corner, J. (2002b) "Performing the Real," *Television and New Media*, 3(3), 255–70.
Costera Meijer, I. and Reesink, M. (eds.) (2000) *Reality Soap! Big Brother en de Opkomst van het Multimediaconcept*, Amsterdam: Boom.
Couldry, N. (2002) "Playing for Celebrity: Big Brother as Ritual Event," *Television and New Media*, 3(3), 283–94.
Cozens, C. (2003) "Round the Clock Reality Arrives," *Guardian*, Tuesday April 29. Online at http://media.guardian.co.uk/realitytv/story/0,7521,945285,00.html (accessed June 27, 2003).
de Leeuw, S. (2001) "*Big Brother*: How a Dutch Format Reinvented Living and Other Stories," unpublished paper.
Dovey, J. (2000) *Freakshows: First Person Media and Factual TV*, London: Pluto.
Ellis, J. (2000) *Seeing Things: Television in the Age of Uncertainty*, London: I.B. Tauris.
Ellis, J. (2002) "A Minister is about to Resign: On the Interpretation of Television Footage," in A. Jerslev (ed.), *Realism and "Reality" in Film and Media*, Copenhagen: Museum Tusculanum Press, pp. 193–210.

Ellis, J. (2003) "Big Debate is Happening Everywhere but on TV," *Broadcast*, June 27, p. 11.

Fetveit, A. (2002) "Reality TV in the Digital Era: A Paradox in Visual Culture," in J. Friedman (ed.), *Reality Squared: Televisual Discourse on the Real*, Brunswick, NJ: Rutgers University Press, pp. 119–37.

Fishman, M. and Cavender, G. (1998) *Entertaining Crime: Television Reality Programmes*, New York: Aldine De Gruyter.

Friedman, J. (ed.) (2002) *Reality Squared: Televisual Discourse on the Real*, Brunswick, NJ: Rutgers University Press.

Goffman, E. (1969) *The Presentation of Self in Everyday Life*, London: Pelican Books (reprint).

Hartley, J. (1999) *Uses of Television*, London: Routledge.

Hawkins, G. (2001) "The Ethics of Television," *International Journal of Cultural Studies*, 4(4), 412–26.

Hill, A. (2000a) "Crime and Crisis: British Reality TV in Action," in E. Buscombe (ed.), *British Television: A Reader*, Oxford: Oxford University Press.

Hill, A. (2000b) "Fearful and Safe: Audience Response to British Reality Programming," *Television and New Media*, 1(2), 193–214.

Hill, A. (2002) "*Big Brother*: The Real Audience," *Television and New Media*, 3(3), 323–40.

Hill, A. and Independent Television Commission (ITC) (2000) "Quantitative Research in Television Audiences and Popular Factual Entertainment," unpublished document, in association with the Broadcasters' Audience Research Board.

Hill, A. and Palmer, G. (2002) "*Big Brother*: Special Issue," *Television and New Media*, 3(3).

Holmes, S. and Jermyn, D. (eds.) (2003) *Understanding Reality TV*, London: Routledge.

ITC/BSC (2003) *Television: the Public's View*, London: Independent Television Commission and Broadcasting Standards Commission.

Kilborn, R. (1994) "How Real Can You Get?: Recent Developments in Reality Television," *European Journal of Communication*, 9, 421–39.

Kilborn, R. (1998) "Shaping the Real: Democratization and Commodification in UK-Factual Broadcasting," *European Journal of Communication*, 13(2), 201–18.

Kilborn, R. (2003) *Staging the Real: Factual TV Programming in the Age of Big Brother*, Manchester: Manchester University Press.

Kilborn, R. and Hibbard, M. (2000) *Consenting Adults?*, London: Broadcasting Standards Commission.

Mathijs, E., Jones, J., Hessels, W., and Verriest, L. (eds.) (2004) *Big Brother International: Critics, Format and Publics*, London: Wallflower Press.

Mikos, L., Feise, P., Herzog, K., Prommer, E., and Veihl, V. (2000) *Im Auge der Kamera: Das Fernsehereignis Big Brother*, Berlin: Vistas.

Montgomery, M. (2001) "Defining 'Authentic Talk,'" *Discourse Studies*, 3(4), 397–405.

Paget, D. (1998) *No Other Way to Tell It*, Manchester: Manchester University Press.

Palmer, G. (2003) *Discipline and Liberty*, Manchester: Manchester University Press.

Phillips, W. (1999) "All Washed Out," *Broadcast*, July 2, 22–3.

Roscoe, J. (2001) "Big Brother Australia: Performing the 'Real' Twenty-Four-Seven," *International Journal of Cultural Studies*, 4(1), 473–88.

Roscoe, J. and Hight, C. (2001) *Faking It: Mock-documentary and the Subversion of Factuality*, Manchester: Manchester University Press.

Schlesinger, P. (1978) *Putting "Reality" Together*, London: Constable.

Van Leeuwen, T. (2001) "What is Authenticity?," *Discourse Studies*, 3(4), 392–7.

Winston, B. (1995) *Claiming the Real: The Documentary Film Revisited*, London: British Film Institute.

Winston, B. (2000) *Lies, Damn Lies and Documentaries*, London: British Film Institute.

CHAPTER TWENTY-FIVE

A Special Audience? Children and Television

David Buckingham

Both in academic research and in popular debates about television, children are frequently identified as a special audience with distinctive characteristics and needs. Although watching television is an everyday activity for people of all age groups, children are singled out for particular attention and concern. Their behavior is closely measured and monitored; they are experimented upon, surveyed, and canvassed for their views; and the "problem" of their relationship with television is frequently a focus of concern for parents, pundits, and politicians.

On one level, this might be seen simply as a response to the relative importance of television in children's lives. Thus, it is frequently pointed out that children today spend more time watching television than they do in school, or indeed on any other activity apart from sleeping. Despite the advent of new media, statistics show that television remains the dominant medium in children's lives, even in countries with high levels of access to computer technology (Livingstone and Bovill, 2001; Roberts and Foehr, 2003). In fact, however, it is elderly people who are the heaviest viewers – and yet there has been little discussion of their viewing habits, either in relation to their needs in relation to television or in terms of their particular vulnerability.

The identification of children as a "special" audience for television is not simply a matter of viewing figures. On the contrary, it invokes all sorts of moral and ideological assumptions about what we believe children – and, by extension, adults – to be. As histories of childhood have shown, the definition and separation of children as a distinct social category is itself a relatively recent development, which has taken on a particular form in Western industrialized societies (Aries, 1973; Hendrick, 1997). This process has been accompanied by a veritable explosion of discourses, both *about* childhood and directed *at* children themselves. The emergence of developmental psychology, and its popularization in advice literature for parents, for example, has been one of the means by which norms about what is "suitable" or "natural" behavior for children have been enforced. Likewise, the production of children's literature and children's toys –

468

and eventually of children's television – has invoked all sorts of assumptions about what it means to be a child.

As these examples imply, this construction of "the child" is both a negative and a positive enterprise: it involves attempts to restrict children's access to knowledge about aspects of adult life (most obviously sex and violence), and yet it also entails a kind of pedagogy – an attempt to "do them good" as well as to protect them from harm. The constitution of children as a television audience, and as objects of research and debate, has been marked by a complex balance between these positive and negative motivations. In the early days of television, for example, one of the primary advertising appeals made by the equipment manufacturers was on the grounds of the medium's educational potential for the young (Melody, 1973); and while some have argued that this pedagogical motivation has increasingly been sacrificed to commercialism, it remains a central tenet of public service provision for children. Likewise, early debates about the role of television in the family, in both the United Kingdom and the United States, were characterized by genuine ambivalence about its potential, as either an attack on family life or as a means of securing domestic harmony (Spigel, 1992; Oswell, 2002).

This definition of the child audience is an ongoing process, which is subject to a considerable amount of social and historical variation. Policies on the regulation of children's programming, for example, often reflect much more fundamental assumptions about the nature of childhood, which vary from one national context to another (Hendershot, 1998; Keys and Buckingham, 1999; Lisosky, 2001). Likewise, the struggle between parents and children over what is appropriate for children to watch and to know is part of a continuing struggle over the rights and responsibilities of children. Indeed, the definition of what is "childish" or "adult" is also a central preoccupation among children themselves, not least in their discussions of television (Buckingham, 1994; Davies, Buckingham and Kelley, 1999).

A History of Concern

It is important to locate the concern about children and television historically, in the context both of evolving definitions of childhood and of recurrent responses to new cultural forms and communications technologies. Fears about the negative impact of the media on young people have a very long history (Starker, 1989). More than 2,000 years ago, the Greek philosopher Plato proposed to ban dramatic poets from his ideal Republic, for fear that their stories about the immoral antics of the gods would influence impressionable young minds. In more recent times, popular literature, music hall, the cinema and children's comics have all provoked "moral panics" which have typically led to greater censorship designed to protect children from their allegedly harmful effects. In this respect, more recent controversies such as the "video nasties" scare of the

469

1980s or the debates about screen violence that followed the killing of James Bulger in Britain in 1993 can be seen as heirs to a much longer tradition (Barker, 1984; Buckingham, 1996). And, of course, these concerns are now being echoed yet again in discussions of children's uses of Internet chatrooms and computer games.

Yet such concerns about the harmful influence of television on children cannot be dismissed as simply irrational scare-mongering. Indeed, different versions of such concerns would seem to recur among many social groups, and right across the political spectrum. The most vocal groups, and perhaps the most influential, are those of the so-called moral majority. The concern here, often motivated by traditional religious beliefs, is essentially with the moral and behavioral impact of television, most obviously in terms of sex, violence, and "bad" language. Yet there are also concerns which might be seen as more "liberal," for example to do with the impact of television viewing on children's academic achievement, their imagination and their capacity for social interaction. Popular books like Marie Winn's symptomatically-titled *The Plug-In Drug* (1985) typically urge parents to encourage more "healthy" viewing behavior and to wean their children away from their "addiction" to television. Meanwhile, on the political Left, it is possible to identify parallel concerns with the negative influence of television – for example, on the grounds that it encourages consumerism, militarism, sexism, racism, and just about any other objectionable ideology one might care to name (e.g. Goldsen, 1977).

Of course, it is important to distinguish between these different areas of concern, and the motivations that underlie them. Yet they share a fundamental belief in the enormous power of television, and in the inherent vulnerability of children. Television has, it would seem, an irresistible ability to brainwash and narcotize children, drawing them away from other, more worthwhile activities and influences. From this perspective, children are at once innocent and potentially monstrous: the veneer of civilization is only skin-deep, and can easily be penetrated by the essentially irrational appeals of the visual media (Barker, 1984). Such arguments often partake of the fantasy of a "golden age" before television, in which adults were able to "keep secrets" from children, and in which innocence and harmony reigned. By virtue of the ways in which it gives children access to the hidden, and sometimes negative, aspects of adult life, television is accused of having caused the "disappearance of childhood" itself (Postman, 1983; and for further discussion, see Buckingham, 2000a).

These arguments reflect a form of displacement that often characterizes popular debates about the media. Genuine, often deep-seated anxieties about what are perceived as undesirable moral or social changes lead to a search for a single causal explanation. Blaming television may thus serve to deflect attention away from other possible causes – causes that may well be "closer to home" or simply much too complicated to understand (Connell, 1985). The symbolic values that are attached to the notion of childhood, and the negative associations of an "unnatural" technology such as television, make this a particularly potent combination for social commentators of all persuasions. Yet they also make it extremely difficult to arrive at a more balanced and less sensationalist estimation of the role of television in children's lives.

Psychological Research: From "Effects" to "Active Audiences"

Research is itself an inextricable part of this process. The production of "scientific" knowledge about children inevitably helps to define what it means to be a child, and thus invokes the kinds of assumptions I have been discussing (Luke, 1990). The discipline of psychology – which, until fairly recently, enjoyed an effective monopoly of the study of children – has been particularly implicated in this process (Burman, 1994).

Broadly speaking, the fundamental aim of most psychological research in this field has been to establish evidence of the existence of negative effects. By comparison, research on positive effects has been a very marginal concern. As new media forms and technologies have been introduced, the same basic questions have tended to recur (Reeves and Wartella, 1985). Thus, the questions researchers were asking about television in the 1960s were very similar to those which had been investigated (and indeed, largely superseded) in relation to film thirty years previously – and which are now recurring yet again in research on computer games and the Internet.

This is most obviously the case with research about the effects of television violence on children – an area which has been exhaustively reviewed elsewhere (some more critical accounts can be found in Freedman, 2002; Gauntlett, 1995; and Rowland, 1983). The problems with this research are partly methodological – for example, to do with the limited validity of laboratory experiments as a guide to real-life behavior, or with the frequent confusion in survey research between correlation and causality. They are also theoretical, reflecting fundamental confusions about the definition of "violence" and the ways in which it is deemed to influence people's behaviour (Barker and Petley, 2001). Above all, however, the key issue is a political one: it is to do with how media research has been used by politicians to deflect attention away from other potential causes of violence that seem to be much more difficult for them to address – for example, by means of gun control. Here, as in many other areas, the focus on children enables campaigners to command assent in a way that would be much harder to achieve were they to propose restrictions on adults (see Jenkins, 1992).

Similar arguments might be made in relation to other key preoccupations of media effects research, such as work on the effects of advertising or of gender stereotyping (for a review, see Buckingham, 1998). Yet in fact, research in these areas has gradually moved away from the cruder form of behaviorism (the so-called "magic bullet theory") that was apparent in some of the early research on children's responses to television violence. While the influence is still seen to flow in one direction, the emphasis now is on the range of "intervening variables" that mediate between the stimulus and the response. In the process, many effects researchers have tended to adopt rather more cautious estimates of the influence of the medium.

For example, research on the place of television in cultivating "sex roles" has increasingly drawn attention to the influence of the broader social context – for example, of family communication patterns, and of relationships within the peer group. Durkin (1985), for example, agrees that television tends to provide "stereotyped" representations of male and female roles, but he rejects the view that these are somehow "burned" into the viewer's unconscious, or that they necessarily have a cumulative effect. He argues that research in this field needs to take much greater account of the developmental changes in children's understanding and use of the medium as they mature, and of how they actively make sense of what they watch.

In this respect, effects research has gradually given way to paradigms that conceive of children as "active viewers." The notion of "activity" here is partly a rhetorical one, and it is often used in rather imprecise ways. Yet what unites this work is a view of children, not as passive recipients of television messages, but as active interpreters and processors of meaning. The meaning of television, from this perspective, is not delivered *to* the audience, but constructed *by* it.

This emphasis is apparent in some research within the "uses and gratifications" tradition. Rosengren and Windahl (1989), for example, paint a complex picture of the heterogeneous uses of television among Swedish adolescents, and its interaction with other media such as popular music. In the process, they challenge many general assertions about media effects, not least in relation to violence, arguing that the socializing influence of television will depend upon its relationship with other influences, and upon the diverse and variable meanings which its users attach to it. Thus, for example, television viewing will have a different significance depending upon the child's orientation toward the school, the family, and the peer group. As they indicate, the influence of variables such as age, gender, and social class means that different children can effectively occupy different "media worlds" – an argument which clearly undermines any easy generalizations about "children" as an homogeneous social group.

This notion of "active viewing" is particularly evident in psychological studies that adopt broadly "constructivist" or cognitive approaches (for reviews of this approach, see Buckingham, 1998; Gunter and McAleer, 1997). Rather than simply responding to stimuli, viewers are seen here as consciously processing, interpreting, and evaluating information. In making sense of what they watch, viewers use "schemas" or "scripts," sets of plans and expectations that they have built up from their previous experience both of television and of the world in general. In studying children's understanding of television, cognitive psychologists have tended to concentrate on the "micro" rather than the "macro" – for example, aspects of mental processing such as attention and comprehension, the understanding of narrative or the ability to distinguish between fantasy and reality. Thus, there is some extremely detailed research about the ways in which particular elements of television "language" (such as camera angles or editing) may "stand in" for internal mental processes or "model" cognitive skills which children do not possess (Salomon, 1979). Much of this work has attempted to map the ways in which children's understanding of television changes along with their general intellectual development.

This research provides many insights into the kinds of mental processing that are involved in understanding television, but it suffers from some limitations that are characteristic of mainstream psychology in general. "Cognition" has largely been considered in isolation, not only from "affect" or emotion, but also from the social and interpersonal aspects of the viewing experience. The focus on the individual's internal mental processes has made it difficult to assess the role of social and cultural factors in the formation of consciousness and understanding. While some psychological researchers do acknowledge these factors in theory, much of the research itself appears to adopt a notion of "the child" that is abstracted from any social or historical context. Ultimately, the relationship between children and television is conceived as a matter of the isolated individual's encounter with the screen. The central questions are about what television does to the child's mind – or, more recently, about what the child's mind does with television. In the process, television itself, and the social processes through which its meaning is established and defined, have tended to be neglected.

Cultural Studies Approaches: A Brief Review

These latter issues have been a particular focus of concern for cultural studies researchers – although the specific consideration of children and television has been a relatively recent phenomenon in this field. In principle, the cultural studies approach would appear to go beyond many of the limitations both of effects research and of "active audience" theories. In particular, it places on the agenda questions about the institutional context of television production, and about the nature of television texts; and it offers a theoretical account of the audience that goes beyond the a-social notion of viewers as isolated "cognitive processors." At least in its original "Birmingham" version, cultural studies is typically conceptualized in terms of an interaction between *institutions*, *texts*, and *audiences* (e.g. Johnson, 1985/6); and it is possible to identify a range of relevant studies in each of these three areas.

Institutions

Accounts of children's television written from within the industry have perhaps inevitably tended towards public relations (e.g. Home, 1993; Laybourne, 1993). Yet the study of the institutional context of children's television production has been a relatively marginal concern within communications research generally.

David Buckingham

Early studies such as Melody (1973) and Turow (1981) adopted a broad "political economy" approach, focusing on questions of ownership, marketing, and regulation. With the exception of Palmer (1986) and Buckingham et al. (1999), there has been very little analysis of producers' assumptions and expectations about the child audience; and while there has been some historical work on the evolution of regulatory policy on children's television (e.g. Hendershot, 1998; Keys and Buckingham, 1999), this area has also remained under-researched.

More recently, however, the increasing commercialization of children's television and the apparent retreat from the public service tradition have generated a growing body of academic research and debate (e.g. Blumler, 1992; Buckingham et al., 1999; Kline, 1993; Seiter, 1993). On the one hand, there has been concern about the decline in factual programming for children, and the extent to which production is increasingly tied in with merchandising; yet there have also been calls for a more positive (or at least less puritanical) account of consumption, and a more thoroughgoing discussion of what is meant by "quality." Hendershot's (2003) study of the children's channel Nickelodeon addresses these issues through a sustained historical analysis of the economic and institutional dimensions of children's television production.

Texts

Of course, children's viewing is not confined to programs that are specifically designed for them; yet the analysis of children's television may provide interesting insights into some of the broader tensions which surround dominant definitions of childhood. Analysis here has focused on the ways in which children's television provides opportunities for "para-social interaction" (Noble, 1975), how it handles the relationship between "information" and "entertainment" (Buckingham, 1995), and how it addresses the child viewer (Davies, 1995). More recently, there have been fruitful discussions of genres such as costume drama (Davies, 2002), children's news programs (Buckingham, 2000b; Banet-Weiser, forthcoming), action-adventure shows (Jenkins, 1999) and pre-school programming (Buckingham, 2002a; Oswell, 2002). As in children's literature, the analysis suggests that the position of television as a "parent" or "teacher" and the process of attempting to "draw in" the child is fraught with difficulties and uncertainties (cf. Rose, 1984).

Perhaps the most interesting work in this area has focused on the widely denigrated area of children's cartoons. As against the continuing use of quantitative content analysis (e.g. Kline, 1995), there have been several studies that have sought to apply semiotics (Hodge and Tripp, 1986; Myers, 1995), psychoanalysis (Urwin, 1995), and postmodernist theory (Kinder, 1991) in qualitative analyses of this apparently simple genre. This work raises interesting hypotheses about the ways in which cartoons offer the potential for "subversive" readings, and enable viewers to explore and manage anxiety, thereby perhaps bringing about more protean forms of subjectivity (Hendershot, 2004b; Nixon, 2002; Wells, 2002).

474

Audiences

Cultural studies offers a perspective on the audience that is significantly more social than that of the research discussed above. At the same time, it has struggled to avoid the danger of regarding individual viewers as simply representatives of given demographic categories. Hodge and Tripp (1986) apply a social semiotic perspective, both to the analysis of children's programming, and also to audience data. In common with constructivists, they regard children as "active" producers of meaning, rather than passive consumers; although they are also concerned with the ideological and formal constraints that are exerted by the text.

This approach has been pursued in my own work, where there is a central emphasis on the ways in which children define and construct their social identities through talk about television (Buckingham, 1993a,b, 1996, 2000a,c). Children's judgments about genre and representation, and their reconstructions of television narrative, for example, are studied as inherently social processes; and the development of knowledge about television ("television literacy") and of a "critical" perspective are seen in terms of their social motivations and purposes. In the process, fundamental questions are raised about the methodological difficulties of research with children, and of interpreting audience data.

In parallel with this work, some researchers have adopted a more strictly ethnographic approach to studying children's viewing, both within the context of the home (e.g. Palmer, 1986; Lindlof, 1987; Richards, 1993) and in the context of the peer group (Wood, 1993). Marie Gillespie's (1995) study of the use of television among a South Asian community in London, for example, combines an analysis of the role of television within the family and the peer group with an account of children's responses to specific genres such as news and soap opera. Television is used here partly as a heuristic means of gaining insight into other cultures, although (as with the work discussed above), there is a self-reflexive emphasis on the role of the researcher, and on the power-relationships between researchers and their child subjects, which is typically absent from mainstream psychological research.

As this brief review suggests, there is now a growing body of research on children and television within the field of cultural studies. Nevertheless, there are particular problems when it comes to integrating the different forms of research identified here. Indeed, I would argue that building connections between research on institutions, texts, and audiences remains a central priority in media research considered more broadly. This is not simply a matter of balancing the equation, and thereby finding a happy medium between the "power of the text" and the "power of the audience." Nor is it something that can be achieved in the abstract. Like many other areas considered in this book, the relationship between children and television can only be fully understood in the context of a wider analysis of the ways in which both are constructed and defined.

David Buckingham

Figure 25.1 John Williams, 5, watches Nickelodeon. Photograph © 1986 by Frederick Williams. Child by Victoria and Fred Williams.

Re-locating the Child Audience

In the final part of this chapter, I would like to provide a brief outline of a recent research project that tried to develop these connections; and to discuss some of the problems that emerged. The project was entitled "Children's Media Culture," although in practice it was largely confined to looking at children's relationship with television, rather than the media more broadly. (Further information about this research can be found in Buckingham, 2002b; Buckingham et al., 1999; Davies, Buckingham and Kelley, 1999, 2000; Kelley, Buckingham and Davies, 2000).

Our starting point with this project was to challenge this category of "the child" and particularly "the child audience." We wanted to make explicit and to deconstruct the assumptions that are made about children – about who children are, about what they need, about what they should and should not see. These are assumptions that derive in turn from a whole range of institutional discourses. We wanted to examine these different (and often contradictory) discourses, and the moral, political, economic, psychological, and educational theories upon which they are based. Our basic research question, therefore, was: How do the media (particularly television) *construct* the child audience? And how do children negotiate with these constructions — how do they define themselves and their needs as an audience? We also wanted to take an historical perspective, to consider how those definitions and constructions have changed historically – and how they do or do not reflect changing social constructions of childhood more broadly.

The key point in terms of my argument here is that these kinds of questions cannot be answered by looking at only one aspect of the picture – for example, just by looking at television itself, or just by looking at the audience. On the contrary, we need to look at the relationships between audiences, texts, and institutions. We need to analyse how these different discourses circulate and are manifested at these different levels – in policy, in production, in regulation, in the practice of research, in scheduling, in choices about content, in textual form, in children's own perspectives on and uses of media, and in how those uses are regulated and mediated within the home.

It is also vital to emphasize that none of these levels is determining: on the contrary, there is an ongoing interaction between them. Each of these relationships involves a form of negotiation, a struggle over meaning, a struggle for fixity and control. Texts position readers; but readers also make meanings from texts. Institutions create policies that are manifested in texts; but policies aren't simply implemented, since producers exercise their own kinds of creativity and professional judgment. Likewise, institutions imagine and target audiences; but audiences are elusive – as Ien Ang (1991) describes them, they are "wild savages" who cannot be definitively known or controlled – and the changing behavior of audiences in turn produces changes in institutions.

Furthermore, all these relationships evolve over time: policies and institutions evolve historically, in response to other forces; texts also bear histories of intertextual or generic relations with other texts, which change; and readers do not come to texts either as blank slates or as wised-up critical viewers – they also have reading histories, histories of engagements with other texts which have enabled them to develop certain kinds of competencies as readers.

Changing Constructions of Childhood: Institutions, Texts, and Audiences

In terms of *institutions*, the production of children's media, we considered three main areas. We looked historically at the evolution of children's television, and the kinds of institutional struggles that went on in attempting to claim and preserve a specific place for children in the schedules; we looked at the contemporary political economy of children's television, and particularly at the fate of public service television in the light of the move toward a more commercial, multi-channel, global system; and we gathered and analyzed instances of policy discourse, in the form of official reports and interviews with policy makers, broadcasters, regulators, lobbyists, and others.

In very broad terms, what we find here is a complex balance between the fear of doing harm (a protectionist discourse) and the attempt to do children good (a pedagogical discourse); and these are discourses that in each case draw on broader discourses about childhood. There are also, obviously, some significant historical shifts, as established traditions and philosophies come under pressure in the changing media environment. At present, for example, older philosophies of child-centeredness, which were very dominant in the United Kingdom in the 1960s and 1970s, are rearticulated through their encounter with more consumerist notions of childhood, and simultaneously with notions of children's rights, which have become a key part of the rhetoric of children's producers over the past ten years or so.

Yet far from enjoying an overweening power to define the child audience, producers and policy makers in fact display a considerable degree of uncertainty about it. Changing economic conditions often appear to have precipitated a much broader set of doubts about the changing nature of childhood. In the 1950s, for instance, the advent of commercial television, and the subsequent dramatic decline in the ratings of the BBC (the public service channel), led to a thoroughgoing process of soul-searching. Those responsible for children's programs were dismayed by their loss of the child audience, and increasingly came to doubt the somewhat middle-class, paternalistic approach they had been adopting. Ultimately, after a period of internal crisis, the BBC's Children's Department was effectively abolished in the early 1960s: it was subsumed into a new Family Department, and when it re-emerged later in the decade, it did so with a much less paternalistic view of its audience.

Similar doubts and uncertainties are apparent in the present situation, as terrestrial broadcasters try to come to terms with the threat of competition from new cable and satellite providers, and (more broadly) with the challenges of globalization and commercialization. In the late 1990s, children in Britain (or at least those whose parents subscribed to pay TV) suddenly gained access to a whole series of new specialist channels; and while the generic range of new programming was comparatively narrow, much of it appeared distinctly fresh and innovative, and there was a great deal more to choose from. As in the 1950s, those children who had the option began to abandon terrestrial television in favor of something that appeared to speak to them more directly, without the patronizing or didactic overtones that still characterized a good deal of British children's television.

Contrasting the publicity material produced by the BBC with that produced by the US-based specialist channel Nickelodeon provides a symptomatic indication of the different definitions of childhood that are at stake here. The BBC still tends to hark back to the past, invoking (or indeed re-inventing) tradition – and in the process, playing to parents' nostalgia for the television of their own childhoods. By contrast, Nickelodeon does not have to achieve legitimacy with parents (and hence secure their continued assent for the compulsory license fee): it can address children directly, and it does so in ways that emphasize their anarchic humor and their sensuality. What we find here, and in the statements of its executives, is a rhetoric of empowerment – a notion of the channel as giving voice to kids, taking the kids' point of view, as the friend of kids. Significantly, children seem to be defined here primarily in terms of being *not adults*. Adults are boring; kids are fun. Adults are conservative; kids are fresh and innovative. Adults will never understand; kids just intuitively *know*. This is a very powerful rhetoric that astutely combines an appeal to children's rights with an assertion of consumer power; and its growing popularity suggests that terrestrial broadcasters urgently need to readjust their conceptions of childhood, at least if they wish to survive.

In terms of *texts*, we were interested in how these assumptions and ideologies of childhood are manifested or negotiated in the practices of producers, and in the form of texts themselves. There were two aspects here. Firstly, we tried to develop a broad view of the range of material that has been offered to children over time, through an audit both of the children's television schedules over the past four decades and of the programs which are most popular with children. The schedules for children's TV in the 1950s obviously embody a very different construction of the space of childhood, and of the nature of children's viewing, as compared with the diversity of material that is on offer today. Our analysis also questions some of the myths of cultural decline that often characterize discussions of children's television: the notion that we once lived in a kind of golden age of quality, and that we are now being swamped by trashy American programming, simply does not hold up in the face of the evidence.

Secondly, we undertook a series of qualitative case studies of particular texts or genres, as well as talking to their producers. We were particularly interested in texts or areas of programming which have a long history, where we can see clear

479

indications of historical change. We looked at how texts address and construct the child viewer – for example, the various ways in which the viewer is spoken to; how the viewer is or is not invited to be involved; the function of children within the programs; how adult-child relations are represented or enacted; and more formal devices – how the visual design of the studio, the camerawork, graphics, and music imply assumptions about who children are, and what they are (or should be) interested in. This analysis is also, of course, about content – about which topics are seen to be appropriate for this audience, and how the perceived interests of the child audience are demarcated from or overlap with those of the adult audience.

The BBC pre-school series *Teletubbies*, and the debates that surrounded it, provide an interesting case study of some of these changes. *Teletubbies*, which began broadcasting in 1997, is an outsourced, independent production, which has generated strong overseas sales and a vast range of ancillary merchandising. It has been accused of abandoning the "great tradition" of educative program-ming, and thereby "dumbing down" its audience; of commercially exploiting children; and (by some overseas critics) of cultural imperialism, in terms of pedagogy and social representation. The controversy it has aroused can be seen as a highly symptomatic reflection of the BBC's current dilemmas, as it attempts to sustain national public service traditions while simultaneously being increas-ingly dependent on commercial activities and global sales.

In terms of both form and content, *Teletubbies* is an amalgam of two historical traditions within children's television – the more didactic (albeit play-oriented), "realist," adult-centered approach of *Playdays* and its predecessor *Playschool* on the one hand, and the more surrealistic, entertaining tradition of many animation and puppet shows on the other. While it is the latter that immediately confounds and surprises many adult critics, it is important to recognize the particular forms of education that are being offered here, and the different ways in which they construct the child viewer. Thus, the "child-centered" pedagogic approach is manifested in documentary inserts shot and narrated from the child's point of view; in the manipulation of knowledge via narrative; and in the slow pace and "parental" mode of address. This contrasts with the more didactic elements, relating to pre-reading and counting skills and the modeling of daily routines.

Teletubbies is extremely popular with its immediate target audience of 2- to 3-year-olds; but it also attained a kind of cult status among older children and among some adults. The program was a frequent topic of conversation in our audience research, although our sample was much older than the target audience. The 6- to 7-year-olds were often keen to disavow any interest in the program, while the 10- to 11-year-olds seemed to relate to it with a kind of subversive irony – although it was often passionately rejected by those with younger siblings. As this implies, the children's judgments about the program reflected their attempts to project themselves as more or less "adult." Combined with more anecdotal information about the program's popularity with older chil-dren and young adults, this suggests that its cult popularity may be symptomatic of a broader sense of irony that suffuses contemporary television culture – and one that reflects ambivalent investments in the *idea* of "childishness." (For further discussion, see Buckingham, 2002a.)

What we find at the level of institutions and texts, then, are some very powerful definitions of the child – definitions which are partly coercive, but also partly very pleasurable, and often quite awkward and contradictory. The obvious question here is how children negotiate with these definitions: that is, how they define themselves as an *audience*. This was the third dimension of the project, and again there was a quantitative and a qualitative dimension.

Audience ratings can clearly tell us a fair amount about how children define themselves as an audience; and however unreliable or superficial they may be, they clearly show (for example) that children are increasingly opting to watch adult programs and not children's programs. At the same time, they do choose to watch particular kinds of adult programs; and it is interesting to look at the versions or aspects of "adulthood" that they choose to buy into, and those they reject or resist.

These kinds of questions were very much the focus of our more qualitative investigations of the child audience, which focused on children aged 6–7 and 10–11. Through a series of focus group discussions and activities, we investigated how children negotiate with these adult definitions of childhood, how they define themselves as children, as young people, and as children of a particular age – and how they do this in different ways in different contexts and for different purposes. In a sense, we were looking again at how "childhood" is defined, at how children collectively and discursively construct the meaning of childhood (and by implication of adulthood). The category "child" and its defining opposite "adult" are thus highly flexible, and also highly charged, parameters of self-definition and of identity formation.

In the children's exploration of what makes a program "appropriate" for children, the strongest arguments were negative ones. Programs featuring sex, violence, and "swearing" were singled out by both groups as being particularly "grown-up." Likewise, children's programs were predominantly defined in terms of absences – that is, in terms of what they do *not* include. One area of our analysis here concerned children's discussions of sex and sexuality on television. On one level, it was clear that "adult" material on television could function as a kind of "forbidden fruit." In discussing this kind of material, the children displayed a complex mixture of embarrassment, bravado, and moral disapproval. Discussions of sex and romance in genres such as dating game shows, soap operas and sitcoms often served as a rehearsal of projected future (hetero-)sexual identities, particularly among girls. Boys were less comfortable here, with the younger ones more inclined to display disgust than fascination; although the older ones were more voyeuristic. Discussion of sexuality was often the vehicle for interpersonal policing among the group, in which girls seemed to enjoy the upper hand.

The children were very familiar with adult definitions of appropriateness, although they were inclined to displace any negative "effects" of television onto those younger than themselves, or onto "children" in general. While some of the youngest children expressed a more censorious rejection of "adult" material, this was much less common among the older children, who aspired to the freedom they associated with the category of the "teenager." Here again, these discussions could serve as a form of mutual policing, particularly among boys. Overall, the analysis here suggests that in discussing their responses to television, the children are performing a kind of "identity work," particularly via claims about their own "maturity." In the process, these discussions serve largely to reinforce normative definitions both of "childhood" and of gender identity (for a fuller discussion, see Kelley, Buckingham, and Davies, 2000).

Another aspect of our investigation here concerned the issue of children's tastes. We were interested to discover whether children have distinctive tastes as an audience, and how these tastes are articulated and negotiated in the context of peer group discussion. We analyzed the social functions and characteristics of children's expressions of their tastes using a set of overlapping paradigmatic oppositions: parents::children; grannies::teenagers; boring::funny; and talk::action. In each case, the children generally favored the former element (associated with children) and disavowed the latter (associated with adults). However, they frequently distinguished here between the tastes attributed to parents in general and those they observed in the case of their *own* parents. The older children were inclined to aspire to the identity of the "teenager," via the display of particular tastes, notably in comedy. By contrast, the tastes of some adults were dismissed as belonging to "grannies," who were parodied as hopelessly "old fashioned" and "uncool." The children were highly dismissive of programs featuring "talk" and enthusiastic about those featuring action – not least action of a violent or otherwise spectacular nature. As this implies, they frequently inverted cultural hierarchies and resisted adult notions of "good taste."

Contemporary debates about children's television have emphasized the need for factual programs, literary adaptations, and socially responsible contemporary drama. Without disputing this, our analysis suggests that there is also a need for entertainment programming – and indeed for programs that a majority of adults would consider "infantile," "puerile," or otherwise "in bad taste." The complex and playful nature of children's judgments of taste and their understanding of taste as "cultural capital" is certainly apparent in the popularity of such self-consciously ironic texts as *South Park* and *Beavis and Butthead*. Nevertheless, children's tastes cannot be defined in an essentialist way, any more than adults' can: both groups are more heterogeneous than is typically assumed (see Davies, Buckingham, and Kelley, 2000).

Conclusion

In this chapter, I have argued that we need to understand audiences in relation both to texts and to institutions. Audiences are indeed "active," but they act under conditions that are not of their own choosing – and to this extent, it would be quite misguided to equate "activity" with *agency* or power. In the case of children, their relationships with television are structured and constrained by wider social institutions and discourses, which (among other things) seek to define "childhood" in particular ways. The child audience – and indeed the "specialness" of that audience – is thus constructed through an ongoing process of social negotiation.

This kind of multifaceted or interdisciplinary approach to the study of television is part of the project of this book. Yet, as in the research project I have described, it is notable that these different aspects of television study are rarely fully integrated in practice. Our publications from the project, for example, fell fairly neatly into those addressing institutions, texts, and audiences, and there was relatively little overlap between them. Some of our audience-focused pieces attempted to take up themes that cross these boundaries. The regulation of children's time, for example, and what is seen as a productive or unproductive way for children to spend their time, is an issue that connects parental regulation in the domestic context with questions about the role of scheduling in television and questions about regulatory policy (see Davies, Buckingham, and Kelley, 1999). Likewise, we attempted to understand children's tastes, not as an expression of their innate characteristics and desires, but (at least in part) as a function of how "childishness" is defined by the media themselves (Davies, Buckingham, and Kelley, 2000). Yet in practice, it is difficult to take account of this "balance of forces" without lapsing into one or other of the two opposing positions outlined earlier. On the one hand, there is a view of childhood (and, by extension, of the subjectivity of children) as somehow inexorably produced by powerful institutional discourses; while on the other is the view that real children somehow automatically and inevitably evade those constructions. Accounting for the real slippages and inconsistencies here – and doing so *in empirical terms*, rather than simply through recourse to a series of "in principle" theoretical qualifications – is a continuing endeavor.

References

Ang, I. (1991) *Desperately Seeking the Audience*, London: Routledge.

Aries, P. (1973) *Centuries of Childhood*, Harmondsworth: Penguin.

Banet-Weiner, S. (2004) "'We Pledge Allegiance to Kids': Nickelodeon and Citizenship," in H. Hendershot (ed.), *Nickelodeon Nation: The History, Politics and Economics of America's Only TV Channel for Kids*, New York: New York University Press, pp. 209–40.

Barker, M. (ed.) (1984) *The Video Nasties*, London: Pluto.

Barker, M. and Petley, J. (eds.) (2001) *Ill Effects: The Media/Violence Debate*, 2nd edn., London: Routledge.

Blumler, J. (1992) *The Future of Children's Television in Britain: An Enquiry for the Broadcasting Standards Council*, London: Broadcasting Standards Council.

Buckingham, D. (1993a) *Children Talking Television: The Making of Television Literacy*, London: Falmer.

Buckingham, D. (ed.) (1993b) *Reading Audiences: Young People and the Media*, Manchester: Manchester University Press.

Buckingham, D. (1994) "Television and the Definition of Childhood," in B. Mayall (ed.), *Children's Childhoods Observed and Experienced*, London: Falmer, pp. 76–96.

Buckingham, D. (1995) "On the Impossibility of Children's Television," in C. Bazalgette and D. Buckingham (eds.), *In Front of the Children*, London: British Film Institute, pp. 47–60.

Buckingham, D. (1996) *Moving Images: Understanding Children's Emotional Responses to Television*, Manchester: Manchester University Press.

Buckingham, D. (1998) "Children and Television: A Critical Overview of the Research," in R. Dickinson, O. Linne and R. Harindranath (eds.), *Approaches to Audiences*, London: Edward Arnold, pp. 19–55.

Buckingham, D. (2000a) *After the Death of Childhood: Growing Up in the Age of Electronic Media*, Cambridge: Polity.

Buckingham, D. (2000b) *The Making of Citizens: Young People, News and Politics*, London: University College London Press.

Buckingham, D. (2002a) "Child-centred Television? Teletubbies and the Educational Imperative," in D. Buckingham (ed.), *Small Screens: Television for Children*, Leicester: Leicester University Press, pp. 38–60.

Buckingham, D. (ed.) (2002b) *Small Screens: Television for Children*, Leicester: Leicester University Press.

Buckingham, D., Davies, H., Jones, K., and Kelley, P. (1999) *Children's Television in Britain: History, Discourse and Policy*, London: British Film Institute.

Burman, E. (1994) *Deconstructing Developmental Psychology*, London: Routledge.

Connell, I. (1985) "Fabulous Powers: Blaming the Media," in L. Masterman (ed.), *Television Mythologies*, London: Comedia/MK Media Press, pp. 88–93.

Davies, H., Buckingham, D., and Kelley, P. (1999) "Kids' Time: Television, Childhood and the Regulation of Time," *Journal of Educational Media*, 24(1), 25–42.

Davies, H., Buckingham, D., and Kelley, P. (2000) "In the Worst Possible Taste: Children, Television and Cultural Value," *European Journal of Cultural Studies*, 3(1), 5–25.

Davies, M. M. (1995) "Babes 'n' the Hood: Pre-school Television and its Audiences in the United States and Britain," in C. Bazalgette and D. Buckingham (eds.), *In Front of the Children*, London: British Film Institute, pp. 15–32.

Davies, M. M. (2002) "Classics with Clout: Costume Drama in British and American Children's Television," in D. Buckingham (ed.), *Small Screens: Television for Children*, Leicester: Leicester University Press, pp. 120–40.

Durkin, K. (1985) *Television, Sex Roles and Children*, Milton Keynes: Open University Press.

Freedman, J. (2002) *Media Violence and its Effect on Aggression: Assessing the Scientific Evidence*, Toronto: Toronto University Press.

Gauntlett, D. (1995) *Moving Experiences: Understanding Television's Influences and Effects*, London: John Libbey.

Gillespie, M. (1995) *Television, Ethnicity and Cultural Change*, London: Routledge.

Goldsen, R. (1977) *The Show and Tell Machine*, New York: Dial Books.

Gunter, B. and McAleer, J. (1997) *Children and Television*, 2nd edn., London: Routledge.

Hendershot, H. (1998) *Saturday Morning Censors: Television Regulation before the V-Chip*, Durham, NC: Duke University Press.

Hendershot, H. (ed.) (2004a) *Nickelodeon Nation: The History, Politics and Economics of America's Only TV Channel for Kids*, New York: New York University Press.

Hendershot, H. (2004b) "Nickelodeon's Nautical Nonsense: The Intergenerational Appeal of Spongebob Squarepants," in H. Hendershot (ed.), *Nickelodeon Nation: The History, Politics and Economics of America's Only TV Channel for Kids*, New York: New York University Press, pp. 182–208.

Hendrick, H. (1997) *Children, Childhood and English Society, 1880–1990*, Cambridge: Cambridge University Press.

Hodge, B. and Tripp, D. (1986) *Children and Television: A Semiotic Approach*, Cambridge: Polity.

Home, A. (1993) *Into the Box of Delights: A History of Children's Television*, London: BBC Books.

James, A. (1993) *Childhood Identities*, Edinburgh: Edinburgh University Press.

Jenkins, H. (1999) "Her Suffering Aristocratic Majesty: The Sentimental Value of *Lassie*," in M. Kinder (ed.), *Kids' Media Culture*, Durham, NC: Duke University Press, pp. 69–101.

Jenkins, P. (1992) *Intimate Enemies: Moral Panics in Contemporary Great Britain*, New York: Aldine de Gruyter.

Johnson, R. (1985/6) "What is Cultural Studies Anyway?," *Social Text*, *16*, 38–80.

Kelley, P., Buckingham, D., and Davies, H. (1999) "Talking Dirty: Childhood, Television and Sexual Knowledge," *Childhood*, *6*(2), 221–42.

Keys, W. and Buckingham, D. (eds.) (1999) "International Perspectives on Children's Media Policy," special issue of *Media International Australia/Culture and Policy 93*.

Kinder, M. (1991) *Playing with Power*, Berkeley: University of California Press.

Kline, S. (1993) *Out of the Garden: Toys and Children's Culture in the Age of TV Marketing*, London: Verso.

Kline, S. (1995) "The Empire of Play: Emergent Genres of Product-based Animations," in C. Bazalgette and D. Buckingham (eds.), *In Front of the Children*, London: British Film Institute, pp. 151–65.

Laybourne, G. (1993) "The Nickelodeon Experience," in G. L. Berry and J. K. Asamen (eds.), *Children and Television: Images in a Changing Socio-Cultural World*, London: Sage, pp. 303–7.

Lindlof, T. (ed.) (1987) *Natural Audiences*, Newbury Park, CA: Sage.

Lisosky, J. (2001) "For *All* Kids' Sakes: Comparing Children's Television Policy-making in Australia, Canada and the United States," *Media, Culture and Society*, *23*(6), 821–45.

Livingstone, S. and Bovill, M. (eds.) (2001) *Children and their Changing Media Environment*, Mahwah, NJ: Lawrence Erlbaum Associates.

Luke, C. (1990) *Constructing the Child Viewer*, New York: Praeger.

Melody, W. (1973) *Children's Television: The Economics of Exploitation*, New Haven: Yale University Press.

Myers, G. (1995) "'The Power is Yours': Agency and Plot in Captain Planet," in C. Bazalgette and D. Buckingham (eds.), *In Front of the Children*, London: British Film Institute, pp. 62–74.

Nixon, H. (2002) "South Park: Not in Front of the Children," in D. Buckingham (ed.), *Small Screens: Television for Children*, Leicester: Leicester University Press, pp. 96–119.

Noble, G. (1975) *Children in Front of the Small Screen*, London: Constable.

Oswell, D. (2002) *Television, Childhood and the Home: A History of the Making of the Child Television Audience in Britain*, Oxford: Oxford University Press.

Palmer, P. (1986) *The Lively Audience*, Sydney: Allen and Unwin.

Postman, N. (1983) *The Disappearance of Childhood*, London: W. H. Allen.

Reeves, B. and Wartella, E. (1985) "Historical Trends in Research on Children and the Media, 1900–1960," *Journal of Communications*, *35*(2), 118–33.

Richards, C. (1993) "Taking Sides? What Young Girls Do with Television," in D. Buckingham (ed.), *Reading Audiences: Young People and the Media*, Manchester: Manchester University Press, pp. 24–47.

Roberts, D. F. and Foehr, U. (2003) *Kids and the Media in America*, New York: Cambridge University Press.

Rose, J. (1984) *The Case of Peter Pan: Or the Impossibility of Children's Fiction*, London: Macmillan.

Rosengren, K. E. and Windahl, S. (1989) *Media Matter: TV Use in Childhood and Adolescence*, Norwood, NJ: Ablex.

Rowland, W. (1983) *The Politics of TV Violence: Policy Uses of Communication Research*, Beverly Hills, CA: Sage.

David Buckingham

Salomon, G. (1979) *Interaction of Media, Cognition and Learning*, San Francisco: Jossey-Bass.

Seiter, E. (1993) *Sold Separately: Parents and Children in Consumer Culture*, New Brunswick, NJ: Rutgers University Press.

Spigel, L. (1992) *Make Room for TV: Television and the Family Ideal in Postwar America*, Chicago: University of Chicago Press.

Starker, S. (1989) *Evil Influences: Crusades Against the Mass Media*, New Brunswick, NJ: Transaction Books.

Turow, J. (1981) *Entertainment, Education and the Hard Sell*, New York: Praeger.

Urwin, C. (1995) "Turtle Power: Illusion and Imagination in Children's Play," in C. Bazalgette and D. Buckingham (eds.), *In Front of the Children*, London: British Film Institute, pp. 127–40.

Wells, P. (2002) "'Tell Me About Your Id, When You Was a Kid, Yah?' Animation and Children's Television Culture," in D. Buckingham (ed.), *Small Screens: Television for Children*, Leicester: Leicester University Press, pp. 61–95.

Winn, M. (1985) *The Plug-In Drug*, Harmondsworth: Penguin.

Wood, J. (1993) "Repeatable Pleasures: Notes on Young People's Use of Video," in D. Buckingham (ed.), *Reading Audiences: Young People and the Media*, Manchester: Manchester University Press, pp. 184–201.

PART EIGHT

Television/ Alternative Challenges

CHAPTER TWENTY-SIX

Local Community Channels: Alternatives to Corporate Media Dominance

DeeDee Halleck

Which is better? One radio programme that reaches one million people with one standard message and language, or one hundred radio programmes that reach ten thousand people each (total, one million), with messages tailored to the local culture and traditions, in the local language and possibly made through a participatory process that involves each community?

Alfonso Gumucio Dagron

There is an abiding myth in the United States that the military and the industries that service that sector are responsible for all the electronic communication tools and toys that we enjoy. Although military budgets (rising to unprecedented levels in the post-9/11 United States) have certainly funded communication research, consumer demand for creative tools has also been a major factor. VCRs that could play back tapes but could not record were never popular. DVDs were designed to be one-way program recordings, but demand for writable DVDs has opened up new markets. Participatory communication tools are now available worldwide and many communities have built infrastructures that support local media production and distribution.

There are sturdy traditions of community-based media throughout the world. In the early days of radio, individuals with homemade transmitters were the first to exchange messages through the "ether" and create tiny radio stations from their bedrooms. When 16 mm and 8 mm film was introduced, Cine Clubs formed in many countries to share amateur productions and to see films that went beyond the mass-market formulae (Stone and Streible, 2003). In the 1970s video collectives pioneered the use of television as social dialog. Much computer hardware and software was developed in the garages and basements of tinkerers and hackers. When the mass media tells stories of "homemade" media tools, narratives are based on tales of individual inventors, who are often mythologized in media history books. What have often been neglected are the stories of how

489

local community organizations around the world have created networks for local and regional media, often using improvised homemade devices, grassroots organizing and unlikely collaborations between artists, technicians, academics, and activists. These examples are rarely studied by communication academics, and are not generally included in standard anthologies or textbooks. From samizdats to pirate radio, there is a dynamic history of oppositional communication tools that have scaled the fortresses of corporate culture and breached the impediments of mass-media censorship. There are also other successful forms of "alternative media" which are not "underground" and rarely combative, but which stand in direct contrast to the corporate media environment which dominate the airwaves and satellite paths of the world.

This chapter will examine some examples of community uses of television throughout the world, and the infrastructures that enable these entities to survive and, in many cases, to flourish, despite the many pressures of commercial broadcasters and government bureaucracies.

Community television has its roots in the radio and 16 mm film movements of the twentieth century. The forms of community media vary not only in the type of technique and hardware that different communities use, but also in the themes that they cover. These themes include women's health issues, human rights, agrarian reform, environmental problems, literacy training, AIDS prevention, youth empowerment, and an entire gamut of culture, art and performance. This type of media can be given a variety of labels: citizens' media, horizontal communication, local media, participatory communication, guerilla video, or alternative media. One of the organizations that promotes this work is called simply "Our Media," which suggests that the "Other" media is "Theirs." However, the use of the word "Our" is more than a proprietary description. This category of citizens/community/alternative media can be defined by the fact that it is made by people who, though often not "professional" or "experts," want to take part in the *production* and *exchange* of the ongoing narratives, cultures, icons, and images of their lives. "Our Media" is not about media they consume as passive audiences. This is media they make. The "Our" is in active mode. And the product is that creative activity, not the resulting media "productions" (Rodriguez, 2001).

The notion of broadcast stations being transmitters from one central place to many "listeners" is not something that is inherent to electronic media. As Bertolt Brecht pointed out in 1936, radio can be from many to many, not just one to many (Brecht, 1964).

Examples of Community Media: Radio

Early radio was very much an interactive affair. In the United States, the thousands of amateurs who set up small transmission stations from their homes were soon displaced by corporations such as Westinghouse and RCA. These corporate interests influenced federal regulation into promoting powerful commercial stations that soon expanded to centralized networks. However, not all countries proceeded in this direction. In postwar Italy, for instance, broadcasting was largely unregulated and the airwaves came alive with thousands of small, unlicensed stations, broadcasting everything from opera to community news. Meanwhile, Bolivian tin miners, often far from their homes for many weeks at a time, set up small radio stations as part of struggles to strengthen union solidarity and lobby for improvements in safety. The radios enabled the miners to communicate their labor issues not only to each other and to their families, but also to the government and the mining concessions. These stations also broadcast greetings to their wives and children (see Beltran, 2000; Dagron and Cajias, 1989; and O'Connor, 1990).

Community radio has also been used to mobilize for reform. In mid-1920s Chicago, radio was used by unions to give information about strikes and working conditions. Alfonso Gumucio Dagron (2001) describes a radio station in the Philippines: "Peasants from Tacunan Community Audio Tower (CAT), in Davao del Norte, in the Philippines, told me that they were certain they could have not obtained electricity, roads and safe water within a couple of years if it were not for their ability to voice their needs through their simple six cone-speaker system."

For communities with ambiguous allegiance to national states, radio can be a "border" issue. Along the border between Tanzania, Rwanda, and Burundi several hundred thousand refugees tune in to Radio Kwize, which provides important information on everything from weather to the current position of border guards.

In the eyes of many small rural communities, national transmissions from urban centers do not address the realities of their lives. The urban settings are far away and isolated, while their own rural community media is in reach, access is available and the content addresses community needs. For example, in El Salvador, Radio Victoria transmits to a small village of people who were displaced during the war. The radio station is popular with teenagers who not only provide DJs to play the music, but also create short radio novellas or soap operas featuring local real-life characters with romantic interests and adolescent fears. Radio Victoria is one of many small stations that fought hard for legalization after the peace accords in the 1990s. These radio stations provide information for local farmers, including the prices of crops and availability of transport, and *avisos* – the equivalent of classified ads. These are only a few of the examples of the varied uses and projects associated with community radio.

Early Examples of Community Media: Film

The use of film as community media also has an extensive history. The distribution of documentaries, experimental, and independent productions was pioneered in the 1920s and 1930s by organizations in Europe and the United States through "Cine Clubs." Although some filmmaker/explorers, such as Robert

Flaherty and Edward Curtis, attempted to gain access to theatrical venues for their documentary material, they had to rely on other exhibition settings. Flaherty's astonishing *Nanook of the North* was screened in Cine Clubs throughout the world, with presentations in cafés and small theaters, in workers' halls and church cellars. Joris Ivens, a Dutch filmmaker, showed his experimental films, *Rain* and *The Bridge*, to Cine Clubs in Europe. Cine Club programs often included modernist films from the Soviet Union, such as Sergei Eisenstein's *Battleship Potemkin* and Dovshenko's *Strike*. In the United States, the Workers Film and Photo League initially started as a Cine Club, but also made their own documentaries showing the Poor March to Washington and confrontations between jobless workers and the police in New York City's Union Square. It might be noted that these documentaries were shot in 35 mm, which made both production and exhibition a cumbersome activity. Some of the filmmakers from the Workers Film and Photo League became part of the Work Projects Administration, which made documentaries such as *The City* and *The River*, as well as other films to promote Roosevelt's New Deal. (See Stange, 1989 for a critical view of both Flaherty and the WPA films.)

After World War II, 16 mm films (roughly half the size of 35 mm) were popularized for institutional and consumer use. War photographers had developed lighter, more portable equipment, including the technology for synchronous sound. After the war, 16 mm films were produced promoting health and safety and marketed to schools and community centers. Public libraries started film collections that could be borrowed by youth clubs and church groups. At the same time, artists began experimenting with 16 mm for documentaries and lyrical films. The more adventurous librarians began to collect these experimental and documentary works produced by people such as Burt Haanstra and Arne Sucksdorf, as well as 16 mm prints of Robert Flaherty's films.

Meanwhile, the National Film Board of Canada was created as a way of promoting Canadian culture. This ambitious filmmaking project, led by Scotsman John Grierson, created a trove of 16 mm films that became staples in school and library collections, not just in Canada, but also throughout the world, with translations into many languages. Norman McLaren, a Canadian animator, and Lotte Reiniger, a World War II refugee living in Canada, created many experimental films, which inspired both European and US filmmakers.

With the development of synchronous sound, *cinema verité* ("film truth") became possible as documentarians strived to capture everyday life. D. A. Pennebaker and Ricky Leacock (who had worked as an assistant to Robert Flaherty) pioneered the use of portable cameras in real-life situations. Their film *Happy Mothers' Day*, about the birth of quintuplets to a Wisconsin farm couple, was one of the first documents of mass-media frenzy, as they focused their cameras not on the quints, but on the hordes of journalists who descended on the family.

During the Vietnam War, several collectives provided alternative views to the official accounts of that conflict. The Newsreel Collective covered US protests and also distributed films from North Vietnam. Wintersoldier was a collective of filmmakers and Vietnam Vets who held hearings in Detroit in 1970. The documentation of the event was edited with gruesome war footage into a feature film entitled *Wintersoldier*, which was shown in GI coffee houses and in film festivals around the world. These attempts to make alternative cinema were the precursors to the early video experiments.

Portable Video Arrives

Although 16 mm cameras had made portable media production possible, film still needed to be processed, edited, and projected. In the 1950s, the Ampex Corporation had developed video tape recorders – but these were huge affairs that used two-inch oxide-coated plastic tape (called "Quad") that was even bigger than 35 mm film. The machines were about the size of refrigerators, laid on their side and sales of these machines were limited to television stations and production studios. In the late 1960s, however, Japan's Sony Corporation revolutionized media making by manufacturing a portable video camera called the Portapak. This was the first video recorder made specifically for the consumer mass market. This small portable machine made it possible to have instant, simultaneous feedback that could be watched on ordinary television sets. In other words, the equipment was "consumer friendly." Many filmmakers and activists in Europe and the United States recognized the potential in this technology for challenging mass-media television and for giving voices to ordinary people. There was no need for professionals to intervene between the production and audience. As a result, video collectives in the United States, France, and Italy proclaimed that television of, by, and for the people was possible. In fact, the notion of a "finished product" itself was questioned. Videotape could be erased and used again and again. In fact, video production could involve a live feedback loop without recording. Thus, many people experimented with television/video as process or dialog. Artists such as Joan Jonas and Nam June Paik used video as a live adjunct to performance. And Peter Campus created installations in art museums that allowed visitors to interact with the art by having their images intermixed with the live video feedback.

Meanwhile, city planners and social workers also began to use video as a way of encouraging dialog. In Bologna, Italy, the leftist government, prodded by Roberto Grandi, a communications professor at the University of Bologna, filmmaker Roberto Faenza, and Eddie Becker, a young video enthusiast from Washington, DC, initiated a project using video for community expression. Booths, set up in the town center, allowed passers-by to voice their reaction to local and national events.

Another example involved a group of prostitutes who had barricaded themselves in the cathedral in Lyon, France, to protest police harassment and the hypocrisy of town officials, many of whom were regular patrons. A feminist group used a video camera to exchange taped messages from the barricaded women and townspeople who wanted to evict them.

In Canada, the National Film Board was quick to grasp the potential of video. Colin Low had initiated a program called Challenge for Change to use film documentation of local Canadian communities, but the turnaround time was cumbersome. NFB's Bonnie Klein and Dorothy Henaut quickly recognized that video was a better tool to allow instant feedback on issues such as fishing rights, housing, and other social welfare matters. The project encouraged communities to communicate with each other and with regional and national authorities through taped messages. George Stoney was one of the facilitators who worked with this project, and took his experiences with Challenge for Change to the United States, where he was one of the founders of the movement for public access television.

While the cost of video recorders and tape was far less than film equipment, video editing equipment was prohibitive for individuals. As a result, several groups in the United States banded together in collectives to share resources and space. Their names resonate with the ideology of Marshall McLuhan, whose aphorisms were the catchwords of the movement: Global Village, Rain Dance, Optic Nerve (Boyle, 1996). Their journal, *Radical Software*, was filled with articles proclaiming a media revolution. Some of the groups experimented not only with making tapes and interacting with "process," but also created their own television transmission. One collective, Videofreex, in the Catskill mountain town of Lanesville, New York, built a small transmitter and began to send live television to their neighbors (Teasdale, 1999). The programs often mimicked network television genres: a cowboy show called *Buckaroo Bart*, and the *News Buggy*, in which the video deck was carried a baby carriage as newscasters covered stories about calf births and fishing contests. The station represented the only broadcast TV available in the narrow valley, which was surrounded by steep mountain escarpments. Rather than being shut down by the Federal Communication Commission (FCC), the station became the model for the FCC's rule-making on Low Power TV in 1980. While activists hoped that a new era of creative alternative TV would develop, unfortunately, the licenses for low-power television stations predominantly went to well-funded evangelical religious groups and Sears and Roebuck. Because of prohibitive overlap restrictions, the only spectrum slots that were available were in underpopulated rural areas. Thus far, the development of alternative low-power television has been somewhat disappointing.

Cable Monopolies and the Justification for Public Access

In the early 1970s, many US cities were approached by cable corporations that wanted to string cables for television reception. Offering more programming than was available over broadcast TV, this service advanced rapidly in large and medium-sized municipalities. Because there was considerable cost involved in building the infrastructure, cable corporations wanted assurances that they would have a monopoly on the service, and most companies drew up contracts with local towns for exclusive rights to their citizens. Towns and cities were reluctant to cede access to information to one corporation, so many drew up franchise agreements, which required the cable corporations to give back both channels and a portion of the fees to the towns. Initially, this process was required by the FCC, whose liberal commissioner, a Lyndon Johnson appointee, Nicholas Johnson, supported community television and ruled that all cities must negotiate to utilize public "rights of way." This form of telecommunications regulation has become what is known as PEG (Public, Educational, and Government) access television. Eventually the federal requirement was dropped, but states and local authorities can still require PEG franchises and many have extensive regulations. There are four levels on which this regulation is justified:

- First, a cable contract grants what is essentially a monopoly to a communications corporation, enabling it to reach a particular market. Therefore, this benefit should be compensated by a public component.
- Second, in order to construct the physical infrastructure (hanging wires, sharing telephone poles, having access to city sewer and tunnels), a cable company needs the cooperation and "right-of-way" from the municipality involved.
- Beyond these two legal justifications, there is the third philosophical argument that citizens have an essential right to information exchange. The assumption here is that if the First Amendment protects free speech in an age of face-to-face argument and print media, these rights are automatically extended into more complicated forms of technology as they are developed.
- There is also an argument that the distribution of most cable programming is via extensive satellite systems which were developed with a heavy investment of public funds which therefore mandates a "pay-back" in the form of public benefits.

All countries have some form of public interest telecommunications regulation, although many of these laws have eroded with the increasing expansion of market-oriented systems. (See Sylvia Harvey's chapter in this volume.) These regulations vary from country to country, but the essential requirement is to ensure that the public benefits from the use of public airwaves and rights of way. Public access in the United States is a creative response to the need to regulate video distribution, and many studies have shown that not only does the public benefit, but the cable corporations receive benefits, as well (Agosta, 1990, p. 20).[1]

Public Access in the United States: Structures and Systems

Public access is implemented in a variety of ways. A public access center might be located at a library, a junior high school, or in a church basement. It could be set up in an abandoned firehouse or a new shopping center. Each municipality works with community leaders and non-profit groups to find a home for the center. The organization of the space and outreach can be quite different from system to system or even within a given system. For example, Dallas, Texas, started with a plan to create centers in many different parts of the city. What they found was that some of the centers flourished and others languished; so they eventually shut those centers that were not active and increased the ones that worked well.

Access centers can become community hubs of activity, especially when the center is in a location where people feel safe and welcome, and where there is ample public transportation (especially at night). Since most cities are divided into neighborhoods along class or ethnic lines, the location of an access center in one specific neighborhood can be seen to favor one group or another. The building of a broad base of community contacts and relationships is essential to the success of access and the location in many ways defines future use.

The administration of public access centers can vary from appointed boards (usually appointed by the mayor and/or the city council) to elected boards (via ballot at the center or by producers and users). The cable corporations themselves administer some access centers. In general, the centers that have some degree of autonomy from the cable corporations seem to be the most successful and creative. There is an inherent conflict of interest between the cable companies (who want the channels for their own purpose and profit) and the access community (especially the community of producers, who are in general independently minded and therefore antagonistic to intrusion or control). The Alliance for Community Media (formerly the National Federation of Local Cable Programmers) is a national trade association for people who work on PEG. Both producers and administrators attend a yearly conference where programming is shared and ideas for improving franchise negotiations are discussed.

One problem with cable has been the fact that although equipment and studios may be available, most people are intimidated by technology, thus it is time-consuming and difficult to make programming. One option has been for people to form programming collectives in order to create weekly series. One example of this phenomenon is Paper Tiger Television (www.papertiger.org), which began in 1981 in Manhattan, providing a weekly critique on various forms of media. The project started with a six-part series, "Herb Schiller Reads the *New York Times*." Since that time, the collective has produced over 300 programs and has served as a model for low-budget, creative production, with subject matter of consequence. Other groups, such as Peoples' Video Network, Whispered Media, and San Diego Stew, have made series especially for cable access (Downing, 2000).

Satellite Distribution

In 1986, Paper Tiger began to distribute programming via satellite to other public access channels in a project named The Deep Dish Network (www.deepdishtv.org). Each year, Deep Dish has produced several series which have been transmitted via rented transponders on commercial satellites. More recently, Deep Dish has worked with Free Speech TV (FSTV), a non-profit organization in Boulder, Colorado, which programs a 24-hour transponder on the Direct Broadcast Satellite (DBS) on the Echo Star Dish Network. The channel is available because activists pressured the FCC to require that all DBS providers have a fixed percentage (4 percent minimum) of non-profit, non-commercial, educational programming. These "set-aside" channels are also used by several universities to provide educational and cultural programming. FSTV transmits documentaries and discussion programs that are not seen on commercial or even public television channels.

Community Television Around the World

Several European cities have versions of public access: Berlin has the Offener Canal, Amsterdam has several channels, and many Scandinavian countries have various forms of community television.[2] Australia has public access in several cities, and there are community facilities in many aborigine communities. In Korea, labor unions have used video for many years and have been able to pressure the government for channel time. There is an elaborate public video facility in Seoul and labor unions have negotiated contracts that include provisions for video documentation of working conditions. In a recent Hyundai contract, for example, the company must pay for a video journalist (chosen by the union) to film weekly reports which are then played on a television set in the workers' lunchroom (http://lmedia.nodong.net).

In Denmark, a group called TV Stop started in 1987 and grew out of the squatters' movement. Their statement declared: "We broadcast news, alternative productions and all kinds of annoying and amusing productions which for political or financial reasons are overlooked by mainstream media . . ." TV Stop has around 30 volunteers at any given time and says that more than 300 people have worked at the station. They say they are called TV Stop "because we want to stop television in principle. This is to say we want people to stop watching TV . . . The ultimate goal for TV Stop is to close down." With a potential audience of a million and a half, the channel regularly attracts around 100,000 viewers at any given time.

In addition, artists and filmmakers have experiments with creative ways to utilize television. Jean Luc Godard has created several series for French television, while Alexander Kluge in Munich became so enthusiastic about television that he gave up filmmaking altogether. Inspired by the potential of Euro-Sat, which can reach the whole of Europe, Kluge began making programs to utilize this potential, including a nightly program featuring intellectuals and cultural figures.

The Internet: Streaming Media

With the development of streaming media for the Internet, television can be watched independent of cable and ordinary broadcast. This feature has been at the heart of the Independent Media Center movement (www.indymedia.org), which began during the demonstrations against the World Trade Organization in Seattle. As clashes between police and demonstrators escalated, ongoing documentation of the situation was provided by combining radio, video, photographs, and print journalism. By the end of the five days of meetings, the indymedia Internet site had hosted over a million separate visitors. The movement has spread to cities across the globe and now more than 130 sites provide information via streaming media to a vast international Internet audience.

Meanwhile, the Superchannel (www.superchannel.org) is a website that originated in Denmark and enables people, organizations, and communities to produce interactive Internet-tv by posting to a given list of topics. The project began in a housing project in Copenhagen to provide a way for tenants to organize for rights. Similar to many of the programs produced for public access channels, it doesn't limit itself to issues like tenant organizing, but includes everything from African braiding experts demonstrating new techniques to groups of teenagers just goofing off. Another significant website is One World TV (http://tv.oneworld.net), which collates programming posts into subject areas. In this case, the subjects and the posts are more carefully moderated and chosen for relevancy and technique. Typically, the subjects include development issues such as AIDS, water, the environment, and war and peace.

Many producers and activists are also attempting to utilize the Internet for the exchange of programming using sharing software that is similar to that used to share music in a "Peer to Peer" (P to P) mode. Because video images need more bandwidth and speed for better resolution and smooth movement, this exchange has been available mostly in countries with highly developed infrastructures. However, as technology advances, it is assumed that this sort of sharing will proliferate in the future.

Strategies for Building Community Media

Aside from sharing video packages, the Internet has allowed community producers around the world to share information and strategies. Because audio files are smaller and more easily streamed, community radio stations in Africa and Asia are able to exchange programming and ideas on a regular basis. In fact, national legislation in many African countries includes explicit support for local media.

While community radio stations, video centers, and computer technology centers have proliferated around the world, the producers and activists who work in these facilities have been active in discussions at the International Telecommunications Union (ITU) and UNESCO to promote regulations and infrastructure for community media. There was a strong presence of grassroots organizers from literally hundreds of media NGOs (non-governmental organizations) at the World Summit on the Information Society (WSIS) held in Geneva in 2003.[3] As corporate media such as Murdoch's Star TV and Disney extend their satellite reach to every corner of the globe, non-profit groups are providing necessary local information and cultural expression. Organizations promoting citizens' media are also working to create regulations on local, regional, and global levels to ensure that public resources such as the airwaves, the public rights of way and geostationary satellite paths are maintained and legislated with provisions to sustain and safeguard space and infrastructure for community media.

Notes

1 This informative study of the use of cable in one state shows that systems of the same size in equal type markets that have public access have more subscribers. Copies may be obtained from Diana Agosta, 180 Claremont Ave #32, NY, NY 10027.
2 A good website that connects to community media worldwide is: http://www.openchannel.se/cat/index.htm.
3 See www.wsis.org for the Civil Society Declaration that demands support for community media.

References

Agosta, D. (1990) "The Participate Report: A Case Study of Public Access Television in New York State," privately published document.

Beltran, L. (2000) *Investigación sobre comunicación en Latinoamérica: inicio, trascendencia y proyección*, La Paz: Universidad Católica Boliviana.

Boyle, D. (1996) *Subject to Change: Guerilla Television Revisited*, New York: Oxford University Press.

Brecht, B. (1964) "The Radio as an Apparatus of Communication," in J. Willet (ed.), *Brecht on Theater*, New York: Hill and Wang, pp. 53–4.

Dagron, A. G. (2001) "Making Waves: Stories of Participatory Communication for Social Change," a report to the Rockefeller Foundation (Job #3184).

DeeDee Halleck

Dagron, A. G. and Cajias, L. (eds.) (1989) *Las radios mineras de Bolivia*, La Paz, Bolivia: CIMCA-UNESCO.

Downing, J. (2000) *Radical Media: Rebellious Communication and Social Movements*, London: Sage.

O'Connor, A. (1990) "The Miners' Radio Stations in Bolivia," *Journal of Communication*, Winter, 102–10.

Rodriguez, C. (2001) *Fissures in the Mediascape: An International Study of Citizen's Media*, Cresskill, NJ: Hampton Press.

Stange, M. (1989) *Symbols of Ideal Life: Social Documentary Photography in America, 1890–1950*, New York: Cambridge University Press.

Stone, M. and Streible, D. (eds.) (2003) "Special Issue: Small-gauge and Amateur Film," *Film History Journal*, 15(2).

Teasdale, P. (1999) *Videofreex: America's First Pirate TV Station and the Catskill Collective That Turned It On*, Hensonville, NY: Black Dome Press.

PART NINE

International Television/Case Studies

CHAPTER TWENTY-SEVEN

Latin American Commercial Television: "Primitive Capitalism"

John Sinclair

Relative to other world regions, Latin America exhibits much in common amongst the 20 or so nations that constitute it. Although long independent from Spain and Portugal, the countries that created them as colonies in the heyday of their expansion after 1492, Latin American nations have inherited considerable cultural similarities. Above all, what makes them "Latin" are the languages of Spanish, dominant in nearly every country of the region, and Portuguese, as spoken in the region's largest nation, Brazil. This relative homogeneity of language and culture has greatly facilitated the region-wide growth of trade in television programs and services, in that this region is as much "geolinguistic" as it is geographic.

On the other hand, what makes Latin America "American" is not just the region's hemispherical continuity with North America, but the political hegemony of the United States. This dates back at least to the Monroe Doctrine of 1823, which sought to warn off European powers from what the United States saw as its "natural" sphere of influence. Apart from its several subsequent military interventions in the region, the United States has been significant for its economic influence via trade and investment, not to mention its *"imperialismo cultural."* However, we shall see in this chapter that although the nations of Latin America have almost universally adopted the commercial model of broadcasting as defined by the United States, the development of television in the region is to be explained in terms of its own "Latin" dynamic, not as a mere, unproblematic extension of direct influence from the United States.

Understanding Television Development in Latin America

Traditional attempts to explain the development of television in Latin America have focused primarily on the influence of the United States, meaning variously its government and the several private sectors with a stake in television development – namely, the networks, the equipment manufacturers, and the transnational advertisers and their agencies. However, other factors need to be taken into account. Clearly, US influence needs to be balanced against that of the governments of the Latin American nations themselves, and the highly variable positions that they have assumed with regard to television at different stages. These have lurched from strict hands-on dictatorial control to the permission of quite lawless environments in which "primitive capitalism" (*capitalismo salvaje*) could flourish. Similarly, a third set of factors is constituted by the role of the broadcasting entrepreneurs and other commercial interests in Latin America, and their influence on their national governments.

In addition, there is an important background factor that is easy to take for granted, and that is the degree to which the establishment of radio in Latin America in the period between the world wars had already cast the mold in which the subsequent development of television was to be formed. Since radio involved the same sets of interests as television – that is, government and private companies, both in the United States and Latin America – we have to take account of the continuities found from one broadcasting medium to the next so as to understand how television became institutionalized as it did in Latin America.

As radio networks began to form in the United States in the 1920s, Latin American entrepreneurs formed partnerships with them, so as to bring the new medium to their own countries. This was in response to the active presence of US radio equipment manufacturers in the region, as well as the encouragement of US officials to institutionalize radio as a commercial, rather than a public, medium (Schwoch, 1990, pp. 96–123). Thus, by the 1930s, commercial radio had become well established in Latin America. At the outbreak of World War II, as one of various initiatives to shore up the ideological defense of the hemisphere, the US government established the Office of Coordinator of Inter-American Affairs under Nelson Rockefeller. From 1940 to 1946, this gave full encouragement to the expansion of the US networks into Latin America, and also supplied programming in Spanish and Portuguese (Fox, 1997, pp. 19–22).

In 1945, the links between the Latin American entrepreneurs and the US networks NBC and CBS were formalized with the establishment of the intercontinental organization AIR (Asociación Interamericana de Radiodifusión – the Interamerican Broadcasting Association). An immediate objective was the establishment of television in the region, and AIR resolved to lobby the various national governments to ensure that, like radio, television would be introduced on a commercial "American" model, rather than a "European" state-operated basis. For this reason, the prevalence of the commercial model in Latin America should be seen to be at least as much due to its adoption by Latin interests, as to the influence of US ones (Fernández Christlieb, 1987, pp. 35–7). Latin American entrepreneurs positioned themselves to take advantage of the fact that there was US official and corporate interest in their region, and supported each other in resisting attempts by national governments to impose regulation. Indeed, the fact that the US government was interested in setting up its own service to Latin America drove NBC and CBS to find common cause with the Latin American entrepreneurs (Fox, 1997, p. 17).

Some of these entrepreneurs were significant figures not only in their own countries, but also elsewhere in Latin America. Goar Mestre, the first president of AIR, had been one of the first exporters of radio programs to the region, including the then innovative Latin commercial genre of the *radionovela*. This was an immediate ancestor of the *telenovela*, a popular cultural form which was to become the dominant export genre of the television era. Mestre was involved with radio and television in Cuba, where he and his brother Abel were backed in their station CMQ with investment from the US network NBC, at that time the broadcast division of the equipment manufacturer RCA. After being exiled from Cuba in the revolution of 1959, Mestre went to Argentina, and became active in television broadcasting and production there, in association with US interests in the form of CBS and Time-Life. He also invested in a Venezuelan television network with these partners, and had a further association with CBS in a Peruvian channel. The next president of AIR was Emilio Azcárraga Vidaurreta, the leading entrepreneur in Mexican radio, where he had built up a chain of stations affiliated with NBC, and then another with CBS. Working through Azcárraga's program distribution connections, NBC also formed affiliates in several other Latin American countries (Fernández Christlieb, 1987, pp. 36–43; Muraro, 1985, p. 80).

Clearly, over this time the various US interests were looking for opportunities and developing strategies to gain a foothold in the region. In his analysis, Silvio Waisbord (1998) discerns three stages in the development of Latin American television, and its relation to US interests is a defining feature of each stage. The first is characterized by the US networks' support for Latin entrepreneurs such as Mestre and Azcárraga, as just outlined, together with the sale of US-manufactured equipment, both for setting up stations and for reception by an audience. Because the US networks were still building up their home market during the early 1950s, when television was still new in the United States itself and Latin markets were too small to be profitable, they had little interest in making direct investment in Latin America. However, the Latin American entrepreneurs were not waiting for this. Television stations were established early, not far behind the US and well ahead of most other world regions. Mexico and Brazil, subsequently the major national markets of the region, set up their first stations in 1950, as did Cuba, which, given its size, had the distinction of being the first nation in the world to extend television to its entire national territory (Bunce, 1976, p. 81). The nations forming today's second echelon of television production and distribution, Argentina and Venezuela, had their first stations in 1951 and 1952 respectively (Waisbord, 1998, pp. 254–5).

After 1959, by which time US corporations had supplied equipment, technical assistance, and in some countries up to 80 percent of programming to Latin American stations, they did begin to take an active interest in direct investment.

505

The newest (but smallest) of the three US networks, ABC, established an international division, Worldvision, which began to set up affiliates in Latin America and other world regions. Meanwhile, the largest, NBC, continued to concentrate more on the sale of management services and equipment, but also exploited the advent of reliable videotape recording around this time by investing in program production (Frappier, 1968, pp. 1–7). As has been noted, CBS was backing Mestre, and had its association with Time-Life which, in its own right, made a most consequential investment in Brazil in 1962 to establish TV Globo, a subsequent regional leader (Marques de Melo, 1988, pp. 13–15).

Based on the same model used by all the US networks to establish themselves on a national basis in their home market, ABC's most characteristic strategy involved the setting up of international networks of affiliates within sub-regions. Thus, in 1960, ABC created the Central American Television Network (CATVN) by signing up affiliates in the five countries involved in a US-sponsored free trade agreement at the time (Bunce, 1976, pp. 81–9). Subsequently, ABC investments in stations from Mexico to the Southern Cone countries of Chile and Argentina were the basis of its LATINO network (Janus and Roncagliolo, 1978, pp. 30–2). However, it is quite possible that these never attracted the advertisers necessary for them to exist as anything more than networks on paper (Fox, 1997, p. 27).

After all, advertisers were the motivating force behind all of this maneuvering to internationalize and commercialize the medium. This was the crucial stage at which US-based corporations were in the process of transforming themselves from national into what were then called the "transnational" (or "multinational") corporations of the 1960s and 1970s, and television offered them access to their prospective markets. However, it was not yet clear just how that was to be achieved. What ABC was trying to do was to provide a transnational medium for transnational advertisers, enabling them to standardize their campaigns. On the other hand, the US-based advertising agencies that were setting up offices in the region at the time were offering their services on a market-by-market basis. It was like a rehearsal for the debate about "global" media for global advertisers that was later to occur in the 1980s (Mattelart, 1991, pp. 48–67).

However, when such strategies proved futile and direct investments did not meet their expectations, the networks began to withdraw, ushering in the second stage of Latin American television as defined by its relation to US interests. The networks were faced with increasing competition in the region, while at home, the US regulatory body, the FCC (Federal Communications Commission), ruled in 1971 that they had to separate their distribution from their syndication activities. Given these circumstances, CBS and NBC sold off their foreign investments (Read, 1976, p. 80), while ABC drastically curtailed its activities, retaining only very minor overseas holdings (Varis, 1978, pp. 16–17). Although the networks continued to supply programs to the region, this period saw a significant maturing of Latin American production companies and the increase of program exports on a regional basis, accompanied by considerable growth in several key domestic markets.

During the era of US network intervention, critics had denounced it as "cultural imperialism" (Schiller, 1969), or, more specifically, "picture-tube imperialism" (Wells, 1972). Such critics were later criticized in turn for having jumped too soon to the conclusion that the dependence upon direct investments and the apparent high levels of programming imported from the United States were going to be permanent structural features of Latin American television, rather than the inevitable compromises of a transitional start-up stage which would be followed by local development (Tunstall, 1977, pp. 38–40). Indeed, this is now seen to be a common pattern in the adoption of television in other world regions.

Nevertheless, the active role of US influence cannot be denied. As in the days of radio, US officials encouraged Latin American governments and media entrepreneurs to adhere to the US commercial model, at the same time as the US networks were selling them the equipment and the management services, and even investing directly. Yet the real conclusion to be drawn is that US influence was significant not because of the incidence of foreign ownership, nor because of the high levels of program imports, neither of which was to last, but because of the institutionalization of the commercial model itself throughout the region (McAnany, 1984, pp. 194–5).

Even where television was state-owned, the commercial model became the norm, the notable exception being Televisora Nacional, Cuba's state-owned monopoly which was formed in 1960 by Castro's revolutionary government, amalgamating the Mestres' CMQ and the other pre-1959 networks (Lent, 1990, pp. 128–9). Columbia developed its own model, going back to radio days, under which the state-maintained ownership of the television channels, but leased broadcasting time to private companies. They, in turn, commercialized the time by supplying programs for it and selling advertising spots and sponsorship (Fox, 1997, pp. 89–100). During the era of the left-wing military regime in Peru, there was a system of majority state ownership of television which prevailed from 1969 until 1981, but even then, the commercial model was not abandoned. In Bolivia and Chile, universities owned the stations, rather than the state, but they were still funded from commercial advertising (Roncagliolo, 1995, pp. 339–40). For a period, Mexico had a mixed system, in which state-owned commercial networks provided nominal competition to the dominant private conglomerate (Sinclair, 1986).

During the second stage in the development of Latin American television, as discerned by Waisbord in the 1970s and 1980s, much international attention was given to patterns of television programming exports and imports. The "flows" of television "traffic" even became one of the issues around which the movement for a New World Information and Communication Order (NWICO) was mobilized in UNESCO during those decades. Part of this mobilization was a movement, particularly amongst the developing countries, to formulate national communication policies so as to defend themselves against the threat of cultural imperialism, which they saw in the news, entertainment, and advertising carried on television. Some studies of television flows from this era had special significance for Latin America.

A wide-ranging comparative study by Tapio Varis confirmed the pattern of flows found in earlier work, that there was a "one-way street" from the United States to the rest of the world, but identified "a trend toward greater regional exchanges," particularly in Latin America (1984). Confirming the same tendency, Everett Rogers and Livia Antola (1985) went further, documenting the very considerable extent to which the *telenovela* had emerged as the preferred commercial genre within the Latin American program trade. This was a trend which Jeremy Tunstall had seen as early as 1977, when he had predicted that "hybrid media forms," such as the *telenovela* (usually described as "the Latin American soap opera") would emerge at an intermediate level in world program trade, between the global and the local (1977, p. 274). In that same year, Ithiel de Sola Pool also argued prophetically that audiences would come to prefer programming which was made in their own language, and had cultural familiarity for them (1977, p. 143).

Now that we can see the development of television in a longer-term and comparative perspective, it appears that the Latin American experience typifies the common pattern of transition through an initial stage of dependence to an eventual maturity of the national market. This process involves not just growth in audience size, but also in domestic program production. The common wisdom amongst researchers now is that audiences prefer television programming from their own country, and in their own vernacular, or if that is not available, from other countries which are culturally and linguistically similar, what Joseph Straubhaar calls "cultural proximity": "audiences will tend to prefer that programming which is closest or most proximate to their own culture: national programming if it can be supported by the local economy, regional programming in genres that small countries cannot afford" (1992, p. 14).

If we look at the hierarchy which has emerged amongst Latin American national markets for television programming and follow Rafael Roncagliolo's classification, we find that the largest and oldest ones, Mexico and Brazil, are pre-eminent as "net exporters" within the region; Venezuela and Argentina are "new exporters," with Colombia, Chile, and Peru seeking to join them, although from far behind; while the rest of the nations in the region, such as the smaller nations of Central America, are "net importers" (1995, p. 337). When cultural proximity is held constant, geographical proximity still matters: these importing nations tend to draw their programming from the nearest exporter, with flows developing on a sub-regional basis – in Central America from Mexico, in Paraguay and Uruguay from Argentina (Fox and Waisbord, 2002, pp. 16–19).

Returning to Waisbord's periodization, the third stage in which the relation to US interests must be assessed in the development of Latin American television is the recent past. New service and content providers, US corporations prominent amongst them, have been attracted by opportunities in the multi-channel environment made possible by "postbroadcast" cable and satellite modes of distribution. Furthermore, the advent of digital direct-to-home (DTH) satellite delivery in particular has encouraged the major Latin American pro-ducers and distributors to enter strategic alliances with US satellite and cable services, bringing Latin American television into the mainstream of globalization.

Even before the end of the 1980s, during which cable channels such as CNN and HBO had achieved national distribution in the United States, they were moving into Latin America. CNN began a news service in Spanish in 1988, and the following year, HBO and ESPN both launched special services for Latin America (Wilkinson, 1992, pp. 13–16). During the 1990s they were joined by several others, such as Discovery and MTV, along with services from the US networks CBS and Fox. The proliferation of channels on the new generation of satellites facilitated this development, allowing for multiple language tracks, but there are also channels carrying programming produced especially for the Span-ish- and Portuguese-speaking worlds. By 1996, 90 percent of television services (that is, satellite and cable signals, as distinct from programs) imported into the Iberoamerican region (that includes Spain and Portugal as well as Latin America) were from the US (Media Research and Consultancy Spain, 1997).

The "trend toward greater regional exchanges" in television program flows observed by Varis (1984) continues, but it is becoming ever more a trade in services rather than of programs, the exchanges are unequal, and it takes place within a globalized context. To take up the case of DTH, the 1990s saw US-based satellite and cable companies combine with the largest Latin American broadcast and satellite-to-cable networks to create multileveled corporate struc-tures able to distribute subscription services throughout the whole region. Thus, a new kind of global corporation was formed in recognition of the strength which the Latin American companies had achieved, not just in their national and regional markets, but also in their whole geolinguistic regions: that is, wherever Spanish or Portuguese was spoken.

More specifically, at the end of 1995, News Corporation announced that it would lead a pan-regional DTH satellite subscription service consortium, DTH Sky. This included not only Televisa of Mexico, but also Globo, its counterpart in Brazil, and the US cable corporation TCI (Tele-Communications Inc, now Liberty Media). This was not so much a new initiative as it was a reaction to the advent of Galaxy, a venture already announced that year by the US-based satellite division of General Motors, Hughes Electronics Corporation. Galaxy, also known as DIRECTV Latin America, incorporated Televisa's cable competi-tor in Mexico, Multivisión, along with Globo's main cable competitor in Brazil, TV Abril, as well as CGC, the owner of Venevisión, a major television producer and distributor based in Venezuela, another key market of the region (Country profile: Mexico, 1996, p. 7). Both services began operating by the end of 1996.

However, the elite market able to pay for DTH has proven too small to sustain the two competing services, with DIRECTV Latin America filing for bankruptcy in March 2003. Furthermore, the fate of DTH in the region is over-shadowed by the fact that News Corporation has acquired DIRECTV from Hughes in the United States. In these circumstances, the most likely outcome is that Sky will merge with or absorb its former competitor in some way. This has already happened elsewhere in the Spanish-speaking world, with the merger of erstwhile competitors Vía Digital and Canal Satélite in Spain (Tapia, 2003). The point in the Latin American case is that although US-based global corporations might need to work through the major regional players, the latter are very much at the mercy of what happens in a bigger game which the former are playing on their home turf – a situation which the theorists of the region used to call "dependency."

The Present Absence of the State

It is important to distinguish state ownership of television, as discussed earlier, from state control. That is, the imposition of the regulation of television by the state, or its relative absence, has consequences for the form which the medium takes as an institution. Whether the state itself owns stations or not is beside the point. Because regimes of regulation differ from one nation to another, and shift over time even within any given nation, along with their political histories as a whole, it is difficult to draw generalizations for the entire region. However, there have been two main phases in which the state has taken an active role in controlling television in the leading nations. The first was in the 1950s, when populist dictators in Argentina, Brazil, and Colombia implemented severe nationalistic policies to assert control over broadcasting and other media; then in the 1970s, in the context of the television flows debate referred to above, several governments took up the NWICO rhetoric of "national communication policies."

This was manifested at first in moves toward public broadcasting in Mexico, Venezuela, and Chile, and then, in the latter part of the decade, in a broader discourse about the need to protect television and other media industries from foreign influence and competition. Yet although the rhetoric affirmed the public service and civic functions of the media, particularly in the construction and defense of national cultures, nowhere does there appear to have been the political will to actually implement national communication policies as such (Fox, 1988). No real tradition of public broadcasting has ever been able to develop in Latin America, because in those instances where the state has asserted control, it has been to serve the interests of dictatorship rather than any public service ideal.

In the most recent era, consolidating itself over the 1980s and 1990s, as elsewhere, the broad trend in the approach of Latin American governments to the television industry has been toward ever more deregulation and privatization. Major privatizations of formerly state-owned networks have occurred in Argentina, Chile, Bolivia, Colombia, and Mexico, while private interests everywhere in the region have been the beneficiaries of the new licenses that have been granted

(Fox and Waisbord, 2002, p. 9). The fact that the commercial model became so institutionalized in the region at the outset has also meant that the private interests who benefited from it then became politically entrenched, so that in spite of government criticism of private control, as in the latter 1970s, the public has learned to be skeptical of the state's capacity to act in the public interest when private interests are at stake. Indeed, because of the general fragility of democratic traditions in the region and disillusion with the concept of national culture after decades of dictatorship, and because state regulation of the media is thus associated historically with authoritarian control in the state's own interests, in some countries the public looks more to the private media than to the state as the legitimate basis for political and cultural leadership (Waisbord, 1998, p. 262).

In addition to such influence as they can wield within the political culture and system of social communication of their domestic markets, Latin American television companies also benefit from the disrepute into which state control has fallen, in that they now operate in a much less regulated environment. It is this situation that is characterized by the term *capitalismo salvaje*, literally "savage" or primitive capitalism. For example, even in the United States, the institutional and ideological home of free enterprise, there are regulatory restraints against monopolization. An instructive case is the FCC ruling referred to above, which obliged US networks to disaggregate their distribution from their syndication activities. In Latin America, by contrast, the state's permission of such vertical integration, especially of production and distribution, is one of the major means by which the region's largest networks have been able to develop their dominant positions in their home markets. Furthermore, Latin American audiences have long been accustomed to the quite intensive commercialization of the medium which has only more recently emerged in other regions, for example, in the excessive proportion of airtime given over to advertising, the egregious insertion of commercial messages into editorial or entertainment content, and other forms of blurring the line between advertising and content, such as product placement (Straubhaar, 1991, pp. 48–9).

The cases of Mexico and Brazil are of particular interest because, in each case, a dominant private network emerged, due in part to the symbiotic relationship maintained over time between government and network. "Favourable government policies" (Rogers and Antola, 1985, p. 27) fostered the pre-eminence of Televisa in Mexico and Globo in Brazil, in return for which the networks gave their ideological support to, respectively, the party that ruled Mexico from the 1930s until the end of the century, and the series of military dictatorships that imposed their "order and progress" upon Brazil from the 1960s until the 1980s. Thus, Televisa and Globo were able to secure competitive dominance over what are respectively the largest Spanish-speaking and Portuguese-speaking domestic markets in the world, and, simultaneously, tie their fortunes to stable regimes and so protect themselves from the vagaries of political instability suffered by the television industry in the rest of the region.

Furthermore, instead of the kind of "arm's-length" relationship between government and media to which the more advanced democracies at least claim to aspire, in Latin America, there is something termed "electronic clientelism." Even a decade after the dictatorship ended in Brazil, a substantial proportion of commercial television broadcasting licenses were still being granted to politicians and former politicians (Fox and Waisbord, 2002, p. 10). In Mexico, the classic case was that of Miguel Alemán Valdés, the president whose decision it was in 1950 that Mexico should have a private commercial system of television, rather than a public model. At the end of his term, Alemán joined the entrepreneur Emilio Azcárraga Vidaurreta in one of the companies which the former president had licensed. Subsequently, in 1972, by then under the control of their sons, this company became the basis for Televisa (Sinclair, 1986, pp. 88–90).

On a regional level, the spread of deregulation and privatization, or the whole overarching ideological disposition that Latin Americans call *neoliberalismo*, has been consistent with the formation of regional free trade blocs in the early 1990s. However, in spite of concerns about the weakening of national cultural industries (Sánchez Ruiz, 1994), Mercosur, the agreement between the Southern Cone countries of Argentina, Brazil, Paraguay, and Uruguay, did not make an issue of audiovisual trade (Galperín, 1999), while in the North American Free Trade Agreement (NAFTA) between the United States, Mexico, and Canada, Mexico has relied on its "natural" protection of linguistic and cultural difference from the rest of North America. The importance of a treaty such as NAFTA is likely to lie more in how it allows the US to enforce intellectual property rights, especially laws against piracy, and the opening up of foreign investment opportunities for it in telecommunications, which includes cable television (Sinclair, 1996). In any event, the severe economic crises of recent years in countries such as Brazil and Argentina seem to be more important than trade agreements or ideology per se in opening them up to foreign investment.

Television Markets of the Latin World

At this point, it would be useful to consider some of the features of the major national markets for television in Latin America. In addition to the three largest shown in Table 27.1, some coverage will also be given to Venezuela, which, although rather smaller in size as a domestic market (with a population of 24.6 million), is exceptional for the relative extent of its international activities.

Brazil

Even though Time-Life had to cancel the financial and management arrangements with which it had supported Roberto Marinho's newspaper and radio group Globo in its venture into television, the financial advantage and, as they say in Portuguese, "American know-how" thus gained by Globo gave it an

Table 27.1 Selected major national markets for television in Latin America, 2002

Country	Population	TV homes	Cable homes	Satellite homes	Imports
Brazil	182,032,604	40 million	2.4 million	1.1 million	23%
Mexico	104,907,991	17 million	2.3 million	1 million	40%
Argentina	38,740,807	9.9 million	5.2 million	430,000	30%

Sources: *Variety* Mipcom supplement, October 7–13, 2002, pp. A20 and A22; and *CIA World Factbook*, http://www.cia.gov/cia/publications/factbook/index.html

unbreachable lead over its competitors. By 1968 it was established as the ratings leader. Similarly, as also mentioned above, Globo enjoyed concrete benefits from the close relationship it established with Brazil's military leaders. First, because it was able to extend itself into a national network by virtue of the telecommunications infrastructure that the government built (Fox, 1997, p. 62); and secondly, because of its concerns over foreign cultural influence on television programming, the government provided Globo with incentives to stimulate Brazilian program production. This became one of the factors enabling Globo to launch an export career (Marques de Melo, 1988, pp. 1–2).

Although new networks have been licensed since the 1980s, notably SBT (Sistema Brasileiro de Televisão) and Manchete, and in spite of having to survive the downfall of the military era at the end of that decade, Globo has continued to dominate. Because Globo's number of stations and audience coverage are not significantly greater than those of SBT, there is constant competition for audience share. In Brazil and the other major markets of the region, ratings are measured by the independent private organization, IBOPE, and although Globo also undertakes its own audience research, both Globo and its competitors constantly adjust their programming in accordance with each other's IBOPE scores. However, by and large, Globo maintains a considerable edge over its competitors through its capacity to provide virtually all of its own programming. Much of this is tailored to suit the more up-market audiences, for which it can charge more for advertising time (Mattelart and Mattelart, 1990, pp. 38–9). As of the end of 2002, Globo's market share was equal to that of its competitors combined, and it was getting 70 percent of the advertising revenue for the sector (Cajueiro, 2002).

Although Globo's dominance of domestic broadcast television seems unassailable, it has encountered stronger competition in the realm of subscription or "pay-TV" services as new modes of transmission have been introduced on this basis. Here it confronts Brazil's other major media conglomerate, Grupo Abril, the largest publishing operation in Latin America. In the early 1990s, Abril began offering broadcast-delivered subscriber services (Abril Group, 1992, pp.

52–3). Globo first met Abril's MMDS (Multipoint Multi-channel Distribution Service) venture with its own service, Globosat, transmitted via Brazil's domestic satellite, Brasilsat (Straubhaar, 1996, pp. 233–4). Globosat then established a cable distribution arm, Net Brasil, and although Abril also then went into cable and satellite, Globo caught up with Abril's early lead and achieved pre-eminence in the subscription television market through Net Brasil (Glasberg, 1995, p. 36B). However, this is not as decisive a lead over its competition as the one it enjoys in broadcast television. As noted, in the DTH era, Abril became allied to the now bankrupt DIRECTV, opposite Globo's luckier association with Sky.

On the international front, Globo and the other Brazilian television producers have fewer and more difficult options in developing overseas markets than the Spanish-language producers. Brazil might be the biggest Portuguese-speaking country in the world, and also the biggest country in Latin America by both area and population, but since it is the only country in its region that speaks Portuguese, it has been more oriented to Europe in its export efforts than to its Spanish-speaking neighbors. Thus, Portugal assumed a strategic significance for Globo, much more than Spain ever has correspondingly for Televisa. For example, Globo took a 15 percent interest, the maximum allowable for a foreign investor, in what it went on to make the leading network in Portugal (Sousa, 1997).

Mexico

In Mexico, the development of television has been very much shaped by the entrepreneurship of the Azcárraga dynasty, and by the emergence of their company, Televisa, as a cross-media conglomerate which, for decades, enjoyed a quasi-monopoly in the Mexican television market. Historically, Televisa has been able to exploit the competitive advantage of its market dominance in the world's largest Spanish-speaking nation as the basis for its international activities. These involve participation in network ownership as well as program export and service provision, not only in Latin America, but also in the United States and Spain.

Televisa's halcyon days in the 1970s and 1980s were thus based on the fact that, because it produces most of what it broadcasts, it always has had vast stocks of programming available for international distribution, programming which has already proven and paid for itself in its domestic market. However, Televisa's success was due not just to this business model, but also to the particular political context in which it operated. For one thing, Televisa enjoyed sympathetic treatment from most of the successive governments over this period, all from the party that ruled Mexico from the 1930s until 2000, the PRI (Partido Revolucionario Institucional). As noted, Televisa was able to maintain its vertical integration of production and distribution, and from 1972 until 1993, faced only nominal competition in the domestic market, although this came from state-owned commercial networks (Sinclair, 1986). For their part, Televisa's owners and management were self-declared supporters of the PRI, and Mexican news and current affairs programs were so strongly oriented to the PRI's interests that they had little credibility outside the country.

However, the state's embrace of privatization ended this golden age when it sold its networks to an electric goods retailer who launched them as TV Azteca in 1993. For Mexican audiences, any alternative was refreshing, and the new company appeared at a time when Televisa was in financial difficulties, as a result of spiraling debt and the devaluation of the peso, so TV Azteca was quickly able to capture audiences. However, it has not been able to push its national audience share much beyond the initial level, which in mid-2003 stood at 30 percent, with a corresponding 33 percent share of television advertising revenue (Tegel, 2003). These levels are more than three times those during the era of state ownership, so Televisa certainly now has to deal with a serious competitor; on the other hand, it is a duopolistic environment, which Televisa still dominates. A true regime of "primitive capitalism" would have allowed many more competitors into the marketplace.

Outside of Mexico, Televisa has had to learn to cope with much less protection, and more regulation, than has been the case at home. In the United States, the FCC turned a blind eye to the fact that the Mexicans set up and ran Spanish-language television for twenty-five years, but eventually, in 1986, it had to enforce foreign ownership provisions against them. As a result, Televisa was reduced to a minority share in its former network, Univisión, its level of ownership since having become far outstripped by Venevisión of Venezuela. Yet this participation still gives both of them a strategic outlet for their programming. Although Univisión's audience share is well ahead of the other Spanish-language networks in the United States, these now include one backed by NBC, as well as a venture by TV Azteca, so its market leadership is coming under more pressure as competition intensifies (Sinclair, 2003).

Argentina

Commercial television in Argentina began with the establishment of three private channels in the late 1950s. However, they were prohibited from forming networks. In the 1970s, they were nationalized, and they did not regain their freedom from direct government control until the 1980s, when democratically elected regimes progressively turned them over to private ownership (Fox, 1997, pp. 101–6). Thus, Argentina is different from the other national industries under consideration here, which have been built on dominant national broadcasting networks, since the historical absence of networks and the relative recency of deregulation has given a special prominence to cable, and latterly, satellite-to-cable and DTH modes of distribution. This range of platforms, in combination with a wave of pro-market reforms in the first half of the 1990s, has produced a volatile competitive environment. In particular, state-owned channels were privatized, cross-media ownership was loosened, satellite distribution and national networks were permitted, and the television industry was opened up to foreign investment (Galperín, 2002, p. 29).

As Table 27.1 shows, more than five million homes, which is half of all television households in Argentina, are cabled. This is nearly half the cabled homes in the entire region, and more than twice as many as in either Mexico or Brazil. This incidence of cable links Argentinian television into the convergence of the media and telecommunications industries and their technologies on a global scale. The leading MSOs (multi-system operators) are: Cablevisión, backed by US investors Liberty Media and Hicks, Muse, Tate & Furst; and Multicanal, owned by Argentinian conglomerate Grupo Clarín (Newbery, 2002).

However, there is also foreign interest in free-to-air television, notably with the Spanish-based telecommunications corporation Telefónica having full ownership of one of the two leading broadcasters, Telefe. Telefónica has emerged as a strong presence in the telecommunications and media industries of the region, with activities in television production as well as distribution. Notable amongst these is Endemol, which, in addition to Argentina, has joint ventures in Mexico and Brazil, with Televisa and Globo respectively (Sutter, 2003). Like Televisa and Globo, the other leading free-to-air broadcaster, Grupo Clarín, is a family company, with interests across a wide range of media industries. In addition to its television activities just mentioned, and its previous participation in the bankrupt DIRECTV DTH venture, it publishes *Clarín*, the newspaper with the largest circulation in the Spanish-speaking world. Since 1999, US-based investment bankers Goldman Sachs have owned 18 percent of the group (Grupo Clarín, 2003). Other foreign investors have come and gone, notably Citicorp Equity Investments, which had extensive holdings across the whole communications sector in the mid-1990s (Galperín, 2002, pp. 31–2).

In spite of the predominance of foreign programming on cable, audiences maintain a strong preference for Argentinian programs. As these attract the best advertising revenues, the national networks have great incentives to produce and distribute their own programming. They are also able to export programs to neighboring countries, and elsewhere in the Americas and beyond, although this operation is on nowhere near the scale of those carried out by either Televisa or Globo (Galperín, 2002, pp. 32–3).

Venezuela

Television broadcasting in Venezuela today is dominated by two strong and internationally active networks – a virtual duopoly, in spite of a series of new channels licensed in a wave of privatization during the 1990s (Mayobre, 2002). The pre-eminent network is Venevisión, which was begun in 1960 by Diego Cisneros, with the help of an initial investment from the US network, ABC. Cisneros built up a diversified industrial group, but since the 1980s, under his son, Gustavo, the concentration has increasingly been on television and new media (Giménez Saldivia and Hernández Algara, 1988, pp. 189–213; Fox, 1997, pp. 72–8). Like Grupo Televisa and Organizacoes Globo, the Cisneros Group of Companies (CGC) is a conglomerate in which media are integrated both vertically and horizontally. CGC also clearly fits a similar pattern of patriarchal, dynastic control and continuity apparent in the history and structure of Televisa and Globo.

Yet unlike its Mexican and Brazilian counterparts, whose pre-eminence derives from their dominance in the largest domestic markets of their geolinguistic regions, CGC has had to make exceptional efforts in terms of international activities. While taking full advantage of the low costs of program production in Caracas, CGC's distribution activities outside of Venezuela are so extensive that it has made Miami the international headquarters of Venevisión. As noted earlier, its participation with Televisa in the leading Spanish-language network in the United States is an important plank in its international strategy. Yet Venevisión has been Televisa's competitor in continental South America, where Cisneros has been the major figure in pushing the expansion of the now bankrupt Galaxy/DIRECTV DTH venture. Cisneros is also an equal partner with AOL Time Warner in AOL Latin America, as it attempts to catch up to Telefónica's Terra Networks and other ISPs (Internet Service Providers) active in the region (Hoag, 2000).

To a much greater extent than its counterparts in Mexico or Brazil, Venevisión faces a substantial competitor both in its domestic market and internationally, in the form of RCTV (Radio Caracas Televisión). As of 2000, Venevisión had 35 percent of the average national audience, compared to the 31 percent held by RCTV (AGB Venezuela, 2000). Very much like Venevisión in its origins, RCTV is owned by a family-based industrial group, Phelps, which has several other horizontally and vertically integrated media and communications companies, as well as other interests (Giménez Saldivia and Hernández Algara, 1988, pp. 180–1; Fox, 1997, p. 77). RCTV's domestic network consists of four national channels, while its international activities are centered on program production and distribution through Coral Pictures Corporation which, like Venevisión International, is based in Miami, "the Hollywood of Latin America" (Sinclair, 2003, pp. 224–5). Thus, just as RCTV offers strong competition to Venevisión in the domestic market, this is also the case in programming export activities. Coral is a major distributor of programming, with its number of hours sold annually comparable to that of Televisa and Globo.

Two final comparative points on the character of the Venezuelan television industry: unlike that of Mexico and Brazil, Venezuela's political history has been too turbulent for there to have been any long-standing relationship between the networks and a particular regime; while, unlike the case of Argentina, there has been an absence of foreign ownership (Mayobre, 2002).

John Sinclair

On the Global Scale

Even if, under the ideological rubric of what the Latin Americans call *neo-liberalismo*, a more competitive environment has emerged out of the wave of deregulation and privatization of the 1990s, the established market leaders have suffered little erosion of their dominance as a result. Their historical and structural advantages in the broadcast market have protected them, and also provided the base from which to launch themselves into new modes of television distribution and other convergent technologies of the new media. Yet for all their market power, by global standards the Latin American television groups are big fish in relatively small ponds, so for them, the savagery of capitalism is felt higher up the food chain. We have seen how their participation in the regional DTH ventures has been subjected to the maneuvers of the global partners. Similarly, drastic currency devaluations as have occurred in Brazil and Argentina in recent years, and also in Mexico at an earlier stage, can cause a debt crisis for the media groups, regardless of their domestic market strength, and make them vulnerable to foreign investment. Until the struggle for economically viable and politically stable democracies is won in Latin America, its television industries will be as much at the mercy of global forces as any other industry or institution in the dependent nations that sustain them.

References

Abril Group (1992) *The Abril Group*, São Paulo: Editora Abril.
AGD Venezuela (2003) AGB Venezuela, 2000: *Promedia share 200 total país*, retrieved September 19 at http://www.agb.com.ve/libro2000/share/20.htm.
Bunce, R. (1976) *Television in the Corporate Interest*, New York: Praeger.
Cajueiro, M. (2002) "Marinho Clan Slaps Sale Sign on Debt-hit Globo," *Variety*, December 2–8, p. 22.
CIA (2003) *World Factbook*, retrieved September 8, at http://www.cia.gov/cia/publications/factbook/index.html.
de Sola Pool, I. (1977) "The Changing Flow of Television," *Journal of Communication*, 27(2), 139–79.
Fernández Christlieb, F. (1987) *Algo más sobre los orígenes de la televisión latinoamericana*. [Something more about the origins of Latin American television], *Dialogos de la Comunicación*, 18 (October), 32–45.
"Country Profile: Mexico," *TV International* (1996) May, pp. 5–8.
Fox, E. (1988) "Media Policies in Latin America: An Overview," in E. Fox (ed.), *Media and Politics in Latin America: The Struggle for Democracy*, London: Sage, pp. 6–35.
Fox, E. (1997) *Latin American Broadcasting: From Tango to Telenovela*, Luton: University of Luton Press.
Fox, E. and Waisbord, S. (2002) "Latin Politics, Global Media," in E. Fox and S. Waisbord (eds.), *Latin Politics, Global Media*, Austin TX: University of Texas Press, pp. 1–21.
Frappier, J. (1968) "US Media Empire/Latin America," *NACLA Newsletter*, 2(9), pp. 1–11.
Galperín, H. (1999) "Cultural Industries in the Age of Free-trade Agreements," *Canadian Journal of Communication*, 24(1), 49–77.

Galperín, H. (2002) "Transforming Television in Argentina," in E. Fox and S. Waisbord (eds.), *Latin Politics, Global Media*, Austin TX: University of Texas Press, pp. 22–37.

Giménez Saldivia, L. and Hernández Algara, A. (1988) *Estructura de los medios de difusión en Venezuela*, Caracas: Universidad Católica Andres Bello.

Glasberg, R. (1995) "Bringing up Brazil," *Multi-channel News International*, April, pp. 6B and 34–6B.

Grupo Clarín (2003) Grupo Clarín SA, retrieved September 18, 2003, at http://www.grupoclarin.com/espanol/grupoclarin/estructura-grupo-clarin-sa.html.

Hoag, C. (2000) "Empire Building: The Slow Track," *Businessweek Online*, September 4, retrieved March 25, 2003, at http://www.businessweek.com:/2000/00_36/b3697159.htm?scriptFramed.

Janus, N. and Roncagliolo, R. (1978) "A Survey of the Transnational Structure of the Mass Media and Advertising," report prepared for the Center of Transnationals of the United Nations, Instituto Latinoamericano de Estudios Transnacionales, Mexico DF.

Lent, J. (1990) *Mass Communications in the Caribbean*, Ames, IA: Iowa State University Press.

McAnany, E. (1984) "The Logic of the Cultural Industries in Latin America: The Television Industry in Brazil," in V. Mosco and J. Wasko (eds.), *The Critical Communications Review Volume II: Changing Patterns of Communications Control*, Norwood, NJ: Ablex, pp. 185–208.

Marques de Melo, J. (1988) *As Telenovelas de Globo*. [The Telenovelas of Globo], São Paulo: Summus Editorial.

Mattelart, A. (1991) *Advertising International: The Privatisation of Public Space*, London and New York: Routledge.

Mattelart, M. and Mattelart, A. (1990) *The Carnival of Images: Brazilian Television Fiction*, New York: Bergin and Garvey.

Mayobre, J. (2002) "Venezuela and the Media: The New Paradigm," in E. Fox and S. Waisbord (eds.), *Latin Politics, Global Media*, Austin, TX: University of Texas Press, pp. 176–86.

Media Research and Consultancy Spain (1997) "La industria audiovisual iberoamericana: datos de sus principales mercados 1997" [The Iberoamerican Audiovisual Industry: Data from Principal Markets, 1997]. Report prepared for the Federación de Asociaciones de Productores Audiovisuales Españoles and Agencia Española de Cooperación Internacional, Madrid, July.

Muraro, H. (1985) *El "modelo" Latinoamericano*. [The Latin American "Model"], *Telos*, 3, 78–82.

Newbery, C. (2002) "Jornadas Confronts the Pay TV Crisis," *Variety*, November 18–24, p. 20.

Read, W. (1976) *America's Mass Media Merchants*, Baltimore, MD: Johns Hopkins University Press.

Rogers, E. and Antola, L. (1985) "*Telenovelas*: A Latin American Success Story," *Journal of Communication*, 35(4), 24–35.

Roncagliolo, R. (1995) "Trade Integration and Communication Networks in Latin America," *Canadian Journal of Communication*, 20(3), 335–42.

Sánchez Ruiz, E. (1994) "The Mexican Audiovisual Space and the North American Free Trade Agreement," *Media Information Australia*, 71, 70–7.

Schiller, H. (1969) *Mass Communications and American Empire*, New York: Augustus M. Kelley.

Schwoch, J. (1990) *The American Radio Industry and its Latin American Activities, 1900–1939*, Urbana and Chicago, IL: University of Illinois Press.

Sinclair, J. (1986) "Dependent Development and Broadcasting: The 'Mexican Formula,'" *Media, Culture and Society*, 8(1), 81–101.

Sinclair, J. (1996) "Culture and Trade: Some Theoretical and Practical Considerations," in E. McAnany and K. Wilkinson (eds.), *Mass Media and Free Trade: NAFTA and the Cultural Industries*, Austin, TX: University of Texas Press, pp. 30–60.

Sinclair, J. (2003) "'The Hollywood of Latin America': Miami as Regional Center in Television Trade," *Television and New Media*, 4(3), 211–29.

Sousa, H. (1997) "Crossing the Atlantic: Globo's Wager in Portugal," Paper presented at International Association for Mass Communication Research conference, Oaxaca, Mexico, July.

Straubhaar, J. (1991) "Beyond Media Imperialism: Assymetrical Interdependence and Cultural Proximity," *Critical Studies in Mass Communication*, 8(1), 39–59.

John Sinclair

Straubhaar, J. (1992) "Asymmetrical Interdependence and Cultural Proximity: A Critical Review of the International Flow of Television Programs," paper presented at Asociación Latinoamericana de Investigadores de la Comunicación conference, São Paulo, Brazil, August.

Straubhaar, J. (1996) "The Electronic Media in Brazil," in R. Cole (ed.), *Communication in Latin America: Journalism, Mass Media, and Society*, Wilmington, DE: Jaguar Books on Latin America, pp. 217–43.

Sutter, M. (2003) "Endemol Soars in South America with a Pan-regional Approach," *Variety*, January 20–26, p. 33.

Tapia, B. (2003) "Digital TV Monopoly Emerges in Spain," Europemedia.net, retrieved April 14, 2003 at www.europmedia.net/shownews.asp?ArticleID=14617.

Tegel, S. (2003) "Battling Billionaire Powers Azteca," *Variety*, May 26–June 1, p. 22.

Tunstall, J. (1977) *The Media Are American: Anglo-American Media in the World*, London: Constable.

Variety (2003) "Mipcom Supplement," *Variety*, October 7–13.

Varis, T. (1978) "The Mass Media TNCs: An Overall View of their Operations and Control Options," Paper prepared for Asian and Pacific Development Administration Centre meeting, Kuala Lumpur, Malaysia.

Varis, T. (1984) "The International Flow of Television Programmes," *Journal of Communication*, *34*(1), 143–52.

Waisbord, S. (1998) "Television in Latin America," in A. Smith (ed.), *Television: An International History*, Oxford and New York: Oxford University Press, pp. 254–63.

Wells, A. (1972) *Picture Tube Imperialism? The Impact of US Television in Latin America*, Maryknoll, NY: Orbis.

Wilkinson, K. (1992) "Southern Exposure: US Cable Programmers and Spanish-language Networks enter Latin America," Paper presented to International Association for Mass Communication Research conference, Guarujá, Brazil, August.

Further Reading

Paxman, A. and Saragoza, A. (2001) "Globalization and Latin Media Powers: The Case of Mexico's Televisa," in V. Mosco and D. Schiller (eds.), *Continental Order? Integrating North America for Cybercapitalism*, Lanham, MD: Rowman & Littlefield, pp. 64–85.

Rockwell, R. and Janus, N. (2003) *Media Power in Central America*, Champaign, IL: University of Illinois Press.

Salwen, M. (1994) *Radio and Television in Cuba: The Pre-Castro Era*, Ames, IA: Iowa State University Press.

Sinclair, J. (1999) *Latin American Television: A Global View*, Oxford and New York: Oxford University Press.

Straubhaar, J. (2001) "Brazil: The Role of the State in World Television," in N. Morris and S. Waisbord (eds.), *Media and Globalization: Why the State Matters*, Lanham, MD: Rowman & Littlefield, pp. 133–53.

CHAPTER TWENTY-EIGHT

Television in China: History, Political Economy, and Ideology

Yuezhi Zhao and Zhenzhi Guo

China's television system is the world's largest, with 1.1 billion regular viewers and 317 million television sets in 2003. As part and parcel of China's unique and rapid transformation from a planned economy to a market economy over the past two-and-a-half decades, television in China has undergone an unprecedented metamorphosis, evolving from a propaganda instrument of the Communist Party to a commercially oriented mass medium within the country's evolving market authoritarian system. While China's broad political economic, technological, and social transformations provide the context for the development of a mass medium that is unsurpassed in geographic reach and population penetration, Chinese television's double articulation of political and commercial propaganda and popular culture has had profound effects on the country's ongoing social transformation. This chapter provides an overview of television's development into the most influential mass medium in China. Following a brief historical analysis of television in pre- and early reform China, ending in 1989, we examine the political economy of the post-1989 industry in transition, analyze its defining ideological and cultural features, and highlight key issues in the current development and future evolution of Chinese television, as an integral political, economic, and cultural institution in a globalizing context.

Pre-reform Chinese Television: Nationalistic Ambitions, Shifting Foreign Models

The analysis of the development of television in China can be broken down into three historical periods. The first runs from the birth of television in China in 1958 to the end of the Cultural Revolution in 1976, when television enjoyed only a marginal existence as a simple political and educational instrument, with little social and cultural impact. The second lasted from 1976 to the political turmoil

521

of the spring of 1989, a period during which television assumed a central role in Chinese politics and culture as political mobilization and ideological ferment sought reform that would bring both economic prosperity and democratic political transformation. The third period was ushered in by the crackdown on the 1989 student movement, culminating in the dramatic events in Tiananmen Square. It is with this suppression of popular demands for political participation in the reform process that the preconditions were created for the consolidation of a market authoritarian social formation (Y. Zhao, 2001). Thus, television assumes a major role in the promotion of a market economy, consumerism, and the nationalistic project of building a "strong and powerful" China, while continuing to serve as a site of discursive contestation.

The first period demonstrated the inextricable linkages between television, the Communist Party's nationalistic ambitions, and Cold War politics. Reports about the planned launch of television by the Nationalist government in Taiwan – on its National Day of October 10, 1958 – led the Communist government to abruptly inaugurate three stations in order to trump its political rival. On May Day 1958, Beijing Television was launched, with Shanghai Television and Harbin Television following later the same year (in fact, television in Taiwan was delayed for four years, till October 10, 1962). In the face of political and economic sanctions imposed by the capitalist world, China was assisted by socialist countries, with transmission technologies from Czechoslovakia, program formats from the Soviet Union and East Germany, and television sets from the Soviet Union. The revolutionary romanticism of the Great Leap Forward in 1958, and the imperative of competing with the capitalist world, led to trial broadcasts in many other provincial capitals shortly after the three inaugural broadcasts. However, inadequate infrastructures, massive economic mismanagement – compounded by the loss of technological and financial assistance from the Soviet bloc as the result of ideological conflicts – led to a profound economic crisis in China in the early 1960s and, with it, a deep retrenchment of the fledgling Chinese television system (Guo, 1991; Hong, 1998).

From its inception, Chinese television was institutionally and ideologically incorporated into the Party's pre-existing media system as a propaganda mouthpiece. Inadequate production capacity limited the activities of early Chinese television to the transmission of existing media content, such as newsreels produced by film studios. Self-produced content was limited to feature stories about communist role models, and live broadcasts of children's and other cultural and educational shows. Entertainment programming in China consisted primarily of feature films. Because of the small audience and the lack of independent economic interests under the planned economy, film studios often showcased new films on television either simultaneously with or even ahead of cinema releases. For example, for the price of a movie ticket, people could collectively watch films or filmed theatrical performances at community cultural centers. In fact, because human interest news stories were non-existent at the time and news content was limited to major political events and state policies, television's entertainment function was hardly explored, and, later, explicitly suppressed. Between 1960 and 1962, Beijing Television tried to provide some diversion by broadcasting a light entertainment variety show, while the population endured the devastating consequences of the Great Leap Forward. However, this popular show was condemned as "vulgar," "superficial," and in "low taste," and eventually canceled, as leftist cultural policies began to dominate the Chinese cultural scene, eventually leading to the period of the Cultural Revolution (1966–76).

The launching of the Cultural Revolution in 1966 was a second major setback to the still-embryonic Chinese television system. The operation of Beijing Television was disrupted for a month at the onset of the Cultural Revolution, while provincial operations were mostly suspended indefinitely. In Shanghai, television became an instrument of factional political struggles. News and cultural programming was broadcast on an ad hoc basis according to the vicissitudes of political necessity. As leftist politics reached its climax, Chinese television also shut its doors to the Western world, canceling a 1963 content exchange agreement with VISNEWS.

Significant technological and infrastructure developments, however, were made during the later stages of the Cultural Revolution. Under the slogan of "carrying the shining image of Chairman Mao to every part of the country," many cities set up stations, interconnecting with each other and transmitting Beijing Television programs via the Ministry of Post and Telecommunication's microwave networks. These interconnected stations provided the initial technological and institutional basis for a national news network. By the end of 1976, with the exception of Tibet, all the provinces and autonomous regions had a television station. Although the national television audience was still extremely small, with one television set for every 1,600 people, the population coverage rate had reached 36 percent (Guo, 1991, p. 123). In an important technological move, the leadership decided to abandon China's efforts to develop a color television system, beginning a trial color broadcast in 1973 using the German PAL system. This choice was made for political reasons, as China's political enemies of the time, the United States and the Soviet Union, used the NTSC and SECOM systems respectively. The breakthrough in US–China relations, symbolized by US president Richard Nixon's visit in 1972, was a pivotal moment for Chinese political history in general and for the Chinese television industry in particular. The professional and technological sophistication of the three US television networks that transmitted live reports of Nixon's visit via satellite back to American audiences had a powerful impression on their Chinese hosts. If it was the Soviet bloc that had helped introduce television to China, it was through US television that the Chinese television industry saw an image of its future.

Yuezhi Zhao and Zhenzhi Guo

From Political Mobilization to Commercial Revolution: Television and the Contradictions of Chinese Reform

The death of Zhou Enlai and Mao Zedong in 1976, and the subsequent political coup by forces opposing the extreme leftists within the Party leadership, marked the beginning of the reform era. Changes in Chinese television were both immediate and drastic. On May Day 1978 Beijing Television was renamed China Central Television, marking the debut of a centrally controlled national television station. Television programs were revamped and those with imported foreign content, from *International News* to *Around the World* to *Dubbed Television Dramas*, quickly became the most attractive fare and key cultural sources for the construction of the Chinese transnational imagination. After many years of isolation, the Chinese audience was eager to learn about the outside world through its new window – television. The visuality of television, as Sun (2002, p. 72) observes, "acquired paramount importance for image-hungry Chinese audiences" in such transitional times.

This window was both structurally shaped and politically tainted by dramatic and far-reaching institutional and ideological transformations. First, advertising was introduced, marking the beginning of Chinese television's historical transformation from a state-subsidized propaganda operation to an entertainment-oriented mass medium with both propaganda and commercial objectives. Although official permission for television advertising was not approved until the end of 1979, Shanghai Television broadcast the first advertisement for a domestic business on January 29, 1979, and the first advertisement for a foreign consumer product, a Swiss watch, on March 15, 1979. Second came the marketized provision of television programs. Economic reforms in the Chinese film industry led state-owned studios, increasingly concerned with economic benefits and competition, to refuse to provide free or cheap films to television stations. Television as a medium was forced to "find its own path" by quickly developing its own productive capacities, and importing foreign programs. Soon afterwards, market mechanisms were introduced into the production and distribution of entertainment television programs. As a result, foreign television imports increased dramatically (Hong, 1998). Japanese cartoons and dramas, complete with advertisements, quickly mesmerized a Chinese audience hungry for any form of entertainment; concurrently, Chinese television became the most effective advertising vehicle for transnational corporations eager to reach the newly opening Chinese markets. Domestic television dramas began to flourish with the support of advertisers and commercial sponsors, and quickly television was exploited as a profit-making medium. This drew film studios and theatrical troupes into the new field of television drama production, leading to the rise of television drama as the most popular audiovisual narrative form (Yin, 2002, p. 30). Thus, a new "cultural revolution" began: television sets became the hottest and most desired consumer item for a population reaping the material benefits of Deng's pragmatic policy of economic development, and a revolution in consumer culture was unleashed in China.

During the initial years of economic reforms the development of Chinese television was astonishing. At the institutional level, 1982 witnessed the establishment of the new Ministry of Radio and Television, under the State Council in the Chinese Cabinet. This strengthened the Chinese state's centralized control of the country's broadcasting operations. The ascendancy of television as the most important means of political communication was formalized during the Party's 12th National Congress in 1982, when central propaganda authorities replaced the 8.00 pm national radio news program with CCTV's 7.00 pm primetime national news program *Xinwen lianbo* as the primary outlet for the day's most important political news. By 1984, when CCTV made a highly successful broadcast of its Chinese New Year Eve gala, television's status as the country's most popular medium had been fully established. By the late 1980s, watching television had joined lighting firecrackers and eating dumplings as an organic component of the Chinese Spring Festival ritual, with the television special garnering a 90 percent share of the national audience.

The Chinese reform program, however, was pregnant with political, economic, and ideological contradictions. Chinese television was not only implicated in these contradictions, but also became one of the primary sites of contestation for different political and social forces (Lull, 1991). On the one hand, television was a key promotional instrument for the Chinese state's modernization program. On the other hand, in the context of a broad intellectual and cultural ferment in the 1980s, television became a powerful forum for promoting the liberal ideas favored by increasing numbers of the Chinese intellectual elite. The relatively liberal political environment of the early and mid-1980s provided the space for reform-minded Chinese intellectuals to explore television as a serious medium of political communication. Yet although commercialism had crept into Chinese television, audience ratings, niche marketing, and advertising contracts were not yet the primary preoccupations of Chinese television. The result was a rare historical window in which television played an unprecedented role in popularizing an elite discourse promoting not only China's modernization and global integration, but also overtly democratic aspirations.

No single media text more forcefully expressed the ideological orientations of this reformist ethos and the liberal intellectual-led "Second Enlightenment" (after the "First Enlightenment" of the "May Fourth Period" in the late 1910s and early 1920s, when Western liberal thought was first introduced into China) than the controversial 1988 six-episode documentary *Heshang* (*River Elegy*). This program, produced by a group of young lecturers, writers, and television producers, assaulted China's so-called "yellow civilization" – river-based agricultural civilization – for its authoritarianism, insularity, and fatalism. By contrast, the program expressed an urgent desire for Western modernity and China's further integration with global capitalism. Rather than promoting conformism and aiming to amuse its audience, this program was an exemplar of the mid-1980s "high-culture fever" (Wang, 1996), drawing upon intellectual sources and iconic images to construct a powerful discourse of national crisis and provoke its audience to ponder the very future of China and Chinese civilization. Articulated as part and parcel of an ongoing debate on the direction of the reform at a critical juncture in Chinese history, this politically charged and rhetorically powerful program was broadcast twice on CCTV within a short period, because of popular demand, and with the support of the reformist fraction of an ideologically divided Party leadership.

The broadcast of *Heshang* and the subsequent "*Heshang* fever" among the urban intelligentsia epitomized the explosive political, economic, and cultural tensions of the late 1980s. This was a time of rising expectations among the urban population, especially intellectuals and students, for democratic participation and the development of an idealized market economy that would bring wealth and national power to China. In reality, however, a dual-track economy created enormous opportunities for profit-seeking activities in which state officials turned political power into economic capital, legally expropriating public property as private property. Rampant official corruption, together with accelerated social stratification at a time of considerable economic inflation, led to deep anxieties about lessening economic opportunities and declining living standards among the urban population. The future direction of the reforms was in the balance as such popular concerns intersected with intensified elite power struggles and ideological divisions between leftist old guards concerned with eroded Party power and a coalition of reformed market authoritarians and liberal democrats (Y. Zhao, 2001). The resulting political drama ended in tragedy in spring 1989, when a student- and intellectual-led urban uprising was suppressed with the violent imposition of martial law in Beijing.

Much of the political drama of 1989 was played out on Chinese television. For their part, journalists and television producers championed a reformist agenda. When the Party's chain of propaganda was paralyzed due to internal division during the height of student protests in May 1989, television helped mobilize and legitimize the movement with its sympathetic coverage of the protests by students and urban residents (Lull, 1991, pp. 188–9). This was the unique historical moment in which Chinese television experienced its freedom – the freedom to report a popular protest movement. It was doubly rare in that media freedom – i.e. freedom from overt Party censorship – became an important rallying cry not just for Chinese journalists, but also for students, intellectuals, and urban citizens in general. This freedom, of course, would be provisional and short lived. In the end, market authoritarian forces, personified by Deng Xiaoping, defeated the reformists led by Party General Secretary Zhao Ziyang, crushed the liberal and democratic inspirations of Chinese students and citizens, and reasserted political control of the media.

The suppression of the 1989 movement marked a turning point in both Chinese history and television. On the one hand, Party control was reinforced and tightened. Party leadership re-emphasized positive propaganda and stressed the role of media in maintaining ideological control. Instead of serving as a forum for public debate, mass media now had to present a unified and politically correct ideological line. The mass media, television in particular, retreated from active engagement with elite political and cultural debates of the day, and shied away from any attempts to assert relative autonomy from the Party state through structural reforms. In 1992, Deng Xiaoping called for accelerated market-oriented economic developments as a response to demise of Communist regimes in the Soviet bloc. Market forces quickly swept through the Chinese media system and transformed it from the inside out (Y. Zhao, 1998). This broad transformation toward market authoritarianism in the Chinese political economy, gained speed in June 1992, when Party officials redefined the mass media and other traditionally state-subsidized ideological and cultural institutions as part of a tertiary industry to be run according to the principles of self-financing (i.e. market-based financing) and profit-making, all in the name of accelerating economic development. Profit was the new name of the game for Chinese television. Meanwhile, the household penetration rate of television accelerated during the economic boom of the early 1990s. Thus, Chinese television achieved unprecedented reach, and was subjugated to the dual imperatives of political control and commercialism.

From Commercialization to Conglomeration: The Evolving Political Economy of Post-1992 Chinese Television

The Chinese television industry is characterized by a unique form of state monopoly capitalism: commercialized operations organized into a hierarchical structure of administrative monopoly. While the Party's Propaganda Department assumes ultimate control over content, operational, administrative, and regulatory control of the country's broadcasting industry is in the hands of the State Administration for Radio, Film, and Television (SARFT – the current incarnation of the Ministry of Radio and Television) and its local counterparts. Before 1983, television was operated at central and provincial levels. In 1983, amidst keen structural reform, the "four-level policy" was implemented. This was, in part, a response to the lack of centralized resources, and a mounting desire to develop a national television infrastructure as quickly as possible. Effectively, television operations were decentralized from the existing national and provincial levels, by allowing municipal and county governments to mobilize local resources in order to launch their own television stations, in part to transmit central and provincial programming, in part to broadcast local programming. This policy played a crucial role in the rapid popularization of television in the country and would plague efforts to reassert central control and more fully commercialize the system.

527

While the term "administrative monopoly" describes the overall structure of the Chinese television industry, it does not fully explain the complexity and dynamics of the system. First, although Party state-owned television still monopolizes broadcasting and news and current affairs production, the provision of entertainment programs has been marketized and partially privatized. Independent and quasi-independent producers, which have flourished since the late 1990s, provide an increasing share of entertainment programming (Y. Zhao, 2004). Private and foreign capital is flowing into the domestic industry through investment in entertainment programming, advertising sponsorships, and co-production arrangements with television stations. The result is a unique system of state and private collaboration: while the Chinese state continues to monopolize television broadcasting and exercise macro-control of television production through mechanisms such as the television drama production permit system and pre-production content clearance (Yin, 2002, pp. 36–7), commercial advertisers and private investors contribute to the system's economic foundation and increasingly shape its content orientation.

Secondly, there is intensive competition in the Chinese television system, both within, and across, different administrative boundaries. At the national level, CCTV has faced increasing pressures from local stations. By 1999, each provincial broadcasting administration had established a general interest satellite channel, which is carried by most local cable networks and has potential national audience reach. At the same time, provincial and municipal broadcasting administrations, seeking to exploit new market opportunities and increase the vitality of the television system, operated multiple terrestrial and cable channels, each of them financially autonomous and competing for similar audiences. In Shanghai, for example, the initial monopoly provider, Shanghai Television, was required to provide start-up technical and personnel support to establish a rival station, Oriental TV, to introduce "friendly competition" within the Shanghai Broadcasting Bureau (Y. Zhao, 1998; Weber, 2002). Competitive market mechanisms have also been introduced into the micro-management of television production, leading to the creation of relatively autonomous programming units within a television station. These units, in turn, have produced innovative programs by borrowing commercially proven popular Western formats, from morning-hour news magazines to investigative documentaries, and talk shows. The result, as Keane (2002a) describes, is a television industry where "as a hundred television formats bloom, a thousand television stations contend."

Advertising is the lifeblood of the Chinese television system. With the exception of television stations in economically underdeveloped areas, state subsidies have ceased to be a significant revenue source for Chinese television. By 1999, commercial revenue accounted for more than 85 percent of the Chinese broadcasting industry's revenues, with television advertising accounting for 90 percent of total advertising revenues. The most profitable television station is CCTV, whose dominant market position is guaranteed by its monopoly status at the central level. Between 1992 and 1999, CCTV's advertising revenue grew nearly eightfold, from 560 million yuan to 4.415 billion yuan (*Zhonghua Renmin Gongheguo Jianshi Bianjibu*, 2001, pp. 304–7). Although it is profitable enough to need no state subsidies, the state insists that CCTV continues to accept its token subsidy in order to maintain its patron role (Lee, 2000, p. 11). The 1990s has also seen the rapid development of television stations in those economically developed coastal provinces and cities that have emerged as regional and local economic powerhouses.

However, just as uneven development is a key feature of the Chinese market economy, so it is in the Chinese television industry, with many stations in economically depressed areas maintaining a marginal existence, due both to the lack of advertising revenues and to the inadequate level of state subsidies. As well, conflict of interests between stations affiliated with different levels of government has been a chronic problem. On the one hand, the SARFT and its affiliated national station CCTV – which serves as the mouthpiece of the central authorities – expect and rely upon local stations to relay CCTV programs to every corner of the country to fulfill central propaganda objectives. On the other hand, CCTV has neither a formal network structure nor advertising-revenue-sharing arrangements with these stations. Local government authorities and stations, with local political and commercial interests in mind, are unwilling to serve simply as CCTV relay stations and are also resentful of its monopolistic and parasitical status. During the height of these tensions in the early to mid-1990s, some local stations replaced CCTV advertising with locally solicited advertising during program transmission. Others even replaced less attractive CCTV programming with locally produced fare, and even pirated programs. In the eyes of the SARFT and CCTV authorities, such local practices were signs of political disobedience and cultural denigration.

By the late 1990s, extensive market competition and audience fragmentation had led central authorities to pursue a policy of re-centralization and market consolidation. On a structural level, the SARFT effectively reversed the "four-level policy" and forced municipal and county broadcasters to limit their self-programming powers and devolve to relay stations of central and provincial programming. Under this new regime, county-level stations were forbidden to run their own independent programming channels. At the same time, the SARFT encouraged provincial broadcasting administrations to merge the existing radio, terrestrial, and cable television stations into a single corporate structure. The merger mania in Western media markets and the increasing pressure of global competition, together with potential domestic competition posed by an aggressive telecommunications industry, which has been increasingly encroaching into cable distribution territory in the context of technological convergence, provided further rationales for this state-mandated industrial consolidation in broadcasting and cable.

Following the establishment of a number of provincial-level broadcasting conglomerates, on December 6, 2001, just a week before China officially joined the World Trade Organization (WTO), the state announced the establishment of the China Radio, Film, & Television Group, a mega-media conglomerate that encompasses CCTV, China National Radio, China Radio International, the China Film Group Corporation, and other related media production and distribution assets. This state conglomerate is supposed to serve as an industry champion ready to face the challenges of transnational media corporations in the global marketplace. As the quintessential expression of "socialism with Chinese characteristics," this group is no ordinary media conglomerate. It has no independent business status. An administrative board, under the leadership of the SARFT's Party committee, serves as its corporate board, with the director of the SARFT, who is also a deputy chief of the Party's Propaganda Department, serving as its chair.

The Ideological and Cultural Dimensions of Contemporary Chinese Television

At this point, it is important to re-emphasize the double articulation of the transformation on display in Chinese television: first, it has become a major platform for capital accumulation, albeit in the context of China's authoritarian market system; secondly, its key role in the struggle over meaning has been redoubled as it is helping to re-establish the ideological hegemony of the post-1989 Chinese state and to mediate social relations amidst rapid social transformation. Thus, television's traditional role as mouthpiece, rearticulated in the official media policy discourse of maintaining a "correct orientation to public opinion," has been rigorously maintained through censorship and self-censorship. In news, this means the subordination of journalists to the Party leadership's propaganda objectives. CCTV's *Xinwen Lianbo*, in particular, defines its market niche by being the official bulletin board of Party leaders' daily activities and an official news outlet of various Party and state organs. In dramatic productions, this policy is implemented through the state-subsidized production of "mainstream melody" shows, which are viewed by political authorities as presenting issues of importance and being reflective of normative behavior and values (Keane, 2002b; Yin, 2002). Typically, these shows express the Party's ideological imperatives through the glorification of the Party's history, the celebration of the Party's current achievements, and the creation of contemporary role models. Meanwhile, Chinese television has assumed new ideological roles in the post-1989 period – namely, reasserting both nationalism and traditional Chinese morality. Two television events in 1990 – at a time when both elite and popular social forces were struggling to come to terms with the military crackdown in 1989 – have been pivotal in this development.

First, the 1990 Asian Games in Beijing offered television the opportunity to eloquently articulate and mobilize a much-needed legitimating ideology for post-1989 China: nationalism. Since then, nationalism has become the dominant ideological framework for Chinese media in general – and for television in particular. By projecting national cultural symbols into Chinese households, staging political and cultural spectacles, invoking popular sentiments, and rallying viewers under the national flag during important events such as Hong Kong's return, Beijing's Olympics bid, and China's WTO entry, television is by far the most powerful site for the construction of an official discourse on nationalism and the mobilization of patriotism. As Sun (2002) observes in the context of Chinese television's coverage of the 2000 Sydney Olympics Games and the reception of this television spectacle by both domestic and overseas Chinese audiences, "the nationalistic agenda of the Chinese state converged with the 'gut feelings' of ordinary Chinese people" (p. 213) and their "spectatorial desire" (p. 191). CCTV's annual Spring Festival Show, in particular, has become more nationalistic – both thematically and symbolically – since the early 1990s. Whether it is to glorify military heroes sacrificing themselves for the people, to celebrate sports stars winning for the nation, to acknowledge the hardships endured by laid-off workers, or to show the entire society's charitable concerns for a terminally ill poor maiden, television aims to address an increasingly stratified Chinese society and create a sense of belonging to a national family, where everybody, regardless of his or her social economic status and localities, is first and foremost Chinese. In this program, the classical Confucian notion of the state as an enlarged family is brought fully into play (B. Zhao, 1998, p. 43). Moreover, as CCTV expands its global reach with satellite footprints in North America, Europe, Australasia, and Asia, the Chinese state is increasingly able to extend the notion of a national family to diasporic Chinese communities in a "global Chinese village" (Sun, 2002).

Secondly, in 1990 Chinese television, which had played such an instrumental role in the pre-1989 political and cultural ferment with the politically and intellectually challenging documentary *Heshang*, turned to mass entertainment. The "culture industry" formally arrived in China with CCTV's broadcast of *Kewang* (*Aspirations*), a 51-episode drama series, considered to be Chinese television's first major successful soap opera. As Yin (2002, p. 32) describes the show, "it was a genuine industrial product: it was made with financial support with non-state sources, and shot in a studio, utilizing artificial indoor scenes, multicamera shots, simultaneous voice recording and post-production techniques." Effectively imitating the story format and conventions of dramatic shows from Asia and Latin America, *Kewang* is the first Chinese drama series that embraced ideological conformism and took popular entertainment, rather than political propaganda, as its explicit aim. Centered around the life of a young and attractive working-class woman living in Beijing, this enormously popular show constructed a "tortured yet appealing tale depicted in the prurient tones of schlock TV" (Barmé, 1999, p. 103) by relegating major political events to the background. Nevertheless, it provided a forum for its audiences to engage with highly emotive issues concerning privilege, education, class background, and self-esteem (Barmé, 1999, p. 102), and addressed the psychological needs of an audience struggling to square its collective identity with tumultuous economic reforms and post-1989 political repression (Guo, Sun, and Bu, 1993). The wild popularity of *Kewang* marked an ideological breakthrough in Chinese television: for the first time, the Party's highest ideological establishments not only accepted, but praised a show centering around a traditional Chinese morality play. With the broadcast of *Kewang*, television's conservative role of creating a morality play of unity and stability through entertainment was officially legitimated and celebrated in China.

If reformist television in the 1980s said farewell to class struggle but ended up being caught in "high culture fever" and taking up an elitist mission to promote the "Second Enlightenment," post-1989 television embraced commercialized mass entertainment wrapped in patriotism, traditional Chinese values, and an increased interest in human interest stories. Thus, when the intellectual elites active in the pre-1989 "high culture fever" recovered from 1989 and returned to the Chinese cultural scene in Spring 1992, they found their platform already taken by the more powerful and prevailing force of the market (B. Zhao, 1998, p. 55). Finally, it seems, Chinese television had caught up with mainstream American television. Rather than undermine the Party, commercialized entertainment has helped to re-establish the Party's hegemonic position in Chinese society. As Barmé (1999) concludes:

> the success of *Aspirations* was a victory for one of the most representative paradigms of American popular culture. With this medium, the message of peaceful evolution – the primary of economics over politics, consumption in place of contention – continued to insinuate itself into the living rooms of millions of mainland TV viewers. It was a development that increasingly served the needs of the Communist authorities while also preparing China's artistic soil for the germination of global culture. Popular culture became the opiate of the consumer, and as it was transmogrified by stylistic improvements inspired by Hong Kong, Taiwan, and international media, it converted areas of contestation into spheres of co-option. (p. 107)

By the late 1990s, intensive competition between proliferating television channels and a vibrant private and quasi-private entertainment production industry had fully developed television's entertainment role. Since 1997, entertainment television, in the form of variety shows, dating games, sitcoms, law and order series, reality television, and the endless parade of emperors and empresses in costume dramas, have filled Chinese television screens. Hong Kong-based Phoenix TV and its mainland imitator, Hunan Satellite Television, carried this entertainment wave to new heights. The popularity of Hunan Satellite Television in the late 1990s confirmed a simple truth to the Chinese television industry: uplifting light entertainment is the safest and fastest means to popularity and capital accumulation. If the "vulgar" was denounced during the Cultural Revolution, "kowtowing to the vulgar" (Barmé, 1999) is the mantra of post-1992 Chinese television. While both liberal and state socialist cultural elitists condemn this development, the affirmation of mundane and commonplace emotions, and the single-minded imperative to delight can also be seen as a "double liberation" in

532

post-1989 China: from both the traditional socialist didactic approach to culture and the elitism and intellectual vanguardism of the 1980s as expressed in *Heshang*.

The ideological impact of television's shift to the distinctive "Chinese commercial model" has been profound in post-1989 China, at a time of enormous social and cultural dislocation, and as the *Falun Gong* movement and widespread worker and farmer protests contest the emerging market authoritarian social order. If advertising aims to mobilize the audience to pursue happiness through the consumption of material goods, entertainment shows mobilize the audience to pursue happiness and Chinese identity affirmation through the consumption of television discourses. Not surprisingly, Hunan Satellite TV's flagship show, *Kuaile dabenying* (*The Grand Happiness Citadel*), is modeled on a Hong Kong television show entitled *Huanle zongdongyuan* (*All Out Mobilization for Joyfulness*). If Maoist propaganda instructs the politicized subject of a socialist state to dedicate his or her transient life to the transcendental cause of "serving the people," commercialized state television, through advertising and what Sun (2002) has termed "indoctritainment" (that is, indoctrination and entertainment in a single package), directs the all-consuming subject/patriotic citizen to devote his or her limited life to the unlimited world of hedonistic personal consumption. And, of course, if this should prove impossible, at least she or he can find happiness through watching television and witnessing nationalistic achievements (Y. Zhao, 2004).

Television's role as the mouthpiece of nationalism and "mass distraction," embedded in the underlying ideological framework of state-led, market-oriented economic development, has been complemented by its role as a "watchdog," albeit one without much bite. This role was pioneered by CCTV's News Commentary Department, a relatively autonomous and financially independent production unit established in 1993 as a result of a complex struggle between political, institutional, and professional imperatives among Party leadership, CCTV top management, and the growing reformist and professional impulses of Chinese journalists (Y. Zhao, 1998, 2000a). Since then, this department has not only produced some of the country's most celebrated watchdog programs, including *Dongfang shikong* (*Oriental Horizons*), *Jiaodian fangtan* (*Focus Interviews*), *Xinwen diaocha* (*News Probe*), and the talk show *Shihua shishuo* (*Tell as It Is*), but also inspired similar programs in provincial and local television stations, and even in the print media. By developing narratives that expose select cases of official corruption, redress economic and social injustices inflicted upon powerless individuals, and reveal various social problems and ethical dilemmas of the reform process, these programs both provide a much-needed forum for the expression of popular concerns and help re-establish the political legitimacy of the post-1989 Party state. Though the economic reforms have vastly improved living standards for a majority of the population, the process has been "socially disruptive and morally distasteful" (Meisner, 1999, p. 248). Moreover, the human suffering brought about by China's capitalistic-style development has been borne disproportionately by the weakest members of society. Within this context, watchdog journalism, which has been praised by both the top leadership and the general public alike, affirms the former's commitment to ordinary Chinese citizens. At the same time, these programs, which even solicit grievances from the general public for story ideas, also offer a state-sanctioned catharsis for those suffering injustices in post-reform China. Although these programs can be best described as "watchdogs on Party leashes" (Y. Zhao, 2000a) and they tend to individualize, localize, and moralize systemic social problems, thus avoiding "a critical interrogation of overall social structure" (Xu, 2000, p. 646), their relative openness and professionalism are both real, and their political and social impact is profound (Li, 2002). By upholding social morality, these programs help to police the political, economic, and social boundaries of an emerging authoritarian market society. Concomitantly, they enhance the credibility of state monopoly television itself and massage the media's ideological role from that of straightforward propaganda to a more subtle form of hegemony (Y. Zhao, 1998; Chan, 2002). While there are inevitable conflicts and tensions between television professionals, Party censors, and the subjects of media exposure, overall, state-sanctioned watchdog journalism has proven to be particularly profitable for both Chinese television, and the management of social tensions.

These different ideological dimensions often intertwine, contributing to the polysemic and hybrid nature of Chinese television discourses. Official propaganda, commercial popular culture, middle-class reformism, and even traces of elite intellectual discourse of the mid-1980s, coexist in Chinese television culture. While the increasing popularity of light entertainment and commercial culture is "jamming" and to some extent dissolving both serious Party propaganda and "the autonomous culture of the intellectual elites" (B. Zhao, 1998, p. 55), these ideological fragments also borrow from and reinforce each other. So there is watchdog journalism delivered in the form of the television drama, or "mainstream melody" propaganda fare in the form of the Hollywood action adventure, replete with suspense, intrigue, love, lust, conspiracy, murder, detectives, car chases, and violence. Likewise, while television dramas about imperial dynasties tend to cultivate a profoundly anti-democratic political culture through the glorification of autocratic power and the traditional values of patriarchy, submission, social hierarchy, their narratives often smuggle in contemporary political and social critiques – e.g. by presenting upright imperial officials to attack contemporary widespread corruption and express popular desires for social justice and a clean and responsible government. Thus, the production and consumption of historical dramas serve as a rich symbolic site for rhetoric contestation among various ideological forces and discursive positions in contemporary China (Yin, 2002). As Sun argues, it is important to understand the complexity and dynamism that mark the relationship between television production, consumption, commercialization, the propagation of state ideology, and the formation of popular subjectivities in contemporary China, thus avoiding "the trap of both the docile audience argument and the theory of the omnipotent state" in understanding Chinese television culture.

Nonetheless, the heavy dependence on advertising revenues, intensive market competition, and also the urban and affluent sociocultural composition of the television labor force skew Chinese television heavily toward the cultural needs and sensibilities of affluent urban consumers. While Chinese television has "virtually" enfranchised the vast majority of the population as audience, the voices and cultural needs of the vast majority, from farmers in the rural areas to ordinary workers and the urban underclass, children and senior citizens, are often ignored and marginalized. The impact of the consumerist discourse via market-driven television culture on women has been particularly paradoxical. Although women's participation in consumption is celebrated as a liberating and modernizing experience, women are increasingly subjected to its commodifying and objectifying gaze. As the pre-reform Maoist discourse of gender equality and repressive sexual mores fade away and love and lust become everyday television fare, women are seen increasingly as objects of male sexual desire. Thus, on the one hand, television has been a powerful promoter of consumer-friendly sexual liberation and the general liberalization of Chinese society. On the other hand, patriarchy has returned with a vengeance in the context of a rising Chinese authoritarian capitalist culture. Thus, in advertising culture, women are cast as all-consuming subjects, while informational and entertainment television, including talk shows, often suggest that "women return home" to ease the pressures of employment and should make "sacrifices" for the benefit of the reforms. While Western-inspired liberal feminist sentiments have found marginal expression on Chinese television, overall, it has been oblivious to the massive social costs of the economic reforms disproportionately borne by lower-class women, from the middle-aged laid-off urban factory women, who makes up the majority of laid-off state enterprise workers, to young female sweatshop workers in the coastal areas, to the desperate rural woman left behind to look after elders, kids, the land, and livestock while their husbands and sons venture into the cities in search of jobs.

With the increasing prominence of market relations and the increasing penetration of capitalist logic into Chinese television, Party control and the inherent social biases of the market are being mutually reinforced. Echoing developments in the West, Chinese television largely ignores both politics and social conflict. If *Heshang*'s intellectual vanguardism and sympathetic coverage of student protests in 1989 helped to foster pressure for political reform, commercialized post-1989 Chinese television, together with other forms of mass media, may have helped to contain opposition by suppressing reports of workers and farmer protests and marginalizing radical perspectives. In short, there is a general exclusion of the issues of those groups that do not constitute media advertisers' most desirable audience, namely the small political and economic elite, and the mostly urban middle class (Zhao and Schiller, 2001).

In China, as elsewhere, mass entertainment, the mobilization of consumption, and business information are both politically safe and financially rewarding. This new regime of ideological domination is based on a compromise between the

Party state and the population, and collusion between the political and media elite, where the state tolerates ideological and moral relaxation in exchange for the population's retreat from making political demands. The media secures political patronage and commercial success while the population gets entertainment and a sense of cultural enfranchisement. Whether such an exchange between television-centered cultural consumption and real citizenship will accommodate the diverse needs of a fractured Chinese society, of course, is by no means guaranteed. The relatively frequent recourse to coercive state intervention, among other reasons, suggests that the current hegemonic bloc of political and economic elites and urban middle-class consumers has not secured a stable regime of ideological domination in the Chinese media (McCormick, 2003).

Chinese Television in the Global Marketplace: Enduring Issues, New Challenges

As Chinese television comes of age in the new century, it faces new political, economic, and cultural challenges. First, pressures for political liberalization continue to build. These manifest themselves most prominently in persistent liberal demands for loosening Party control on media reporting of sensitive political and social issues. The new Party leadership, under Hu Jintao, has responded by initiating some changes. One highly symbolic move was to explicitly order CCTV to reduce the amount of broadcast time devoted to the ritualistic display of political power by national leaders in *Xinwen lianbo*. Under intensive international and domestic pressures, after the initial media cover-up of the SARS outbreak in Spring 2003 and in the context of an unfinished power transition, the new leadership also decided to encourage the media coverage of SARS. How far this political momentum of change can be maintained remains an open question. In the context of China's deeply entrenched political economic structure and power relations, it would be naïve to place high hopes for media liberalization on the new leadership.

Second, the regulatory and structural reforms of Chinese television continue to face enormous challenges. While its mouthpiece role has enabled the industry to grow as a commercial monopoly in the reform era, it has also prevented Chinese television operators from achieving independent corporate status, thus limiting its ability to absorb outside capital, or to use market mechanisms to rationalize industry structure. Conflict among various bureaucratic interests both inside and outside the industry, including state broadcasting versus telecommunication authorities, broadcasting bureaus and television stations, not to mention central, provincial, municipal, and county levels, remains acute (Y. Zhao, 2000b, 2004; Hu, forthcoming). The state's campaign for re-centralization and market consolidation has not only run into considerable resistance at the local level; it has also failed to produce the desired ends of increased market efficiency and profitability. To date, the Chinese state has entertained neither privately owned television nor any notion of public broadcasting in the Western sense. Reformist efforts to establish a stable regime of regulation for the television industry continue to be frustrated by the Party state's arbitrary directives (Guo, forthcoming). Thus, a formidable challenge remains to (re)-invent a sustainable institutional form for television under "socialism with Chinese characteristics."

Thirdly, the challenges of globalization have assumed a new sense of urgency in the context of China's WTO entry. Although China's WTO accession agreements did not include the opening of its broadcasting and cable industries to foreign investment, the state has increased China's film import quota, and permitted transnational media corporations to expand their operations in cinema services, audiovisual co-production, distribution, and retailing. Transnational satellite television, meanwhile, has significantly expanded its reach in China. In addition to Phoenix TV, which since 1997 has been the most localized and influential foreign-invested television channel in China, the state has also approved limited operations of two major Mandarin channels, one owned by Star TV, the other jointly owned by Time Warner and Tom Group. Although none of these channels has a national audience reach and profitability remains only a remote possibility, their competitive pressures and effectiveness in attracting affluent urban market niches are profound. The critical issue is not whether China's domestic television industry will survive "an invasion of foreign culture" after WTO entry. While cultural nationalism remains a strong force, the industry is becoming smarter by localizing foreign formats and borrowing branding and management strategies from transnational corporations. From this perspective, some observers remain optimistic about the prospects for China's audiovisual industry in the global marketplace (Keane, 2002c). From the point of view of social communication, however, the most crucial questions are how will the new television environment address the diverse communicative needs of an increasingly fractured Chinese society? Will it further intensify the existing social biases of the Chinese television system (Zhao, forthcoming)? Will television provide the Chinese with an adequate *Dictionary of Happiness* (*Kaixin Cidian*, the Chinese imitation of *Who Wants to Be a Millionaire?*) in their *Expedition into Shangrila* (*Zourun Xianggelida*, the Chinese imitation of *Survivor*) as China's integration with the global market system accelerates, and as the domestic and international contradictions of this integration intensifies in the new millennium?

As the intertwining logics of state, domestic, and international market forces negotiate the terms of domination over Chinese television, other social forces continue to contest the current regime of control. Inside the Chinese television industry, socially conscious professionals continue to explore new institutional and symbolic spaces for more autonomous expression. Outside the television industry, politically, economically, and culturally disenfranchised Chinese continue to stake their claims over a television system that is still run under pretense of "serving socialism, serving the people." Their struggles have taken many forms. *Falun Gong* practitioners hacked into satellite and cable television systems to tell their side of the story. Countrywide, victims of power abuses flood the reception area of CCTV headquarters on a daily basis and beg it to use its investigative and exposing powers in order to promote social justice. A nascent ad hoc women's media watchdog group has lodged protests against sexism on television. Even television audiences have overcome their passivity and protested against excessive advertisements during their favorite shows. If television has been an "agent of modernization" in China (Lull and Sun, 1988), certainly these groups are not satisfied with what nearly three decades of state-organized Chinese modernization has brought them. As Chinese television continues its search for the appropriate institutional and cultural forms in a globalizing context, the unfolding struggles over control of television in China will only intensify.

References

Barmé, G. (1999) *In the Red*, New York: Columbia University Press.

Chan, A. (2002) "From Propaganda to Hegemony: Jiandian Fangtan and China's Media Policy," *Journal of Contemporary China*, *11*(30), 35–51.

Guo, Z. Z. (1991) *Zhongguo dianshi shi (A History of Chinese Television)*, Beijing: Zhongguo renmindaxue chubanshe.

Guo, Z. Z. (forthcoming) "Play the Game by the Rules? Television Regulation in China," *Javnost*.

Guo, Z. Z., Sun, W. S., and Bu, W. (1993) "Jingyan, Rentong yu Manzhu: Dui Kewang Shouzhong Laixin de Neirong Fenxi" (Experience, Identification, and Gratification: A Content Analysis of Letters from *Kewang* Audience), *Proceedings of the China–Canada Television and Film Cooperation Conference*, Beijing.

Hong, J. H. (1998) *The Internationalization of Television in China: The Evolution of Ideology, Society, and Media Since the Reform*, Westport, CT: Praeger.

Hu, Z. R. (forthcoming) "The Post-WTO Restructuring of the Chinese Media Industries and the Consequences of Capitalization," *Javnost*.

Keane, M. (2002a) "As a Hundred Television Formats Bloom, A Thousand Television Stations Contend," *Journal of Contemporary China*, *11*(30), 5–16.

Keane, M. (2002b) "Television Drama in China: Engineering Souls for the Market," in T. J. Craig and R. King (eds.), *Global Goes Local: Popular Culture in Asia*, Vancouver: UBC Press, pp. 120–37.

Keane, M. (2002c) "Facing Off on the Final Frontier: The WTO Accession and the Rebranding of China's National Champions," *Media International Australia*, *105*, 130–47.

Lee, C. C. (2000) "Chinese Communication: Prisms, Trajectories, and Modes of Understanding," in C. C. Lee (ed.), *Power, Money, and Media: Communication Patterns and Bureaucratic Control in Cultural China*, Evanston, IL: Northwestern University Press, pp. 3–44.

Li, X. P. (2002) "'Focus' (Jiaodian Fangtan) and the Changes in the Chinese Television Industry," *Journal of Contemporary China*, *11*(30), 17–34.

Lull, J. (1991) *China Turned On: Television, Reform, and Resistance*, London and New York: Routledge.

Lull, J. and Sun, S. W. (1988) "Agent of Modernization: Television and Urban Chinese Families," in J. Lull (ed.), *World Families Watch Television*, Newbury Park, CA: Sage Publications.

McCormick, B. L. (2003) "Recent Trends in Mainland China's Media: Political Implications of Commercialization," *Issues & Studies*, *38*(4)/*39*(1), 175–215.

Meisner, M. (1999) "China's Communist Revolution: A Half-Century Perspective," *Current History*, September, pp. 243–8.

Sun, W. N. (2002) *Leaving China: Media, Migration, and Transnational Imagination*, Lanham, Boulder, New York, and Oxford: Rowman & Littlefield.

Wang, J. (1996) *High Culture Fever: Politics, Aesthetics, and Ideology in Deng's China*, Berkeley, CA: University of California Press.

Weber, I. (2002) "Reconfiguring Chinese Propaganda and Control Modalities: A Case Study of Shanghai's Television System," *Journal of Contemporary China*, *11*(30), 53–75.

Xu, H. (2000) "Morality Discourse in the Marketplace: Narratives in the Chinese Television News Magazine *Oriental Horizon*," *Journalism Studies*, *1*(4), 637–49.

Yin, H. (2002) "Meaning, Production, Consumption: The History and Reality of Television Drama in China," in S. H. Donald, M. Keane and Yin, H. (eds.), *Media in China: Consumption, Content, Crisis*, London: Routledge/Curzon, pp. 28–39.

Zhao, B. (1998) "Popular Family Television and Party Ideology: The Spring Festival Eve Happy Gathering," *Media, Culture & Society*, *20*, 43–58.

Zhao, Y. Z. (1998) *Media, Market, and Democracy in China: Between the Party Line and the Bottom Line*, Urbana and Chicago: University of Illinois Press.

Zhao, Y. Z. (2000a) "Watchdogs on Party Leashes? Contexts and Limitations of Investigative Journalism," *Journalism Studies*, *1*(4), 577–97.

Zhao, Y. Z. (2000b) "Caught in the Web: The Public Interest and the Struggle for Control of China's Information Superhighway," *Info*, *2*(1), 41–65.

Zhao, Y. Z. (2001) "Media and Elusive Democracy in China," *Javnost*, *8*(2), 21–44.

Zhao, Y. Z. (2004) "The State, the Market, and Media Control in China," in P. Thomas and Z. Nain (eds.), *Who Owns the Media? Global Trends, Local Resistance*, Penang: Southbound and Zed Books.

Zhao, Y. Z. (forthcoming) "Transnational Capital, the Chinese State, and China's Communication Industries in a Fractured Society," *Javnost*.

Zhao, Y. Z. and Schiller, D. (2001) "Dances with Wolves? China's Integration with Digital Capitalism," *Info*, *3*(2), 137–51.

Zhonghua Renmin Gongheguo Jianshi Bianjibu (2001) *Gaige Kaifangzhong de Guangbodianshi, 1984–1999 (Broadcasting in the Reform Period, 1984–1999)*, Beijing: Zhongguo Guoji Guangbo Chubanshe.

CHAPTER TWENTY-NINE

Japanese Television: Early Development and Research

Shunya Yoshimi

Open-Air Television in the Mid-1950s

In postwar Japan, most of the early TV sets were placed not in private homes, but on street corners, in railway stations, and in parks and shrines – locations that attracted large numbers of people. Thus, Japanese television in the mid-1950s was something like an "open-air theater" (*Gaitou Terebi*). This phenomenon of open-air television has its origins in the marketing strategy of Nihon Terebi (NTV), the first commercial TV station, which was established in 1953, around the same time as NHK, the national broadcasting system. At that time, a basic television set cost about 20,000 Yen. Since the starting salary for a university graduate was only 8,000 Yen a month, televisions were still expensive luxury items that were well beyond the reach of most households. NTV installed 220 large-screen televisions at 55 locations, in front of railway stations and other public places. Even though most people could not afford a television, there was still widespread interest in the new medium. NTV's intention was to encourage this interest and to instill in people the desire to buy their own television sets, while simultaneously boosting the station's advertising revenue.

It must be emphasized that the popularity of open-air television occurred at a time when few households possessed TV sets. When NHK started broadcasting in 1953, only 866 households had contracts to receive the broadcasts. Although this number had risen to 50,000 by 1955, until the late 1950s the number of people watching television out-of-doors in a crowd far exceeded the number of people watching at home. In September 1953, NTV conducted on-site surveys at the places where open-air television sets were located. These revealed that there were on average a total of 9,200 viewers at any given time, each of whom spent an average of 30 minutes standing in front of the television. It was estimated that the average number of daily viewers numbered in excess of 100,000. Thus, in those early days of television broadcasting, watching open-air television was far more significant than viewing within the home (Hoso Bangumi Iinkai, 2003, p. 28).

The location of each open-air television set was monitored carefully by the company. Matsutaro Shoriki, the president of NTV, required his employees to write daily observations about the audiences in front of each open-air television. Shoriki read these "diaries" carefully and ordered changes in the location if the television set did not attract enough audience members (Sano, 1994, p. 474). Under Shoriki's rigorous inspection, employees of NTV always had to find better locations for their television sets around the densely populated districts of Tokyo. Once they had found a good location, they also had to ask the neighbors to accept the placement of the television set without payment. Documents also indicate that, in many cases, NTV did not even pay for the electricity charges for the televisions.

One of the most interesting features of these early open-air television sets was their geographical distribution. In 1950s Tokyo, open-air televisions were located not only in places of popular leisure and entertainment ("*sakariba*"), such as Hibiya Park, Ginza Yonchome and the west side of Shinjuku, but also in the grounds of temples and shrines frequented by the working class. These latter sites included Koto Rakutenchi, Asakusa Kannon, Suitengu, Sugamo Togenuki Jizo, and Kagurazaka Bishamonten. The railway stations where open-air televisions were installed were concentrated in the eastern part of the city, near the coast where factories were concentrated. These included Ueno, Shinagawa, Asakusa, Gotanda, Omori, and Kawasaki, as well as 16 stations on the Keisei line, and four on the Keihin line. By contrast, in the middle-class residential districts of western Tokyo, there were surprisingly few open-air televisions. There was, therefore, a major regional gap between the areas where open-air television was concentrated in the early phase of the development of television and the areas that led the way during the later expansion of domestic indoor television. In other words, open-air television in 1950s Japan seems not to have been a part of the newly developing middle-class culture in suburban houses, but a part of the traditional working-class culture that had developed on the street corners.

Performing "Japanese-ness" on the TV Screen

Open-air television was important not only as the stimulus for the quantitative growth of television viewing. It also reflected a linkage between television and the popular imagination characteristic of a certain postwar period. During this period the most popular programs on open-air television were not dramas or documentaries, but the relaying of sports events, and especially professional wrestling matches featuring the well-known wrestler, Rikidouzan. Due to the limitations of camera technology at the time, action unfolding over a wide space, such as on a baseball field, was difficult to broadcast. By contrast, professional wrestling's high-speed action, confined to a small ring, was ideally suited as a genre for television. Naoki Inose succinctly describes the relationship between wrestling and TV in postwar Japan: "[the] existence of television was dependent on professional wrestling, and the existence of professional wrestling depended on television. Rikidouzan became a hero thanks to television, and television gained an audience by giving birth to such heroes" (Inose, 1990, p. 288).

Rikidouzan's showmanship was American-inspired (he had learned the skills of professional wrestling in the United States) and he was always conscious of the television camera. On the TV screen, he played out the heroic role of a brave Japanese wrestler confronting underhand giant American wrestlers. This formidable performer skillfully established postwar Japan's convoluted pseudo-anti-Americanism in the medium of television. Whereas other professional wrestlers based in the Osaka and Kyushu regions saw wrestling as nothing more than a combative sport, Rikidouzan realized that it could be viewed as a national symbolic drama performed in front of an audience of several million people gathered in front of the television. In this way, he gained his great success by aligning his own brilliant performance with people's convoluted feelings toward things American. This was exemplified by his first tag match: battling the Sharp brothers in February 1954, he knocked down his giant American opponent with a karate chop, to a groundswell of cheers from the excited crowds.

From the point of cultural history, the most significant irony in the popular fandom toward Rikidouzan's performances was the fact that Rikidouzan himself was North Korean, not ethnically "Japanese" in the narrow postwar sense. Born in a poor rural village near the east coast of North Korea in the early 1920s, he was recruited to Japan to be a Sumo wrestler in 1940. So Rikidouzan's life had been rooted deeply in traditional Korean society during the colonial period. Although Rikidouzan played actively as a Sumo wrestler during the 1940s, a "Korean" Sumo wrestler could reach only a minor status in the hierarchy of traditional Sumo society. Finally, he quit Sumo and applied for the registration of "Japanese" at a local government office. From that point on, he constructed a fictional "Japanese" identity of himself separated from his actual origins (Lee, 1998). Soon after, he decided to become a professional wrestler and moved to the United States to learn American wrestling skills.

This complex story was carefully concealed until the late 1970s in Japan, even after his sudden death. So in the 1950s, most Japanese knew nothing about his Korean origins. With an awareness of the very ambiguous sentiments that existed toward other nationalities in Japanese society, Rikidouzan himself consciously continued to play a "Japanese hero" who actively fought against violent American wrestlers.

The popularity of professional wrestling in the 1950s can also be seen in the contemporaneous fad of imitative "wrestling games" among children. This phenomenon developed into a social problem, leading to injuries and even deaths. In November 1955, for instance, a sixth-grade elementary school pupil in Maebashi (Gunma Prefecture) died as the result of being kicked while pretending to be a professional wrestler. As such incidents became more frequent and public criticism of professional wrestling grew, Rikidouzan made repeated appeals to the effect that wrestling was a "professional sport valuing sportsmanship," required intensive training, and was not a children's game.

Despite the number of these incidents, professional wrestling retained its popularity throughout the 1950s. Newspapers continued to report numerous cases of disturbances associated with the broadcasting of professional wrestling. These included an incident in which people in a public bath got into a fight over seats for watching the broadcasts, and a case in which so many people packed into a house to watch a match that the floor collapsed. Even as late as the early 1960s, by which time televisions in the home had become increasingly common, wrestling matches still had far higher TV ratings than any other type of programming. For example, in October 1960, the audience rating for the professional wrestling broadcast was 50.3 percent, far ahead of the second most popular program (which was an American program, *A Dog Called Lassie*) at 36.7 percent (Minpou and Kenkyukai, 1966, p. 22).

From the late 1950s, open-air television began to lose its dominant position to indoor and domestic television. Realizing that the success of open-air television could be used to attract more customers, owners of shops and restaurants began installing television sets inside their premises or in their windows. Gradually, television moved from the open air to the indoors of commercial establishments, and, finally, to the home. This shift in the "place for watching television" led to a change in the social significance of television. In 1955 the *Asahi Shimbun* had already pronounced that the popularity of open-air television had passed its peak, and that the place of television was shifting "from the outdoors to the home." The same newspaper article stated: "Around the time that commercial stations were setting up open-air televisions in great numbers, the sponsors did not think that the people gathered there would become good television set buyers. The new stratums of TV viewers, however, are well within the bounds of prospective buyers." As this trend continued, the focus of television programming shifted from sports broadcasts to programs that were more suited to "relaxed domestic viewing," such as dramas and studio programs.

Indeed, the broadcasting of professional wrestling matches was clearly out of step with the development of television as a medium for domestic entertainment. In the early 1960s, a survey of commercial television viewers showed that professional wrestling matches were the type of program most often considered to be "trashy" and "unfit for domestic viewing" (Kokumin Seiji Kenkyukai, 1964). Although such programs were still overwhelmingly popular in terms of their ratings, they had also become the object of intense criticism, being regarded as deviant and unsuitable for domestic television. News coverage of wrestling moved toward a discussion of cases emphasizing the ill effects of the broadcasts on people. A typical case occurred in April 1962, when it was widely reported that two elderly people watching a Rikidouzan match "died of shock" at the cruelty of what was being shown. In reality, the two individuals concerned were later found to have suffered from existing heart conditions, and there was no proof of any causal link between the content of the program and their deaths. Nevertheless, newspaper coverage placed a particular emphasis on the "cruelty" of professional wrestling. Reports of violent acts by professional wrestlers further damaged the reputation of the sport, and its image became sullied with the stain of deviance. When a gangster stabbed Rikidouzan to death in December 1963, the indifferent attitude displayed in the press toward the death of this postwar "hero" hardly came as any surprise.

Postwar "Sacred Treasures" in the Living Room

From the early 1960s, television became a prominent medium linking the family with the state, defining the national consciousness, and dominating people's imaginative views of both the past and the present. Professional wrestling, which had been the mainstay of the early years of Japanese television broadcasting, had to be pushed to the margins of this new form of television as a national medium. During the 1960s, new forms of television culture emerged from within the sphere of family life into which tiny television sets had been introduced, even though they were still too expensive for the majority of the Japanese population.

In parallel with this change in the location of television, the role of the hero(ine) in Japanese TV culture was transferred from "Riki" (Rikidouzan) to "Micchie." "Micchie" was the nickname given to Michiko Shouda, who had become engaged to Prince Akihito, the oldest son of Emperor Hirohito. The announcement of their engagement in 1958 led to a popular fad for the new princess across the country. Furthermore, the emerging television networks used all of their resources to mobilize this fad. The image on the television screen with which people now wanted to identify was no longer so much the "Karate Chop" delivered by a small Japanese wrestler against a monstrous American opponent, but instead the "Happy Smile" of the new beautiful princess who was said to have been chosen from among the "ordinary people."

By April 1958, the number of televisions registered in Japan had exceeded one million, but the number increased even further later that year when the engagement was announced. It exceeded two million in April 1959, and had reached three million by October 1959. A major factor in the rapid popularization of television was the broadcast of the marriage parade in April 1959, and the highlighting of a series of "fevers," in which three stations (NHK, Nihon Terebi, and KRT) competed with one another. Each station used around forty television cameras along the roadside and sent approximately ten broadcast vehicles to strategic places. The cameras pursued the passing horse-drawn carriage by moving parallel with it, trying to take close-up photographs of the Crown Prince and Princess's expressions. Each station laid down rails for carrying trucks that would transport television cameras anywhere along the roadside, and also devised a plan to film the procession from a bird's-eye view by placing a camera on the top of a large-scale crane more than five meters in height.

Akira Takahashi et al. (1959) conducted a study of the connection between television and the popular classes with regard to the marriage parade. They emphasize that the television screen structures an image that clearly surpasses the actual sightseeing experience along the roadside, and exerted a powerful influence on the viewers' "feelings about the emperor system":

> While the sightseers watching the parade from the roadside come in contact with the "real thing," that experience does not go beyond one point along the total distance of 8.8 kilometers. In contrast, "the site as it is reconstructed" on television connects close-ups of the hero and heroine with the motion of the frenzied crowd, thus giving nothing short of an impression of going beyond the momentary contact of sightseeing. In addition, the television commentary, by building a bridge from one screen to another, instantly projects the entire structure and mood of the parade.

Takahashi's research team conducted a sample survey of television viewing among approximately 600 households along the route of the marriage parade. According to this study, the majority "gave up their chance that was bestowed by geographical proximity, and turned to 'the site as it is reconstructed.'" However, when the viewers gave a reason for their choice, they did not cite the "passive" reason of the unprecedented crowding of the roadsides predicted by television; rather, they stated that television enabled them to see the whole picture, while only one part could be seen from the roadside, or that more than what is "seen by the eye" could be understood by watching television because of the presence of announcers and commentators. Takahashi et al. suggest that the dramatization of the marriage event by the media translated into a victory over the desires of the people for direct experience (Takahashi et al., 1959, pp. 3–13).

This study shows that for many Japanese people, the marriage ceremony of the Crown Prince had been a television experience wherein they individually received within their homes a festive portrait edited for their consumption. In fact, the number of people who watched the April 10 parade on television is estimated to have reached nearly 15 million, while the number of people who actually went to the parade did not go beyond 500,000. As an event, the marriage of the Crown Prince symbolized the birth of the popular emperor system, materialized primarily within the television screen. Furthermore, in line with the structural principles of the space of imagery, the key image was the close-up of the Crown Prince and Princess within the horse-drawn carriage. Each television company invested a great deal of effort in shooting close-ups of the couple's cheerful expressions. For example, NTV distributed its cameras, and adopted the slogan "The Up Couple!" as the pillar of its entire coverage of the parade. The close-up image itself, cut off from the surrounding setting and accompanied by sound effects, was made into a snapshot ready for consumption.

In the course of the shift from the time when large crowds had thronged around open-air television to the period when innumerable eyes concentrated on the happy smile of the princess on the TV placed in the living room, a structural transformation had taken place in the definition of Japanese television. This change of definition can be compared to the change in the role of movies in US national culture that occurred in the 1930s, when movies were transformed in the national consciousness from a disreputable urban entertainment for the poor laboring classes to a special emblem of American national culture. Throughout the 1960s, the concept of television as Japan's "National Medium" gained particular prominence, as Japanese people re-constructed their national memory and identity, and also their image of the future. After the 1960s, within this concept of television being a "National Medium," there was no longer any space for professional wrestling to retain the centripetal role it had played in the formation of early TV culture.

What, then, did the TV set placed in each Japanese household after the 1960s mean for the ordinary Japanese? The most familiar image of the TV set at the time was that of being one of the principal "*kaden*," or domestic electrical appliance. However, in the mid-1950s, the words *kaden* (domestic electrical appliance) and *katei denka* (home electrification) had not yet even come into use. In addition, those appliances that became the major "*kaden*" in later years – such as washing machines, refrigerators, and vacuum cleaners – had not necessarily been key products in the first half of the 1950s. According to a survey on the proliferation of electrical appliances in households conducted by Kansai Electric Power Company in 1951, the top electrical appliances in use at the time were the radio (in 67 percent of households), the electric iron (38 percent), and the electric kitchen stove (23.5 percent) (Yamada, 1983). Therefore, the electrical appliances that people most wanted to buy in the near future were not refrigerators or washing machines, nor, of course, televisions.

However, from the middle of the 1950s a new image appeared. This symbolic change in the image of home electric appliances starting from the mid-1950s, is reflected in the expression "*Sanshu no Jingi*" (Three Sacred Treasures). This expression referred specifically to the washing machine, the refrigerator, and the black-and-white TV. Although these electric appliances were quite expensive for the standard Japanese family at that time, they gained rapidly in popularity. In 1955, only 4 percent of households owned washing machines, and the figures for television sets and refrigerators were both less than one percent. However, the use of these appliances increased rapidly, and by 1960, 45 percent of Japanese homes had acquired washing machines, 54 percent had TVs, and 15 percent had refrigerators. By the 1970s, the figures for all three appliances had exceeded 90 percent.

Although there are various arguments about the origins of the expression *Sanshu no Jingi*, there is no doubt that it first appeared in the context as the economic boom that began in 1955 – the so-called "Jinmu Boom." (Jinmu was the name of a mythological emperor in the ancient period. The next economic boom in the 1960s was similarly called the "Izanagi Boom," Izanagi being the name of another mythological figure.) The phrase has something to do with the prevailing trend since the late 1950s to talk about the economy in terms of images taken from Japanese mythological antiquity, which was reinforced by nationalistic discourses. Therefore, what is important for our analysis is the mythical meaning the expression inherently holds. The original "Three Sacred Treasures," consisting of the sword, the jewel and the mirror, are national symbols for authenticating the position of the emperor as the ruler of the Japanese archipelago, and they were emphasized during the formation of the modern Japanese nation–state. However, this expression has been used in the private domain since the late 1950s. It has been used as a symbol for authenticating the identity of the individual household as a "modern family."

Besides this, according to the testimonies of many people who lived at the time, television was not something that they "bought," but something that "came" to their home. For example, the Association of Japanese Commercial Broadcasting Networks received 454 letters from the general audience recording their memories of the role of television in their lives. Among these, one woman, who was only 10 years old in 1960, wrote that she had cried with excitement when her family bought a small second–hand television and had exclaimed: "TV has come!" She also wrote that many schoolchildren talked about the "coming" of television while playing at school (Jinsei Dokuhon, 1983). To these people, television was viewed as something much more significant than an artifact or a machine. Perhaps, we can say that it seemed more like a respected visitor. Such people had been viewed as "living Gods" by Japanese in the countryside before the modern period. These developments can perhaps be seen in the same light as events that occurred in the 1890s: the Japanese government distributed the Emperor's portrait photograph ("*Goshinei*") throughout the country, and each village and school celebrated the "coming" of the photograph. There was a similar mentality among postwar Japanese when they welcomed the "coming" of the television to their home.

The symbolic importance of television to the ordinary postwar Japanese family is also reinforced by the fact that up to the 1970s Japanese families tended to have a more passionate attachment to their televisions than was recorded in comparable American families. In the early 1970s, NHK's Broadcasting Culture Research Institute surveyed the attitudes of Japanese and Americans toward some important consumer goods. They asked the respondents what they would choose if they could only bring one item from the following – TV, car, newspaper, telephone and refrigerator – to a new life in an out-of-the-way place. According to this survey, 41 percent of Americans answered that they would bring the car, 38 percent the refrigerator, 11 percent the telephone, while only 5 percent chose the television. By contrast, 37 percent of Japanese respondents answered that they would bring the television, and only 12 percent of Japanese said they would bring the car (NHK Housou Bunka Kenkyujo, 2002). Thus, at least until the 1970s, television had acquired a special symbolic value for ordinary Japanese people.

Shunya Yoshimi

Early Television Studies in Postwar Japan

As we have mentioned above, around the end of the 1950s and the beginning of the 1960s, television viewing in Japan was undergoing a transition from being an outdoor activity that occurred in public places to being a domestic activity located in the home. It was at precisely this time that postwar Japanese television studies began developing in a distinctive way. By the mid-1950s, the field of mass communications studies had been introduced to Japan, including most of the major theoretical approaches – most especially that of American "effect studies." However, until the early 1960s Japanese television studies did little more than simply apply American mass communications theories. There had been extensive debate on various aspects of journalism and mass communications in Japan since the 1920s. Building on this foundation, there thus already existed the potential for distinctive developments in the field.

To illustrate this point, we may consider a special issue of the journal *Shisou* on the theme of "Television as Mass Communication," which was published in November 1958, only five years after the advent of television broadcasting in Japan. Although it is an extremely early example of Japanese television studies, the quality and depth of the research published there is such as to surpass much subsequent work in the field. All of the principal areas of research found in later television studies are represented: television and political consciousness, the daily life of viewers and their reception of programs, television as entertainment or art, the organization and management of broadcasting stations, television and cinema, comparisons with various other forms of media such as publishing, etc. Numerous leading postwar scholars and intellectuals contributed to this special issue, including Ikutaro Shimizu, Rokuro Hidaka, and Hiroshi Minami, alongside younger scholars based at those research institutes which were later to become the main centers for mass communications research in Japan, such as the Institute of Journalism and Communication Studies at the University of Tokyo, the Institute of Social Psychology at Hitotsubashi University, and the Institute of Twentieth-Century Studies founded by Ikutaro Shimizu. Most of these contributors were in their twenties or thirties at the time, giving a fresh and lively start to television studies in Japan.

The leading figure in early Japanese television research was Ikutaro Shimizu, whose numerous publications provided theoretical direction to the work of other younger scholars. In a book called "Socio-psychology" (*Shakaishinrigaku*, Iwanami Shoten, 1951), Shimizu convincingly assimilated the work of Walter Lippman, as well as early twentieth-century sociological theory, into the contemporary Japanese social context. This book became extremely influential as a demonstration of the direction to be taken by postwar Japanese communication studies in general. Over a short period of time at the end of the 1950s, Shimizu also produced a number of theoretical observations dealing specifically with the phenomenon of television.

Thus, in the opening contribution to the above-mentioned special issue of *Shisou*, Shimizu gave a clear exposition of the direction to be taken by television studies, setting out numerous issues of great contemporary relevance, such as the media-historical genealogy of television's origins, the significance of the transition to televisual media for structures of human sensation, the transformation of social reality by television, and the power relations on which television was predicated. In this, Shimizu was able to show television's position of "newness" in media history, and present the various sociological issues raised by its advent. He demonstrated that the development of television had significance far greater than simply the addition of one more form of media to the string of media inventions since the nineteenth century. Although "the thousands-of-years-old monopoly of writing, and the hundreds-of-years-old monopoly of printing had ended" with the invention of radio and cinema at the turn of the twentieth century, this fact was not immediately evident. Not until half a century later, "after the appearance of television, did people realize retrospectively that the monopoly of print had ended with the nineteenth century." Indeed, it was still very difficult to grasp the full significance of the ending of the age of print – the new horizons opened up for human consciousness, the possibilities for further development of televisual media, and the potential for social transformation.

Shimizu began his observations with a comparison of television with print media, paying attention to the particular form of social reality produced as a result of the direct image presentation of television. With print media, the image comes as a final result of the reading process, but with television the image is thrust before the viewer immediately. People are therefore freed from the labor of assembling reality for themselves. At the same time, however, great effort is required in order to extract oneself from the already presented reality. "Even before television programs developed much sophistication – in other words, while people still had the sense that they were not being entirely absorbed – television was already in fact absorbing people entirely." Thus, television produced the phenomenon of "reality excess." In opposition to the contemporary argument that television would stimulate a revival of domestic conviviality, Shimizu claimed instead that television's introduction into ordinary households would bring about a situation in which "the mass production of symbols has progressed to the same stage as electricity, gas and water, thus driving people into the home . . . Whenever they have a need, all people have to do is turn a tap or flip a switch. The water from the tap is sterilized, and so is the water from the 'cultural water works'" (Shimizu, 1958).

The tone of these comments by Shimizu shows the influence of his earlier work on "Rumor" (*Ryugen Higo*, Nihonshinbunsha, 1937). Rumors were a response to the paucity of information induced by state control – a vibrant collective practice in which people spun out alternative dissenting realities using unofficial channels. What then could be people's response to the situation of boundless "reality excess"? Shimizu maintained that the more technologically developed media became, the more reactionary and conservative their content would be. From this it followed that "the transition from the age of print to the age of images is also a transition from an age when we grasped evidence ourselves to an age when evidence is taken out of our hands." This is a condition of some wretchedness, leaving little or no scope for resistance. If "strong" networks of personal communication had been maintained as they had existed previously in the use of rumor against state control, then there would still have been some possibility of relativizing the reality provided by television. However, now that this social base had been lost, there was no longer any means of resisting television's "reality excess."

At about the same time, Shimizu produced a number of other notable works on the subject of television. For example, he produced a series entitled "Lectures on Contemporary Mass Communications" (*Kouza – Genzai Masu Komyunikeshon*, Kawade Shobou Shinsha, 1960), which he edited together with Hidaka, Minami, and Shirobe. The second volume of this series takes the "television age" as its main theme, and includes various analyses on such topics as television and mass manipulation, the structure of visual images, audiences and reception, the organization of commercial television, and comparisons of television, radio and cinema as forms of media. At the very beginning of the series, Shimizu expressed misgivings about the television research being conducted by younger scholars who were increasingly coming under the influence of American mass communication theory. What he found lacking in such work was a direct confrontation with "more basic issues," by which he meant issues of "thought." For Shimizu, such issues of "thought" came before "media" issues in the study of contemporary communications. In his view, these matters could not be reduced to questions of "culture." As had previously been stated in his work on rumor,

> amid the towering structures of the mass media, the apparently insignificant force of rumor gained surprising strength from the fact that it touched inconspicuously on areas of the mind entirely untouched by other media, bringing to light the darkest suspicions, and at the same time whispering responses to those suspicions.

In other words, Shimizu's notion of "thought" referred to intellectuals' attempts to grapple with the everyday collective consciousness of the masses.

Shimizu criticized as "bureaucratic (intellectuals') arrogance" the view that people's preference for entertainment-oriented television over newspapers and political debate was an expression of "backwardness" or "political indifference" on the part of the masses. Instead, he argued that "the masses sit in front of their TV sets not because they find satisfaction in doing so, but because they have nothing else to do . . . However unsophisticated television programs may be, they do recognize to some degree the existence of things which are egotistical, sordid, detestable and dangerous mixed up together with what is good and sublime within human beings." The makers of television programs take this recognition of mass indifference as their starting point. No television research can resist the new power structures encompassing television unless it examines this dimension.

It is also worthy of note that Shimizu was at the same time very active in the introduction of foreign television studies to Japan. For instance, the first volume of a series on "mass leisure" (*Masu Reja Sousho*, Kinokuniya Shoten, 1961) directed jointly by Shimizu, Hidaka and Minami, was on the theme of the "Merits and Demerits of Television." This introduced Japanese readers to ten examples of foreign television research, in which Shimizu himself collaborated in the translation along with Susumu Hayashi and Masaki Takezawa. A key contribution was the translation of Theodor Adorno's famous work on "Television and the Typologies of Mass Culture." In this paper, Adorno expressed his reservations about the social psychological "effect studies" of television, which had developed mainly in the United States. Rather than "communication surveys limited only to present conditions," he emphasized the need for "careful and thorough investigation into the development and background of modern mass media" and for psychoanalytical approaches focusing on the multilayered structures of meaning built up through television viewing. According to Adorno, it was necessary to "examine the social psychological stimuli inhering in television images both descriptively and on the psycho-mechanical level, and analyze the presuppositions of those stimuli and their overall pattern." Evaluation of "television's effects" is possible only after such an analysis has been accomplished. Adorno's views resonate with Shimizu's own concerns about the research being conducted by younger scholars at the time. By the end of the 1950s, there was already an awareness among Japanese television researchers of the limitations of television studies conducted only from the standpoint of mass communication studies.

As described above, by the late 1950s/early 1960s television studies in postwar Japan was beginning to develop in a diverse and distinctive way. Nevertheless, from the mid-1960s, the paradigm of effects theory imported from the United States became the dominant influence in the academy. From that time onwards, rather than deepening its own distinctive theoretical approach, Japanese television studies joined the worldwide trend of amassing social psychological surveys on the effect of television. During the same period, some of the distinguished scholars who had developed original perspectives toward early television left television studies after the late 1960s, while television itself became a more and more prevalent reality in everyday life in Japan. During the 1970s, Japanese television studies became empirical surveys on the social perception of television programs, or on new social uses of cable television (especially after the late 1970s).

The Invention of National TV Time and Narrative

Meanwhile, the significance of television in Japanese society since the 1960s has developed even further than its importance as one of the *"Sanshu no Jingi"* decorating the living room. Besides being a symbolic object linked to nation and gender in specific ways, television is now also seen as a flow of information structuring the horizons of people's bodily sensations and experience through its broadcasts. Considering this dimension, we see that one of the most important influences of television has been the introduction of a nationwide time structure into the domestic sphere. Television inherited this function from radio. In the process of putting out a continuous stream of daily programming, broadcasters arrange programs according to a particular way of thinking, thus creating a particular structure of time. Besides the direct mechanism of the "time signal," broadcasting has fulfilled the role of providing a uniform national time, such that the rhythm of people's daily lives has come to be organized according to the programming timetable.

The most remarkable manifestation of this time structuring effect of television is the phenomenon of "Golden Time." This is the term used in Japan to refer to the three hours – from 7 pm to 10 pm – during which television viewing was concentrated from the 1960s to the 1980s. The experience of many Japanese households during those hours became extremely uniform. This would usually begin with the news at 7 pm, followed in the same hour by children's cartoons and family quiz programs. After 8 pm, the focus would then shift to drama and variety programs, with drama then becoming the dominant genre from 9 pm to 10 pm. Professional baseball games were also usually broadcast between 7 and 9 pm. These were particularly popular with adult male viewers. Such television viewing patterns remained broadly constant throughout the 1960s, 1970s, and 1980s. The fact that these evening hours were given the name "Golden Time," rather than "prime time" as in many other countries, is an indication of how important this "living-room time" (*ochanoma no jikan*) created by television was in the overall temporal lifestyle of the Japanese family.

In addition to the "Golden Time" described above, television led to the creation of at least two more national time zones in the 1960s. One of these was the morning period between 7 am and 9 am. In the households of urban salaried employees, this coincides with the busy period immediately before the commute to work or school. Since the 1960s, television stations have oriented their programming during these hours to the tastes of female viewers, with such programs as NHK's serialized dramas (known as the *terebi shosetsu*, or "television novel"), and the morning shows of the commercial stations. This time zone thus contrasts with the evening "Golden Time" when the selection of which channel to watch is usually a male prerogative. Through the creation of these morning and evening time zones, television became a place of mediation where the real-life time of the household was synchronized with the postwar national time consciousness. Another time zone created by television was the period between 11 am and 2 pm when drama serials (*merodorama*, literally "melodrama") and talk shows (*waidosho* or "wide shows") are the dominant genres. Whereas the first two time zones mentioned have had the role of synchronizing the rhythm of family life to the tune of national time consciousness, the latter period has given rise to a gendered reality, in a manner similar to that of women's magazines, since around this time of the day most viewers are housewives. Until the 1980s,

Japanese television broadcasting was structured around these three contrasting time zones of "morning," "midday," and "evening." Broadcasting at other times of day was relatively unstructured and formed a margin around these core time zones.

The important point here is that the very meaning attributed to different times of day, whether it be 7 pm, 8 am or 12 noon, came to be experienced through its relation with television, as a result of the creation of time zones such as the three described above. This in turn led to a restructuring of the everyday forms of social groups. For example, according to a nationwide survey carried out by NHK on the temporal aspects of daily life, there is a striking correlation between the time flows of "people watching television" and "people eating." However, this does not necessarily mean that people started watching television during their already established meal times. This may be true to some extent, but it is also quite likely that people have adopted the habit of eating their meals during times when they are watching television. This is also true of people's experience of time itself. It is impossible to say for sure whether people watch the NHK news because the time is 7 pm, or whether the time 7 pm itself exists for people as the experience of watching the NHK news. Television is certainly not an external factor constraining our experience of time and lifestyle habits from the outside. On the contrary, television is a key structural constituent enabling those very experiences and habits.

It is therefore entirely inappropriate, at least as far as concerns postwar Japan, to view the family as a closed pre-existing social unit into which television has intervened from the outside. The intimate sphere of the Japanese household in its postwar form was itself created on a national scale through the medium of television. Indeed, according to an NHK survey carried out in 1979 and 1980, there is a positive correlation between the frequency of communication among family members and the amount of time spent watching television. In the same questionnaire, the number of people who answered that television had increased the opportunity for discussion among family members (50 percent) was far higher than those who replied that television had had a negative effect on the opportunity for discussion among family (19 percent). When asked whether television was "necessary for a harmonious family life," affirmative responses (46 percent in Tokyo, 61 percent outside the capital) far outnumbered negative ones (15 percent in Tokyo, 10 percent elsewhere). When asked whether they actually had the television on during times the family spent together, 67 percent in Tokyo and 77 percent in the regions, answered affirmatively. The particular form of family life characteristic of postwar Japan, whether described in terms of domestic harmony or communication, has depended on the existence of television, and was itself created through the medium of television.

Television's catalytic role is not restricted to temporal organization and relations within the family. With respect to the television programs as texts, since the 1960s television has played a central role in linking the time of domestic memory with the time of national history. This is well illustrated by the "serialized television novel" (*renzoku terebi shosetsu*), which has become a fixture of morning broadcasting on NHK since it began in 1961. The first such serial was a drama-tized adaptation of a novel by Shishi Bunroku called *My Daughter and I* (*Musume to Watashi*). The original novel portrays the life history of the author's own family from the end of the Taisho Era, through the war and into the postwar period, ending with the daughter's departure for Europe in 1951. In order to make this story into a television serial to be shown continuously from April 1961 until March of the following year, the scriptwriter, Yamashita Yoshiichi, at-tempted to "recreate in dramatic form the memory of the most difficult period for the Japanese people, the War." This was something not mentioned directly in the original text of the novel. Yamashita explained that he "thought it was necessary to portray *My Daughter and I* from a contemporary perspective amid the turbulent waves of the times." As a result of this adaptation, the "rebuilding of a family's good fortunes" after the wartime age of upheaval became the central theme of the drama serial as a whole. Thus began the 40-year history of the morning "television novel," which continues to the present day. Its stereotypical style became established from the very first serial, *My Daughter and I*. All sub-sequent serials, including the highly popular *Ohanasan* (shown in 1966), and the unprecedented international success of *Oshin* (first shown in 1983), followed the same pattern of portraying the life story of a woman attaining happiness after passing through the travails of the prewar and wartime eras.

From what perspective did these morning drama serials, with their obvious repetitious pattern, narrate the popular historical memory of the Japanese in the period of high economic growth? Being shown every morning around breakfast time, their audience was predominantly female. They were most often set in the historical period straddling World War II (the 1920s to the 1950s), since this was the period that attracted the highest viewer ratings. This was in sharp contrast to the historical dramas described below, which were set mostly in the sixteenth and seventeenth centuries and became particularly popular among male viewers. For postwar Japanese women, the most popular storylines were based on ex-periences of World War II. The television dramas repeatedly presented the Japa-nese war experience from the perspective of "Japanese women as the victims," although men do appear in the dramas as soldiers sent away to the battlefields or returning home wounded after the war. There is, however, no direct portrayal of the invasions of the Asian continent and Pacific islands, nor of the harsh condi-tions experienced by soldiers at the front. Women remain the main characters of the drama, experiencing military oppression and American attacks as victims, suffering great loss even as they finally survive the ordeal. History is presented here in the form of an individual person's memory from the perspective of the average Japanese woman, leaving no room for the memories and viewpoints of those who were colonized or invaded.

The historical periods and themes on which the historical dramas shown weekly on Sunday evenings on Japanese television differ markedly from the morning serials described above. The first of these historical dramas, *Hana no*

Shogai, was shown on NHK in 1963, and depicted the tragedy of the fall of the Shogunate in the mid-nineteenth century. The program producers' stated objective was to "create a large-scale work of healthy entertainment," rather than to make a serialized historical drama. It nevertheless became a representative expression of popular historical narrative among postwar Japanese, following the second series called *Akou Roshi*, and the third, *Taikouki*, portraying the unification of the country under Toyotomi Hideyoshi in the late sixteenth century. By the 1970s the basic pattern for these highly popular programs became established. The most popular series were those set in the sixteenth and seventeenth centuries. The dramatized exploits of Japanese warlords in the age of transition from the "Warring States" period (*Sengoku Jidai*) to the era of "unification" (*Tenka Toitsu*) has consistently gained the greatest popular following among viewers, as illustrated by *Ten to Chi to* in 1969 (viewer rating: 32.4 percent) and *Tokugawa Ieyasu* in 1983 (viewer rating: 37.4 percent). Besides providing a narrative of nation building, this focus on the historical period at the end of the Middle Ages when the Japanese state was unified, also had the significant effect of divorcing this history from the modern history of Japanese imperialism and invasion of the Asian continent. Although drama series have not infrequently been set in the nineteenth century, these have never achieved the same degree of popularity as those set in the period of the "Warring States." The age of Japanese imperialism, from the Sino-Japanese War and Russo-Japanese War until World War II, has not once been covered in these historical drama series. This is in stark contrast to the historical dramas shown on Korean television. Although Korean Television has modeled the pattern of its historical dramas to some extent on those of NHK, its most popular series have been precisely those dealing with the Korean resistance to Japanese imperialism.

Another factor in the popularity of these weekly-dramatized representations of the historical narrative of civil disorder and unification is that the view of society portrayed in them resonated with the accepted notions of liberalism and capitalism. The virtues of those warlords who survived the era of disorder had something in common with the ethos of those successful entrepreneurs from the competitive society of the age of high economic growth. On one hand, there was the narrative of the Japanese state's unification obscuring the history of modern imperialism and expansionism, while on the other hand, the entrepreneurial spirit of capitalist society was narrated through the metaphor of medieval warlords. In some senses, there was something in common here with the morning drama serials like *Oshin*. The latter portray in one detailed episode after another how women exercised their wit and perseverance in order to survive the war, and "managed" their households in a way similar to an entrepreneur, bringing economic success to their families. For a period of more than thirty years – from the 1960s to the 1990s – this simple narrative structure characterized NHK's high-profile drama programs.

The two types of NHK drama described above show a pattern in which the history of Japan as a nation is narrated metaphorically from the standpoint of

Shunya Yoshimi

"Japanese women as victims" and "businessmen as medieval warlords." Although this pattern was certainly prevalent during the period of high economic growth, it did not entirely dominate television broadcasting. Different dominant patterns can be seen in the contemporaneous drama programs of the commercial stations, where stories tended to be centered around mother/daughter relations (or mother-in-law/daughter relations), or on the exploits of eccentric heroes bringing salvation to the community (including, in many cases, the feudal regime of the Edo Shogunate). Taken as a whole, the narrative of television is always multifaceted, and cannot be reduced to a single structure. Nevertheless, as a conclusion to the above discussion, one can say that television from the 1960s until the beginning of the 1980s was closely linked to postwar national identity and the particular constructs of time and history associated with that identity. This is true whichever of the three dimensions we consider: television as a symbolic object; television as a factor organizing the horizons of everyday experience; or the world of textual meaning according to which programs and genres of programs are structured. Until the 1980s, the link between television and nationalism was also highly stable and largely unconscious. This stable structure was predicated firstly on a very clear-cut gender division (engineers vs housewives, "Golden Time" vs "Morning Time," Sunday historical drama vs morning drama serials), and, secondly, on the transition in the social location of the television set from the public space of the open air to the privacy of the household.

References

Amano, M. and Atsushi, S. (1992) *Mono to Onna no Sengoshi*, Tokyo: Yushindo.
Aramata, H. (1997) *TV Hakubutsushi*, Tokyo: Shougakukan.
Asada, T. (1987) *Waido Show no Genten (Birth of Japanese TV Show)*, Tokyo: Shinsen-sha.
Carter, E. (1997) *How German Is She?*, Ann Arbor, MI: University of Michigan Press.
Fujitake, A. (1985) *Telebi Media no Shakairyoku*, Tokyo: Yuuhikaku.
Hoso Bangumi Iinkai (2003) *Terebi 50nen wo Kangaeru*, Tokyo: Hoso Bangumi Kojo Kyogikai.
Inose, N. (1990) *Yokubou no Media (Media of Desire)*, Tokyo: Shougakkan.
Inui, N. (1990) *Gaikoku Telebi Film Seisuishi (History of Foreign Television Film)*, Tokyo: Shobunsha.
Ito, M. and Mafum, F. (1999) *Television Polyphony*, Kyoto: Sekai Shisousha.
Ito, M. (ed.) (2002) *Media no Kenryoku Sayou (Power with Media)*, Tokyo: Serica Shobou.
Jinsei Dokuhon Terebi (1983) Tokyo: Kawade Shobou Shinsha.
Kitamura, H. and Osamu, N. (eds.) (1983) *Nihon no Telebi Bunka (Television Culture in Japan)*, Tokyo: Yuuhikaku.
Kobayashi, N. and Yoshitaka, M. (2003) *Terebi wa Doumiraretekitanoka*, Tokyo: Serika Shobou.
Kokumin, S. K. (1964) *Shufu wa Terebi wo Doumiruka*, monthly report.
Lee, S. (1998) *Mouhitori no Rikidozan*, Tokyo: Shougakukan.
Minpou, G. and Kenkyukai, C. (1966) *Nihon no Shichousha*, Tokyo: Seibundo Shinkosha.
NHK Hoso Bunka Kenkyujo (1981) *Kazoku to Telebi (Family and Television)*, Tokyo: NHK Publishing.
NHK Hoso Bunka Kenkyujo (2003) *Shichosha kara Mita "Terebi 50nen" (50 Years of Television in the View of the Audience)*, Tokyo: NHK Broadcasting Culture Research Institute.

Okamura, M. (ed.) (2002) *Rikidouzan to Nihonjin*, Tokyo: Seikyu-sha.

Sano, S. (1994) *Kyokai-Den*, Tokyo: Bungei Shunju.

Spigel, L. (1992) *Make Room for TV*, Chicago: University of Chicago Press.

Takahashi, A., et al. (1959) "Telebi to 'Kodokuna Gunshu'" ("Television and 'Lonely Crowd'"), *Housou to Senden*, June, pp. 3–13.

Uratan, T., et al. (1983) *Jinsei Dokuhon: Teleb I (Television)*, Tokyo: Kawadeshobou Shinsha.

Yamada, S. (1983) *Kadan Konjaku Monogatari*, Tokyo: Sanseido.

Yoshimi, S. (1999) "'Made in Japan': The Cultural Politics of 'Home Electrification' in Postwar Japan," *Media, Culture and Society, 21*(2), 149–71.

CHAPTER THIRTY

Change and Transformation in South African Television

Ruth Teer-Tomaselli

The South African television landscape today consists of three nationally owned and operated channels: SABC1, with the largest footprint, broadcasts in the Nguni group of languages (Zulu and Xhosa, some Venda and Ndebele) as well as English; SABC2 broadcasts in Sotho, Tswana, Afrikaans, and English; and SABC3 broadcasts in English. While the first two channels carry the bulk of the public service mandate, the latter is aimed at a more cosmopolitan, sophisticated audience made up of all races. Entertainment Television, or e-TV, is a free-to-air commercial channel, licensed in 1998. M-Net is an encrypted subscription channel, owned by MultiChoice, which also operates a subscription-based direct-to-home satellite bouquet, DSTV, offering nearly 40 channels of local and global television.

This chapter is organized in two parts: the first section deals with the history and structure of South African television, then the second part describes a few program highlights of locally produced television shows.

History and Structure

National broadcasting in South Africa was inaugurated in 1936, following the formation of the South African Broadcasting Corporation (SABC). At first, only radio services were provided, divided along language and racial lines. These served to draw, and then reinforce, the social attitudes of a segregated apartheid society (Tomaselli, Tomaselli, and Muller, 1989). 1976 is most frequently remembered as the year of the June uprising by Soweto schoolchildren, a highly significant date signaling the first crisis of apartheid. However, January of that year witnessed a profound change of another kind when television was introduced into South Africa, despite disputes within the ruling National Party over its possible deleterious effects. On January 5 that year, then-Prime Minister B. J.

Vorster appeared on the screen, grim with foreboding: "Television brings the world into our living rooms, bringing certain advantages, but also certain responsibilities . . . Already we can see how easy it is to instil unfair views of countries . . . South Africa has long been the victim of such misrepresentation" (quoted by Bauer, 1996).

At the outset, the SABC, the only television broadcaster, dominated the airwaves. Initially, only one channel was in operation for five hours each evening, and the broadcast time was equally divided between English and Afrikaans programs. A second channel was introduced in 1982, carrying TV2 and TV3 as a split signal. Each channel developed its own character, style, and position in the market. TV2 broadcast in Nguni languages – Zulu and Xhosa – while TV3 broadcast in the Sotho family of languages – North and South Sotho and Pedi. The two channels shared a frequency, but were beamed to different regions of the country. In the late 1980s, the television structure was changed. TV1 remained purely English (57 percent) and Afrikaans (43 percent). Local productions made up the majority of programming broadcast (65 percent of the total).

The encrypted channel M–Net was introduced in 1986. Initially, it was owned jointly by a consortium of all four of the press houses, and in time, National Press bought out the channel. The programming content was dominated by imported films and series. M–Net was allowed an "open window" of two hours unencrypted time at earlier prime time (17:00–19:00), during which time it broadcasts its local soapie, *Egoli* (see below).

In 1993, TV2/3/4 became a single integrated channel, named Contemporary Cultural Values (CCV). Its programming was in English (52 percent), the Sotho group of languages (24 percent), and the Nguni Group (24 percent). The precise allocation of equal minutes for these two groups underscores the strongly ideological rationale of the broadcaster, determined to structure the programming along "equal" lines. This was to change later, in 1994, when the language policy of the SABC insisted on "equitable" rather than "equal," introducing an element of proportional language allocation. Although there was still a significant amount of African-language programming, with a transmitter split to accommodate the news in different languages to different parts of the country, the "glue" which held together the programming of CCV was English, since, according to the station's manager, English was the only common bond between all the peoples of South Africa.

In 1992 a third channel, National Network Television (NNTV), was established as a non-commercially funded educational and public service channel. A year later, the designation of the channel was changed to TSS (Top Sport Surplus). TSS was designed specifically to carry extra sporting programming, together with "hardcore" public service offerings such as education, actuality, documentaries and religion, broadcast predominantly in English (96 percent) with a small amount of Afrikaans (4 percent). Here, more than half (57 percent) of programming was local (see various SABC *Annual Reports*, 1976–83).

In the 1980s the SABC explicitly supported the then government in its effort to combat what was represented as the "Total Onslaught" of "revolutionary forces," supposedly spearheaded by the African National Conference (ANC) in exile (Teer-Tomaselli and Tomaselli, 1996). This was done most particularly through its news coverage and current affairs programming. *Annual Reports*, as well as internally circulated in-house documents, provide evidence that the philosophy of the Corporation during the mid-1980s was self-consciously based on the principles of national security in order to combat revolutionary forces (Teer-Tomaselli, 1993, 1995). From 1985 until 1990 the country experienced a series of States of Emergency, accompanied by stringent media censorship and overt propaganda (see Teer-Tomaselli, 1993 for a detailed account). In response, the SABC's programming was characterized by self-censorship and crude National Party propaganda. In particular, news had little legitimacy, demonizing any opposition to the government as a "terrorist threat," and discursively connecting the exiled liberation movement, the African National Congress, and its internal supporters as "communists" threatening not only to establish a Marxist state within South Africa, but also as working in concert with other movements internationally. Warned the SABC radio programme, *Comment*, "Americans have learned – in Nicaragua, Cuba and elsewhere – what happens with broad coalitions in which Leninists have control. They establish Marxist Governments" (*Comment*, January 10, 1987, quoted in Teer-Tomaselli, 1993).

Entertainment consisted mainly of American imports. Sanctions ensured that most European countries (with the exception of France and Germany) refused to allow their television programming to be shown in South Africa. Although English-speaking white South Africans shared strong historical and cultural connections with Britain, crucially, the British Actor's Union, Equity, boycotted South Africa, leading to a situation in which the majority of programs on the screen were imported from America.

Local drama was not contextualized into the realities of South African society and politics, but included a great many historical plays, lavish costume dramas, and a decidedly Eurocentric attitude to programming.

Changing Direction: The Late 1980s

Since the introduction of commercial radio in 1955, the SABC had been financed by a fixed funding system, in which the major portion of income, ranging from 70 percent to the approximately 80 percent at present, was commercially derived from advertising, sponsorship and entrepreneurial activity, while 20 to 30 percent came from licenses. Direct government monies have been earmarked for specific projects only – typically in support of school curriculum programming and health-education programming. This state of affairs led to the realization that the Corporation was chronically schizophrenic: in its attempt to be a fully fledged public broadcaster, it was forced to rely more and more on commercial logic, which meant providing inexpensive programming that appealed to the largest possible segment of the attractive high-end audience. This seriously compromised the public mandate of the Corporation.

Toward the end of the 1980s the government stranglehold over the SABC began to slip. Even before the Berlin Wall came down in 1989, bringing with it the symbolic end to the Cold War in which any thought of South Africa's continued protection in the slipstream of American foreign policy's fight against the cloak of communism was untenable, the SABC had begun to anticipate the change. Ironically, it was because Riaan Eksteen, the director-general at the time (note the very civil service nature of his designation), failed to toe the party line sufficiently, that he was replaced in 1987 by Wynand Harmse. Styling himself as the group chief executive, Harmse was an accountant by profession, and when he became head of the SABC, the focus of the organization shifted from pro-paganda to the pursuit of financial stability. In January 1991, the SABC was reorganized into a number of separate "business units," each with its own re-sponsibility as a profit-making entity (Currie, 1993, pp. 48–9; Collins, 1993, p. 87).

On-screen, significant changes were apparent. The overtly propagandistic newscast, *Network*, restyled itself as *Agenda*, and debate and engagement were the currency of the day. Once banned films – most famously, *Guess Who's Coming to Dinner?*, a drama featuring the ultimate apartheid taboo, inter-racial romance – were shown to an amazed viewing public. "The SABC had taken a long, hard look at its twisted reflection, and decided, like a smoker with emphy-sema, it was time to adapt or die" (Bauer, 1996) and dying wasn't a good alternative.

With the general transformation of South African political imperatives, being perceived as the voice of the government was no longer a viable option: it had become a political and commercial liability. Thus, from the late 1980s, a process of restructuring began in which pragmatism, rather than propaganda, became the dominant ethos. Furthermore, by the end of the 1980s, the SABC was feeling the competitive effect of M-Net, the subscription television channel, which, with its wealthier viewership, was beginning to draw advertising away from the SABC.

The transition period was driven by forces both internal and external to the SABC. Internally, the SABC became increasingly commercially driven, in what Harmse referred to as the "most extensive process of change in the Corporation's hitherto 56 year history" (SABC *Annual Report*, 1992, pp. 9–10). The Corpora-tion was divided into business units with a system of internal cost recovery, and the top-heavy administration of human resources and financial management was decentralized to each of the business units. At the management level, remunera-tion was linked to performance, while at all levels significant staff reductions were implemented. In the financial year between 1990 and 1991, the surplus was doubled, an achievement which was marred by a major industrial conflict in 1992.

561

In 1991 the SABC, under the direction of the board chairperson, Professor Christo Viljoen, set up a Task Group to study the issue of broadcast regulation. The report, published in August of that year, was popularly known as the "Viljoen Report." It covered the role of the national broadcaster, local program-ming, educational broadcasting, the liberalization of broadcasting through the issuing of new licenses, and, most significantly, the establishment of a regulatory body for broadcasting (see Louw, 1993, for further details). Despite the crit-icisms leveled against the Task Group, particularly over the manner of its appointment and composition, many of the Viljoen Report's recommendations were similar to those made by more radical bodies.

External pressure on the SABC came in the form of numerous pressure groups, headed by the Film and Allied Workers Organization (FAWO) and the Campaign for Open Media (COM). For three years, from the beginning of 1991 to the end of 1993, numerous workshops, seminars, position papers, and public protests brought together a very catholic collection of interests – students, media workers, and political activities inside the country together with sectors of the ANC in exile. This "moment" in the popular struggle for the ownership of the media and its contents have been well documented elsewhere (Currie, 1991, 1993; Louw, 1993; Jabulani, 1991).

Prior to 1994, the top management of the SABC was entirely male, white and predominantly Afrikaans. The Board of Directors, appointed directly by the president, was also largely male and Afrikaans, and practically all white. Many, if not most of the SABC top management and board belonged to the "super Afrikaner" organization known as the Broederbond (Tomaselli, Tomaselli and Muller, 1986). Since 1994, there has been a comprehensive overhaul of most of the legislative and regulatory mechanisms governing the country. However, even before the elections, broadcasting was identified as an area of immediate concern. It was feared by progressive forces that any election that was undertaken while the national broadcaster, the SABC, was under the control of the apartheid government, would be neither free nor fair. It was imperative, therefore, that an independent regulator for broadcasting be established as a matter of urgency. It was against this background that the Independent Broadcasting Authority was established in 1993 (RSA, 1993).[1]

The direct result of the popular campaigns was the appointment of the first democratically nominated Board of Governors, after a lengthy process of public hearings. This was rapidly followed by the passing the Independent Broadcast-ing Authority (IBA) Act (1993), and further public nominations and hearings to select the inaugural five Councillors of the authority. The installation of the new SABC Board, and the reorganization that it was able to drive, was the beginning, rather than the end of the process. Ivy Matsepe Casaburri, then chairperson of the SABC, and now the Minister of Communication, opined that "The old SABC is not yet dead, and . . . the new broadcaster we want to be, is not yet born. The SABC has to "re-invent" itself: it is a vision of the SABC as the national public broadcaster, to deliver services of value to all South Africa's people in more creative and unique ways than ever before" (RSA, 1994).

Triple Enquiry

One of the first tasks of the new IBA was to undertake the "Triple Enquiry" by convening hearings into, and formulating policy on: the viability of the public broadcaster; local content; and cross-media ownership and control. A year of national and regional public hearings resulted in the report to Parliament. The Triple Enquiry attempted to present the broadcasting landscape as a whole, rather than looking simply at the SABC. The outcome was "a tightly argued ecological approach to broadcasting with shared opportunities and obligations on all broadcasters" (Gillwald, 2002, p. 37). A primary purpose of the Enquiry was to break the monopoly of the SABC and to open up the airwaves to a multiplicity of voices and economic opportunities. A three-tier system of broadcasting licenses was advocated: public, private and community. In order to stimulate the "private" category, and counter-act the criticism of unfair advantage of the SABC receiving both license-fee monies as well as advertising revenue, the Enquiry Report's recommendations to Parliament included the paring down of the SABC to two television channels, and the licensing of an independent commercial channel.

The years between 1994 and 1996, can be seen as "the golden season of public broadcasting in South Africa" and saw significant changes to the internal organization of the SABC, as well as the programming fare broadcast. In its submission to the Triple Enquiry, the SABC argued, inter alia, that "SABC programming can deliver value by nurturing and reflecting cultural identity, meeting basic needs, developing human resources, building the economy and democratising the state and society" (SABC, 1995) and to a large extent the Corporation made good on this promise. The Charter to the SABC, which is found as an Appendix to the Broadcasting Act (RSA, 1999), provides a neat outline of the SABC's public service mandate.[2]

The reorganization following on the heels of the Triple Enquiry included a "relaunch" of all three television stations. At an extravagant relaunch party, rumored to have cost between R4 million and R9 million (probably around R6 million), and broadcast simultaneously on all three channels, SABC television heralded its new beginning. The channels were reconfigured according to the outline at the beginning of this chapter. Afrikaans, from being a co-equal language with English, was downgraded to a minor language, and given a greatly reduced allocation of broadcast time, on a par with other African languages. SABC3 was reinvented as an all-English-language channel, designed to meet the needs of the urban, educated audiences of all races.

The unforeseen outcome of these moves was to change the demographics of television viewing substantially. After an initial precipitous decline in the viewing audiences, overall viewing figures recovered to return to their pre-launch levels; however, the balance between channels changed noticeably. A significant number of "CIW," advertising-speak for Coloured–Indian–White, or non-black audiences, migrated to the pay channel M-Net, but this was compensated for by in increase in the number of black viewers (*Financial Mail*, August 30, 1996). In turn, this change impacted on the profitability of the SABC, since it was the wealthier viewers with greater spending power who had deserted the corporation. This realization led to a repositioning and strengthening of the commercial emphasis for SABC3 in an attempt to re-attract the wealthier audience and regain advertising revenue. In part, the channel has been successful, aiming now at a broad, multiracial middle class, enormously strengthened by the economic changes that have occurred since the early 1990s.

The McKinsey Process

In 1996, the SABC experienced a financial crisis. This was a combination of many factors: increased competition in the form of new radio stations; a migration of wealthy television viewers to the subscription channel, M-Net, leading to a fall in revenue resulting from the loss of significant sections of the well-to-do Afrikaans audience; and the provision of local programming in a plethora of languages, way beyond the resource capacity of the Corporation. The following year the SABC sought to cut costs in a bid to turn its finances around. This it did by undergoing extensive reorganization along lines recommended by the international consultants, McKinsey and Associates.[3]

As a result of the McKinsey recommendations, the local content component in programming was reduced significantly (Duncan, 2002, p. 127). Local production is relatively costly, and much of this was replaced with cheaper imported programming, chiefly from the United States. Although the cutting back on local production has been regarded as detrimental to the public service mandate of the corporation, it alleviated the cost constraints in two ways: by saving money on acquisition, and, ironically, by attracting more money in the way of advertising revenue through the broadcast of popular sitcoms and dramas aimed at the "lowest common denominator" audience.

A major change was implemented in the reorganization of the News Division. Previously divided along media lines (radio and television), the News Division was reconfigured as a "bi-media" operation in order to streamline production and cut costs. While the move succeeded in cutting costs, overall it was unsuccessful. Different media require significantly different approaches to news gathering and presentation, and radio coverage particularly suffered from being an afterthought consideration. The staff experienced a significant intensification of work pressure to accommodate the needs of producing more "outputs" in more formats than ever before. The bi-media project was scrapped late in 2001, and a return to the logic of separate, but cooperative, radio and television News

Divisions was implemented. There was also less money for foreign correspondents, and most of the SABC's foreign bureaus were closed down. Some of these were not re-established until after 2002, and then only sparingly. In order to redeem itself from the situation, the SABC required R1.8 billion from government, claiming that without the money, the Corporation would not be able to provide anything more than "ordinary news on current events" (*Sunday Argus*, May 31, 1997).[4]

The Independent Broadcasting Authority Act (1993) failed to provide specifically for satellite broadcasting. The IBA was responsible for the regulation of all terrestrial broadcasting. While aspirant entrants to the satellite sector waited in vain throughout the 1990s, the terrestrial subscription broadcaster, M-Net, used the opportunity and the gap in policy prescription to declare itself exempt from the need to apply for license for satellite activity. In 1999 it established a direct-to-home digital satellite service, DSTV (Gillwald, 2002, p. 43). The success of the satellite service has been, to some extent, at the expense of the terrestrial pay channel, M-Net, as the holding company of both, MultiChoice, has encouraged the migration of its considerable subscriber base from M-Net to DSTV. The SABC's own attempt to establish an analog satellite service initiative, under the project name of Astrasat, failed miserably, and was scrapped at great expense.

The most significant source of competition has come from the free-to-air terrestrial commercial channel, e-TV, which, after protracted negotiation and threatened litigation, was licensed in 1998. e-TV is jointly owned by a local empowerment consortium, dominated by trade union interests, with a 20 percent foreign ownership by Time Warner. At present, the channel commands around 10 percent of audience share, and 23 percent of television ad spending (SAARF, 2002).

The Period of Regulated Liberalization, 1998–2002

The McKinsey Report did have legislative repercussions, since its recommendations indirectly informed the 1999 Broadcasting Act (RSA, 1999), repealing the existing legislation of 1976 (RSA, 1976). In terms of the new Act, the SABC was to be governed by a Charter, clearly spelling out a clear set of objectives for the Corporation.[5] Most significantly, the SABC was "incorporated" into a limited liability company, which was to consist of two separate "operational entities": a "public service broadcaster" and a "public service commercial broadcaster." Cross-subsidization of some kind would take place between the two entities. Thus, the SABC would now be transformed into a corporate share structure, with the Minister of Communications acting on behalf of the government as the sole shareholder. A number of powers not previously available to the government are associated with this arrangement. As well as the traditional obligation to monitor annual and quarterly reports, the shareholder (in the person of the minister), engaged in a "Shareholder Compact" with the Corporation, the details of which were to be set out in an Appendix to the Act.

Having only recently emerged from the McKinsey process, the SABC now went into a second round of consultancy-driven restructuring, this time under the guidance of Gemini Consulting,[6] at an estimated cost of US$9 million. The *Gemini Report* criticized the SABC management for allowing a set of unwritten practices to override the stated values and vision of the Corporation (SABC, 2000). According to the report, the Corporation was top-heavy, with too many levels of management hindering the decision-making process, while management performance and evaluation was neglected. Staff were said to suffer from low morale, in part because of the lack of clear roles and responsibilities. Areas of particular weakness singled out for castigation were in the business and financial management, and a lack of overt strategy. Internal competition between divisions was also cited as a problem area. However, the Corporation was seen to have a competent staff component, with strong broadcasting skills and commitment. The restructuring process would be undertaken to reconfigure the Corporation as a "flat" organization, with a horizontal management structure that would eliminate duplication.

Five options were offered for television,[7] most of which offered some permutation of regional splits. On an experimental basis, three splits were instituted for an hour three times a week, but after less than two years the experiment was terminated for lack of funds.

Broadcast Amendment Bill 2002[8]

In the early 2000s the South African media environment was particularly tight, with no advertising growth to speak of. All broadcasters were faced with lower advertising incomes, and higher production costs. Fiscal restraint and careful management were required in order to make ends meet, let alone undertake significant structural changes. The Broadcast Amendment Bill (2002) was the second major piece of legislation in five years, and consolidation and stability seemed to be ethereal objectives. While the Act (1999) was not without its difficulties, with the introduction of the Amendment Bill (2002), new issues arose.[9] A fundamental change introduced by the bill was the creation of two management boards which are intended to oversee the public broadcasting services (PBS) and the public commercial broadcasting services (PCBS) divisions of the organization. The specific functions of the management boards would be set by the minister in the *Memorandum and Articles of Agreement*, a contractual agreement entered into between the broadcaster and Minister of Communication.

A second important aspect of the Broadcasting Bill was the stipulation that the SABC should establish two regional television channels, one in the "north," loosely defined, and the second in the "south" (RSA, 2002, section 32a). The primary purpose of these stations was to increase the amount of television programming broadcast in indigenous African languages, thus strengthening delivery against the language mandate of the SABC, as well as fulfilling its local content requirement. It would appear that the original idea was to have the new regional stations up and running in time for the prelude to the national elections of April 2004. This did not happen, and at the time of writing in mid-2004, plans were afoot to proceed with the establishment of these channels. The proposed channels are something of an anomaly: although they are to be run by the SABC, they were to be under the ownership and direct control of the Department of Communications. From the outset, the SABC was loath to take on the responsibility for these enterprises, noting that while the Corporation "supports any sustainable measures directed towards providing services to marginalised official languages (however) in the absence of having detailed feasibility plan" this would be very difficult to implement (SABC, 2003). There were a number of logistical problems associated with the fast-tracking of the channels' establishment.[10] Thus, the proposed establishment of these regional stations would mean an enormous restructuring exercise on the part of the SABC, in terms of infrastructure development; the rollout of programs and commissioning of new material; raising start-up financing; and human resources transfer at a time when there was a dearth of trained personnel.

Programming on television

If programming in the 1980s had depicted a schizophrenic world of political turmoil in the news programs and happy, head-in-the-sand drama programming, the structural transformations of the 1990s had an equally strong impact on the programming of the period. It is now a truism to point out that in South Africa the dominant political imperative since 1994 "has been to find new ways of building and promoting an inclusivist model of political integration which combines diversity within overall norms of unity" (Barnett, 1998, p. 553). In an age of global media penetration into all markets, in which the great majority of markets are dominated by American products, there is a tension between external ideologies, images and programming and internal attempts to create a rhetoric of nation building. This tightrope between economic liberalism (foreign investment, significant inflows of foreign media material, particularly in the audiovisual field) and ideological localism has been managed to some extent through policy making and media regulations, such as that of the local content quotas. Clive Barnett identifies three sometimes contradictory understandings of the connection between broadcasting and national identity: firstly, the idea that radio and television are useful in the dissemination of "symbolic representations of national unification and reconciliation"; secondly, that the media and communications sector are strategic and potentially profitable "in the process of economic development and reconstruction"; and, thirdly, the mass media have been identified as having a crucial role to play in extending the process of political participation (Barnett, 1998, p. 553). All three of these processes are visible on the screens of South African television, and can be seen in programs across the spectrum – those that entertain, educate, and inform.

Following the programming transformation in the mid-1990s, there was a tendency to overplay the "political" hand of serious drama. *The Line*, the first offering of this new "political turn," covered the internecine fighting between followers of the ANC and the Zulu-nationalist party, the Inkatha Freedom Party. The program, which can be seen as an unparalleled take on violence with superb production values and comment on the rotten state, was popular among viewers, but lambasted by the Inkatha party. Another drama, *Homeland*, dealt with the pain experienced by soldiers of the South African Defence Force in a rather stilted and politically correct manner. Adaptations of books and short stories were popular, notably *The Principal*, adapted from short stories written by Alan Paton,[11] recounting his years as the head of a reformatory school for wayward boys in the 1930s.

Deafening Silence, a screen adaptation of John Miles' novel *Kroniek uit Doofpot*, received excellent critical reviews as a "story of abuse and discrimination but also featured the best cast yet, led by our brightest talents Vusi Kune and Nthathi Moshesh" (Janet Smith and Andrew Worsdale, *Mail and Guardian*, February 12, 1998, p. 3). From the late 1990s, serious dramas became less overtly political, as exemplified by the outstanding production, *Natural Rhythm*. Again, this story was an adaptation from a novel, this time by Alison Lowry, featuring a stellar cast of South African stage actors/actresses, in which the political situation was a "subtext" to the plot which dealt with more contemporary matters such as health, friendships, marriage and home lives, sexual orientation and only then, political activism. Noted producer Johan Van den Berg: "bigotry, guilt and intransigence still exist, but instead of being didactic, we're trying to be more subtle with the issues . . . our approach to telling the story is to sow one tiny little seed rather than spilling the whole packet" (Smith and Worsdale, 1998, p. 3).

Each of the five terrestrial television channels have a "flagship" soap opera, which accounts for a large percentage of its advertising revenue, and fulfills a significant portion of its "drama" quota. Inevitably, these programs are among the three most popular slots on the channel, frequently taking top spot. The popularity of the programs may be attributed not only to the format – soap operas, which enjoy widespread popularity across the world – but also to the ease with which audiences are able to identify with the characters, contexts, and situations portrayed. Both soap operas and sitcoms have been the subject of a number of scholarly investigations. Magrietha Pitout (1996, p. 2) has noted that since 1992 when the African National Congress (ANC) and other political parties were unbanned, existing white power structures within the country began to disintegrate, leading to an "Africanization" which found resonance in popular cultural products, including television series. Seen in this light, entertainment programs can serve as a source of information concerning different cultural groups in South Africa.

The highest viewership of any program in the country is claimed by *Generations* (SABC1), a soap opera in the grand style of *Dallas*, depicting the fortunes and tribulations of an upwardly mobile black family as they make their fortunes in the "new" and multiracial South Africa. Its enormous popularity is attributed to the positive aspirational messages embodied in the program.

7de Laan (SABC2) is the only Afrikaans-language soap opera. Set in suburban Johannesburg, the show is peopled with a multi-ethnic, multicultural cast, linked by the places in which they work, shop, and gossip. Broadcast since April 2000, it is the second most popular soap opera after *Generations*, with a viewership of more than 1.5 million persons per episode, more than 70 percent of the Afrikaans-speaking audience. While the series is primarily in Afrikaans, English subtitles make it accessible to a wider audience.[12]

Isidingo (SABC3, repeated on SABC1) is the top-rated show on the channel in terms of audience viewership. The South African version of the program is multilingual, and an all-English version is distributed widely throughout Africa. Created by Gray Hofmeyer, a veteran of South African television, and produced by Endemol SA, in partnership with the global company of the same name, the show is set in a fictitious compound of Horizon Deep Gold Mine, and various parts of glitzy Johannesburg. Since its introduction in 1989, the storyline has moved its focus from the plebeian world of the compound to the wealthier milieu of Johannesburg's corporate world. The current plot structure revolves around the creation of a new television station, allowing the scriptwriters unimpeded access to a series of trials, tribulations, romances, passions, and intrigues for the characters.

Egoli, Place of Gold (M-Net) is the longest running of all current soap operas, having started in 1991. At time of writing this chapter, the soap has run for close on 15 years, or more than 3,000 episodes. The program is a mix of Afrikaans (approximately 60 percent) and English, aimed mainly at women viewers, but with a wide crossover in terms of race and ethnicity.

Backstage (e-TV) is a youth-targeted soap opera. Set in a college for performing arts and a nearby nightclub in Johannesburg, the storyline revolves around young aspirants' attempts to enter the world of music, dance, and theater, a concept which owes much to *Fame*, but which has been indigenized into the African milieu. The genre is pure soap opera, with generous doses of jealousy, ambition, heartache, and all the accoutrements of melodrama. Enthuses the publicity material: "kids who fall in love, who get hurt, who get high, kids who fall down and who manage to pick themselves up again."[13] The programmers have included guest appearances by a range of up-and-coming artistes and cultural figures, including local hip hop talent, R&B idols, DJ legends, dancers, fashion designers and that particularly South African music form: kwaito.

In contrast to the stereotypical American soap, South African drama productions, both soaps and sitcoms, avoid an overreliance on studio settings. Both *Egoli* and *Isidingo* (SABC3) carry out a significant proportion of filming on location, a factor which gives the subject a flavor of reality. South African soapies tend to be anchored in current reality to a greater extent than their American counterparts. Topical storylines are included, with references to topical events such as World Cup soccer and rugby, Valentine's Day, or the Olympic Games. Many productions also feature guest stars, both local and international. These local "personalities," and the raw nature of some of the subject matter, including "rape, HIV/Aids, traffic accidents and racism [which] feature in local shows without any discernable adverse reaction from the majority of viewers" (Addison, 2004, p. 24), have created a specifically "South African" type of drama. What differentiates these programs from the gratuitous violence and sensationalism of their imported counterparts is the strong engagement with social issues that is embedded into the narrative. *Yizo-Yizo*, the follow-on series, *Gazlam* (SABC1), and also to a lesser extent *Backstage* (e-TV), all deal with gritty, sometimes terrifying circumstances facing the younger generation of viewers, while at the same time providing positive role models and coping strategies for those facing gang violence, substance addiction, domestic abuse, teenage pregnancy, and the other challenges of a fast-developing society in transition.

While soap operas provide the latitude to deal with these issues in a dramatic genre, sitcoms allow for humor as the route to reconciliation. Sitcoms with strong social realism have proliferated from the mid-1980s, and show no sign of becoming redundant. The genre is seen as the flipside of paranoia and, in the face of great tragedy and uncertainty in the country, "laughter is emerging as a positive response to fear, as gallows humour uses personal experiences, not just faceless stereotypes" (Roome, 1987, p. 72). Dorothy Roome argues that "comic realism" on television can be conceptualized as part of the national drive toward transformation and reconciliation, since comedy is able to lampoon society in a way which is both cathartic and unthreatening.

Focusing on *Going Up*, a sitcom which ran for four seasons in the mid-1980s, Roome (1997, 1998) analyzed the success of South African sitcoms to address changing cultural standards and operate as a "site of negotiation of cultural change and difference." Using a standard technique within sitcoms, *Going Up* is located within a small workplace situation – in this instance, a law firm – which allows the scriptwriters and producers to explore the interactions of characters, often drawn as stereotypes, who occupy different positions within the social milieu, and experience the political, economic, and cultural changes of post-apartheid South Africa in unexpected and sometimes stressful ways. Numerous other sitcoms have used similar devices. The multilingual, mostly Zulu *Emzini Wwzinsizws* (SABC1), based in a mining hostel, touches on some political issues, but does so in a light-hearted manner, approaching the subject matter satirically, rather than didactically. *Scoop-Scombie* (e-TV) recounts the adventures of a irresponsible white journalist hack working for a small suburban newspaper, sharing an office with the usual suspects of post-apartheid life: the elegantly nouveau-riche young black accountant, the ditzy but sexy black receptionist, the sharp-tongued white female reporter and the fatherly black senior editor. In this show the humor lies in the manner in which expected roles are reversed, stereotypes are both reinforced (providing reassurance on the part of the audience) and inverted (providing a strong sociopolitical satire).

Suburban Bliss, a English–Afrikaans production, was the top attraction for several consecutive years. It told the story of two neighboring families, the Moloi's, who are black, and the Dwyers, who are white. The families work together in a furniture factory, and serendipitously move into neighboring houses (de Klerk, 2002). *Madam and Eve* (e-TV), based on a well-loved comic strip first syndicated in 1992, depicts the interdependency of Gwen, a white suburban housewife and the "madam" of the title, and Thando, the streetwise, caustic and ironic "maid." Other characters include the gin-swilling 60-something mother, the sometime gardener and handyman, and assorted other regulars. Over the period, the characters have evolved in parallel with the changing national circumstances. In a recent comic strip, Madam's mother is shown being questioned about racism based on old strips in which she features. She defends herself, saying: "That's not fair. It's an old cartoon. I've changed since then." The mother was introduced in 1994 "because Madam was getting too liberal and the contrast between her and Eve was dropping off" noted one of the writers (*Pretoria News*, March 19, 2004), but now she also displays a tolerance and humanity which threatens her role as the old colonial archetype, and requires newer, meaner characters to be introduced in order to provide the comic foil.

In essence, suggests Roome (1997, p. 77), these comedies "aid transformation and cultural reconciliation since [they] confirm tolerance of, and respect for, other human beings, stressing individual liberty and social justice." Roome identifies four key "consensual values" in these sitcoms: social mobility; morality; national reconciliation; and (anti-) racism. Using these as building blocks, Marize de Klerk (2002, p. 87) developed a hierarchy of indicators of domination and subordination within *Suburban Bliss*, in order to track the way in which these entertainment programs were able to "express values, beliefs, myths and ideologies which are peculiar to the culture in which were produced," and to track the way in which these are negotiated through the interaction of different groups in society.

Most soap operas and sitcoms are multilingual, using English as an anchor language, with generous additions of other South African languages interspersed. Despite the popularity, however, the majority of scripts are still written in English, by white writers, and then translated into various languages. Gray Hofmeyer, an independent producer of one of the most successful South African productions, *Isidingo*, bemoans the fact that good black writers are often attracted away to advertising copywriting for better pay, leaving mainly white writers to formulate the daily trauma (Addison, 2004, p. 24). The deeply South African nature of the soapies, and sitcoms, make them difficult to on-sell into other markets, except in Africa, while even there, there are difficulties. The use of indigenous languages exacerbates this difficulty – Egoli, with a 60 percent Afrikaans content, and only 40 percent in English, produced an "export" version for a while, translating the whole of the Afrikaans soundtrack into English.

Finally, this project was abandoned, since the sales revenue was not commensurate with the excessive translation costs.

However, not all programming on South African television is purely entertainment. The current crisis in education in South Africa presents a grave challenge. Television has been identified as a powerful medium, permeating the daily lives of significant numbers of people. Putting these two circumstances together, the hope has been expressed that television broadcast-based distance education may be a successful way to instruct a vast number of students simultaneously. Only the national broadcaster, the SABC, has a specific requirement to broadcast educational programming.

Takalani Sesame Street[14] is the South African version of the now famous Sesame Street project, targeting areas of basic numeracy, literacy and life skills.[15] "Takalani" is a TshiVenda word for "be happy," and conveys the spirit of happiness and innocence that is supposed to characterize the project. The Takalani version is a joint project of the broadcaster, foreign aid, and private sponsorship.[16] The partnership as a whole drives the project – from the development of the concept, the research and curriculum adaptation, through to the outreach, marketing and support, and, importantly, the commissioning and monitoring of the production. Internationally, the Sesame project has two didactic elements – television programming and printed support/outreach material. In South Africa, a third medium, radio, has been developed. The radio programs are the only such audio-only in the Sesame stable worldwide. In South African television programming, 50 percent of the content is lifted from the American product, and 50 percent is originated locally.[17] A recent addition to the series has been the world's first HIV-positive puppet, Kami, who has since been appointed by Unicef as a global "champion for Children," in order to highlight "areas where children are particularly vulnerable – from illiteracy to disability and abuse – in ways that are gentle, honest and compassionate" (UNICEF, 2003).[18]

Curriculum-based educational programming is best exemplified by the *The Liberty Life Learning Channel*, a secondary-school-level and youth development program broadcast daily, with an interactive phone-in omnibus on Saturday mornings. The program concentrates on the four subjects most commonly identified as problematic by high school students: English language and literature; mathematics; physical science and chemistry; and biology. It is used in the classroom, either in the form of a direct, real-time broadcast, or as delayed videotapes, as well as by students in their own homes. All programs are supported with printed materials. During the academic year, material is presented in an incremental fashion, while toward the end of the year, emphasis is placed on revision and examination preparation. A recent study reports that the teachers, as well as students themselves, find the project extremely helpful (Ivala, 2004).

Soul City (SABC1 and SABC2) is perhaps the best-known of all the "social marketing" entertainment-education programs in the country. Begun as a non-governmental organization (NGO) in 1992, its main mission was to use mass media as an ongoing vehicle for health promotion and social change (Singhal and

Rogers, 1999, p. 213; Tufte, 2001, p. 30).[19] The entertainment education project deals with a number of issues, including HIV/AIDS, tuberculosis, violence, abuse, alcohol, and gender inequality. Its multi-media approach uses television fiction, radio drama and print materials, which are distributed through newspapers. The story is set in the fictional town called Soul City and centers on the Masakhane clinic (Tufte, 2002). The radio drama uses the same messages as the television series, but includes different characters and storylines (Parker et al., 2000, p. 28). The different media have reached audiences in excess of 16 million people in a season, and have been the recipients of several national and international awards.[20]

Yizo-Yizo (SABC1), translated as "it's the real thing," was a three-season series, following the lives of students at a township high school, with the overt aim of "highlight[ing] the crisis in education" (Smith, 2003, p. 248). The extensive use of popular township music, especially kwaito, contributed to the popularity and success of the show among black youth, with approximately two million people watching each episode (excluding re-runs and repeats). The characters and situations used in the drama were all based on prior research, and gender-based harassment (including rape) were depicted in the series. Unlike the majority of television programs, which attract limited attention for only a short period, and then are relegated to the dustbin of nostalgia, *Yizo-Yizo* made a significant impact on the national public consciousness. Claiming both supporters and detractors, the highly controversial programming attracted a great deal of negative criticism and produced considerable public debate: "Newspapers all over the country carried letters from irate parents and community leaders blaming the programme for setting antisocial trends, reviewers raved about the realism and the black teenagers nodded their heads that someone understood the dilemmas they faced each day both inside and outside their classrooms" noted one commentator (Oppelt, 1999, p. 4). Despite the lack of consensus around the series, its immense popularity among the youth, the large number of international awards, and consciousness-raising effect of the content, all resulted in raising questions around youth culture and the crisis in education to the top of the national agenda.

Documentary programs have long been a forte of South African television. In the early 1990s long-running features on the Truth and Reconciliation Commission were aired for three years. However, as discussed earlier in this chapter, documentaries have been vulnerable to the vagaries of budget cuts, and at the present time the offerings are slim. The SABC broadcasts two regular weekly slots: *Special Assignment* (SABC3) in English, and *Fokus* (SABC2) in Afrikaans. Both programs have dedicated research and production teams.[21] Despite the numerous accolades and awards notched up between the two, they receive relatively little support from the Corporation, possibly because of the uncomfortable approach that often offends public dignitaries, and their reputation for regarding no one as "untouchable." Both programs provide a mix of full-length documentaries and shorter investigations into "the stories behind the stories."

Carte Blanche (M-Net), with a longer history since 1988, is regarded as an institution in South African television. Unusually for such a program, the original anchors are still with the show.[22] One of the first stories concerned a day in the life of two ordinary people living in an African township – at the time, an extraordinary subject. The flagship current affairs programming is *3rd Degree*, a personality-driven exposé program hosted by Debora Patta. Publicity material for the channel describes the program in lurid terms: "this seasoned journalist routinely investigates recent incidents that have made news headlines. The more shocking the story, the more reason for Debora to tackle the controversy head on."[23] Over the years these four programs have aired a great many stories covering the ordinary to the truly extraordinary: exposés with apartheid spies, violence in the workers' hostels, the first mixed marriages, racial inter-mixing in schools and public places; corruption in politics and business; celebrity interviews with presidents and film stars – the range of material has covered the significant (the Truth and Reconciliation Commission) to the outrageous, at least by conservative South African standards (transvestite prostitution and sex change operations).

No account of South African television is complete without a mention of sport. Each of the three broadcasters has a fully constituted sports division. *SABC Sport* produces 16 programs – or 14 hours a week – over three channels. *Supersport* is the trade name for M-Net and MultiChoice's sportbrand; Multi-Choice has three 24-hour sports channels on DSTV, and many hours per week on M-Net; e-TV fronts e-Sport. The regulator, ICASA, has drawn up a list of "sports of national interest," including rugby football, soccer, cricket, and athletics, which must be aired on free-to-air channels in certain set proportions.

As is the case internationally, the cost of sports rights has risen exponentially in the past few years, and the competition between the various broadcasters becomes a matter of careful balance: broadcasters are required – both by their audiences, and by law – to carry heavy amounts of sports coverage, while at the same time, the economics of sports broadcasting means that the large flagship events (such as World Cup cricket, or the Olympics) seldom make a profit (see Evans, 2003, for detailed survey and analysis).

Local content is extremely popular. Weekly audience figures, as measured by the South African Audience Research Foundation (SAARF), consistently indicate that local content is as popular, if not more popular, than the imported programming. This is despite the complaint that the quality of production is uneven – while there are some excellent programs, at the same time, much of what is produced is below the best international standards. Eight of the ten most popular programs are South African. However, despite the preference for local programming, the majority of television and radio fare remains foreign, and South Africans are deeply integrated into the global audience. Mention has already been made of the financial constraints currently affecting broadcasters, particularly those who rely on commercial funding (and this includes those nominally "public" broadcasters whose main income is derived from commercial activities, such as the SABC). Since local programming can cost up to twenty times as much as the equivalent genre of imported programming, it is to be expected that a large proportion of programming would be imported, thus exposing the audience to foreign, global programming. Worldwide, most of this is American in origin. In South Africa, this trend was exacerbated during the 1980s. Throughout the period of the cultural boycott, South African television relied on massive amounts of imported American programming, since European nations refused to supply television programming to South Africa. Thus, most of the imported programming was American. This created a certain "taste" for American programming, bred on the basis of familiarity. Furthermore, this was intensified by the introduction of subscription television (M-Net), which had, and continues to have, a very low commitment to local content. The introduction of direct-to-home television (DSTV) in 1995 was a further intensification of the process (Mytton, Teer-Tomaselli, and Tudesq, 2004). People in the upper-income groups who have access to satellite television are able to draw on global television in the real sense, having access to 40 channels from around the world. One way around the dilemma of the expense of local content versus the preference for foreign programming in certain sectors of the audience has been to introduce dubbing in order to add "local flavor" to an essentially foreign import. While this trend has been reversed in recent years, it is still practiced in relation to children's programming.

Conclusion

Television broadcasting in South Africa is a particularly interesting case study, since it is presently at the confluence of a number of different currents that underline the contradictions in which broadcasting finds itself internationally. A survey of broadcasting in this country illustrates a comprehensive process of broadcast re-regulation, which, since the early 1990s, has ensured the break-up of the SABC's monopoly; the struggle to move from a state broadcaster to a public service broadcaster; and the difficulties of upholding the mandates of the public broadcaster in the face of unrelenting commercial competition. On-screen, the television speaks eloquently of the changes in the society around it, acting as a mediator and catharsis to a rapidly changing nation.

Notes

1 Three years later, the country's Constitution (RSA, 1996) ratified the value placed on social diversity in its reference to the establishment of an Independent Authority to regulate broadcasting in the public interest, and to ensure fairness and diversity of views broadly representing South African society (RSA, 1996, Section 192).

2 The Corporation is required to "Make services available to South Africans in all official languages; Reflect the country's unity, diversity and multilingualism; Ensure quality across all

services; Provide significant news and current affairs programming, which is fair, unbiased, impartial and independent from government, commercial and other interests; Enrich South Africa's cultural heritage; Strive to offer a broad range of services targeting, particularly, children, women, the youth and the disabled; Include both in-house and commissioned programmes; include national sports programming, including developmental and minority sports."

3   The same group that reconfigured a number of other international public broadcasters, including the BBC.

4   The money was requested in successive tranches: R215 million in 1997; R441 million in 1998; R572 million in 1999 (the year of the election); and R539 million in 2000. The money is to cover regional broadcasts, coverage of the Truth and Reconciliation Commission (TRC) broadcasts, the 1999 elections, and also live parliamentary broadcasts. In the event, the TRC and the elections were covered, but the other two projects were dropped.

5   The Charter contents included detailed prescriptions for the SABC covering, among other issues, national objectives; performance, new services; board appointment procedures; lines of accountability; and enforcement of the Charter. These were to be monitored and compliance enforced by ICASA.

6   Gemini South Africa was the result of a management buyout, under the auspices of Metra Holdings, of the local operation of the Global company, Cap Gemini Group.

7   Option 1: SABC2 becomes a public channel with 5 regional splits; SABC1 and SABC3 become commercial channels; Option 2: SABC2 and SABC3 become public service channels; SABC1 becomes a commercial channel; Option 3: SABC1 and SABC2 become public service channels; SABC3 becomes a commercial channel; Option 4 SABC 2, with regional splits, and SABC1, become public service channels; SABC 3 becomes a commercial channel (this was the preferred option); Option 5: SABC2, with regional splits, and a modified SABC3 (with increased language and local content represents) become PBS channels. The regional splits occur in the same manner as option 1 and SABC3 will provide mass audience appeal for PBS.

8   Details here are taken from Parliamentary Portfolio Committee on Communications, September 17, 2002.

9   Selected areas of concern of the Amendment Bill. Included: Corporate governance; Independence of the SABC from the organs of the state; Conversion of the SABC from a wholly owned state enterprise into a limited liability entity, registered under the Companies Act; Reorganization of the SABC along the lines of Public Broadcasting Services and Commercial Public Broadcasting Services; the establishment of regional television services; and the television licence fee allocation. There were also numerous constitutional issues at stake.

10 Specifically, SABC expressed concerns about the following issues: Timing; Infrastructure rollout; Financial viability; Funding [if the stations belong to the DOC, and the SABC were the "management partners," who would pay for their infrastructure, programming costs, personnel, etc.?]; Licensing process [the timeframe did not allow sufficient time for the usual licensing process to be lodged with ICASA in time for the permission to be granted and the stations to be established . . . need a fully-worked out feasibility plan, a financial model, a proposed schedule of programs etc.]. Impact on the present channels of the SABC had not been determined.

11 Paton is best known for his novel *Cry the Beloved Country*, London: Jonathan Cape, 1948, reprinted several times by Penguin.

12 www.7delaan.co.za.

13 www.etz.co.za.

14 www.sabcedusation.co.za.

15 Initiated by the Children's Television Workshop, the project has been renamed Sesame Workshop (SW) based in New York, USA. Several international partnerships and country-specific programs are now produced.

16   Partners include the Sesame Workshop, the South African Department of Education (DOE) and the South African Broadcasting Corporation (SABC), with financial sponsorship from the insurance giant, SANLAM and the US overseas development agency, USAID. The SABC donates the airtime and supports the marketing budget of the project. Together, these partners constitute the Takalani Sesame partnership, with a secretariat housed on the premises of the SABC, alongside the SABC's Education Department.

17   In the radio series, the entire 100 percent is originated locally. The television and radio series have been commissioned by the Education Division of the SABC, and are produced by the Kwasukasukela production house. The radio programs were produced by Veluka Productions, an independent radio production company based in Durban.

18   Reported on www.iafrica.com.

19   The Soul City project was established in 1992 by two medical doctors, Dr Garth Japhet and Dr Shereen Usdin.

20   "The television series consistently achieved top audience ratings, winning six coveted Avanti Awards, including the prize for South Africa's Best Television Drama" (Singhal et al., 2003, p. 16).

21   *Special Assignment* is produced by Jacques Pauwand Anneliese Burgess (see www.sabcnwew.com/specialassignment). *Fokus* is produced by Freek Robinson and Ida Jooste (www.sabcnew.com/fokus).

22   Afrikaans newsreader (ex-SABC) Ruda Landman, and ex-sports commentator, Derek Watts (English).

23   www.etz.co.za.

References

Addison, G. (2004) "Soapies-R-Us," *The Media: Independent Industry Intelligence*, April, pp. 22–6.

Barnett, C. (1998) "The Contradictions of Broadcasting Reform in Post-apartheid South Africa," *Review of African Political Economy*, 78, 551–70.

Collins, R. (1993) "Reforming South African Broadcasting," in P. E. Louw (ed.), *South African Media Policy: Debates of the 1990s*, Bellville: Anthropos, pp. 85–92.

Comment (1987) SABC radio editorial comment, mimeograph, October 1.

Currie, W. (1991) "The Control of Broadcasting: Transition Period," in *Jabulani! Freedom of the Airwaves!*, Amsterdam: African European Institute, pp. 9–12.

Currie, W. (1993) "The People shall Broadcast! The Battle of the Airwaves," in E. P. Louw, *South African Media Policy: Debates of the 1990s*, Bellville: Anthropos, pp. 40–70.

de Klerk, M. (2002) "Die Raming van Dominasie en Ondergeskiktheid in *Suburban Bliss*," PhD thesis, Potchefstroom University for Higher Christian Education.

Duncan, J. (2002) *Broadcasting and the National Question: South African Broadcast Media in an Age of Neo-liberalism*, Johannesburg: Forum for the Freedom of Expression (FXI); Amsterdam: Netherlands Institute for Southern Africa.

Evans, I. (2003) "ICC Cricket World Cup 2003: Sports Broadcasting in South Africa, National Interest and Money," MA thesis, University of Natal, Durban.

Gillwald, A. (2002) "Experimenting with Institutional Arrangements for Communication Policy and Regulation: The Case of Telecommunications and Broadcasting in South Africa," in *The Southern African Journal of Information and Communication*, 2(1), 34–70.

Ivala, E. (2004) "The Uses of Television Broadcast-based Distance Education: A Case Study of the *Liberty Life Learning Channel*," PhD thesis, Durban: University of KwaZulu-Natal.

Jabulani (1991) *Jabulani! Freedom of the Airwaves!*, Amsterdam: African European Institute.

Louw, E. (1993) *South African Media Policy: Debates of the 1990s*, Chicago: Lake View Press.

Mytton, G., Teer-Tomaselli, R. E., and Tudesq, J.-P. (2004) "Transnational Television World-wide: Towards a New Media Order," in Jean Chalaby (ed.), *Transnational Television Worldwide Globalisation and Transnationality in the Media*, London: I.B. Tauris.

Oppelt, P. (1999) "Interpreting the Reel World," *Sunday Times Magazine*, October 31, pp. 12–14.

Parker, W., Dalrymple, L., and Durden, E. (2000) *Communicating Beyond AIDS Awareness*, Pretoria: National Department of Health, Beyond Awareness Campaign, chapters 1 and 2.

Pitout, M. (1996) "*Televisie en Resepsiestduie: 'n Analyise van Kykersinterpretasie van die Seep-opera Egoli – Plek van Goud*," PhD thesis, University of South Africa, Pretoria.

Republic of South Africa (RSA) (1976) *Broadcasting Act*, No. 73 of 1976, Pretoria: Government Printer.

Republic of South Africa (1993) *Independent Broadcasting Authority Act*, No. 53 of 1993, Pretoria: Government Printer.

Republic of South Africa (1999) *Broadcasting Act*, No. 4 of 1999, Pretoria: Government Printer.

Republic of South Africa (2002) *Broadcasting Amendment Act*, No. 64 of 2002, Pretoria: Government Printer.

Republic of South Africa (IBA) (1994) "Issues Paper: The Protection and Liability of Public Broadcasting Services," Johannesburg: Independent Broadcasting Authority.

Roome, D. (1997) "Transformation and Reconciliation: 'Simunye', a Flexible Model," *Critical Arts*, 1(1/2), 66–94.

Roome, D. (1998) "Humour as 'Cultural Reconciliation' in South Africa: An Ethnographic Study of Multicultural Female Viewers," PhD thesis, University of Natal, Durban.

SAARF (2002) *South African Advertising Research Foundation Audience Report*, Johannesburg: SAARF.

SAARF (2003) *South African Advertising Research Foundation Audience Report*, Johannesburg: SAARF.

SABC (1976) *Annual Report of the South African Broadcasting Corporation*, Johannesburg: SABC.

SABC (1991) *Annual Report of the South African Broadcasting Corporation*, Johannesburg: SABC.

SABC (1995) "Delivering Value, Vol. One, Submission to the Independent Broadcasting Authority on the Viability of the Public Broadcaster," Johannesburg: SABC.

SABC (2000) "Gemini Consulting: Overarching Analysis and Summary Recommendations," Johannesburg: SABC/Gemini Consulting.

SABC (2003) "Presentation to Parliamentary Portfolio Committee," Cape Town, August.

Singhal, A. and Rogers, E. M. (1999) "Lessons Learned about Entertainment-Education," in A. Singhal and E. M. Rogers (eds.), *Entertainment-Education: A Communication Strategy for Social Change*, London: Lawrence Erlbaum Associates, pp. 205–27.

Singhal, A., Cody, M. J., Rogers, E. M., and Sadibo, M. (2003) *Entertainment-Education and Social Change: History, Research and Practice*, London: Lawrence Erlbaum Associates.

Smith, J. and Worsdale, A. (1998) Comment, *Mail and Guardian*, December 2, p. 3.

Smith, R. (2003) "*Yizo Yizo* and Essentialism: Representations of Women and Gender-based Violence in a Drama Series based on Reality," in H. Wasserman and S. Jacobs (eds.), *Shifting Selves: Post-apartheid Essays on Mass Media, Culture and Identity*, Cape Town: Kwela Books.

Teer-Tomaselli, R. E. (1993) "The Politics of Discourse, and the Discourse of Politics: Images of Violence and Reform on the South African Broadcasting Corporation's Television News Bulletins, July 1985–November 1986," PhD dissertation, University of Natal.

Teer-Tomaselli, R. E. (1995) "Moving Towards Democracy: The South African Broadcasting Corporation and the 1994 Election," *Media, Culture and Society*, 17(4), 577–601.

Teer-Tomaselli, R. E. (1996) "DEBI does Democracy: Recollecting Democratic Voter Education in the Electronic Media prior to the South African Elections," in G. Marcus (ed.), *CONNECTED: Engagements with the Media*, Chicago: University of Chicago Press, pp. 377–423.

Tomaselli, R. E., Tomaselli, K., and Muller, J. (1989) *Broadcasting in South Africa*, Chicago: Lake View Press.

Teer-Tomaselli, R. E. and Tomaselli, K. G. (1996) "Reconstituting Public Service Broadcasting: Media and Democracy during Transition in South Africa," in M. Brun-Andersen (ed.), *Media and Democracy*, Oslo: University of Oslo Press.

Tufte, T. (2001) "Entertainment-Education and Participation: Assessing the Communication Strategy of Soul City," *Journal of International Communication*, 7(2), 21–50.

Tufte, T. (2002) "Edutainment in HIV/AIDS Prevention: Building on the Soul City Experience in South Africa," *Approaches to Development Communication*, Paris: UNESCO.

UNICEF (2003) Statement, reported at www.iafrica.com, November 24.

CHAPTER THIRTY-ONE

Television in the Arab East

Nabil H. Dajani

Introduction

The Arab East, or "al-Mashrik al Arabi" as it is commonly known in Arabic, sits where Africa, Asia, and Europe meet. It represents three Arab regions that fall between the eastern Mediterranean, the Arabian (Persian) Gulf, the Red Sea, and the Indian Ocean. These regions were under Ottoman occupation for over 400 years (from 1516 to 1918).

Egypt is the first of these regions and is considered to be a leader in the development of Arab broadcasting. Because of its distance from Istanbul, the seat of the old Ottoman Empire, it was relatively free from direct imperial control and thus enjoyed more freedom than the other two Arab regions. The second region falls on the eastern shores of the Mediterranean Sea and extends from Turkey and Iran in the north to Egypt in the south and the Arabian Peninsula and the Red Sea to the east. As it directly borders present-day Turkey, the center of the Ottoman Empire, it was under the strict control of the Ottoman rulers. This region was historically known as Bilad ash-Sham, sometimes referred to as the Fertile Crescent or Greater Syria, and includes the present-day states of Syria, Lebanon, Jordan, Iraq, and Palestine. The third region represents the Arabian Peninsula, or Arabian Gulf, a largely desert area that includes the largest mass of sand on the planet but which is, nevertheless, an important ancient route to India and East Asia. Because of its desert environment, it was difficult for the Ottomans to control this region directly. It includes the present states of Saudi Arabia, Yemen, Oman, Qatar, Bahrain, the Arab Emirates, and Kuwait.

When Napoleon landed in Egypt, in 1798, and advanced into Palestine, the French and the British soon recognized the strategic importance of the Arab East region, and it was later the site of struggles between these two countries and the Ottoman Empire to establish spheres of influence. This rivalry ended at the conclusion of World War I with the French and British occupation of the areas of all these three regions, except Saudi Arabia, which is 95 percent desert and difficult to occupy.

While all of the Eastern Arab states, with the exception of Palestine, are independent today and while all share very similar cultural backgrounds and recent history, their present state institutions, including the media organizations, were affected differently by the differential political and administrative controls exercised on them by the Ottomans and/or British and French occupiers.

The modern development of the institutions of the Arab West (al-Maghreb al-Arabi) followed a different pattern to those experienced by the Arab East countries. The Arab West, which occupies the northern African Mediterranean and Red Sea coasts, experienced a different type of occupation under the French and the Italians.

The colonial administrators placed broadcasting under governmental control from the start and sought to exploit it as an instrument of colonial rule. In contrast to the popular print media that was introduced in the Arab East by men of letters and social reformers, radio broadcasting was introduced, in the late 1930s, by the ruling British and French administrations. This action was taken mainly to combat the Nazi and Fascist propaganda that was being beamed to this part of the world by the Arabic programs of Radio Berlin in Germany and Radio Bari in Italy.

Thus, it is clear that early Arab broadcasting institutions began as institutions managed, financed, and run by the government. Arab television broadcasting continued this pattern of being established by governments for the purpose of advocating their policies. Little effort was given to the professional development of the medium. However, those Arab countries that have witnessed revolutions, such as Egypt, Iraq, Syria, and Yemen, have tended to give higher priority to the development of television. They paid considerable attention to the use and exploitation of this medium as an instrument to mobilize the masses for both political and developmental goals. Arab revolutionary governments have generally made more effort than other governments in the region to make television programming convey political and developmental messages to the masses. The politicized programming generally has revolutionary overtones, advocating change at home or abroad (Rugh, 1979, p. 118). However, such change as was advocated was limited to that which agreed with the ideology of the ruling revolutionary authority. On the other hand, during normal times the majority of material broadcast on the electronic channels of the Arab revolutionary regimes was entertainment-based and there was little political content. Egypt is considered to be the production hub for television programs in the Arab world. It has enough talent to produce a high percentage of Arab television programs and stands out as the leader in the production of television programs that address Arab developmental and political issues. Syria has recently followed Egypt's lead and, perhaps, has shown itself to be bolder in producing programs that address sensitive social and political issues. Egypt and the Fertile Crescent regions are traditionally known for their production of television programs and the Arab Gulf region has generally been seen as the market for such production.

Non-revolutionary Arab countries, particularly those of the Arab Gulf, Lebanon, and Jordan, have focused on entertainment programs and have been less interested in using television as an instrument for social change. Of these states, Lebanon is perhaps the regional leader in producing quality entertainment programs that are aired widely across the Arab world. Political programs produced by television authorities in non-revolutionary Arab states were more often aimed at preserving the status quo rather than at bringing about political awakening and change. Broadcast personnel in these countries generally avoided controversial political issues, and only rarely did they air views that were opposed to the ruling authorities.

Broadcasting in the Arab East

Television broadcasting in all countries of the Arab East – with the exception of Lebanon – was initially developed as a government function. Television stations operated within the framework of official government institutions that were, in most cases, not technically or professionally adequate for this venture. For many Arab governments, a television station was more a national status symbol than a communication channel to address the people.

While these individual Arab countries have had different experiences with television because of the unique local factors that influenced their media, they also pool their experiences and cooperate with one another through their membership in the Arab League's "Arab States Broadcasting Union" (ASBU), that was established in 1969 and based in Cairo. While ASBU has not been successful in developing inter-Arab cultural cooperation and programming, it has had some success in promoting news exchanges and cooperative technical efforts. Regional exchanges and cooperation have prospered primarily in the Arab Gulf states (Boyd, 1999, p. 348).

All of the Arab television authorities had difficulty in recruiting and training qualified personnel, because they continued to practice the authoritarian colonial tradition of treating broadcasting as an instrument of the state. Arab broadcasting gatekeepers were usually (and still are) given instructions on their political programming by government officials. Lack of concern for the nature, orientation, and interests of the audience is also a common feature of Arab domestic television authorities. Although ASBU promotes Arab cooperative research and training, little valid audience research is conducted and programming is largely in accordance with the whims of the authorities in charge.

In June 1955, the first television station in the Arab East was introduced in Saudi Arabia; however, it was not an Arab station. It was established by the US Air Force to broadcast to American personnel stationed at the US air command at the Dhahran Air Force Base, and also to US Arab American Oil Company

(Aramco) personnel (Boyd, 1999, p. 147). At this time, Saudi religious groups fought intensely to block the launching of television into the Kingdom – an attitude to the medium that helps to explain the delay in its official introduction into Saudi Arabia.

Iraq was the first Arab state to establish a government-operated television station. The station started broadcasting in May 1956 as an irregular service using a modest 500-watt facility devoted largely to entertainment programs. The facility was originally brought to Iraq for a commercial exhibition (Boyd, 1999, p. 130; Rugh, 1979, p. 118). The service became regular and grew in importance after the July 14, 1958 revolution that overthrew the Iraqi monarchy. The revolutionary government boosted the power of the station's signal and increased the political content of its programs. Within a few years, every government in the region had its own domestic television broadcasting service. Most of the governments lacked the necessary infrastructure or technical know-how, but plunged into this venture mainly for reasons of national prestige. Official news and canned entertainment programs formed the bulk of their programs.

Prior to the spread of satellite broadcasting it was difficult to make a clear distinction between domestic and regional television broadcasting in the region, since domestic programs could be received in neighboring countries. Egyptian, Lebanese, and Syrian television signals could be picked up in neighboring countries, particularly during the warm weather conditions that were characteristic for nine months of the year. It was a common sight to see Arab television homes displaying tall rotating antenna towers that were not necessary to receive local television; they were there to receive signals from regional neighbors.

A regional Arab satellite system, "Arabsat," became operational in 1985, following several years of planning by 22 Arab countries. It serves member states in the Arab League, and its main purpose was for Arab states to be self-sufficient in the area of satellite use and also to strengthen news exchanges among Arab television stations. However, misuse of this exchange system, together with the lack of professionalism evident in Arab broadcasting, had resulted in the decline of the effectiveness and credibility of joint Arab broadcasting efforts. Having the most material resources in the area, Saudi Arabia may be considered the Arab state that has contributed greatly to the development of Arab satellite broadcasting.

After a slow start, the Arab region embarked on a rapid development of satellite channels. This new phenomenon is transforming the media environment and creating a need for the media situation in this region to be re-examined. The Council of Islamic Information Ministers, which includes representatives from many African and Asian countries, has also called for a series of satellite channels that can be used to influence global public opinion.

Since the 1980s, satellite television and the spread of digital multi-channel satellite platforms and networks have made a significant impact on the Arab world. Practically, this impact began in the early 1980s, when the two Saudi-owned newspapers, *al-Sharq al-Awsat* and *al Hayat*, used satellite transmissions from London to print their papers in several locations in the Arab world (Amin, 2000). Prior to that date, there had been no attempts to utilize the services of Arabsat satellites to initiate direct broadcast services (DBS). To that point, broadcasting through Arabsat was used only for the exchange of local news and public relations between Arab state-owned national television stations. National television channels, regardless of their differing state political ideologies, were mere pressrooms for government officials.

Four Arab countries – Jordan, Egypt, Lebanon, and the United Arab Emirates – announced their intention to initiate plans to develop Free Media Zones where freedom of operation is to be assured and no limitations are to be placed on legitimate media enterprises. Egypt and the Arab Emirates have already implemented these plans. These media zones, however, have been received cautiously by the Arab nations, who fear that this would invite influences that would threaten Arab and Islamic values. In addition to the increasing attention being paid to the regulatory framework for international broadcasting, there is also concern about technical issues.

The following section briefly discusses the development of broadcasting in the three Arab East regions.

The Egyptian region

Located in the northeastern corner of Africa and including the Asian Sinai Peninsula, the Arab Republic of Egypt is the most populous country in the Arab East, with a current population close to 75 million people. While Egypt was not the first nation to introduce television in the Arab world, it was the principal driving force behind the development of Arab broadcasting and has influenced radio and television development in the region. Historically, Egypt is the center of Arab theatre and performing art. Its capital, Cairo, is considered the Hollywood of the Arab World and is the primary television program producer in the Arab region (Boyd, 1999, p. 50).

Egypt began its television system in 1960 during the regime of Gamal Abdul Nasser, who gave considerable attention to the use of the mass media as means to reach the public and to support his government's political positions and development programs. During the early years of television in the country, the Egyptian government subsidized and installed hundreds of sets in rural and urban cultural centers (Boyd, 1999, p. 52). President Nasser used the mass media, mainly radio and television, to instill enthusiasm for the political and social campaigns he was promoting and also to launch attacks on his enemies. The Egyptian government built TV facilities on a grand scale; fitting out its studios with the latest equipment and hiring an army of well-trained technical staff who were able to produce the most effective, subtle, and well-executed political programming in the Arab world. The coming to power of President Anwar Sadat in 1973 heralded a period when imported films and programs – notably from the United States – were increasingly seen on Egyptian television. This pattern continues under the

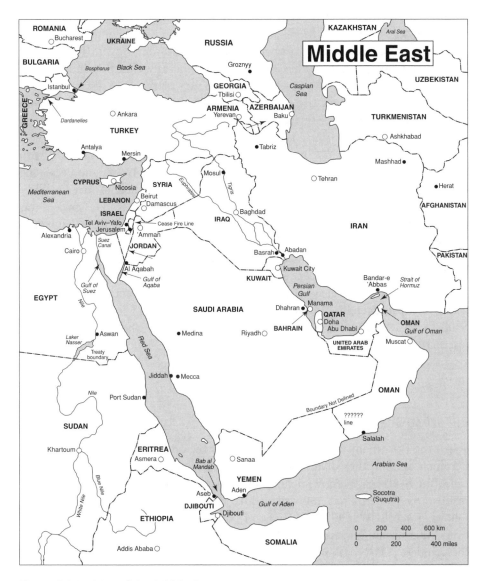

Figure 31.1 Map of the Middle East

current administration of President Husni Mubarak. At the end of the twentieth century, a plan was announced to decentralize TV in Egypt by introducing a number of local channels.

In 1990 the government approved the establishment of Cable News Egyptian (CNE), a cooperative 25-year arrangement with CNN. The main purpose of CNE was to retransmit CNN international in Egypt. At the end of 1994, CNE rebranded itself as Cable Network Egypt, and then made an agreement with a South Africa-based company to market CNE in Egypt. Within the framework of this agreement, this company began selling a new decoder and introduced additional services such as MTV and M-Net.

Arab satellite broadcasting also made its first entry in Egypt. This was in December 1990, just before the Gulf War, when the Egyptian Satellite Channel began transmissions. Its role was to hasten re-entry of Egypt to the Arab world after a period of isolation because it had signed a unilateral peace treaty with Israel. A second satellite service, Nile TV International, started experimental broadcasting in October 1993. This second satellite service was aimed at promoting the image of Egypt in Europe and to attract tourism (Amin, 2000). In April 1998, having been excluded from the Arab partnership in Arabsat, and realizing the importance of digital satellite broadcasting, Egypt launched its own satellite – Nilesat – that beams down more than 100 digital TV channels. The project was financed by the Egyptian Radio and Television Union (ERTV), Egyptian banks, and foreign investors. Nilesat 102 was launched in August 2000 and carries approximately 159 TV stations and more than 60 radio stations. ERTV makes use of most Nilesat channels, which gives Egypt the chance to produce and broadcast its own specialized channels. ERTV currently broadcasts the Nile TV specialized package. In addition, it has made distribution agreements with the Saudi Arab Radio and Television (ART) and also with Showtime. To attract international investors and media companies to Egypt, a free media zone, the Egyptian Media Production City, was introduced by the Egyptian government in 1992.

Bilad Ash-Sham region

When the British and French moved into this region following the defeat of the Ottoman Empire, they divided it between themselves in a diplomatic settlement called the Sykes–Picot Agreement. This agreement established a framework for a mandate system that was imposed by a League of Nations decision in the years following World War I. The League of Nations awarded Britain the mandates over Transjordan (later renamed Jordan), Palestine, and Iraq. France was given the mandate over Syria and Lebanon.

Television in Iraq

Located in the northeast of the Arabian Peninsula, bordering Saudi Arabia, Kuwait, Iran, Turkey and Syria, the Republic of Iraq has witnessed a number of changes in its system of government during the past century. Iraq was part of the Ottoman Empire until Britain occupied it during the course of World War I. In 1920, it was placed under the mandate of the British government, which helped set up Iraq as a kingdom in 1921, with Prince Faisal, the son of the Hashemite Sharif Hussein of Mecca, as its king. Iraq became independent in 1932 and the kingdom was abolished by a bloody coup in 1958. A series of coups ensued and several military dictators have ruled the country since then, the latest being Saddam Hussein. In 2003, Iraq was occupied by the United States, the United

Kingdom, and a number of allied countries. At the time of writing (November 2004), the United States governs the country aided by an appointed provisional government.

The Iraqi broadcast media was established, financed, managed, and controlled by the government through the jurisdiction of the Ministry of Information and Culture. During its early period, 1956 to 1958, Iraqi television was mainly a modest service devoted to entertainment. The 1960s and 1970s witnessed the expansion of broadcasting in Iraq, the installation of powerful transmitters, and the modernization of studios. Broadcasts were mainly focused on political programs and documentaries with little Western programming.

Prior to the occupation of Iraq in 2003, Iraqi TV comprised two national channels that were transmitted from five local stations each located in one of the Iraqi provinces of Kirkuk, Mosul, Basrah, Missan and al-Muthana. The first channel, the older, focused on news, entertainment and documentaries while the second included cultural and educational programs as well as a number of Western programs.

The old ruling Ba'ath party in Iraq viewed the role of the mass media as the channels that could transmit the party principles and the president's ideologies, reinforcing the faith in the revolution, and combating those opposing the Ba'ath party. While the government controlled most of the information available to Iraqis, it did sometimes tolerate criticisms of the corruption of some state officials.

Satellite broadcasting in Iraq began in 2000 and focused on communicating the viewpoints of the Iraqi officials, and reporting the achievements of the Iraqi president. The Iraqi satellite service comprised two channels: the Iraqi Satellite Channel (ISC) that included transmissions from the two national channels; and Qanat ash-Shabab (Youth Channel) that targeted youth with a variety of programs.

In 2000, a third satellite service was introduced by one of the Iraqi Kurdish factions in Sulaymanieh, the capital of Iraqi Kurdistan. A fourth satellite service, al-Hurriyah (Liberty) Satellite, was introduced in 2001 by the Iraqi opposition with US support. It was the first to broadcast programs outlawed in Iraq and its programs were a mixture of news, talk shows, and documentaries. However, this service was discontinued in 2002 owing to a lack of funds.

With the invasion of Iraq by the United States and its allies, the Iraqi television installations were bombed and the television service went off the air on April, 8, 2003. Within a few days, on April 10, the US occupation forces in Iraq had established a new Arabic-language television channel, al-Hurriyah (Freedom) Television that is managed by the Kurdish Nation Union. In mid-May 2003, a nationwide Iraqi television network, the Iraqi Media Network (IMN), was set up by US broadcasters and Iraqi opposition groups with US funding. Other regional channels were introduced in the provinces of Mosul, Karkuk, Mosul, and Turkoman. Television programs on all channels are subject to review by US-appointed officials.

Television in Syria

Modern Syria had a rather long history of political struggle. After a prolonged period of Ottoman rule, the Hashemite Prince Faisal was able to govern a large part of present-day Syria and Lebanon between October 1918 and June 1920 when the French army ended his rule and forced him to flee to Iraq where the British proclaimed him king. The French divided Syria into several states but, faced by strong popular opposition, in 1925 they were forced to declare the state of Syria and, in 1936, to expand this state to its present boundaries. France recognized the independence of Syria in 1943 and removed its military presence in 1946.

Television in both Syria and Egypt was inaugurated on the occasion of the anniversary of the Egyptian revolution on July 23, 1960, during the brief period when both countries were united under the leadership of President Abdul Nasser. Syrian broadcasters benefited from the advanced Egyptian broadcasting experience. Television transmission was introduced first in the capital city, Damascus and subsequently introduced across most of the Syrian provinces. By 1967, television stations across all the major cities of Syria were connected by a microwave link and transmission links were set up in most of the Syrian regions. Television studios were also set up in most of the Syrian provinces. These studios contributed to the direct broadcast operations of Syrian television. In 1985, Syrian television began broadcasting on a second channel that concentrated more on cultural programs (Dajani and Najjar, 2003).

The Arab Syrian Television is an independent government authority that is financed directly by the government and by the sale of advertising time. It operates within the General Directorate of Radio and Television whose board is chaired by the Minister of Information. In July 1995, Syria began its experimental satellite TV operation.

Syrian television had a wealth of technical and artistic talent, but in its early years of operation, it lacked "enthusiasm for creative programming or innovative production techniques." Although Egypt has traditionally led the way in Arab popular culture with radio and television dramas, "Syria is now taking the mantle, enthralling [Arab audiences] with livelier and more realistic scripts. [While] the abilities of Syrian writers and poets have long been highly regarded, many note with surprise that Syria . . . has come to lead the region in the modern realm of sophisticated TV series" (Peterson, 2002).

Syrian soap operas deal boldly with historical themes and are full of clever satire. They address many of the current social problems that trouble the Arab masses and that other Arab television channels are reluctant to tackle.

Television in Lebanon

Lebanon has much in common with Syria, sharing close historical, economic, and cultural links. Both gained their independence from the French in 1943.

However, having experienced more stable governments since 1948, it follows a more liberal political philosophy and the average Lebanese person enjoys more political and economic freedom than in any other country in the region.

Unlike all other Arab countries, television in Lebanon was introduced by business executives rather than by the government. Lebanese television began with two private commercial television stations that were licensed by the government. The first station, the first commercial television station in the Arab world, was backed by the French communication network (Sofirad), and began transmission in May 1959. It transmitted television signals on two channels – one devoted to Arabic programs or programs with Arabic subtitles and the other for foreign, mainly French, programs. The second station was backed by a US network (ABC) and began broadcasting on one channel in May 1962.

During the Lebanese civil war (from 1975 to 1989), the warring militias occupied the television stations and several of the broadcasting installations were shelled and destroyed. By the end of the second year of the war, in 1976, both companies were on the verge of bankruptcy and appealed to the government for help. In December 1977, the government merged the two ailing television institutions into a new company in which the government provided liquidity by purchasing 50 percent of its shares.

With the civil war escalating further and the central government becoming weaker and weaker, some of the warring factions established their own pirate television stations. Several of these pirate stations proved to be more successful than the government-subsidized station. They also managed to draw considerable advertising revenues.

By the end of the civil war some ten pirate stations were broadcasting and 36 others were planning to go on the air (Dajani and Najjar, 2003). This situation generated public debate and consequently in 1994 the Lebanese parliament passed an audiovisual law organizing broadcasting in the country. In September 1996, the government granted licenses to four television stations in addition to the existing licensed station. The four belonged to members of the government or their relatives. This action caused a public uproar because it was based on political and sectarian prejudices rather than professional considerations. In the face of the objections, the government granted four additional licenses.

The early 1990s witnessed an acceleration of satellite broadcasting in Lebanon and satellite dishes became among the fastest-selling commodities in Lebanon. Lebanese business people and television stations saw an excellent profit potential in satellite broadcasting. Two Lebanese stations ventured into this field – in October 1994 and April 1996. The success of both Lebanese satellite stations prompted most of the existing television institutions to start satellite channels.

Today, cable connection is widespread in Beirut and the major Lebanese cities because cable subscription fees are inexpensive. Unlicensed satellite TV cable distribution companies have increased rapidly in the past few years and have penetrated almost two-thirds of Lebanon's households.

Television in Jordan

The Kingdom of Jordan is the outgrowth of the emirate of Transjordan that did not accurately correspond to a political entity but was traditionally part of the Ottoman Wilayet (administrative unit) of Damascus. It was established by the British to pacify the Hashemite emirs of Hijaz, who were promised the rule of Syria in return for their support in the war against the Ottomans. Consequently, Emir Faisal was declared king of Iraq and his brother, Emir Abdullah, was proclaimed ruler of Transjordan in April 1921. This new entity included the Syrian regions of Ajloun, Balqa and Ma'ab, and the Hijazi (Saudi) regions of Ma'an and Tabuk (Aqaba). Britain formally recognized the Emirate of Transjordan as a state under the leadership of Emir Abdullah in May 1923. In March 1946, the British ended their mandate over Transjordan but they maintained a military presence there in return for providing financial subsidy to the new state. Emir Abdullah was proclaimed king in May 1946 and the country was named the Hashemite Kingdom of Jordan.

Television transmission in Jordan began in April 1968 with one single channel. This transmission was government-financed and -operated and was under complete government control. Its purpose was to advocate the government's policies and decisions, and it concentrated mainly on entertainment programs. Channel 2 was added in 1972 to broadcast news and programs in English, and French programming was added in 1978.

Jordan Radio, established in 1956, merged with television to form Jordan Radio and Television Corporation (JRTV). In January 2001, the corporation underwent major restructuring. Channel 2 specializes in sports and offers some coverage of parliamentary debates (transmission is usually interrupted when discussions get intense). Channel 3, which is operated with the private sector, offers a morning cartoon channel and an evening movie channel.

JRTV has an appointed eight-member board of directors headed by the Minister of Information and including representatives from the Ministry of Religious Affairs as well as from the Armed Forces. Jordan started satellite broadcasting in 1993, but its programs are quite unassuming and thus Jordanians turn to other Arab satellite services (Dajani and Najjar, 2003).

The Arabian Peninsula

The Arab Gulf region has few cinemas and nightclubs, and those that do exist do not generally appeal to the home and family-centered culture of this region. Television is quite popular as it has relatively little competition (Boyd, 1999, p. 8).

Television in Saudi Arabia

The Kingdom of Saudi Arabia covers about three-fourths of the Arabian Peninsula. It was established in 1932 by King Abdul Aziz Bin Rahman al-Saud as a Moslem monarchy dedicated to following the Wahhabi fundamentalist interpretation of Islam. The introduction of television to the kingdom was strongly resisted by Wahhabi religious leaders who prohibit employing images and pictures of living beings.

Because of these developments, Saudi Arabia was late to introduce television. The first Saudi television station was started as a state-run medium in 1965 with the help of the American network, NBC. The station is under the direct control of the Ministry of Information and provides four channels over a large network of stations that covers the kingdom. The first Saudi television channel is an Arabic-language channel; the second is in English with a few French programs, and the third is a sports channel. A fourth channel, dedicated to news reporting, was added in January 2004.

Although Saudi Arabia has no cinemas or nightclubs and television is the only available entertainment medium, Saudi television, with its controlled programs, has little appeal in the kingdom and the Saudis generally resort to watching satellite broadcasts from more liberal Arab regimes. This, perhaps, explains the haste of a number of members of the Saudi Royal family and their business associates to establish, with Saudi government blessing, Arab satellite television stations that broadcast from Europe. The first of the Saudi satellite services is the Middle East Broadcasting Centre (MBC), headed by a Saudi executive, Sheikh Walid Ibrahimi, a brother-in-law of King Fahd. It began transmission from London in September 1991 and later moved to Dubai Media City. MBC is quite professional in comparison to Arab national channels and provides a blend of news and public affairs programming, along with sports, movies, and other entertainment content. According to a London-based Arab journalist, Abd al-Jabbar Atwan, the debates broadcast by MBC "seemed like an ideological revolution that drew vast numbers of viewers if only because participants spoke with relative freedom" (*Daily Star*, 2003).

The second Saudi satellite service is Arab Radio and Television (ART), established in January 1994 by a former partner in MBC, Sheikh Saleh Kamel, a Saudi businessman who is related (by marriage) to the Saudi Crown Prince Abdallah. ART began transmission from the Telespazio in Fucino, Italy, providing four channels: a movie channel, a sport channel, a variety channel, and a children's channel. It has developed into a global stage for more than 20 channels with many of its programs being filmed and produced in several Arab and European cities.

The third satellite system is Orbit, the world's first fully digital, multi-channel, multilingual, direct-to-home, pay-TV and radio satellite service, with more than 30 television and radio channels. It was launched in May 1994 by the Saudi investment group al-Mawared. Orbit transmits from Rome and it offers more non-Arabic language programs than either of its Saudi competitors. Orbit was the first to include in its package of channels an exclusive Arabic news channel that was produced by the BBC Arabic news World Television Service. However, this news operation was suspended when it became troubling to Saudi officials. Many Arab staff members of this service later joined the Qatari Al-Jazeera service.

Although the Saudi government banned satellite receiving equipment in 1994, it did not implement this ban, and satellite dishes are widespread throughout the kingdom. The kingdom contributed the largest share of the cost of the Arabsat system and was the major force behind this project. It is also among the top ten largest shareholders in the International Telecommunications Satellite Organization (Intelsat). According to the Saudi Ministry of Information, Saudi Arabia "has risen to seventh position in the world, and to second position in Asia after Japan, in the use of satellites" (Rampal, 1994, p. 252).

Television in Bahrain

Bahrain is a tiny sheikhdom that was declared a monarchy in 1973. It comprises a group of 35 minute and mostly uninhabited desert islands in the Arabian Gulf which are situated between the west coast of Qatar and the east coast of Saudi Arabia. Most of the population lives on the principal and largest island of Bahrain, which is connected to Saudi Arabia by a 15-mile bridge. Unlike its two neighbors, Saudi Arabia and Qatar, which adhere to the strict Wahhabi interpretation of Islam, Bahrain has a relatively liberal attitude to Islam. In recent years, Bahrain has been experimenting with democratic reforms and these changes have had a direct effect upon the level of media and press freedom in the country.

Bahrain is the smallest of all the Arab states, yet because of its strategic location, it attracted foreign domination. The Portuguese were the first to occupy the islands in the sixteenth century. The British placed it under its protection from 1861 to 1971 when it was declared an independent state.

Although Bahrain is viewed as an important telecommunication center in the Arab Gulf, its television services had a modest beginning. In 1972 the government gave a concession to an American firm, RTV International, that was headed by a former director of Radio Liberty, a service that "is known to have been started and then operated by the US Central Intelligence Agency (CIA)" (Boyd, 1999, p. 95). Although this service introduced color signals, it had primitive studio facilities and provided little local production apart from news broadcasts. The bulk of its programs were canned and imported from outside, mainly the United States. This situation, together with financial difficulties that RTV faced during its first two years of operation, prompted the Bahraini government to take over the station in 1976.

The government-run television service began with one Arabic channel. A second channel, broadcasting in English, was added in 1981 and three more channels were added in the early 1990s. Two of these provide the terrestrial broadcast of the regional satellite services of the Egyptian Satellite Television and the Saudi Middle East Broadcasting Center (MBC) service; the third broadcasts the BBC satellite World Television Service. The programs of the original English (second) channel were later expanded to include the broadcast of CNN news and programs. A satellite service was added in the late 1990s. In addition, Bahrain Radio and TV runs a subscription service that provides 27 additional channels for a fee.

Bahrain has succeeded to some extent in inviting Arab talents to produce programs for Arab distribution, but it has been unable to become a significant broadcasting center in its region.

Television in Kuwait

Kuwait is a small sheikhdom between Saudi Arabia and Iraq at the head of the Arabian Gulf in the northern corner of the Arabian Peninsula. Like the other small states on the Arabian Gulf, a large proportion of its population is foreign. Kuwait was a British protectorate from 1899 until June 1961 when it was declared independent. The country is now ruled by an "emir" whose authority is decisive, although the state does have an elected parliament.

Kuwait is one of the more liberal states in the Arabian Gulf. The government allows a margin of press freedom and in 2002 it promised to introduce democratic reforms and to amend the present media law and end the practice of prior restraint. Kuwait makes no restrictions on the installation and use of satellite dishes or receivers. However, the Ministry of Information censors imported media material that is considered to be politically and morally offensive. Internet providers at the Ministry of Communication use filtering technology that allows the blocking of certain sites that are considered politically and morally unacceptable. Broadcasting in Kuwait falls under the authority of the Ministry of Information with broadcasting stations being owned, managed, and financed by the government. There are no radio or television receiver license fees.

Television began in Kuwait in 1957 as a private enterprise run by a local RCA television receiver dealer who wanted to promote set sales (Boyd, 1999, p. 139). When it achieved independence in 1961, Kuwait's television station was transferred to the government. At present, Kuwait operates three television network services. The first is a national network that includes four television channels. KTV1, the main station, broadcasts Arabic educational and information programs; the second, KTV2, is an English-language channel that was started in 1978 and mainly broadcasts movies, shows, and soaps; KTV3 transmits sports programs in both English and Arabic; finally, the fourth channel is the Kuwait Television Entertainment Channel, which airs programs in both Arabic and English. There are 13 television channels in Kuwait. The Ministry of Information is currently considering the possibility of changing the present broadcasting structure in Kuwait and permitting the introduction of private television stations.

The second television network service is cable/satellite television and it consists of the Kuwait Space Channel and the Kuwait TV Satellite Channel. The third service offers two pay-TV facilities: one offering the Orbit Satellite Television channels, and the other the Western family entertainment channels of Showtime Network Arabia.

593

Television in Oman

The Sultanate of Oman is strategically located in the southeastern part of the Arabian Peninsula, at the tip of the small Musadam Peninsula on the southern shore of the Strait of Hormuz. It was a regional commercial power in the nineteenth century that held territories along the coast of East Africa and on the coast of the Arabian Sea. When its possessions were lost, Oman withdrew into isolationism and became one of the less developed nations in the region until 1970 when a bloodless coup ended the rule of an ultraconservative sultan and the installation of his Western-educated son.

Television was introduced into the capital city, Muscat, on November 1974 and a year later broadcasting was extended to Salalah in the Dhofar area. In June 1979, the two stations at Muscat and Salalah were joined by microwave and satellite links to form a unified broadcasting service. This service is transmitted to the Omani regions through a network of stations, some 120 transmitters and relays, spread across the country. The service is also transmitted by satellite. Broadcasting in Oman is managed and financed by the government and no license fees for broadcast services are levied.

Television in Qatar

Qatar is a peninsula and a number of islands in the Arabian Gulf, bordering Saudi Arabia and the United Arab Emirates. In common with many other states in the region, the country was ruled by the Ottomans, and then the British, before gaining its independence in 1971. Its population of less than a million was noted mainly for pearl fishing before the discovery of significant oil and natural gas resources.

The state-owned television Arabic broadcasting system was introduced to Qatar in August 1970. A second channel broadcasting in English, "Qatar Television Two," was added in 1984. Cable television broadcasting is supplemented by three networks: Orbit Satellite Television, Showtime Network Arabia, and QCV – Qatar Cable Vision, a service owned and operated by Qatar public telecommunications.

Being a tiny state with little influence in its region, Qatar sought to influence the Arab world by adopting a mediating role in regional disputes and by setting up an all-news satellite television channel. A Qatari satellite news channel, Al-Jazeera (the Peninsula), was thus introduced in 1997 and quickly became the most outspoken and controversial Arabic news channel in the region. (Several Arab critics accuse Al-Jazeera of being a pro-Western medium that advocates the normalization of relations with Israel before ending its occupation of Palestinian territories, while simultaneously Western critics accuse it of being anti-Western, especially in its coverage of recent events in Afghanistan and Iraq.)

Because it covers topics that other media in the region do not, Al-Jazeera has grown quickly in popularity, overtaking the government-run stations of the region and Arab satellite networks, as well as those Western satellite news channels that are received in the Arab world. The majority of Al-Jazeera's staff has had experience within the Western media, especially the BBC. All of its employees are Arab, but only the administrative staff and a few of the technicians are Qataris. The editorial staff constitutes a mixture of talent from all over the world that represents a variety of opinions and political backgrounds.

Television in the United Arab Emirates
The United Arab Emirates (UAE) is located in the eastern part of the Arabian Peninsula, along the southern coast of the Arabian Gulf. It borders Saudi Arabia to the west and south, Qatar to the north, and Oman to the east. It was established in 1971 as a federation of six of the Trucial States of the Arabian Gulf coast that were protectorates of the United Kingdom: Abu Dhabi, Ajman, Fujairah, Sharjah, Dubai, and Umm al Qaiwain. These were joined in 1972 by Ra's al-Khaimah. Bahrain and Qatar, the two remaining Trucial States, opted to stay out of this merger. The UAE is the most wired and technologically advanced in the region, but also the least populated state, with over 85 percent of its population being foreign by birth.

Television was first introduced in Abu Dhabi in August 1969. After the establishment of the UAE, this station became the national channel, owned and operated by the federal government (Boyd, 1999, p. 192). A second national television channel, primarily an English-language channel, was added in the late 1980s. Other member emirates of the UAE set up their own television systems. Today Dubai, Sharjah, Ajman, and Abu Dhabi have their own local and satellite systems, in addition to the federal system, and also a pay-TV cable system that provides the channels of the Saudi Orbit Satellite Television and Showtime Network Arabia. Following the success of Qatar's Al-Jazeera television, Abu Dhabi started its own satellite news service in 2003.

Television in Yemen
The Republic of Yemen is located in the southwestern part of the Arabian Peninsula, on the Gulf of Aden and the Red Sea between Oman and Saudi Arabia. It is the outcome of the merger of the socialist South Yemen state with the state of North Yemen. The former was a British protectorate until 1967. The latter was a former primitive monarchy that became independent of the Ottoman Empire in 1918. The two countries were formally unified as the Republic of Yemen in 1990.

Broadcasting in both North and South Yemen was the responsibility of two government institutions. These were merged in 1990 to form the Public Corporation for Radio and Television (PCRT), which operates as a separate body within the Ministry of Information.

The electronic media in Yemen are a state monopoly, comprising two national television channels and seven radio stations. Channel one was originally the northern television service, and began broadcasting in September 1975. The former television service in the south was established in September 1964 and now operates as Channel two.

In Yemen, the possession of satellite dishes is not restricted. These are widely used due to the poor quality of the programs available on local channels.

Satellite Broadcasting

Two events fueled the expansion of satellite television in the Arab world in the late 1980s and early 1990s. The first event occurred in the fall of 1989, when CNN began to transmit, via an unsteady Soviet satellite, which has a footprint covering part of the Arab world. Later, CNN moved on to Arabsat, as per an agreement between Egypt and CNN, to broadcast terrestrially as CNE (Cable Network Egypt). By December 1990 Egypt had begun the first Arab satellite television broadcasting through its Egyptian Satellite Channel. The second key event was the Iraqi invasion of Kuwait in 1990 when Egyptian television began to broadcast free and uncensored CNN coverage of the war. CNN broadcasts of the Gulf War were also carried on Lebanese and other Arab television channels. Meanwhile, Saudi television aired taped and censored CNN coverage to their domestic audiences (Foote and Amin, 1993).

A few months after the CNN coverage of the Gulf War in September 1991, the Saudis launched from Europe their first satellite television system, MBC. This was followed by two other satellite broadcasting systems, ART and Orbit. All three were privately owned systems that are actually controlled by Saudi Arabian business interests linked to members of the Saudi royal family.

In 1996, two Lebanese stations, LBC and Future, developed satellite delivery to the Middle East. With their relaxed and informal approach, these channels had an immediate impact on viewing patterns in the Gulf and Saudi Arabia. By February 2003, a total of 45 major Arabic-language satellite channels were on the airwaves (*Daily Star*, 2003). Satellite channels became a matter of status for Arab governments. The motivation of these governments, however, was more to acquire the "prestige" of getting into satellite broadcasting than to produce and air quality programs. Several of the Arab government satellite channels simply rebroadcast the same programs aired on their terrestrial channels.

As we mentioned earlier, after several unsuccessful Arab attempts to establish a Pan-Arab satellite news channel similar to CNN, in 1995 Qatar launched Al-Jazeera, the first successful Arab all-news and public affairs satellite channel. The channel formed a free forum in the region.

Although the Arab satellite channels were originally established by Arab regimes to act in their own interest, they have now become seen as increasingly dangerous challenges to these very regimes. According to Sharabi (2003), Arab satellites have broken old taboos and they can now "criticize state policies, attack corruption in government and call for political and social change." Sharabi argues that "nothing in the modern history of the Arabs, including the revolutionary ideologies of the sixties and seventies, parallels the power of the Arab satellite networks in altering consciousness and attitude on a mass scale."

With the increasing accessibility of satellite television systems, there is grow-ing concern expressed by both Arab and non-Arab officials, including the United States, about the content of programs aired by Arab satellites. On the one hand, there is concern about a possible radical political reaction of anti-Western fac-tions to disorienting anti-Arab cultural materials broadcast directly via satellites by Western media. On the other hand, there is concern about the liberal news coverage of events in the region by Arab news satellite stations like Al-Jazeera that display an unfavorable view of the US and Arab regimes.

The possibility of radical satellite broadcasts by hostile movements or coun-tries launching satellite stations that would air programs undermining moderate Arab governments is another concern. Israel inaugurated an Arabic-language satellite channel in 2002, Iran has two Arabic-language channels (al-Alam and Sahar), and in February 2004 the US government launched a $100-million Arabic-language news channel (al-Hurra) that operates from the Washington DC area (Springfield, Virginia) with a broadcast center in Dubai. It broadcasts on satellite across the Middle East and plans are underway to make it available terrestrially within Iraq (*The Daily Star*, 2003; CropWatch, 2004).

For the first time in the history of broadcasting in the region Arab audiences have the luxury of selecting news from a menu of major news networks such as CNN, MBC, Nile News, BBC and Al-Jazeera.

Television Content

Arab television news content today focuses mainly on political events and cheap entertainment at the expense of educational and cultural programs. Political content in Arab television programs is not intended to promote political edu-cation. It is more in the nature of a political circus or an entertainment show. Political programs usually avoid the discussion of public issues that are embar-rassing to the ruling establishments. The media usually avoid the discussion of real problems that concern the average citizen. Representation of the Arab citizen is largely absent on television screens; however, one observes more freedom and more transparency on satellite systems than on state-owned television networks.

Due to the lack of good scripts, acting talents, and scarce funding, most Arab television entertainment programs lack originality and creativity. They are usu-ally copied from the western channels and fail to relate to Arab heritage, history, or identity.

Arab audiences were introduced to different television content in the early 1990s. This period witnessed an influx of Mexican, Venezuelan, and Brazilian *telenovelas* that are dubbed into Arabic. These are usually soap operas that run for months on a daily basis. For some, this kind of content was a welcome cheap alternative that could fill long airtime hours. However, the popularity of these Latin American productions on the Arab screens started to decline with the spread of Arab satellite services with their specialized channels.

By the end of the 1990s, satellite channels, and particularly the financially affluent Saudi ones, utilized the services of Egyptian and Syrian talents to produce a variety of programs, many of which became very popular both in the Arab world and also abroad. Today, Arab television audiences have access to a broad mix of news, drama, entertainment, sports, and music. There are also a number of specialized channels that are dedicated to a specific content, such as drama or music.

While Arab local television channels reflect the official point of view of the political authority or group managing or sponsoring it, the development of communication technology and the spread of satellite broadcasting have contributed to the relaxation of many of the restrictions that had hitherto prevented broadcasters from addressing sensitive political and social issues. In the new environment, significant events can no longer be concealed from the public.

The success of satellite television compelled local Arab television systems to focus their news content more on factual reporting, and to rely increasingly on live broadcasts from the scene of events. Arab audiences are thus able to see the coverage of major events that have a direct bearing on their livelihood, such as the Palestinian Intifada, the invasion of Iraq, and the ensuing resistance activities there. The boom in news coverage was accompanied by another phenomenon in current affairs and political programs, mainly talk shows. Arab audiences were introduced to live talk shows and audience call-ins. Several television stations and satellite channels air programs that host opposing advocates with call-ins from the audience.

Television drama and soap operas in the Arab world are traditionally produced in three countries: Egypt, Syria, and Lebanon. Egyptian studios were able to fill the airtime of many television stations and satellite channels in the Arab world. Syrian production houses have recently added a "new" flavor to Arab television screens and became a serious competitor to the dominant Egyptian drama with their historical and social soap operas and comedies. Many specialized channels air around the clock programming, mainly from Egypt and Syria. In the 1970s and 1980s, Lebanon was one of the active production centers of Arab television drama, but it has recently shifted to focus on entertainment and musical programs. Television channels generally air their best drama and soap operas during the holy month of Ramadan when Arab audiences stay up late at night.

Entertainment and musical programs have recently come to dominate television content in the Arab world. Currently, these programs dominate the prime time of many Arab television channels. A number of satellite channels specializing in entertainment are proving particularly popular. These channels are seen as an Arabic alternative to the phenomenon of MTV – the global channel of music. Lebanon and Egypt are the main source of Arab world televised entertainment.

Sports programming is also becoming increasingly popular in the Arab world. Millions of enthusiastic Arab viewers watch sports programs on local stations and satellite channels, and the channels themselves attract a great deal of advertising revenue.

Conclusion

Facing the exploding number of Arab satellite channels, one finds that in most countries of the region Internet facilities are available only to the elite. One also finds restrictions on news gathering, tough enforcement of broad licensing of journalists, and extensive censorship systems that have the aim of protecting the government from criticism.

While there are substantial differences between the media laws and policies in operation across the region, the processes of media regulation in all states are not transparent and their media systems are politically charged. Arab journalists are usually restrained from criticizing friendly nations or challenging the state's position on many sensitive national issues.

Most Arab states require approval to be obtained before any politically sensitive broadcast can be carried on cable systems. Some states even control the access to transponders in order to limit or affect the content of satellite programs. National security and anti-terrorist provisions are also frequently used as a basis for media repressions or controls.

The growing international Arab media and Arab media based in and operating from Europe have had a considerable influence on state-based Arab media, and hence on the legal framework affecting them. They set a liberal standard for the Arab media by initiating debates on Arab affairs that engage diverse points of view. The level of freedom practiced by these media is undermining the communication regulatory frameworks of most Arab states, since their own media are being placed at a competitive disadvantage.

A major achievement of the pan-Arab satellite channels is that they are helping to culturally unite the Arab world. Another achievement has been in sparking a new buoyant mood and a degree of boldness among the Arab general public. This boldness may be witnessed in public political discourse and liberal cultural codes. A few pan-Arab satellite channels have been trendsetters in many areas of life in the Arab societies. For example, Al-Jazeera challenged forbidden public political discourse and the Lebanese channels did the same with traditional cultural sensitive codes.

Although originally established by the Arab regimes to act in their interest, Arab satellite channels have now become perceived as dangerous challenges to these regimes. The success of these satellite channels has prompted several government-owned TV channels to remodel themselves. As a result, live TV programs, talk shows, and game shows flourish because they are usually less expensive to produce. Game shows offering the chance to win cash prizes have gained social acceptance. Lottery on TV is no longer considered a taboo;

especially during the holy month of Ramadan such programs attract thousands of callers.

Of course, the success of one type of program on one TV channel does not mean that it can be replicated on another. A major outcome of the forceful presence of the pan-Arab satellite channels is the emergence of a brand personality among a selected few, a differentiation that was impossible in the local government-owned TV environment. The new trend in Arab television viewing has been to see a shift from "channel-led" viewing to a "program-led" viewing.

Pan-Arab satellite channels have been trendsetters in many areas of life in the Arab world. Because of their novelty appeal, they have attracted large audiences. However, the magnitude of their social, cultural, and political impact needs to be researched further.

Acknowledgments

The author wishes to acknowledge the help of Dr. Mahoud Tarabay, who made valuable suggestions on the "Television Content" section.

References

Abbas, H. (2001) "The Gloomy Picture of the Syrian Press: Ink, Paper, and Eternal Slogans," *An-Nahar*, Cultural Supplement, June 25.

Alterman, J. (1998) *New Media, New Politics: From Satellite Television to the Internet in the Arab World*, Washington, DC: The Washington Institute for Near East Policy.

Alterman, J. B. (2002) "The Effects of Satellite Television on Arab Domestic Politics," *Transnational Broadcasting Studies Journal*, 9, Fall/Winter (tbsjournal.com).

Amin, H. (2000) "The Current Situation of Satellite Broadcasting in the Middle East," *Transnational Broadcasting Studies Journal*, Fall, retrieved at tbsjournal.com.

Amin, H. (1992) "The Development of SpaceNet and its Impact," in R. E. Weisenborn (ed.), *Media in the Midst of War*, Cairo: Adham Center Press, pp. 16–24.

Amin, H. (1998) "Pay TV: World Overview," paper delivered at the Fourth International Radio and Television Festival, Cairo, Egypt, July.

Ayalon, A. (1995) *The Press in the Arab Middle East: A History*, Oxford: Oxford University Press.

Ayish, M. (2002) "The Impact of Arab Satellite Television on Culture and Value Systems in Arab Countries: Perspectives and Issues," *Transnational Broadcasting Studies Journal*, retrieved at tbsjournal.com.

BBC (2004) "Country Profile: Report on Saudi Arabia," retrieved at http://news.bbc.co.uk/1/hi/world/middle_east/country_profiles.

Boyd, D. A. (1999) *Broadcasting in the Arab World*, 3rd edn., Ames, IA: Iowa State University Press.

CropWatch (2004) "Information Warfare or Yesterday's News: Pentagon Media Contractor . . . ," retrieved January 6 at http://www.cropwatch.org/issues/PID.jsp?articleid=9508.

Daily Star (2003) "Columnist Views 'Battle of Influence' in Arab Satellite TV," *Daily Star*, March 5.

Dajani, N. (1989) *The Analysis of the Press in Four Arab Countries*, Paris: The Vigilant Press, UNESCO.

Dajani, N. (1993) *Disoriented Media in a Fragmented Society: The Lebanese Experience*, Beirut: American University Press.

Dajani, N. (1997) "The Cultural and Communication Dimension of the New World Order" (in Arabic), *al-Mustakbal al-Arabi*, *20*(224), 58–66.

Dajani, N. (2001) "Lebanese Television: Caught Between Government and the Private Sector," in J. Atkins (ed.), *Journalism as a Mission: Ethics and Purpose from an International Perspective*, Ames, IA: Iowa State University Press, pp. 123–42.

Dajani, N. and Naijar, O. (2003) "Status of the Press in Syria, Lebanon and Jordan," in D. H. Johnston (ed.), *Encyclopedia of International Media and Communication*, vol. 4, San Diego, CA: Academic Press, pp. 301–16.

Fakhreddine, J. (2000) "Pan Arab Satellite Television: Now the Survival Part," *Transnational Broadcasting Studies Journal* retrieved at tbsjournal.com.

Foote, J. and Amin, H. (1993) "Global TV News in Developing Countries: CNN's Expansion to Egypt," *Ecquid Novi*, *14*(2), 153–78.

Ghareeb, E. (2000) "New Media and the Information Revolution in the Arab World: An Assessment," *The Middle East Journal*, *54*, 395–418.

Hafez, K. (2001) *Mass Media, Politics & Society in the Middle East*, Cresskill, NJ: Hampton Press.

Human Rights Watch (2003) "Saudi Arabia, 2003 Report," retrieved at http://www.hrw.org/wr2k3/mideast6.html.

Kalb, M. and Socolovsky, J. (1999) "The Emboldened Arab Press," *Harvard International Journal of Press Politics*, *4*(3).

Kamalipour, Y. and Mowlana, H. (eds.) (1994) *Mass Media in the Middle East: A Comprehensive Handbook*, Westport, CT: Greenwood Press.

Khazen, J. (1999) "Censorship and State Control of the Press in the Arab World," *Harvard International Journal of Press Politics*, *4*(3), 87–92.

Peterson, S. (2002) "Syrian Soaps Grab Arabs' Prime Time," retrieved at http://csmonitor.com/durable/1997/10/02/intl/intl.4.html.

Rampal, K. R. (1994) "Telecommunication Policy and Development in Saudi Arabia," in Y. Kamalipour and H. Mowlana (eds.), *Mass Media in the Middle East: A Comprehensive Handbook*, Westport, CT: Greenwood Press.

Rugh, W. (1979) *The Arab Press*, Syracuse: Syracuse University Press.

Sakr, N. (2002) "Arab Satellite Channels Between State and Private Ownership: Current and Future Implications," *Transnational Broadcasting Studies Journal*, retrieved at tbsjournal.com.

Schleifer, A. (1995) "Media Explosion in the Arab World: The Pan Arab Satellite Broadcasters," paper presented at the Broadcast Education Association conference, Las Vegas, NV, April 8.

Schleifer, A. (2000) "Does Satellite TV Pay in the Arab World Footprint: Exploring the Economic Feasibility?," *Transnational Broadcasting Studies Journal*, retrieved at tbsjournal.com.

Sharabi, H. (2003) "The Political Impact of Arab Satellite Television on Post-Iraq War Arab World," *Transnational Broadcasting Studies Journal* (Fall-Winter), retrieved at tbsjournal.com.

Index

Index

Index

Index

Index

Index

Index

Index

Index

Index

Index

Index